Mechanisms and Robots Analysis with MATLAB®

Dan B. Marghitu

Mechanisms and Robots Analysis with MATLAB®

Springer

Dan B. Marghitu, Professor
Mechanical Engineering Department
Auburn University
270 Ross Hall
Auburn, AL 36849
USA

ISBN 978-1-84996-799-0 e-ISBN 978-1-84800-391-0
DOI 10.1007/978-1-84800-391-0
Springer Dordrecht Heidelberg London New York

British Library Cataloguing in Publication Data
A catalogue record for this book is available from the British Library

Cover design: eStudioCalamar, Figueres/Berlin

Printed on acid-free paper

Springer is part of Springer Science+Business Media (www.springer.com)

to Stefania, to Daniela,
to Valeria, to Emil

Preface

Mechanisms and robots have been and continue to be essential components of mechanical systems. Mechanisms and robots are used to transmit forces and moments and to manipulate objects. A knowledge of the kinematics and dynamics of these kinematic chains is most important for their design and control. MATLAB® is a modern tool that has transformed the mathematical calculations methods because MATLAB not only provides numerical calculations but also facilitates analytical calculations using the computer. The present textbook uses MATLAB as a tool to solve problems from mechanisms and robots. The intent is to show the convenience of MATLAB for mechanism and robot analysis. Using example problems the MATLAB syntax will be demonstrated. MATLAB is very useful in the process of deriving solutions for any problem in mechanisms or robots. The book includes a large number of problems that are being solved using MATLAB. The programs are available as appendices at the end of this book.

Chapter 1 comments on the fundamentals properties of closed and open kinematic chains especially of problems of motion, degrees of freedom, joints, dyads, and independent contours. Chapter 2 demonstrates the use of MATLAB in finding the positions of planar mechanisms using the absolute Cartesian method. The positions of the joints are calculated for an input driver angle and for a complete rotation of the driver link. An external m-file function can be introduced to calculate the positions. The trajectory of a point on a link with general plane motion is plotted using MATLAB. In Chap. 3 the velocities and acceleration are examined. MATLAB is a suitable tool to develop analytical solutions and numerical results for kinematics using the classical method, the derivative method, and the independent contour equations. In Chap. 4, the joint forces are calculated using the free-body diagram of individual links, the free-body diagram of dyads, and the contour method. MATLAB functions are applied to find and solve the algebraic equations of motion. Problems of dynamics using the Newton–Euler method are discussed in Chap. 5. The equations of motion are inferred with symbolical calculation and the system of differential equations is solved with numerical techniques. Finally, the last chapter uses computer algebra to find Lagrange's equations and Kane's dynamical equations for spatial robots.

Contents

Chapter 1
Introduction

1.1 Degrees of Freedom and Motion

The *number of degrees of freedom* (DOF) of a mechanical system is equal to the number of independent parameters (measurements) that are needed to uniquely define its position in space at any instant of time. The number of DOF is defined with respect to a reference frame.

Figure 1.1 shows a rigid body (RB) lying in a plane. The distance between two particles on the rigid body is constant at any time. If this rigid body always remains in the plane, three parameters (three DOF) are required to completely define its position: two linear coordinates (x, y) to define the position of any one point on the rigid body, and one angular coordinate θ to define the angle of the body with respect to the axes. The minimum number of measurements needed to define its position are shown in the figure as x, y, and θ. A rigid body in a plane then has three degrees of freedom. The particular parameters chosen to define its position are not unique. Any alternative set of three parameters could be used. There is an infinity of sets of parameters possible, but in this case there must always be three parameters per set, such as two lengths and an angle, to define the position because a rigid body in plane motion has three DOF.

Six parameters are needed to define the position of a free rigid body in a three-dimensional (3-D) space. One possible set of parameters that could be used are

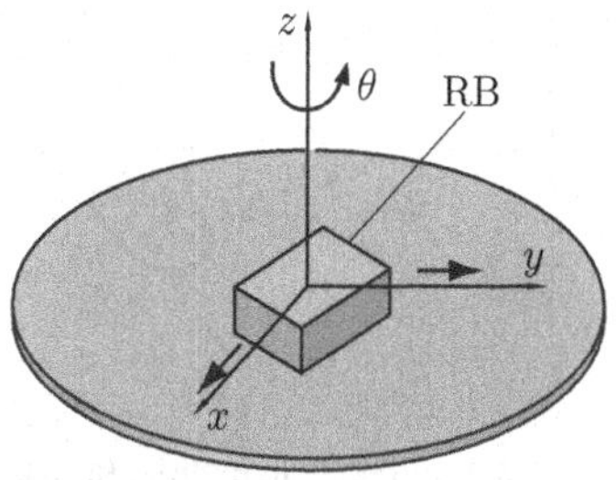

Fig. 1.1 Rigid body in planar motion with three DOF: translation along the x-axis, translation along the y-axis, and rotation, θ, about the z-axis

three lengths, (x, y, z), plus three angles $(\theta_x, \theta_y, \theta_z)$. Any free rigid body in three-dimensional space has six degrees of freedom.

A rigid body free to move in a reference frame will, in the general case, have complex motion, which is simultaneously a combination of rotation and translation. For simplicity, only the two-dimensional (2-D) or planar case will be presented. For planar motion the following terms will be defined, Fig. 1.2:

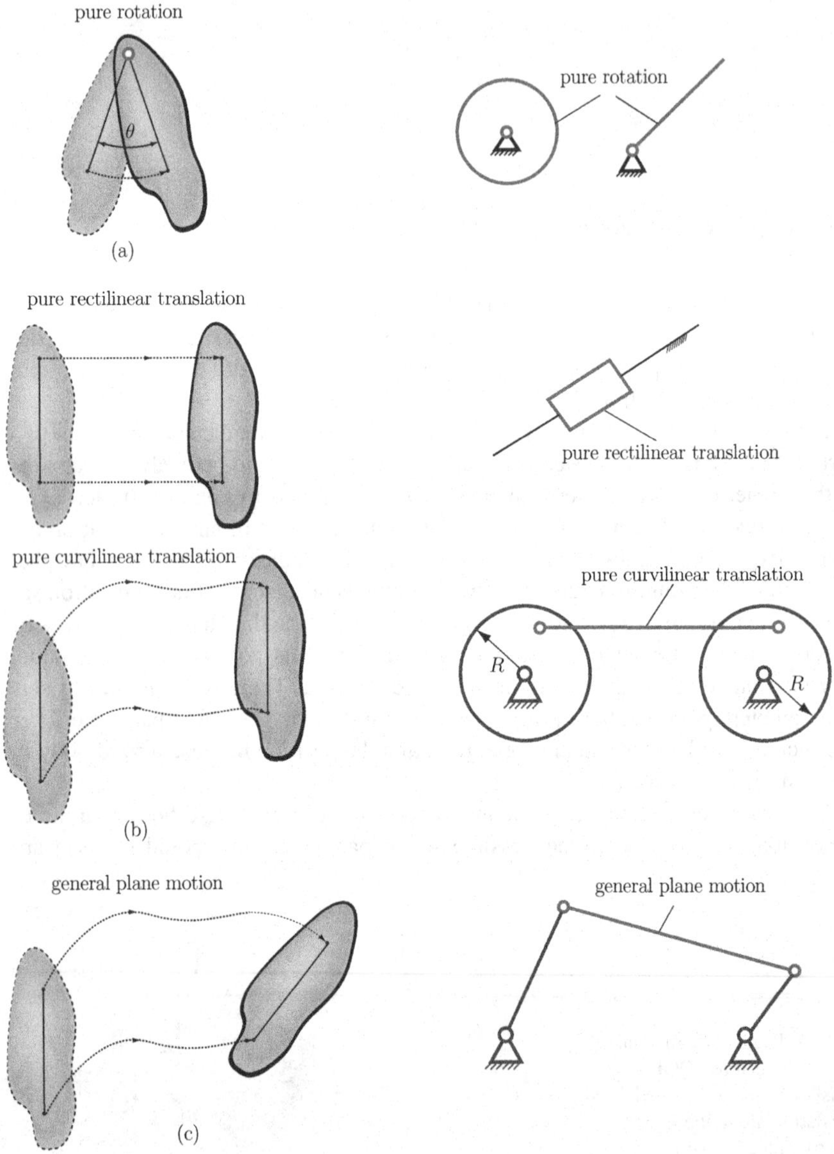

Fig. 1.2 Rigid body in motion: (a) pure rotation, (b) pure translation, and (c) general motion

1. *pure rotation* in which the body possesses one point (center of rotation) that has no motion with respect to a "fixed" reference frame, Fig. 1.2a. All other points on the body describe arcs about that center;
2. *pure translation* in which all points on the body describe parallel paths, Fig. 1.2b;
3. *complex or general plane motion* that exhibits a simultaneous combination of rotation and translation, Fig. 1.2c.

With general plane motion, points on the body will travel non-parallel paths, and there will be, at every instant, a center of rotation, which will continuously change location.

Translation and rotation represent independent motions of the body. Each can exist without the other. For a 2-D coordinate system, as shown in Fig. 1.1, the x and y terms represent the translation components of motion, and the θ term represents the rotation component.

1.2 Kinematic Pairs

Linkages are basic elements of all mechanisms and robots. Linkages are made up of links and joints. A link, sometimes known as an element or a member, is an (assumed) rigid body that possesses nodes. Nodes are defined as points at which links can be attached. A joint is a connection between two or more links (at their nodes). A joint allows some relative motion between the connected links. Joints are also called kinematic pairs.

The number of independent coordinates that uniquely determine the relative position of two constrained links is termed the *degree of freedom* of a given joint. Alternatively, the term *degree of constraint* is introduced. A kinematic pair has the degree of constraint equal to j if it diminishes the relative motion of linked bodies by j degrees of freedom; i.e. j scalar constraint conditions correspond to the given kinematic pair. It follows that such a joint has $(6 - j)$ independent coordinates. The number of degrees of freedom is the fundamental characteristic quantity of joints. One of the links of a system is usually considered to be the reference link, and the position of other RBs is determined in relation to this reference body. If the reference link is stationary, the term frame or ground is used.

The coordinates in the definition of degree of freedom can be linear or angular. Also the coordinates used can be absolute (measured with regard to the frame) or relative.

Figures 1.3a and 1.3b show two forms of a planar, one degree of freedom joint, namely a rotating pin joint and a translating slider joint. These are both typically referred to as full joints. The one degree of freedom joint has 5 degrees of constraint. The pin joint allows one rotational (R) DOF, and the slider joint allows one translational (T) DOF between the joined links.

Figure 1.4 shows examples of two degrees of freedom joints, which simultaneously allow two independent, relative motions, namely translation (T) and rotation (R), between the joined links. A two degrees of freedom joint is usually referred to

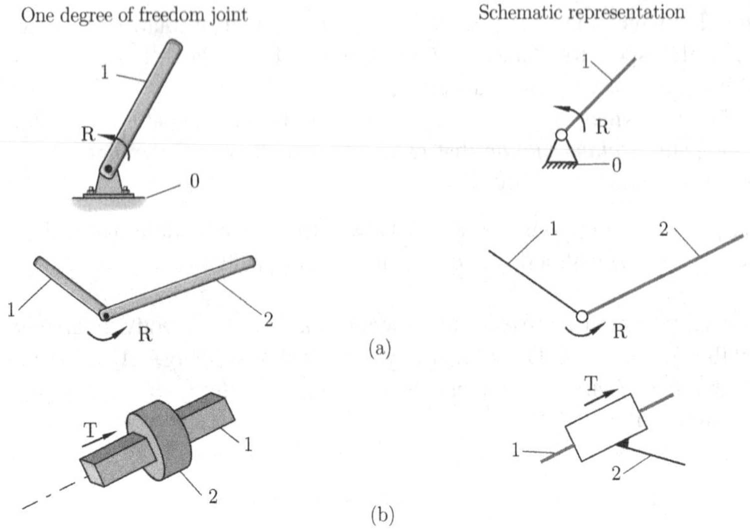

Fig. 1.3 One degree of freedom joint, full joint (c_5): (a) pin joint, and (b) slider joint

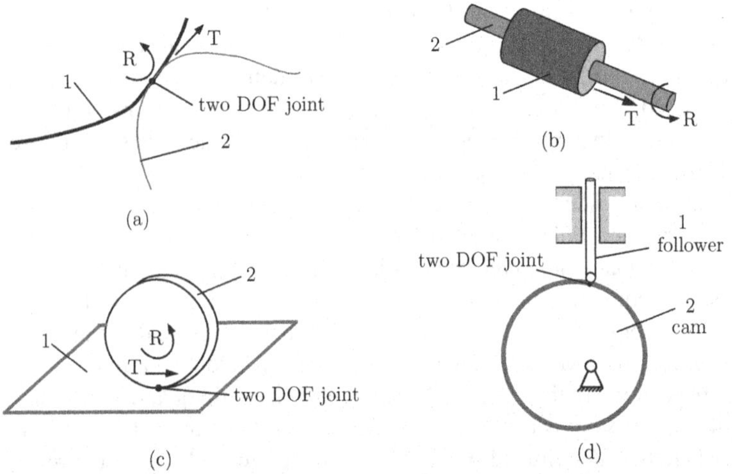

Fig. 1.4 Two degrees of freedom joint, half-joint (c_4): (a) general joint, (b) cylinder joint, (c) roll and slide disk, and (d) cam-follower joint

as a half-joint and has 4 degrees of constraint. A two degrees of freedom joint is sometimes also called a roll-slide joint because it allows both rotation (rolling) and translation (sliding).

Figure 1.5 shows a joystick, a ball-and-socket joint, or a sphere joint. This is an example of a three degrees of freedom joint (3 degrees of constraint) that allows three independent angular motions between the two links that are joined. Note that to visualize the degree of freedom of a joint in a mechanism, it is helpful to “mentally

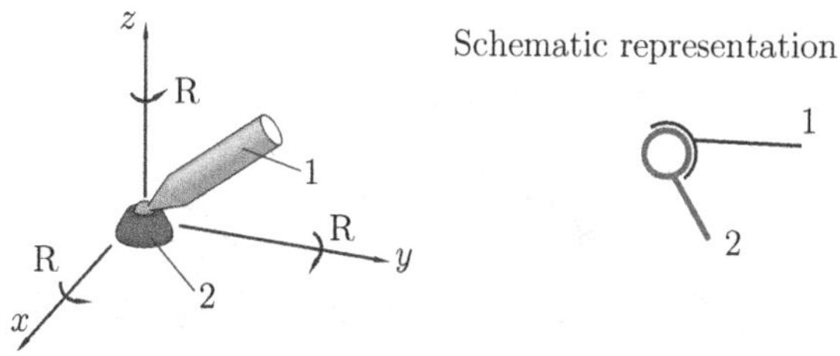

Fig. 1.5 Three degrees of freedom joint (c_3): ball and socket joint

disconnect" the two links that create the joint from the rest of the mechanism. It is easier to see how many degrees of freedoms the two joined links have with respect to one another.

The type of contact between the elements can be point (P), curve (C), or surface (S). The term lower joint was coined by Reuleaux to describe joints with surface contact. He used the term higher joint to describe joints with point or curve contact.

The *order of a joint* is defined as the number of links joined minus one. The combination of two links has order one and it is a single joint, Fig. 1.6a. As additional links are placed on the same joint, the order is increased on a one for one basis, Fig. 1.6b. Joint order has significance in the proper determination of overall degrees of freedom for an assembly. Bodies linked by joints form a *kinematic chain.* Kinematic chains are shown in Fig. 1.7. A *contour* or *loop* is a configuration described by a polygon consisting of links connected by joints, Fig. 1.7a.

The presence of loops in a mechanical structure can be used to define the following types of chains:

- *closed kinematic chains* have one or more loops so that each link and each joint is contained in at least one of the loops, Fig. 1.7a;

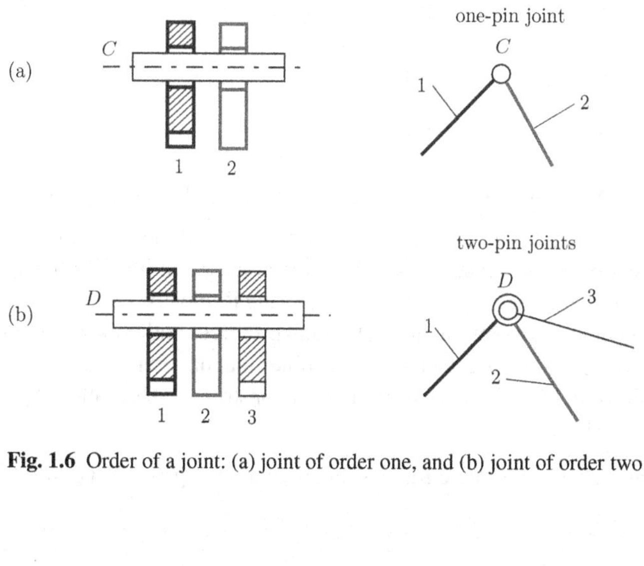

Fig. 1.6 Order of a joint: (a) joint of order one, and (b) joint of order two (multiple joints)

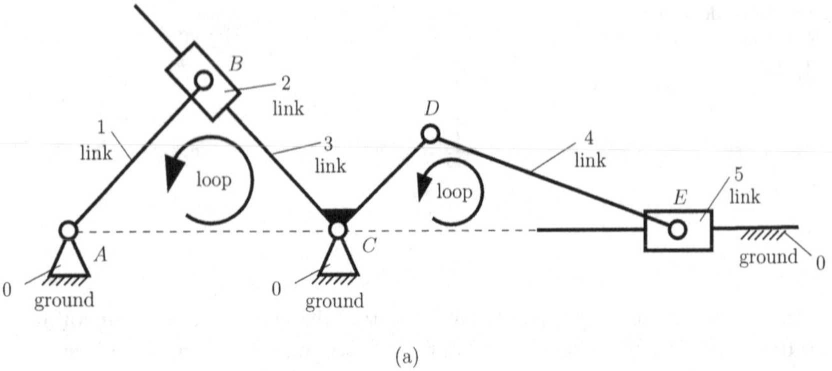

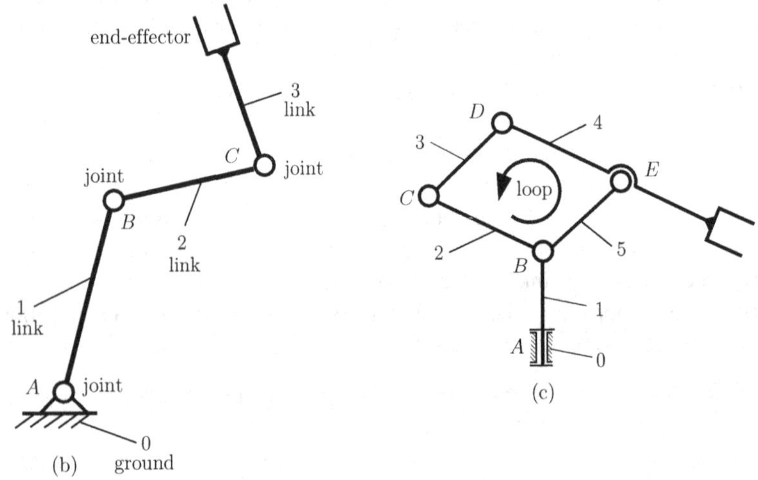

Fig. 1.7 Kinematic chains: (a) closed kinematic chain, (b) open kinematic chain, and (c) mixed kinematic chain

- *open kinematic chains* contain no closed loops, Fig. 1.7b. A common example of an open kinematic chain is an industrial robot;
- *mixed kinematic chains* are a combination of closed and open kinematic chains.

Figure 1.7c shows a robotic manipulator with parallelogram hinged mechanism.

A *mechanism* is defined as a kinematic chain in which at least one link has been "grounded" or attached to the frame, Figs. 1.7a and 1.8. Using Reuleaux's definition, a *machine* is a collection of mechanisms arranged to transmit forces and do work. He viewed all energy, or force-transmitting devices as machines that utilize mechanisms as their building blocks to provide the necessary motion constraints. The following terms can be defined, Fig. 1.8a:

- a *crank* is a link that makes a complete revolution about a fixed grounded pivot;

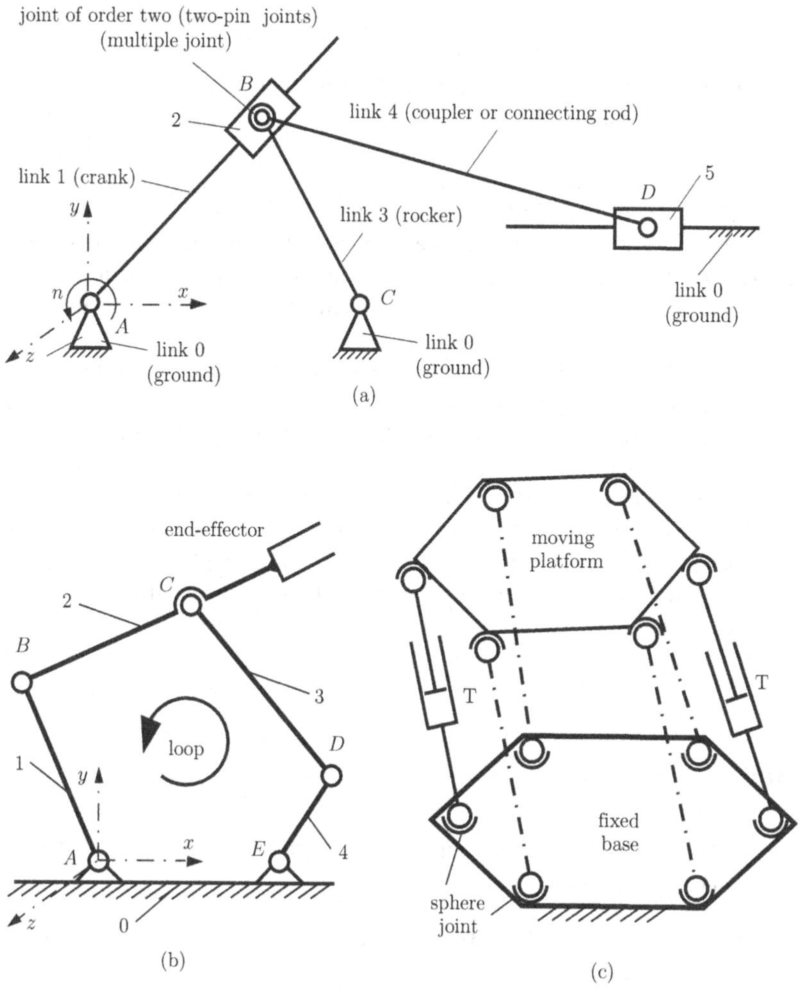

Fig. 1.8 (a) Mechanism with five moving links, (b) parallel link robot, and (c) Stewart mechanism

- a *rocker* is a link that has oscillatory (back and forth) rotation and is fixed to a grounded pivot;
- a *coupler* or connecting rod is a link that has complex motion and is not fixed to ground.

Ground is defined as any link or links that are fixed (non-moving) with respect to the reference frame. Note that the reference frame may in fact itself be in motion.

Figure 1.8b illustrates a five-bar linkage consisting of five links, including the base link 0, connected by five joints. The mechanism can be viewed as two link arms (1, 2 and 3, 4) connected at a point C. It is a closed kinematic chain formed by the five links. The position of the end-effector is determined if two of the five joint angles are given. Figure 1.8c shows the Stewart mechanism, which consists of a moving platform, a fixed base, and six powered cylinders connecting the moving platform to the base frame. The position and orientation of the moving platform are determined by the six independent actuators. This mechanism has spherical joints (three degrees of freedom joints).

The concept of *number of degrees of freedom* is fundamental to the analysis of mechanisms. It is usually necessary to be able to determine quickly the number of DOF of any collection of links and joints that may be used to solve a problem.

The number of degrees of freedom or the *mobility* of a system can be defined as: the number of inputs that need to be provided in order to create a predictable system output, or the number of independent coordinates required to define the position of the system.

The *class* f of a mechanism is the number of degrees of freedom that are eliminated from all the links of the system.

Every free body in space has six degrees of freedom. A system of class f consisting of n movable links has $(6-f)\,n$ degrees of freedom. Each joint with j degrees of constraint diminishes the freedom of motion of the system by $j-f$ degrees of freedom. The number of joints with k degrees of constraint is denoted as c_k. A *driver* link is that part of a mechanism that causes motion. An example is a crank. The number of driver links is equal to the number of DOF of the mechanism. A *driven* link or *follower* is that part of a mechanism whose motion is affected by the motion of the driver.

1.3 Dyads

For the special case of planar mechanisms (f=3) the number of degrees of freedom of the particular system has the form

$$M = 3n - 2c_5 - c_4, \tag{1.1}$$

where n is the number of moving links, c_5 is the number of one degree of freedom joints, and c_4 is the number of two degrees of freedom joints.

There is a special significance to kinematic chains that do not change their degrees of freedom after being connected to an arbitrary system. Kinematic chains defined in this way are called *system groups* or *fundamental kinematic chains*. Connecting them to or disconnecting them from a given system enables given systems to be modified or structurally new systems to be created while maintaining the original degrees of freedom. The term system group has been introduced for the classification of planar mechanisms used by Assur and further investigated by Artobolevski. Limiting to planar systems from Eq. 1.1, it can be obtained as

$$3n - 2c_5 = 0, \tag{1.2}$$

according to which the number of system group links n is always even. In Eq. 1.2 there are no two degrees of freedom joints because a c_4 joint (two degrees of freedom joint) can be substituted with two one degree of freedom joints and an extra link.

The simplest fundamental kinematic chain is the binary group with two links (n=2) and three one degree of freedom joints ($c_5 = 3$). The binary group is also called a *dyad*. The sets of links shown in Fig. 1.9 are dyads and one can distinguish the following classical types:

1. rotation rotation rotation or dyad RRR as shown in Fig. 1.9a;
2. rotation rotation translation or dyad RRT as shown in Fig. 1.9b;
3. rotation translation rotation or dyad RTR as shown in Fig. 1.9c;
4. translation rotation translation or dyad TRT as shown in Fig. 1.9d;
5. translation translation rotation or dyad RTT as shown in Fig. 1.9e.

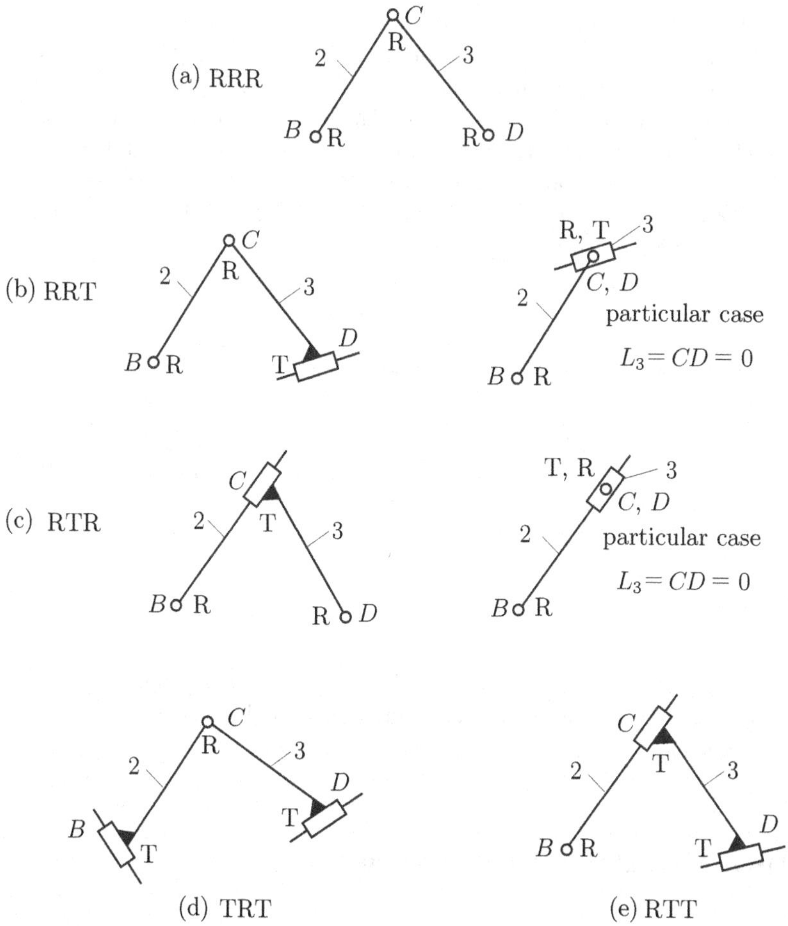

Fig. 1.9 Types of dyads: (a) RRR, (b) RRT, (c) RTR, (d) TRT, and (e) RTT

The advantage of the group classification of a system lies in its simplicity. The solution of the whole system can then be obtained by composing partial solutions.

1.4 Independent Contours

A contour is a configuration described by a polygon consisting of links connected by joints. A contour with at least one link that is not included in any other contour of the chain is called an *independent contour*. The number of independent contours, N, of a kinematic chain can be computed as

$$N = c - n, \tag{1.3}$$

where c is the number of joints, and n is the number of moving links.

Planar kinematic chains are presented in Fig. 1.10. The kinematic chain shown in Fig. 1.10a has two moving links, 1 and 2 ($n = 2$), three joints ($c = 3$), and one independent contour ($N = c - n = 3 - 2 = 1$). This kinematic chain is a dyad. The kinematic chain shown in Fig. 1.10b has three moving links, 1, 2, and 3 ($n = 3$), four joints ($c = 4$), and one independent contour ($N = c - n = 4 - 3 = 1$). A closed chain with three moving links, 1, 2, and 3 ($n = 3$), and one fixed link 0, connected by four joints ($c = 4$) is shown in Fig. 1.10c.

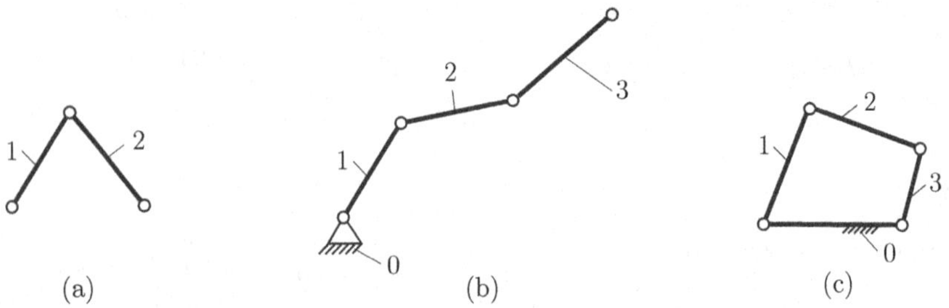

Fig. 1.10 Planar kinematic chains with contours

This is a four-bar mechanism. In order to find the number of independent contours, only the moving links are considered. Thus, there is one independent contour ($N = c - n = 4 - 3 = 1$).

1.5 Planar Mechanism Decomposition

A planar mechanism is shown in Fig. 1.11. This kinematic chain can be decomposed into system groups and driver links. The number of DOF for this mechanism is $M = 3n - 2c_5 - c_4 = 3n - 2c_5$. The mechanism has five moving links ($n = 3$).

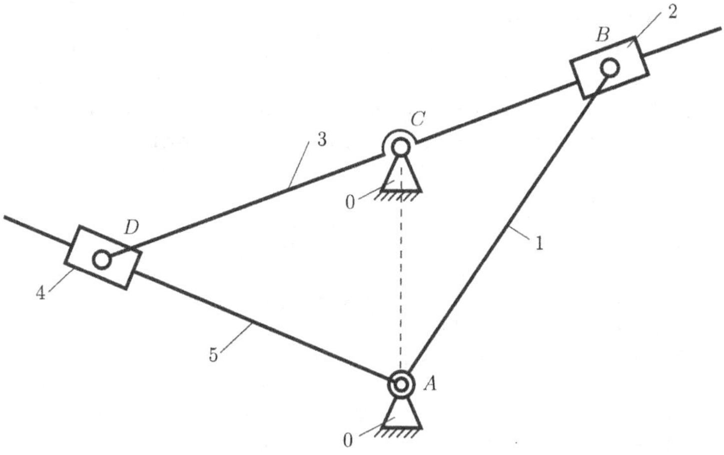

Fig. 1.11 Planar R-RTR-RTR mechanism

To find the number of c_5 a *connectivity table* will be used, Fig. 1.12a. The links are represented with bars (two node links) or triangles (three node links). The one degree of freedom joints (rotational joint or translation joint) are represented with a cross circle. The first column has the number of the current link, the second column shows the links connected to the current link, and the last column contains the graphical representation. The link 1 is connected to ground 0 at A and to link 2 at B, Fig. 1.12a. The link 2 is connected to link 1 at B and to link 3 at B. Next, link 3 is connected to link 2 at B, link 0 at C, and link 4 at D. Link 3 is a ternary link because it is connected to three links. At B there is a joint between link 1 and link 2 and a joint between link 2 and link 3. Link 4 is connected to link 3 at D and to link 5 at D. The last link, 5, is connected to link 4 at D and to 0 at A. In this way the table in Fig. 1.12a is obtained. At A there is a multiple joint, two rotational joints, one joint between link 1 and link 0, and one joint between link 5 and link 0.

The *structural diagram* is obtained using the graphical representation of the table connecting all the links Fig. 1.12b. The c_5 joints (with cross circles), all the links, and the way the links are connected are represented on the structural diagram. The number of one degree of freedom joints is given by the number of cross circles. From Fig. 1.12b it results that $c_5 = 7$. The number of DOF for the mechanism is $M = 3(5) - 2(7) = 1$. If $M = 1$, there is just one driver link. One can choose link 1 as the driver link of the mechanism. Once the driver link is taken away from the mechanism the remaining kinematic chain (links 2, 3, 4, 5) has the mobility equal to zero. The dyad is the simplest system group and has two links and three joints. On the structural diagram one can notice that links 2 and 3 represent a dyad and links 4 and 5 represent another dyad. The mechanism has been decomposed into a driver link (link 1) and two dyads (links 2 and 3, and links 4 and 5).

Another graphical construction for the connectivity table, shown in Fig. 1.12a, is the *contour diagram*, that can be used to represent the mechanism in the following way: the numbered links are the nodes of the diagram and are represented by circles, and the joints are represented by lines that connect the nodes. Figure 1.12c shows the contour diagram for the planar mechanism. The maximum number of independent contours is given by $N = c - n = 7 - 5 = 2$, where $c = 7$ is the number of joints and $n = 5$ is the number of moving links. The connectivity table, the structural diagram,

link	connected to	representation
1	2 0	A 1 B
2	1 3	B 2 B
3	0 2 4	3 D B C
4	3 5	D 4 D
5	0 4	D 5 A

(a)

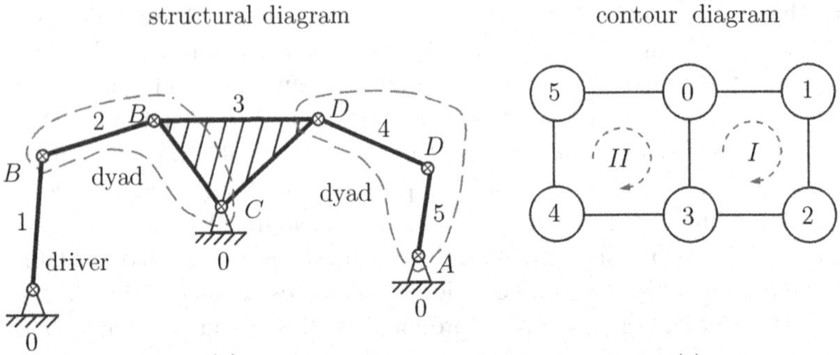

(b) (c)

Fig. 1.12 Connectivity table, structural diagram, and contour diagram for R-RTR-RTR mechanism

and the contour diagram are not unique for this mechanism. Using the structural diagram the mechanism can be decomposed into a driver link (link 1) and two dyads (links 2 and 3, and links 4 and 5). If the driver link is link 1, the mechanism has the same structure no matter what structural diagram is used.

Next, the driver link with rotational motion (R) and the dyads are represented as shown in Fig. 1.13. The first dyad (BBC) has the length $l_2 = l_{BB}$ equal to zero,

$l_{BB} = 0$, Fig. 1.13b. The second dyad (DDA) has the length $l_4 = l_{DD}$ equal to zero, $l_{DD} = 0$, Fig. 1.13c.

Using Fig. 1.13b, the first dyad (BBC) has a rotational joint at B (R), a translational joint at B (T), and a rotational joint at C (R). The first dyad (BBC) is a rotation translation rotation dyad (dyad RTR). Using Fig. 1.13c, the second dyad (DDA) has a rotational joint at D (R), a translational joint at D (T), and a rotational joint at A (R). The second dyad (DDA) is a rotation translation rotation dyad (dyad RTR). The mechanism is a R-RTR-RTR mechanism.

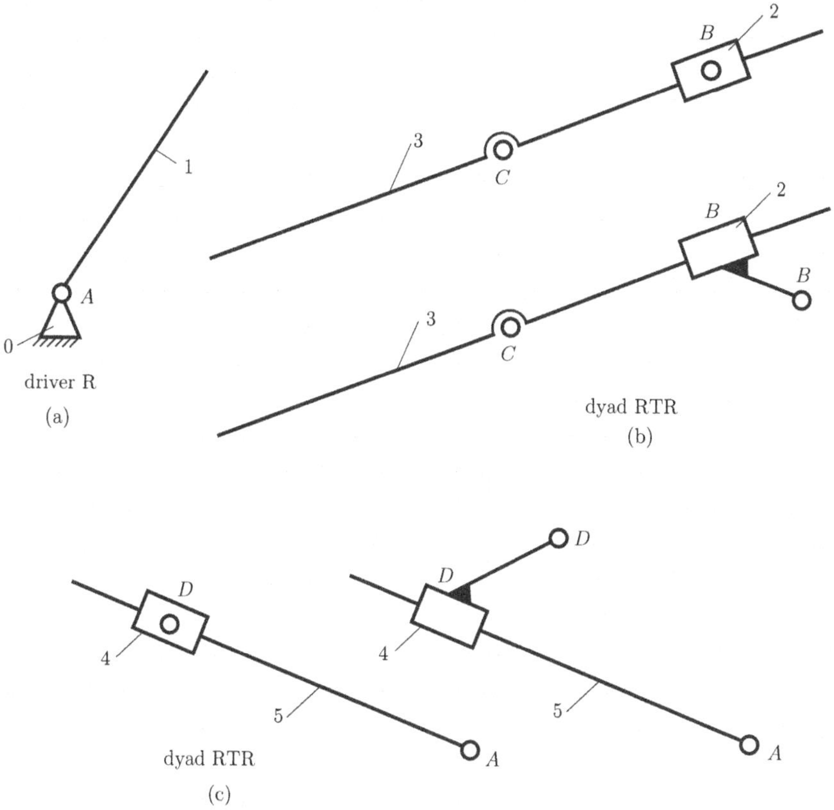

Fig. 1.13 Driver link and dyads for R-RTR-RTR mechanism

Chapter 2
Position Analysis

2.1 Absolute Cartesian Method

The position analysis of a kinematic chain requires the determination of the joint positions, the position of the centers of gravity, and the angles of the links with the horizontal axis. A planar link with the end nodes A and B is considered in Fig. 2.1. Let (x_A, y_A) be the coordinates of the joint A with respect to the reference frame xOy, and (x_B, y_B) be the coordinates of the joint B with the same reference frame. Using Pythagoras the following relation can be written

$$(x_B - x_A)^2 + (y_B - y_A)^2 = AB^2 = L_{AB}^2, \tag{2.1}$$

where L_{AB} is the length of the link AB. Let ϕ be the angle of the link AB with the horizontal axis Ox. Then, the slope m of the link AB is defined as

$$m = \tan\phi = \frac{y_B - y_A}{x_B - x_A}. \tag{2.2}$$

Let n be the intercept of AB with the vertical axis Oy. Using the slope m and the intercept n, the equation of the straight link, in the plane, is

$$y = mx + n, \tag{2.3}$$

where x and y are the coordinates of any point on this link.

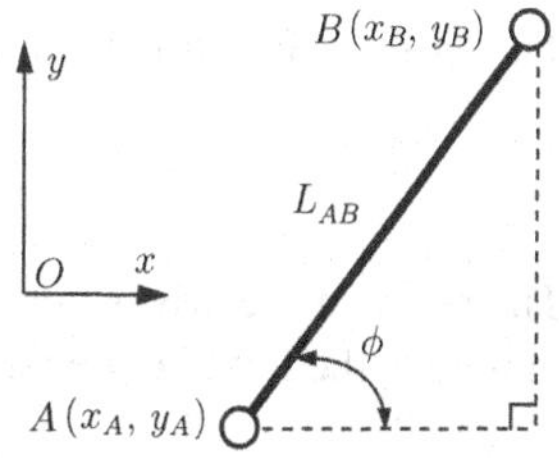

Fig. 2.1 Planar rigid link with two nodes

2.2 Slider-Crank (R-RRT) Mechanism

Exercise
The R-RRT (slider-crank) mechanism shown in Fig. 2.2a has the dimensions: $AB = 0.5$ m and $BC = 1$ m. The driver link 1 makes an angle $\phi = \phi_1 = 45°$ with the horizontal axis. Find the positions of the joints and the angles of the links with the horizontal axis.

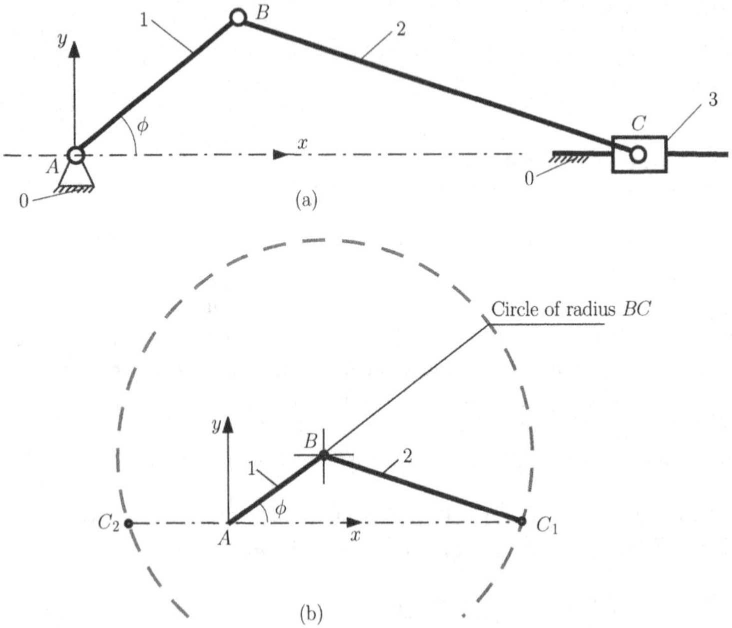

Fig. 2.2 (a) Slider-crank (R-RRT) mechanism and (b) two solutions for joint C: C_1 and C_2

Solution
The MATLAB® program starts with the statements:

```
clear all % clears all variables and functions
clc  % clears the command window and homes the cursor
close all % closes all the open figure windows
```

The MATLAB commands for the input data are:

```
AB=0.5;   BC=1.;
```

The angle of the driver link 1 with the horizontal axis $\phi = 45°$. The MATLAB command for the input angle is:

```
phi=pi/4;
```

where `pi` has a numerical value approximately equal to 3.14159.

Position of Joint A

A Cartesian reference frame xOy is selected. The joint A is in the origin of the reference frame, that is, $A \equiv O$,

$$x_A = 0, \; y_A = 0,$$

or in MATLAB:

```
xA=0; yA=0;
```

Position of Joint B

The unknowns are the coordinates of the joint B, x_B and y_B. Because the joint A is fixed and the angle ϕ is known, the coordinates of the joint B are computed from the following expressions:

$$\begin{aligned} x_B &= AB \cos\phi = (0.5)\cos 45^\circ = 0.353553\,\text{m}, \\ y_B &= AB \sin\phi = (0.5)\sin 45^\circ = 0.353553\,\text{m}. \end{aligned} \tag{2.4}$$

The MATLAB commands for Eq. 2.4 are:

```
xB=AB*cos(phi);
yB=AB*Sin(phi);
```

where `phi` is the angle ϕ in radians.

Position of Joint C

The unknowns are the coordinates of the joint C, x_C and y_C. The joint C is located on the horizontal axis $y_C = 0$ and with MATLAB:

```
yC=0;
```

The length of the segment BC is constant

$$(x_B - x_C)^2 + (y_B - y_C)^2 = BC^2, \tag{2.5}$$

or

$$(0.353553 - x_C)^2 + (0.353553 - 0)^2 = 1^2.$$

Equation 2.5 with MATLAB command is:

```
eqnC='(xB-xCsol)^2+(yB-yC)^2=BC^2';
```

where `xCsol` is the unknown. To solve the equation, a specific MATLAB command will be used. The command:

```
solve('eqn1','eqn2',...,'eqnN','var1','var2',...'varN')
```

attempts to solve an equation or set of equations `'eqn1','eqn2',...,'eqnN'` for the variables `'eqnN','var1','var2',...'varN'`. The set of equations are symbolic expressions or strings specifying equations. The MATLAB command to find the solution `xCsol` of the equation:

```
eqnC='(xB-xCsol)^2+(yB-yC)^2=BC^2'
```

is

```
solC=solve(eqnC,'xCsol');
```

Because it is a quadratic equation two solutions are found for the position of C. The two solutions are given in a vector form: `solC` is a vector with two components `solC(1)` and `solC(2)`. To obtain the numerical solutions the `eval` command has to be used:

```
xC1=eval(solC(1));
xC2=eval(solC(2));
```

The command `eval(s)`, where `s` is a string, executes the string as an expression or statement. The two solutions for x_C, as shown in Fig. 2.2b, are:

$$x_{C_1} = 1.289 \text{ m} \quad \text{and} \quad x_{C_2} = -0.5819 \text{ m}.$$

To determine the correct position of the joint C for the mechanism, an additional condition is needed. For the first quadrant, $0 \le \phi \le 90°$, the condition is $x_C > xB$. This MATLAB condition for `xC` located in the first quadrant is:

```
if xC1 > xB xC = xC1; else xC = xC2; end
```

The general form of the `if` statement is:

`if` *expression statements* `else` *statements* `end`

The x-coordinate of the joint C is $x_C = x_{C_1} = 1.2890$m. The angle of the link 2 (link BC) with the horizontal is

$$\phi_2 = \arctan\frac{y_B - y_C}{x_B - x_C}.$$

The MATLAB expression for the angle ϕ_2 is:

```
phi2 = atan((yB-yC)/(xB-xC));
```

The statement `atan(s)` is the arctangent of the elements of `s`. The numerical solutions for B, C, and ϕ_2 are printed using the statements:

```
fprintf('xB = %g (m) \n', xB)
fprintf('yB = %g (m) \n', yB)
fprintf('xC = %g (m) \n', xC)
fprintf('yC = %g (m) \n', yC)
fprintf('phi2 = %g (degrees) \n', phi2*180/pi)
```

The statement `fprintf(f,format,s)` writes data in the real part of array `s` to the file `f`. The data is formated under control of the specified `format` string. The results of the program are displayed as:

```
xB = 0.353553 (m)
yB = 0.353553 (m)
xC = 1.28897 (m)
yC = 0 (m)
phi2 = -20.7048 (degrees)
```

The mechanism is plotted with the help of the command `plot`. The statement `plot(x,y,c)` plots vector `y` versus vector `x`, and `c` is a character string. For the R-RRT mechanism two straight lines AB and BC are plotted with:

```
plot([xA,xB],[yA,yB],'r-o',[xB,xC],[yB,yC],'b-o')
```

The line AB is a red (`r` red), solid line (`-` solid), with a circle (`o` circle) at each data point and the line BC is a blue (`b` blue), solid line with a circle at each data point. The graphic of the mechanism obtained with MATLAB is shown in Fig. 2.3. The x-axis and y-axis are labeled using the commands:

```
xlabel('x (m)')
ylabel('y (m)')
```

and a title is added with:

```
title('positions for \phi = 45 (deg)')
```

On the figure, the joints A, B, and C are identified with the statements:

```
text(xA,yA,' A'),...
text(xB,yB,' B'),...
```

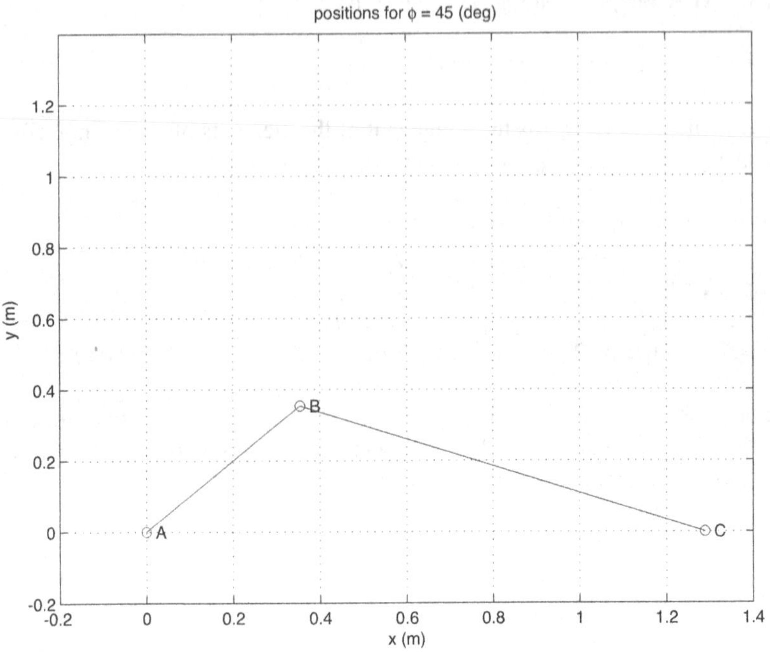

Fig. 2.3 MATLAB graphic of R-RRT mechanism

```
text(xC,yC,' C'),...
axis([-0.2 1.4 -0.2 1.4]),...
grid
```

The commas and ellipses (`...`) after the command are used to execute the commands together. Otherwise, the data will be plotted, then the labels will be added and the data replotted, and so on.

The statement `axis([xMIN xMAX yMIN yMAX])` sets scaling for the x and y axes on the current plot. To improve the graph a background grid was added with the command `grid`.

The MATLAB program for the positions is given in Appendix A.1.

2.3 Four-Bar (R-RRR) Mechanism

Exercise

The considered four-bar (R-RRR) planar mechanism is shown in Fig. 2.4. The driver link is the rigid link 1 (the element AB) and the origin of the reference frame is at A. The following data are given: AB=0.150 m, BC=0.35 m, CD=0.30 m, CE=0.15 m,

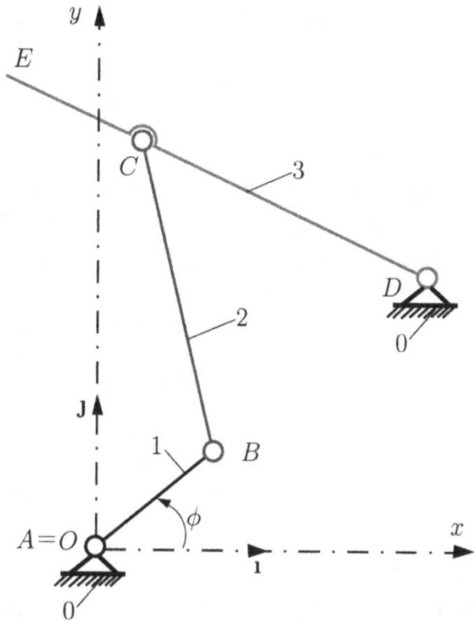

Fig. 2.4 Four-bar (R-RRR) mechanism

x_D=0.30 m, and y_D=0.30 m. The angle of the driver link 1 with the horizontal axis is $\phi = \phi_1 = 45°$. Find the positions of the joints and the angles of the links with the horizontal axis.

Solution

The Cartesian reference frame *xyz* with the unit vectors $[\mathbf{ı}, \mathbf{ȷ}, \mathbf{k}]$ is shown Fig. 2.4. Since the joint A is the origin of the reference system $A \equiv O$ the coordinates of A are $x_A = 0$, $y_A = 0$ and the position vector of A is $\mathbf{r}_A = x_A\,\mathbf{ı} + y_A\,\mathbf{ȷ}$. The position vectors $\mathbf{r}_A$ and $\mathbf{r}_D$ are introduced in MATLAB as:

```
rA = [xA yA 0];
rD = [xD yD 0];
```

In the MATLAB environment, a three-dimensional vector `v` is written as a list of variables `v = [ x y z ]`, where `x`, `y`, and `z` are the spatial coordinates of the vector `v`. The first component of the vector `v` is `x=v(1)`, the second component is `y=v(2)`, and the third component is `z=v(3)`.

Position of Joint B

The unknowns are the coordinates of the joint B, x_B and y_B. Because the joint A is fixed and the angle ϕ is known, the coordinates of the joint B are computed from the following expressions:

$$x_B = AB\cos\phi = 0.106\,\text{m}, \quad y_B = AB\sin\phi = 0.106\,\text{m}.$$

The position vector of B is $\mathbf{r}_B = x_B\mathbf{\imath} + y_B\mathbf{\jmath}$. The MATLAB program for this part is:

```
xB = AB*cos(phi); yB = AB*sin(phi); rB = [xB yB 0];
```

Position of Joint C
The unknowns are the coordinates of the joint C, x_C and y_C. Knowing the positions of the joints B and D, the position of the joint C can be computed using the fact that the lengths of the links BC and CD are constants

$$(x_C - x_B)^2 + (y_C - y_B)^2 = BC^2,$$
$$(x_C - x_D)^2 + (y_C - y_D)^2 = CD^2,$$

or

$$(x_C - 0.106)^2 + (y_C - 0.106)^2 = 0.350^2,$$
$$(x_C - 0.300)^2 + (y_C - 0.300)^2 = 0.300^2. \quad (2.6)$$

Equations 2.6 consist of two quadratic equations. Solving this system of equations, two sets of solutions are found for the position of the joint C. These solutions are

$$x_{C_1} = 0.0401\,\text{m}, \; y_{C_1} = 0.4498\,\text{m} \;\text{ and }\; x_{C_2} = 0.4498\,\text{m}, \; y_{C_2} = 0.0401\,\text{m}.$$

The MATLAB program for calculating the coordinates of C_1 and C_2 is:

```
eqnC1 = '( xCsol - xB )^2 + ( yCsol - yB )^2 = BC^2';
eqnC2 = '( xCsol - xD )^2 + ( yCsol - yD )^2 = CD^2';
solC = solve(eqnC1, eqnC2, 'xCsol, yCsol');
xCpositions = eval(solC.xCsol);
yCpositions = eval(solC.yCsol);
% first component of the vector xCpositions
xC1 = xCpositions(1);
% second component of the vector xCpositions
xC2 = xCpositions(2);
% first component of the vector yCpositions
yC1 = yCpositions(1);
% second component of the vector yCpositions
yC2 = yCpositions(2);
```

The points C_1 and C_2 are the intersections of the circle of radius BC (with the center at B) with the circle of radius CD (with the center at D), as shown in Fig. 2.5. To determine the correct position of the joint C for this mechanism, a constraint condition is needed: $x_C < x_D$. Because $x_D = 0.300\,\text{m}$, the coordinates of joint C have the following numerical values:

$$x_C = x_{C_1} = 0.0401\,\text{m} \quad \text{and} \quad y_C = y_{C_1} = 0.4498\,\text{m}.$$

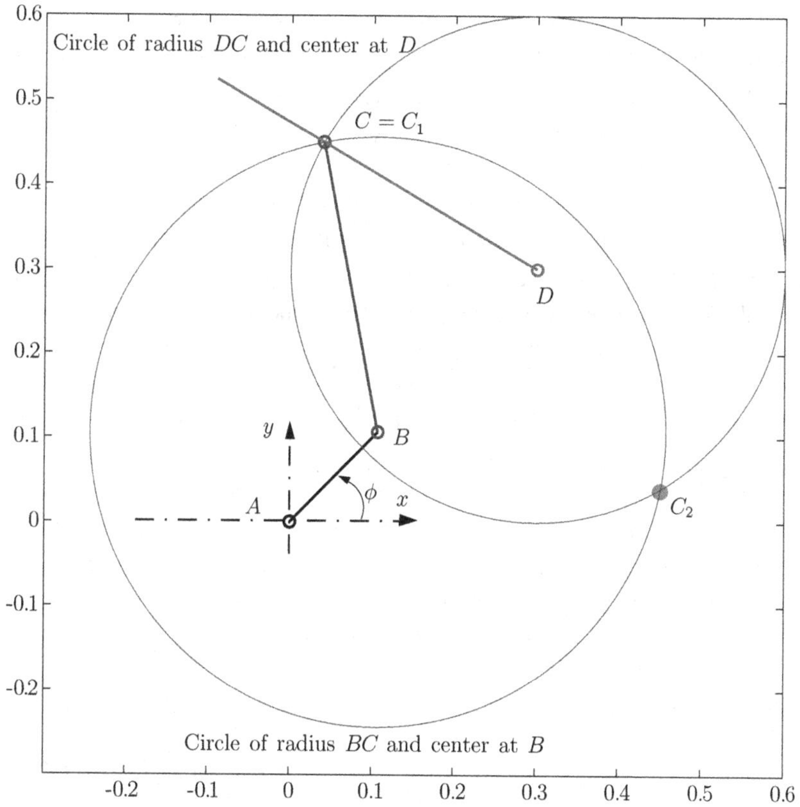

Fig. 2.5 Two solutions for the position of joint C

The MATLAB program for selecting the correct position of C is:

```
if xC1 < xD
    xC = xC1; yC=yC1;
else
    xC = xC2; yC=yC2;
end
rC = [xC yC 0]; % Position vector of C
```

Position of Point E

The unknowns are the coordinates of the point E, x_E and y_E. The position of the point E is determined from the equation

$$(x_E - x_C)^2 + (y_E - y_C)^2 = CE^2, \tag{2.7}$$

or

$$(x_E - 0.0401)^2 + (y_E - 0.4498)^2 = 0.15^2.$$

The joints D, C and E are located on the same straight element DE. For these points, the following equation can be written

$$\frac{y_D - y_C}{x_D - x_C} = \frac{y_E - y_C}{x_E - x_C}, \tag{2.8}$$

or

$$\frac{0.300 - 0.4498}{0.300 - 0.0401} = \frac{y_E - 0.4498}{x_E - 0.0401}.$$

Equations 2.7 and 2.8 form a system from which the coordinates of the point E can be computed. Two solutions are obtained, Fig. 2.6, and the numerical values are

$$x_{E_1} = -0.0899\,\text{m},\ y_{E_1} = 0.5247\,\text{m},$$
$$x_{E_2} = 0.1700\,\text{m},\ \ y_{E_2} = 0.3749\,\text{m}.$$

The MATLAB program for calculating the coordinates of E_1 and E_2 is:

```
eqnE1 = '( xEsol - xC )^2 + ( yEsol - yC )^2 = CE^2 ';
eqnE2 = '(yD-yC)/(xD-xC)=(yEsol-yC )/(xEsol-xC)';
solE = solve(eqnE1, eqnE2, 'xEsol, yEsol');
```

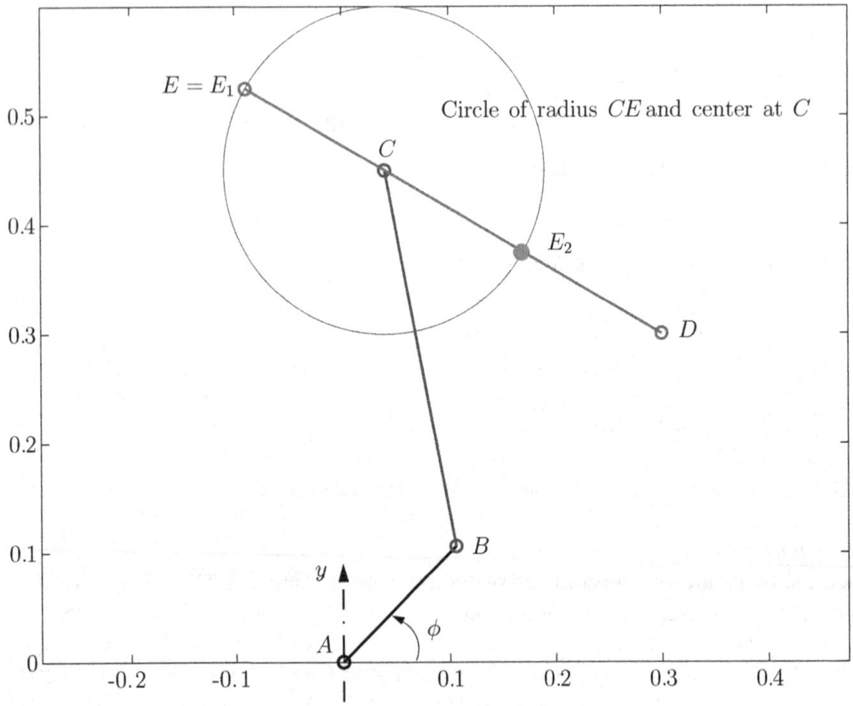

Fig. 2.6 Two solutions for the position of point E

```
xEpositions=eval(solE.xEsol);
yEpositions=eval(solE.yEsol);
xE1 = xEpositions(1); xE2 = xEpositions(2);
yE1 = yEpositions(1); yE2 = yEpositions(2);
```

For continuous motion of the mechanism, a constraint condition is needed, $x_E < x_C$. Using this condition, the coordinates of the point E are

$$x_E = x_{E_1} = -0.0899\,\mathrm{m} \quad \text{and} \quad y_E = y_{E_1} = 0.5247\,\mathrm{m}.$$

The MATLAB program for selecting the correct position of E is

```
if xE1 < xC
    xE = xE1; yE=yE1;
else
    xE = xE2; yE=yE2;
end
rE = [xE yE 0]; % Position vector of E
```

The angles of the links 2, 3, and 4 with the horizontal are

$$\phi_2 = \arctan\frac{y_B - y_C}{x_B - x_C}, \quad \phi_3 = \arctan\frac{y_D - y_C}{x_D - x_C},$$

and in MATLAB

```
phi2 = atan((yB-yC)/(xB-xC));
phi3 = atan((yD-yC)/(xD-xC));
```

The results are printed using the statements:

```
fprintf('rA = [ %g, %g, %g ] (m) \n', rA)
fprintf('rD = [ %g, %g, %g ] (m) \n', rD)
fprintf('rB = [ %g, %g, %g ] (m) \n', rB)
fprintf('rC = [ %g, %g, %g ] (m) \n', rC)
fprintf('rE = [ %g, %g, %g ] (m) \n', rE)
fprintf('phi2 = %g (degrees) \n', phi2*180/pi)
fprintf('phi3 = %g (degrees) \n', phi3*180/pi)
```

The graph of the mechanism using MATLAB for $\phi = \pi/4$ is given by:

```
plot([xA,xB],[yA,yB],'k-o','LineWidth',1.5)
hold on % holds the current plot
plot([xB,xC],[yB,yC],'b-o','LineWidth',1.5)
hold on
plot([xD,xE],[yD,yE],'r-o','LineWidth',1.5)
```

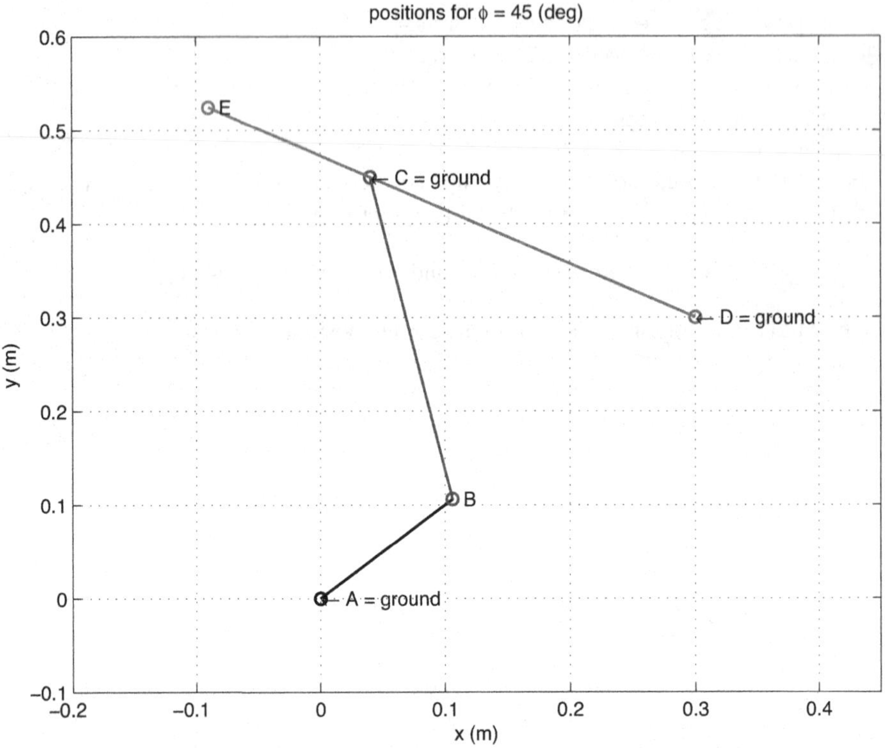

Fig. 2.7 MATLAB graphic of R-RRR mechanism

```
% adds major grid lines to the current axes
grid on,...
xlabel('x (m)'), ylabel('y (m)'),...
title('positions for \phi = 45 (deg)'),...
text(xA,yA,'\leftarrow A = ground',...
'HorizontalAlignment','left'),...
text(xB,yB,' B'),...
text(xC,yC,'\leftarrow C = ground',...
'HorizontalAlignment','left'),...
text(xD,yD,'\leftarrow D = ground',...
'HorizontalAlignment','left'),...
text(xE,yE,' E'), axis([-0.2 0.45 -0.1 0.6])
```

The graph of the R-RRR mechanism using MATLAB is shown in Fig. 2.7. The MATLAB program for the positions and the results is given in Appendix A.2.

2.4 R-RTR-RTR Mechanism

Exercise

The planar R-RTR-RTR mechanism considered is shown in Fig. 2.8. The driver link is the rigid link 1 (the link AB). The following numerical data are given: $AB = 0.15$ m, $AC = 0.10$ m, $CD = 0.15$ m, $DF = 0.40$ m, and $AG = 0.30$ m. The angle of the driver link 1 with the horizontal axis is $\phi = 30°$.

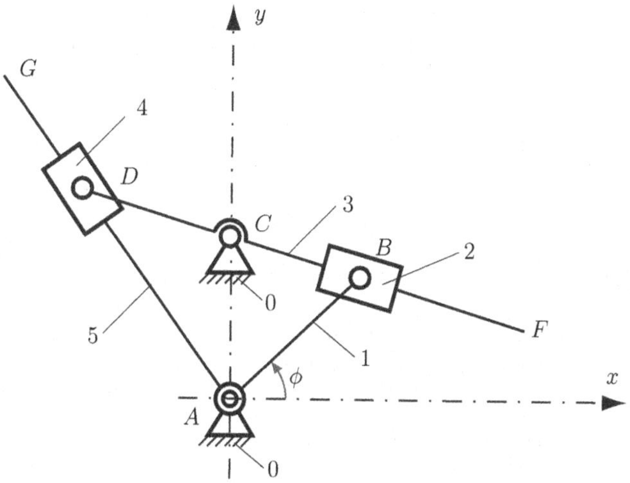

Fig. 2.8 R-RTR-RTR mechanism

Solution

The MATLAB commands for the input data are:

```
AB=0.15; AC=0.10; CD=0.15;    %(m)
phi=pi/6;   %(rad)
DF=0.40; AG=0.30;    % (m)
```

A Cartesian reference frame xOy is selected. The joint A is in the origin of the reference frame, that is, $A \equiv O$, $x_A = 0$, $y_A = 0$.

Position of Joint C

The position vector of C is $\mathbf{r}_C = x_C\,\mathbf{ı} + y_C\,\mathbf{ȷ} = 0.1\,\mathbf{ȷ}$ m.

Position of Joint B

The unknowns are the coordinates of the joint B, x_B and y_B. Because the joint A is fixed and the angle ϕ is known, the coordinates of the joint B are computed from the following expressions:

$$x_B = AB\cos\phi = 0.15\cos 30^\circ = 0.1299\,\text{m},\ y_B = AB\sin\phi = 0.15\sin 30^\circ = 0.075\,\text{m},$$

and $\mathbf{r}_B = x_B\mathbf{\imath} + y_B\mathbf{\jmath}$. The MATLAB statements for the positions of the joints A, C, E, and B are:

```
xA = 0 ; yA = 0 ; rA = [xA yA 0] ;  % Position of A
xC = 0 ; yC = AC ; rC = [xC yC 0] ; % Position of C
% Position of B
xB=AB*cos(phi); yB=AB*sin(phi); rB=[xB yB 0];
```

Position of Joint D
The unknowns are the coordinates of the joint D, x_D and y_D. The length of the segment CD is constant:

$$(x_D - x_C)^2 + (y_D - y_C)^2 = CD^2, \tag{2.9}$$

or

$$(x_D - 0)^2 + (y_D - 0.10)^2 = 0.15^2.$$

The points B, C, and D are on the same straight line with the slope

$$m = \frac{(y_B - y_C)}{(x_B - x_C)} = \frac{(y_D - y_C)}{(x_D - x_C)}, \tag{2.10}$$

or

$$\frac{(0.075 - 0.1)}{(0.1299 - 0.0)} = \frac{(y_D - 0.1)}{(x_D - 0.0)}.$$

Equations 2.9 and 2.10 form a system from which the coordinates of the joint D can be computed. To solve the system of equations the MATLAB statement `solve` will be used:

```
eqnD1='( xDsol - xC )^2 + ( yDsol - yC )^2 = CD^2 ';
eqnD2='(yB - yC)/(xB - xC)=(yDsol - yC)/(xDsol - xC)';
solD = solve(eqnD1, eqnD2, 'xDsol, yDsol');
xDpositions = eval(solD.xDsol);
yDpositions = eval(solD.yDsol);
% first component of the vector xDpositions
xD1 = xDpositions(1);
% second component of the vector xDpositions
xD2 = xDpositions(2);
% first component of the vector yDpositions
yD1 = yDpositions(1);
% second component of the vector yDpositions
yD2 = yDpositions(2);
```

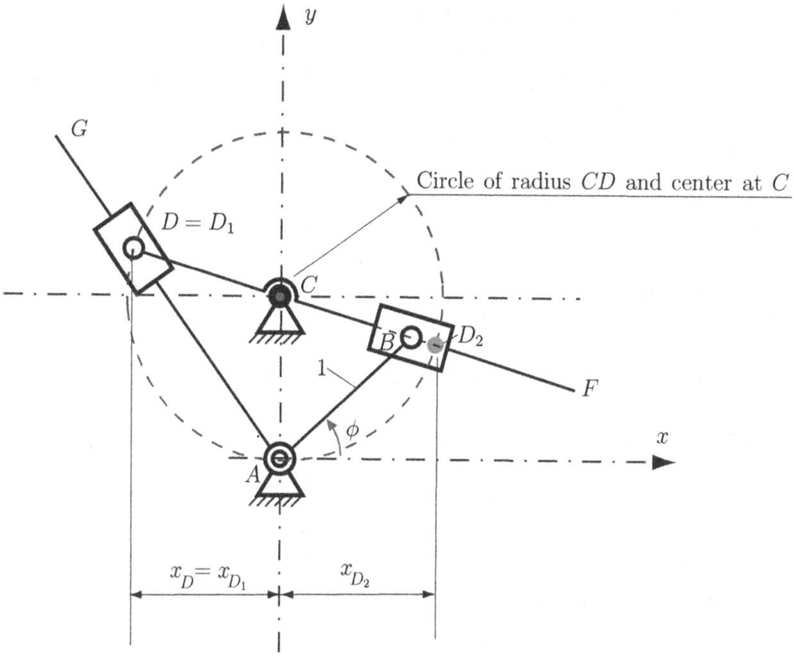

Fig. 2.9 Graphical solutions for joint D

These solutions D_1 and D_2 are located at the intersection of the line BC with the circle centered in C and radius CD (Fig. 2.9), and they have the following numerical values:

$$x_{D1} = -0.1473 \text{ m}, \; y_{D1} = 0.1283 \text{ m},$$
$$x_{D2} = 0.1473 \text{ m}, \; y_{D2} = 0.0717 \text{ m}.$$

To determine the correct position of the joint D for the mechanism, an additional condition is needed. For the first quadrant, $0 \leq \phi \leq 90°$, the condition is $x_D \leq x_C$. This condition with MATLAB is given by:

```
if xD1 <= xC
   xD = xD1; yD=yD1;
else
   xD = xD2; yD=yD2;
end
rD = [xD yD 0]; % Position of D
```

Because $x_C = 0$, the coordinates of the joint D are:

$$x_D = x_{D1} = -0.1473\,\text{m} \quad \text{and} \quad y_D = y_{D1} = 0.1283\,\text{m}.$$

The angles of the links 2, 3, and 4 with the horizontal are

$$\phi_2 = \arctan\frac{y_B - y_C}{x_B - x_C}, \quad \phi_3 = \phi_2, \quad \phi_4 = \arctan\frac{y_D}{x_D} + \pi, \quad \phi_5 = \phi_4,$$

and in MATLAB:

```
phi2 = atan((yB-yC)/(xB-xC));
phi3 = phi2;
phi4 = atan(yD/xD)+pi;
phi5 = phi4;
```

The points F and G are calculated in MATLAB with:

```
xF = xD + DF*cos(phi3) ; yF = yD + DF*sin(phi3) ;
rF = [xF yF 0]; % Position vector of F
xG = AG*cos(phi5) ; yG = AG*sin(phi5) ;
rG = [xG yG 0]; % Position vector of G
```

The results are printed using the statements:

```
fprintf('rA = [ %g, %g, %g ] (m) \n', rA)
fprintf('rC = [ %g, %g, %g ] (m) \n', rC)
fprintf('rB = [ %g, %g, %g ] (m) \n', rB)
fprintf('rD = [ %g, %g, %g ] (m) \n', rD)
fprintf('phi2 = phi3 = %g (degrees) \n', phi2*180/pi)
fprintf('phi4 = phi5 = %g (degrees) \n', phi4*180/pi)
fprintf('rF = [ %g, %g, %g ] (m) \n', rF)
fprintf('rG = [ %g, %g, %g ] (m) \n', rG)
```

The graph of the mechanism in MATLAB for $\phi = \pi/6$ is given by:

```
plot([xA,xB],[yA,yB],'k-o','LineWidth',1.5)
hold on % holds the current plot
plot([xD,xC],[yD,yC],'b-o','LineWidth',1.5)
hold on
plot([xC,xB],[yC,yB],'b-o','LineWidth',1.5)
hold on
plot([xB,xF],[yB,yF],'b-o','LineWidth',1.5)
hold on
plot([xA,xD],[yA,yD],'r-o','LineWidth',1.5)
hold on
plot([xD,xG],[yD,yG],'r-o','LineWidth',1.5)
grid on,...
xlabel('x (m)'), ylabel('y (m)'),...
title('positions for \phi = 30 (deg)'),...
```

```
text(xA,yA,'\leftarrow A = ground',...
'HorizontalAlignment','left'),...
text(xB,yB,' B'),...
text(xC,yC,'\leftarrow C = ground',...
'HorizontalAlignment','left'),...
text(xD,yD,' D'),...
text(xF,yF,' F'), text(xG,yG,' G'),...
axis([-0.3 0.3 -0.1 0.3])
```

The MATLAB program for the positions and the results for the R-RTR-RTR mechanism for $\phi = 30^\circ$ is given in Appendix A.3.

2.5 R-RTR-RTR Mechanism: Complete Rotation

For a complete rotation of the driver link *AB*, $0 \leq \phi \leq 360^\circ$, a step angle of 60° is selected. To calculate the position analysis for a complete cycle the MATLAB statement `for` *var=startval:step:endval, statement* `end` is used. It repeatedly evaluates *statement* in a loop. The counter variable of the loop is *var*. At the start, the variable is initialized to value *startval* and is incremented (or decremented when *step* is negative) by the value *step* for each iteration. The *statement* is repeated until *var* has incremented to the value *endval*. For the considered mechanism the following applies:

```
for phi=0:pi/3:2*pi, Program block, end;
```

2.5.1 *Method I: Constraint Conditions*

Method I uses constraint conditions for the mechanism for each quadrant. For the mechanism, there are several conditions for the position of the joint *D*. For the angle ϕ located in the first quadrant $0^\circ \leq \phi \leq 90^\circ$ and the fourth quadrant $270^\circ \leq \phi \leq 360^\circ$ (Fig. 2.10), the following relation exists between x_D and x_C:

$$x_D \leq x_C = 0.$$

For the angle ϕ located in the second quadrant $90^\circ < \phi \leq 180^\circ$ and the third quadrant $180^\circ < \phi < 270^\circ$ (Fig. 2.11), the following relation exists between x_D and x_C:

$$x_D \geq x_C = 0.$$

The following MATLAB commands are used to determine the correct position of the joint *D* for all four quadrants:

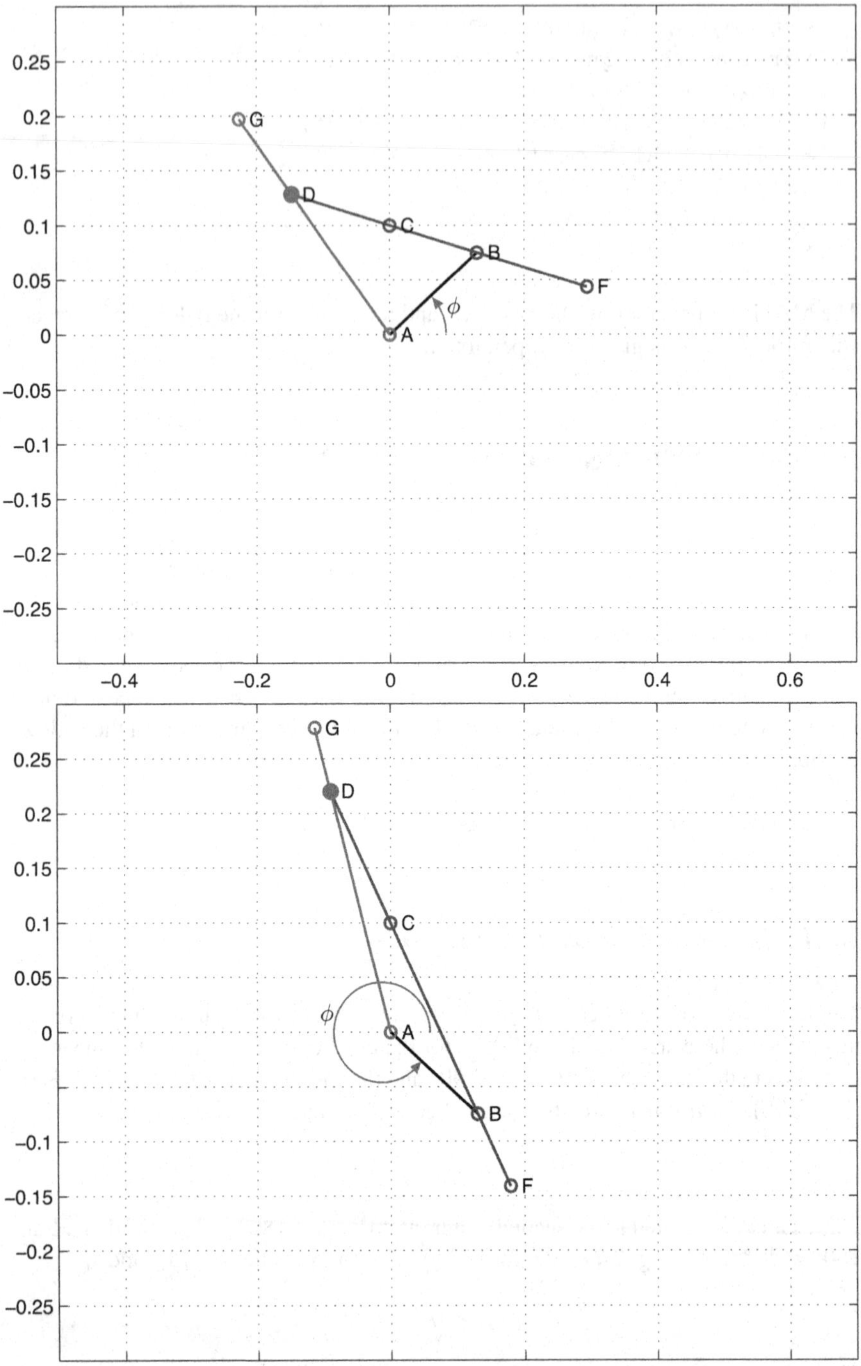

Fig. 2.10 R-RTR-RTR mechanism for $0^\circ < \phi \leq 90^\circ$ and $270^\circ \leq \phi \leq 360^\circ$

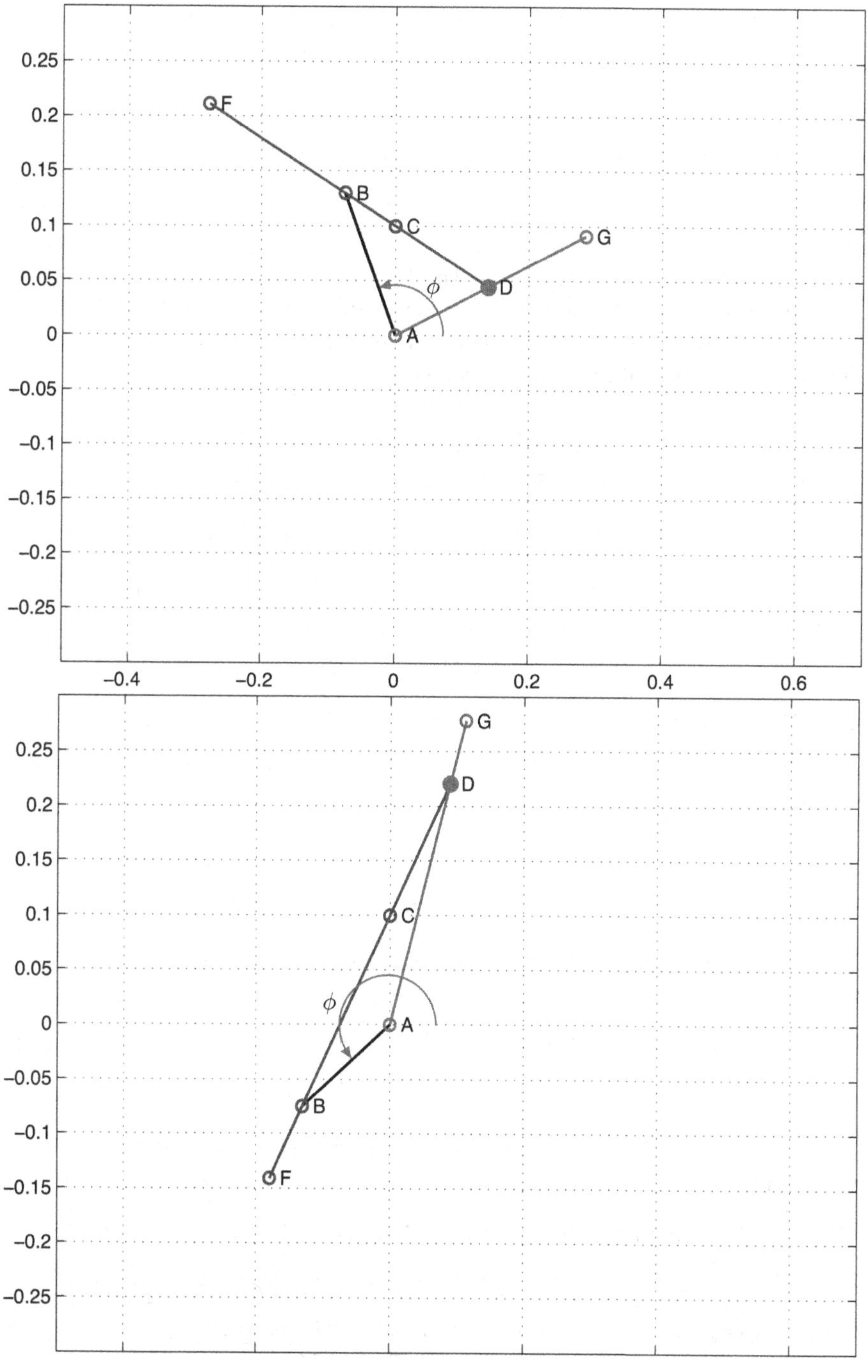

Fig. 2.11 R-RTR-RTR mechanism for $90^\circ < \phi \leq 180^\circ$ and $180^\circ \leq \phi \leq 270^\circ$

```
if (phi>=0 && phi<=pi/2)||(phi >= 3*pi/2 && phi<=2*pi)
if xD1 <= xC xD = xD1; yD=yD1; else xD = xD2; yD=yD2;
end
else
if xD1 >= xC xD = xD1; yD=yD1; else xD = xD2; yD=yD2;
end
end
```

where || is the logical OR function. The MATLAB program and the results for a complete rotation of the driver link using method I is given in Appendix A.4. The graphic of the mechanism for a complete rotation of the driver link is given in Fig. 2.12. To simplify the graphic the points E and G are not shown on the figure.

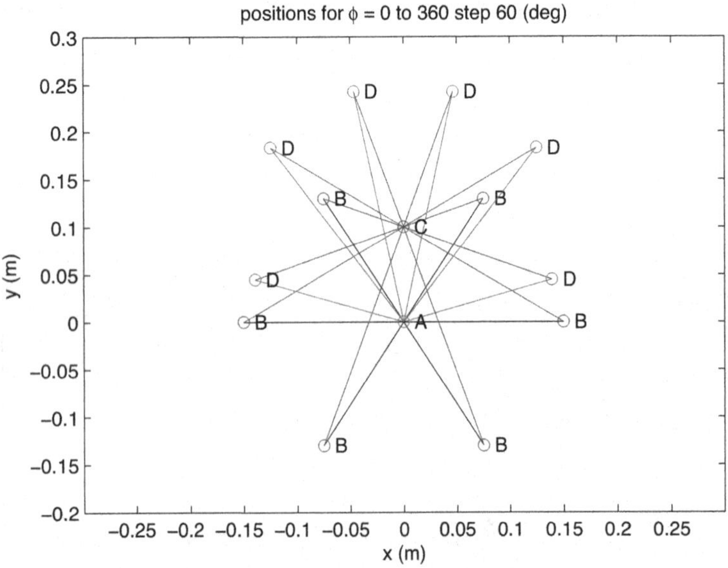

Fig. 2.12 MATLAB graphic of R-RTR-RTR mechanism for a complete rotation of the driver link $0° \leq \phi \leq 360°$

Another way of plotting the simulation of the mechanism for a complete rotation of the driver link is:

```
plot([xA,xB],[yA,yB],'k-o',[xB,xC],[yB,yC],'b-o',...
[xC,xD],[yC,yD],'b-o',[xD,xA],[yD,yA],'r-o'),...
hold off % resets axes properties to their defaults
text(xA,yA,'  A'), text(xB,yB,'  B'),...
text(xC,yC,'  C'), text(xD,yD,'  D'),...
axis([-0.3 0.3 -0.2 0.3]),grid,...
pause(0.8)
```

The MATLAB command `hold off` resets the axes properties to their defaults before drawing new plots and the command `pause(T)` pauses execution for `T` seconds before continuing.

2.5.2 Method II: Euclidian Distance Function

Another method for the position analysis for a complete rotation of the driver link uses constraint conditions only for the initial value of the angle ϕ. Next for the mechanism, the correct position of the joint D is calculated using a simple function, the Euclidian distance between two points P and Q:

$$d = \sqrt{(x_P - x_Q)^2 + (y_P - y_Q)^2}. \tag{2.11}$$

In MATLAB, the following function is introduced with a m-file (Dist.m):

```
function d=Dist(xP,yP,xQ,yQ);
d=sqrt((xP-xQ)^2+(yP-yQ)^2);
end
```

For the initial angle $\phi = 0°$, the constraint is $x_D \leq x_C$, so the first position of the joint D, that is, D_0, is calculated for the first step $D = D_0 = D_k$. For the next position of the joint, D_{k+1}, there are two solutions D^I_{k+1} and D^{II}_{k+1}, $k = 0, 1, 2, \ldots$. In order to choose the correct solution of the joint, D_{k+1}, the distances between the old position, D_k, and each new calculated positions D^I_{k+1} and D^{II}_{k+1}. The distances between the known solution D_k and the new solutions D^I_{k+1} and D^{II}_{k+1} are d^I_k and d^{II}_k are compared. If the distance to the first solution is less than the distance to the second solution, $d^I_k < d^{II}_k$, then the correct answer is $D_{k+1} = D^I_{k+1}$, or else $D_{k+1} = D^{II}_{k+1}$ (Fig. 2.13).

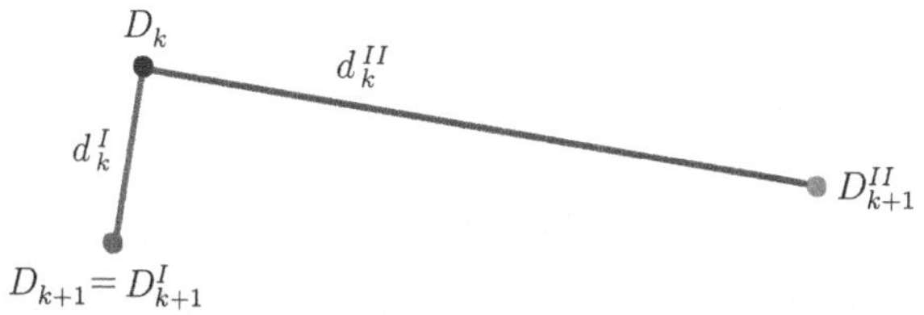

Fig. 2.13 Selection of the correct position: $d^I_k < d^{II}_k \Rightarrow D_{k+1} = D^I_{k+1}$

The following MATLAB statements are used to determine the correct position of the joint D using a single condition for all four quadrants:

```
% at the initial moment phi=0 => increment = 0
increment = 0 ;

% the step has to be small for this method
step=pi/6;
for phi=0:step:2*pi,

xB = AB*cos(phi); yB = AB*sin(phi); rB = [xB yB 0];
fprintf('rB =  [ %g, %g, %g ] (m)\n', rB)
eqnD1='( xDsol - xC )^2 + ( yDsol - yC )^2=CD^2';
eqnD2='(yB-yC)/(xB-xC)=(yDsol-yC)/(xDsol-xC)';
solD = solve(eqnD1, eqnD2, 'xDsol, yDsol');
xDpositions = eval(solD.xDsol);
yDpositions = eval(solD.yDsol);
xD1 = xDpositions(1); xD2 = xDpositions(2);
yD1 = yDpositions(1); yD2 = yDpositions(2);

% select the correct position for D
%    only for increment == 0
% the selection process is automatic
%    for all the other steps

if increment == 0
   if xD1 <= xC xD=xD1; yD=yD1; else xD=xD2; yD=yD2;
   end
else
  dist1 = Dist(xD1,yD1,xDold,yDold);
  dist2 = Dist(xD2,yD2,xDold,yDold);
  if dist1 < dist2 xD=xD1; yD=yD1; else xD=xD2; yD=yD2;
   end
end
xDold=xD;
yDold=yD;

increment=increment+1;

rD = [xD yD 0];
end
```

At the beginning of the rotation the driver link makes an angle `phi=0` with the horizontal and the value of counter `increment` is 0. The MATLAB statement:

```
increment=increment+1;
```

specifies that 1 is to be added to the value in `increment` and the result stored back in `increment`. The value `increment` should be incremented by 1.

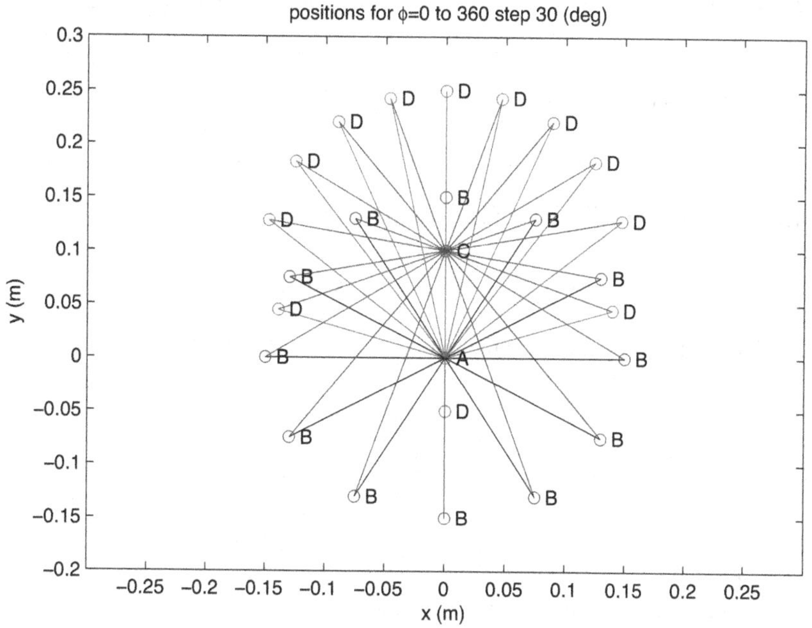

Fig. 2.14 MATLAB graphic of R-RTR-RTR mechanism for a complete rotation of the driver using the Euclidian distance

With this algorithm the correct solution is selected using just one constraint relation for the initial step and then, automatically, the problem is solved. In this way, it is not necessary to have different constraints for different quadrants.

For the Euclidian distance method the selection of the step of the angle ϕ is very important. If the step of the angle has a large value the method might give wrong answers and that is why it is important to check the graphic of the mechanism.

The MATLAB program for a complete rotation of the driver link using the second method is given in Appendix A.5. The graph of the mechanism for a complete rotation of the driver link (the step of the angle is 30°) is given in Fig. 2.14 (the points E and G are not shown).

2.6 Path of a Point on a Link with General Plane Motion

Exercise: R-RRT Mechanism

The mechanism shown in Fig. 2.2a has $AB = 0.5$ m and $BC = 1$ m. The link 2 (connecting rod BC) has a general plane motion: translation along the x-axis, translation along the y-axis, and rotation about the z-axis. The mass center of link 2 is located at C_2. Determine the path of point C_2 for a complete rotation of the driver link 1.

Solution
The coordinates of the joint B are

$$x_B = AB\cos\phi \quad \text{and} \quad y_B = AB\sin\phi,$$

where $0 \leq \phi \leq 360°$. The coordinates of the joint C are

$$x_C = x_B + \sqrt{BC^2 - y_B^2} \quad \text{and} \quad y_C = 0.$$

The mass center of the link 2 is the midpoint of the segment BC

$$x_{C_2} = \frac{x_B + x_C}{2} \quad \text{and} \quad y_{C_2} = \frac{y_B + y_C}{2}.$$

The MATLAB statements for the coordinates of C_2 are:

```
AB = .5; BC = 1; xA = 0; yA = 0; yC = 0;
incr = 0;
for phi=0:pi/10:2*pi,
  xB = AB*cos(phi); yB = AB*sin(phi);
  xC = xB + sqrt(BC^2-yB^2);
  incr = incr + 1;
  xC2(incr)=(xB+xC)/2; yC2(incr)=(yB+yC)/2;
end % end for
```

For the complete rotation of the driver link AB, $0 \leq \phi \leq 360°$, a step angle of $\pi/10$ was selected. For the coordinates of C_2 two vectors:

```
xC2=[xC2(1) xC2(2)  ...  xC2(incr)  ... ]
yC2=[yC2(1) yC2(2)  ...  yC2(incr)  ... ]
```

are obtained. The first components `xC2(1)` and `yC2(1)` are calculated for `phi=0` and `incr=1`. The path of C_2 is obtained by plotting the vector `yC2` in terms of `xC2`:

```
plot(xC2, yC2, '-ko'),...
xlabel('x (m)'), ylabel('y (m)'),...
title('Path described by C2'), grid
```

Figure 2.15 shows two plots: the mechanism for $0 \leq \phi \leq 360°$ and the closed path described by the point C_2 on the link 2 in general plane motion. The plots are obtained using the program in Appendix A.6.

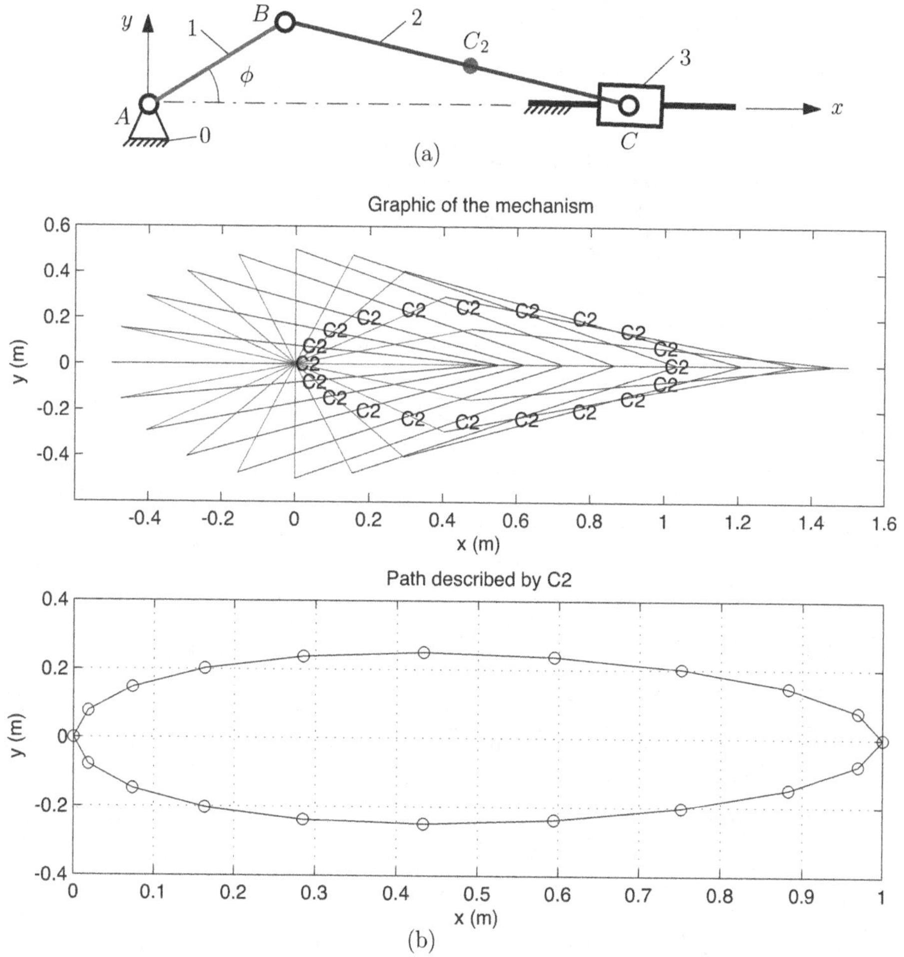

Fig. 2.15 (a) R-RRT mechanism, $AB = 0.5$ m, $BC = 1.0$ m, and $BC_2 = C_2C$; (b) MATLAB plots: mechanism for $0 \leq \phi \leq 360°$ and closed path described by point C_2

R-RRR Mechanism

The mechanism shown in Fig. 2.4 has the dimensions given in Sect. 2.3. The link 2 (link *BC*) has a general plane motion. The positions of the mechanism for $0 \leq \phi \leq 360°$ and the closed path described by the mass center C_2 of the link 2 are shown in Fig. 2.16. The plots are obtained using the program in Appendix A.7.

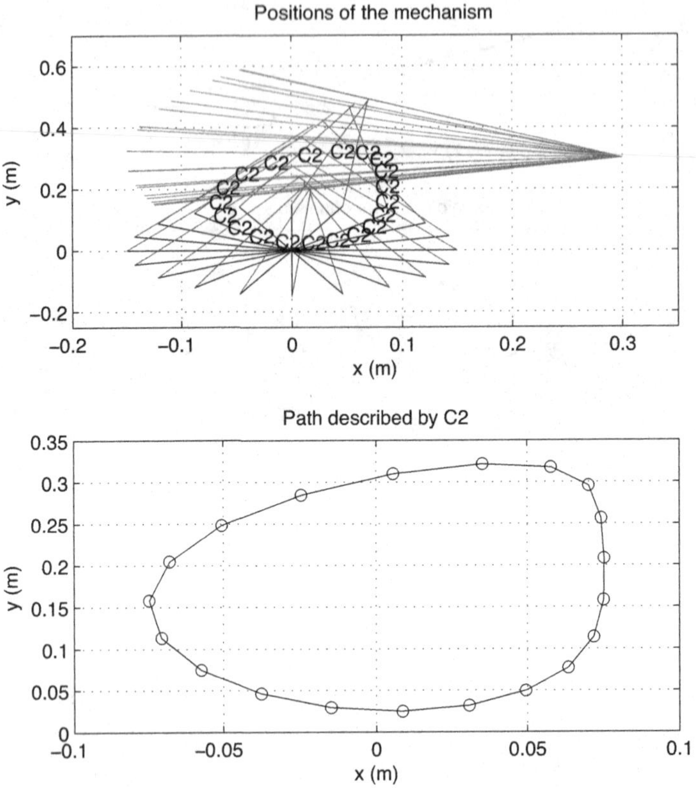

Fig. 2.16 Positions of the R-RRR mechanism for $0 \leq \phi \leq 360°$ and closed path described by the mass center C_2 of link 2.

2.7 Creating a Movie

The R-RTR-RTR mechanism shown in Fig. 2.8 has the dimensions given in Sect. 2.4. This example illustrates the use of movies to visualize the positions of the mechanism for $0 \leq \phi \leq 360°$.

The statement `moviein` is used to create a matrix large enough to hold 12 frames:

```
M = moviein(12);
```

The program has the structure

```
AB=0.15; AC=0.10; CD=0.15; %(m)
xA = 0; yA = 0; xC = 0 ; yC = AC;
% allocate/initialize the matrix to have 12 frames
M = moviein(12);
```

```
incr = 0;
for phi=0:pi/180:2*pi,
xB = AB*cos(phi); yB = AB*sin(phi);
eqnD1='(xDsol-xC)^2+(yDsol-yC)^2=CD^2';
eqnD2='(yB-yC)/(xB-xC)=(yDsol-yC)/(xDsol-xC)';
solD = solve(eqnD1, eqnD2, 'xDsol, yDsol');
xDpositions = eval(solD.xDsol);
yDpositions = eval(solD.yDsol);
xD1 = xDpositions(1); xD2 = xDpositions(2);
yD1 = yDpositions(1); yD2 = yDpositions(2);
if(phi>=0 && phi<=pi/2)||(phi >= 3*pi/2 && phi<=2*pi)
 if xD1 <= xC xD=xD1; yD=yD1; else xD=xD2; yD=yD2;
 end
 else
 if xD1 >= xC xD=xD1; yD=yD1; else xD=xD2; yD=yD2;
 end
end

plot([xA,xB],[yA,yB],'k-o',...
       [xB,xC],[yB,yC],'b-o',...
       [xC,xD],[yC,yD],'b-o',...
       [xD,xA],[yD,yA],'r-o'),...
text(xA,yA,'  A'), text(xB,yB,'  B'),...
text(xC,yC,'  C'), text(xD,yD,'  D'), grid;

% xlim([Xmin Xmax])
% sets the x limits to the specified values
xlim([-0.3 0.3]);
% ylim([Ymin Ymax])
% sets the x limits to the specified values
ylim([-0.3 0.3]);

incr = incr + 1;

M(:,incr) = getframe; % record the movie

end % end for

movie2avi(M,'RRTRRTR.avi');
```

The statement, `getframe` returns the contents of the current axes, exclusive of the axis labels, title, or tick labels. After generating the movie, the statement, `movie2avi(M,'filename.avi')` creates the AVI movie `filename` from the MATLAB movie `M`. The `filename` input is a string enclosed in single quotes. In this case the name of the movie file is `RRTRRTR.avi`.

Chapter 3
Velocity and Acceleration Analysis

3.1 Introduction

The motion of a rigid body (*RB*) is defined when the position vector, velocity and acceleration of all points of the rigid body are defined as functions of time with respect to a fixed reference frame with the origin at O_0.

Let $\mathbf{\imath}_0, \mathbf{\jmath}_0$, and $\mathbf{k}_0$, be the constant unit vectors of a fixed orthogonal Cartesian reference frame $x_0y_0z_0$ and $\mathbf{\imath}, \mathbf{\jmath}$ and $\mathbf{k}$ be the unit vectors of a body fixed (mobile or rotating) orthogonal Cartesian reference frame xyz (Fig. 3.1). The unit vectors $\mathbf{\imath}_0$, $\mathbf{\jmath}_0$, and $\mathbf{k}_0$ of the primary reference frame are constant with respect to time.

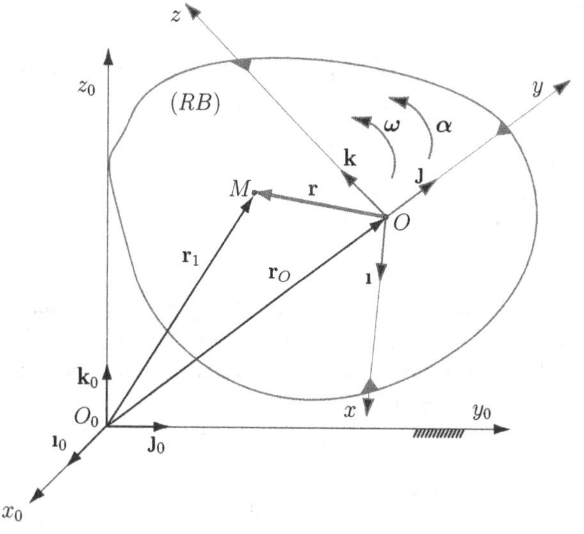

Fig. 3.1 Fixed orthogonal Cartesian reference frame with the unit vectors $[\mathbf{\imath}_0, \mathbf{\jmath}_0, \mathbf{k}_0]$; body fixed (or rotating) reference frame with the unit vectors $[\mathbf{\imath}, \mathbf{\jmath}, \mathbf{k}]$; the point M is an arbitrary point, $M \in (RB)$

A reference frame that moves with the rigid body is a *body-fixed* (or rotating) reference frame. The unit vectors $\mathbf{\imath}, \mathbf{\jmath}$, and $\mathbf{k}$ of the body-fixed reference frame are not constant, because they rotate with the body-fixed reference frame. The location of the point O is arbitrary.

The position vector of a point M, $M \in (RB)$, with respect to the fixed reference frame $x_0y_0z_0$ is denoted by $\mathbf{r}_1 = \mathbf{r}_{O_0M}$ and with respect to the rotating reference frame $Oxyz$ is denoted by $\mathbf{r} = \mathbf{r}_{OM}$. The location of the origin O of the rotating reference frame with respect to the fixed point O_0 is defined by the position vector $\mathbf{r}_O = \mathbf{r}_{O_0O}$. Then, the relation between the vectors $\mathbf{r}_1$, $\mathbf{r}$ and $\mathbf{r}_0$ is given by

$$\mathbf{r}_1 = \mathbf{r}_O + \mathbf{r} = \mathbf{r}_O + x\mathbf{\imath} + y\mathbf{\jmath} + z\mathbf{k}, \tag{3.1}$$

where x, y, and z represent the projections of the vector $\mathbf{r} = \mathbf{r}_{OM}$ on the rotating reference frame $\mathbf{r} = x\mathbf{\imath} + y\mathbf{\jmath} + z\mathbf{k}$.

The magnitude of the vector $\mathbf{r} = \mathbf{r}_{OM}$ is a constant as the distance between the points O and M is constant, $O \in (RB)$, and $M \in (RB)$. Thus, the x, y and z components of the vector $\mathbf{r}$ with respect to the rotating reference frame are constant. The unit vectors $\mathbf{\imath}, \mathbf{\jmath}$, and $\mathbf{k}$ are time-dependent vector functions. The vectors $\mathbf{\imath}, \mathbf{\jmath}$ and $\mathbf{k}$ are the unit vector of an orthogonal Cartesian reference frame, thus one can write

$$\mathbf{\imath} \cdot \mathbf{\imath} = 1, \quad \mathbf{\jmath} \cdot \mathbf{\jmath} = 1, \quad \mathbf{k} \cdot \mathbf{k} = 1, \tag{3.2}$$

$$\mathbf{\imath} \cdot \mathbf{\jmath} = 0, \quad \mathbf{\jmath} \cdot \mathbf{k} = 0, \quad \mathbf{k} \cdot \mathbf{\imath} = 0. \tag{3.3}$$

3.2 Velocity Field for a Rigid Body

The velocity of an arbitrary point M of the rigid body with respect to the fixed reference frame $x_0y_0z_0$, is the derivative with respect to time of the position vector $\mathbf{r}_1$

$$\begin{aligned} \mathbf{v} &= \frac{d\mathbf{r}_1}{dt} = \frac{d\mathbf{r}_{O_0M}}{dt} = \frac{d\mathbf{r}_O}{dt} + \frac{d\mathbf{r}}{dt} \\ &= \mathbf{v}_O + x\frac{d\mathbf{\imath}}{dt} + y\frac{d\mathbf{\jmath}}{dt} + z\frac{d\mathbf{k}}{dt} + \frac{dx}{dt}\mathbf{\imath} + \frac{dy}{dt}\mathbf{\jmath} + \frac{dz}{dt}\mathbf{k}, \end{aligned} \tag{3.4}$$

where $\mathbf{v}_O = \dot{\mathbf{r}}_O$ represent the velocity of the origin of the rotating reference frame $O_1x_1y_1z_1$ with respect to the fixed reference frame $Oxyz$. Because all the points in the rigid body maintain their relative position, their velocity relative to the rotating reference frame xyz is zero, i.e., $\dot{x} = \dot{y} = \dot{z} = 0$.

The velocity of point M is

$$\mathbf{v} = \mathbf{v}_O + x\frac{d\mathbf{\imath}}{dt} + y\frac{d\mathbf{\jmath}}{dt} + z\frac{d\mathbf{k}}{dt} = \mathbf{v}_O + x\dot{\mathbf{i}} + y\dot{\mathbf{j}} + z\dot{\mathbf{k}}.$$

The derivative of the Eqs. 3.2 and 3.3 with respect to time gives

$$\frac{d\imath}{dt}\cdot\imath = 0, \quad \frac{d\jmath}{dt}\cdot\jmath = 0, \quad \frac{d\mathbf{k}}{dt}\cdot\mathbf{k} = 0, \tag{3.5}$$

and

$$\frac{d\imath}{dt}\cdot\jmath + \imath\cdot\frac{d\jmath}{dt} = 0, \quad \frac{d\jmath}{dt}\cdot\mathbf{k} + \jmath\cdot\frac{d\mathbf{k}}{dt} = 0, \quad \frac{d\mathbf{k}}{dt}\cdot\imath + \mathbf{k}\cdot\frac{d\imath}{dt} = 0. \tag{3.6}$$

For Eq. 3.6 the following notation is used

$$\frac{d\imath}{dt}\cdot\jmath = -\imath\cdot\frac{d\jmath}{dt} = \omega_z,$$

$$\frac{d\jmath}{dt}\cdot\mathbf{k} = -\jmath\cdot\frac{d\mathbf{k}}{dt} = \omega_x,$$

$$\frac{d\mathbf{k}}{dt}\cdot\imath = -\mathbf{k}\cdot\frac{d\imath}{dt} = \omega_y, \tag{3.7}$$

where ω_x, ω_y and ω_z may be considered as the projections of a vector $\boldsymbol{\omega}$

$$\boldsymbol{\omega} = \omega_x\imath + \omega_y\jmath + \omega_z\mathbf{k}.$$

To calculate $\frac{d\imath}{dt}, \frac{d\jmath}{dt}, \frac{d\mathbf{k}}{dt}$ the relation for an arbitrary vector $\mathbf{v}$ will be used

$$\mathbf{v} = v_x\imath + v_y\jmath + v_z\mathbf{k} = (\mathbf{v}\cdot\imath)\imath + (\mathbf{v}\cdot\jmath)\jmath + (\mathbf{v}\cdot\mathbf{k})\mathbf{k}. \tag{3.8}$$

Using Eq. 3.8 and the results from Eqs. 3.5 and 3.6 one can write

$$\begin{aligned}\frac{d\imath}{dt} &= \left(\frac{d\imath}{dt}\cdot\imath\right)\imath + \left(\frac{d\imath}{dt}\cdot\jmath\right)\jmath + \left(\frac{d\imath}{dt}\cdot\mathbf{k}\right)\mathbf{k}\\ &= (0)\imath + (\omega_z)\jmath - (\omega_y)\mathbf{k}\\ &= \begin{vmatrix}\imath & \jmath & \mathbf{k}\\ \omega_x & \omega_y & \omega_z\\ 1 & 0 & 0\end{vmatrix} = \boldsymbol{\omega}\times\imath,\end{aligned}$$

$$\begin{aligned}\frac{d\jmath}{dt} &= \left(\frac{d\jmath}{dt}\cdot\imath\right)\imath + \left(\frac{d\jmath}{dt}\cdot\jmath\right)\jmath + \left(\frac{d\jmath}{dt}\cdot\mathbf{k}\right)\mathbf{k}\\ &= (-\omega_z)\imath + (0)\jmath + (\omega_x)\mathbf{k}\\ &= \begin{vmatrix}\imath & \jmath & \mathbf{k}\\ \omega_x & \omega_y & \omega_z\\ 0 & 1 & 0\end{vmatrix} = \boldsymbol{\omega}\times\jmath,\end{aligned}$$

$$\begin{aligned}\frac{d\mathbf{k}}{dt} &= \left(\frac{d\mathbf{k}}{dt}\cdot\imath\right)\imath + \left(\frac{d\mathbf{k}}{dt}\cdot\jmath\right)\jmath + \left(\frac{d\mathbf{k}}{dt}\cdot\mathbf{k}\right)\mathbf{k}\\ &= (\omega_y)\imath - (\omega_x)\jmath + (0)\mathbf{k}\end{aligned} \tag{3.9}$$

$$= \begin{vmatrix} \mathbf{\imath} & \mathbf{\jmath} & \mathbf{k} \\ \omega_x & \omega_y & \omega_z \\ 0 & 0 & 1 \end{vmatrix} = \boldsymbol{\omega} \times \mathbf{k}.$$

The relations

$$\frac{d\mathbf{\imath}}{dt} = \boldsymbol{\omega} \times \mathbf{\imath}, \quad \frac{d\mathbf{\jmath}}{dt} = \boldsymbol{\omega} \times \mathbf{\jmath}, \quad \frac{d\mathbf{k}}{dt} = \boldsymbol{\omega} \times \mathbf{k}. \tag{3.10}$$

are known as *Poisson formulas* and $\boldsymbol{\omega}$ is the angular velocity vector. Using Eqs. 3.4 and 3.10 the velocity of the point M on the rigid body is

$$\mathbf{v} = \mathbf{v}_O + x\boldsymbol{\omega} \times \mathbf{\imath} + y\boldsymbol{\omega} \times \mathbf{\jmath} + z\boldsymbol{\omega} \times \mathbf{k} = \mathbf{v}_O + \boldsymbol{\omega} \times (x\mathbf{\imath} + y\mathbf{\jmath} + z\mathbf{k}),$$

or

$$\mathbf{v} = \mathbf{v}_O + \boldsymbol{\omega} \times \mathbf{r}. \tag{3.11}$$

Combining Eqs. 3.4 and 3.11 it results that

$$\frac{d\mathbf{r}}{dt} = \dot{\mathbf{r}} = \boldsymbol{\omega} \times \mathbf{r}. \tag{3.12}$$

Using Eq. 3.11 one can write the components of the velocity as

$$\begin{aligned} v_x &= v_{Ox} + z\,\omega_y - y\,\omega_z, \\ v_y &= v_{Oy} + x\,\omega_z - z\,\omega_x, \\ v_z &= v_{Oz} + y\,\omega_x - x\,\omega_y. \end{aligned}$$

The relation between the velocities $\mathbf{v}_M$ and $\mathbf{v}_O$ of two points M and O on the rigid body is

$$\mathbf{v}_M = \mathbf{v}_O + \boldsymbol{\omega} \times \mathbf{r}_{OM}, \tag{3.13}$$

or

$$\mathbf{v}_M = \mathbf{v}_O + \mathbf{v}_{MO}^{\text{rel}}, \tag{3.14}$$

where $\mathbf{v}_{MO}^{\text{rel}}$ is the relative velocity, for rotational motion, of M with respect to O and is given by

$$\mathbf{v}_{MO}^{\text{rel}} = \mathbf{v}_{MO} = \boldsymbol{\omega} \times \mathbf{r}_{OM}. \tag{3.15}$$

The relative velocity $\mathbf{v}_{MO}$ is perpendicular to the position vector $\mathbf{r}_{OM}$, $\mathbf{v}_{MO} \perp \mathbf{r}_{OM}$, and has the direction given by the angular velocity vector $\boldsymbol{\omega}$. The magnitude of the relative velocity is $|\mathbf{v}_{MO}| = v_{MO} = \omega\, r_{OM}$.

3.3 Acceleration Field for a Rigid Body

The acceleration of an arbitrary point $M \in (RB)$ with respect to a fixed reference frame $O_0x_0y_0z_0$, represents the double derivative with respect to time of the position vector $\mathbf{r}_1$

$$\mathbf{a} = \ddot{\mathbf{r}}_1 = \dot{\mathbf{v}} = \frac{d\mathbf{v}}{dt} = \frac{d}{dt}(\mathbf{v}_O + \boldsymbol{\omega} \times \mathbf{r}) = \frac{d\mathbf{v}_O}{dt} + \frac{d\boldsymbol{\omega}}{dt} \times \mathbf{r} + \boldsymbol{\omega} \times \frac{d\mathbf{r}}{dt}$$
$$= \dot{\mathbf{v}}_O + \dot{\boldsymbol{\omega}} \times \mathbf{r} + \boldsymbol{\omega} \times \dot{\mathbf{r}}. \tag{3.16}$$

The acceleration of the point O with respect to the fixed reference frame $O_0x_0y_0z_0$ is

$$\mathbf{a}_O = \dot{\mathbf{v}}_O = \ddot{\mathbf{r}}_O. \tag{3.17}$$

The derivative of the vector $\boldsymbol{\omega}$ with respect to the time is the angular acceleration vector $\boldsymbol{\alpha}$ given by

$$\alpha = \frac{d\boldsymbol{\omega}}{dt} = \frac{d\omega_x}{dt}\mathbf{\imath} + \frac{d\omega_y}{dt}\mathbf{\jmath} + \frac{d\omega_z}{dt}\mathbf{k} + \omega_x\frac{d\mathbf{\imath}}{dt} + \omega_y\frac{d\mathbf{\jmath}}{dt} + \omega_z\frac{d\mathbf{k}}{dt}$$
$$= \alpha_x\mathbf{\imath} + \alpha_y\mathbf{\jmath} + \alpha_z\mathbf{k} + \omega_x\boldsymbol{\omega} \times \mathbf{\imath} + \omega_y\boldsymbol{\omega} \times \mathbf{\jmath} + \omega_z\boldsymbol{\omega} \times \mathbf{k}$$
$$= \alpha_x\mathbf{\imath} + \alpha_y\mathbf{\jmath} + \alpha_z\mathbf{k} + \boldsymbol{\omega} \times \boldsymbol{\omega} = \alpha_x\mathbf{\imath} + \alpha_y\mathbf{\jmath} + \alpha_z\mathbf{k}, \tag{3.18}$$

where $\alpha_x = \frac{d\omega_x}{dt}$, $\alpha_y = \frac{d\omega_y}{dt}$, and $\alpha_z = \frac{d\omega_z}{dt}$. In the previous expression the Poisson formulas have been used. Using Eqs. 3.16–3.18 the acceleration of the point M is

$$\mathbf{a} = \mathbf{a}_O + \boldsymbol{\alpha} \times \mathbf{r} + \boldsymbol{\omega} \times (\boldsymbol{\omega} \times \mathbf{r}). \tag{3.19}$$

Using Eq. 3.19 the components of the acceleration are

$$a_x = a_{Ox} + (z\alpha_y - y\alpha_z) + \omega_y(y\omega_x - x\omega_y) + \omega_z(x\omega_x - x\omega_z),$$
$$a_y = a_{Oy} + (x\alpha_z - z\alpha_x) + \omega_z(z\omega_y - y\omega_z) + \omega_x(x\omega_y - y\omega_z),$$
$$a_z = a_{Oz} + (y\alpha_x - x\alpha_y) + \omega_x(x\omega_z - z\omega_x) + \omega_y(y\omega_z - z\omega_y).$$

The relation between the accelerations $\mathbf{a}_M$ and $\mathbf{a}_O$ of two points M and O on the rigid body is

$$\mathbf{a}_M = \mathbf{a}_O + \boldsymbol{\alpha} \times \mathbf{r}_{OM} + \boldsymbol{\omega} \times (\boldsymbol{\omega} \times \mathbf{r}_{OM}). \tag{3.20}$$

In the case of planar motion

$$\boldsymbol{\omega} \times (\boldsymbol{\omega} \times \mathbf{r}_{OM}) = -\omega^2\,\mathbf{r}_{OM},$$

and Eq. 3.20 becomes

$$\mathbf{a}_M = \mathbf{a}_O + \boldsymbol{\alpha} \times \mathbf{r}_{OM} - \omega^2\,\mathbf{r}_{OM}. \tag{3.21}$$

Equation 3.21 can be written as

$$\mathbf{a}_M = \mathbf{a}_O + \mathbf{a}^{\text{rel}}_{MO}, \tag{3.22}$$

where $\mathbf{a}^{\text{rel}}_{MO}$ is the relative acceleration, for rotational motion, of M with respect to O and is given by

$$\mathbf{a}^{\text{rel}}_{MO} = \mathbf{a}_{MO} = \mathbf{a}^{n}_{MO} + \mathbf{a}^{t}_{MO}. \tag{3.23}$$

The normal relative acceleration of M with respect to O is

$$\mathbf{a}^n_{MO} = \boldsymbol{\omega} \times (\boldsymbol{\omega} \times \mathbf{r}_{OM}), \tag{3.24}$$

is parallel to the position vector $\mathbf{r}_{OM}$, $\mathbf{a}^n_{MO} || \mathbf{r}_{OM}$, and has the direction towards the center of rotation, from M to O. The magnitude of the normal relative acceleration is

$$|\mathbf{a}^n_{MO}| = a^n_{MO} = \omega^2 r_{OM} = \frac{v^2_{MO}}{r_{OM}}.$$

The tangential relative acceleration of M with respect to O

$$\mathbf{a}^t_{MO} = \boldsymbol{\alpha} \times \mathbf{r}_{OM}, \tag{3.25}$$

is perpendicular to the position vector $\mathbf{r}_{OM}$, $\mathbf{a}^t_{MO} \perp \mathbf{r}_{OM}$, and has the direction given by the angular aceeleration $\boldsymbol{\alpha}$. The magnitude of the normal relative acceleration is

$$|\mathbf{a}^t_{MO}| = a^t_{MO} = \alpha\, r_{OM}.$$

Remarks:

1. If the orientation of a rigid body (RB) in a reference frame RF_0 depends on only a single scalar variable ζ, there exists for each value of ζ a vector $\boldsymbol{\omega}$ such that the derivative with respect to ζ in RF_0 of every vector $\mathbf{c}$ fixed in the rigid body (RB) is given by

$$\frac{d\mathbf{c}}{d\zeta} = \boldsymbol{\omega} \times \mathbf{c}, \tag{3.26}$$

where the vector $\boldsymbol{\omega}$ is the rate of change of orientation of the rigid body (RB) in the reference frame RF_0 with respect to ζ. The vector $\boldsymbol{\omega}$ is given by

$$\boldsymbol{\omega} = \frac{\dfrac{d\mathbf{a}}{d\zeta} \times \dfrac{d\mathbf{b}}{d\zeta}}{\dfrac{d\mathbf{a}}{d\zeta} \cdot \mathbf{b}}, \tag{3.27}$$

where $\mathbf{a}$ and $\mathbf{b}$ are any two non-parallel vectors fixed in the rigid body (RB). The vector $\boldsymbol{\omega}$ is a free vector, i.e. is not associated with any particular point. With the help of $\boldsymbol{\omega}$ one can replace the process of differentiation with that of cross multiplication. The vector $\boldsymbol{\omega}$ may be expressed in a symmetrical relation in $\mathbf{a}$ and $\mathbf{b}$

$$\boldsymbol{\omega} = \frac{1}{2}\left(\frac{\dfrac{d\mathbf{a}}{d\zeta} \times \dfrac{d\mathbf{b}}{d\zeta}}{\dfrac{d\mathbf{a}}{d\zeta} \cdot \mathbf{b}} + \frac{\dfrac{d\mathbf{b}}{d\zeta} \times \dfrac{d\mathbf{a}}{d\zeta}}{\dfrac{d\mathbf{b}}{d\zeta} \cdot \mathbf{a}} \right). \tag{3.28}$$

2. The first derivatives of a vector $\mathbf{p}$ with respect to a scalar variable ζ in two reference frames RF_i and RF_j are related as follows

$$\frac{^{(j)}d\mathbf{p}}{d\zeta} = \frac{^{(i)}d\mathbf{p}}{d\zeta} + \boldsymbol{\omega}_{ij} \times \mathbf{p}, \tag{3.29}$$

where $\boldsymbol{\omega}_{ij}$ is the rate of change of orientation of RF_i in RF_j with respect to ζ and $\frac{^{(j)}d\mathbf{p}}{d\zeta}$ is the total derivative of $\mathbf{p}$ with respect to ζ in RF_j.

3. The angular velocity of a rigid body (RB) in a reference frame RF_0 is the rate of change of orientation with respect to the time t

$$\boldsymbol{\omega} = \frac{1}{2}\left(\frac{\frac{d\mathbf{a}}{dt} \times \frac{d\mathbf{b}}{dt}}{\frac{d\mathbf{a}}{dt} \cdot \mathbf{b}} + \frac{\frac{d\mathbf{b}}{dt} \times \frac{d\mathbf{a}}{dt}}{\frac{d\mathbf{b}}{dt} \cdot \mathbf{a}}\right) = \frac{1}{2}\left(\frac{\dot{\mathbf{a}} \times \dot{\mathbf{b}}}{\dot{\mathbf{a}} \cdot \mathbf{b}} + \frac{\dot{\mathbf{b}} \times \dot{\mathbf{a}}}{\dot{\mathbf{b}} \cdot \mathbf{a}}\right).$$

The direction of $\boldsymbol{\omega}$ is related to the direction of the rotation of the rigid body through a right-hand rule.

4. Let RF_i, $i = 1, 2, ..., n$ be n reference frames. The angular velocity of a rigid body r in the reference frame RF_n, can be expressed as

$$\boldsymbol{\omega}_{rn} = \boldsymbol{\omega}_{r1} + \boldsymbol{\omega}_{12} + \boldsymbol{\omega}_{23} + ... + \boldsymbol{\omega}_{r,n-1}.$$

Proof

Let $\mathbf{p}$ be any vector fixed in the rigid body. Then,

$$\frac{^{(i)}d\mathbf{p}}{dt} = \boldsymbol{\omega}_{ri} \times \mathbf{p}$$
$$\frac{^{(i-1)}d\mathbf{p}}{dt} = \boldsymbol{\omega}_{r,i-1} \times \mathbf{p}.$$

On the other hand,

$$\frac{^{(i)}d\mathbf{p}}{dt} = \frac{^{(i-1)}d\mathbf{p}}{dt} + \boldsymbol{\omega}_{i,i-1} \times \mathbf{p}.$$

Hence,

$$\boldsymbol{\omega}_{ri} \times \mathbf{p} = \boldsymbol{\omega}_{r,i-1} \times \mathbf{p} + \boldsymbol{\omega}_{i,i-1} \times \mathbf{p},$$

as this equation is satisfied for all $\mathbf{p}$ fixed in the rigid body

$$\boldsymbol{\omega}_{ri} = \boldsymbol{\omega}_{r,i-1} + \boldsymbol{\omega}_{i,i-1}. \tag{3.30}$$

With $i = n$, Eq. 3.30 gives

$$\boldsymbol{\omega}_{rn} = \boldsymbol{\omega}_{r,n-1} + \boldsymbol{\omega}_{n,n-1}. \tag{3.31}$$

With $i = n-1$, Eq. 3.30 gives

$$\omega_{r,n-1} = \omega_{r,n-2} + \omega_{n-1,n-2}. \tag{3.32}$$

Substitute Eq. 3.32 into Eq. 3.31

$$\omega_{rn} = \omega_{r,n-2} + \omega_{n-1,n-2} + \omega_{n,n-1}.$$

Next use Eq. 3.30 with $i = n-2$, then with $i = n-3$, and so forth.

3.4 Motion of a Point that Moves Relative to a Rigid Body

A reference frame that moves with the rigid body is a body-fixed reference frame. Figure 3.2 shows a rigid body (*RB*) in motion relative to a primary reference frame with its origin at point O_0, $x_0y_0z_0$. The primary reference frame is a fixed reference frame or an Earth-fixed reference frame. The unit vectors $\mathbf{ı}_0, \mathbf{ȷ}_0$, and $\mathbf{k}_0$ of the primary reference frame are constant.

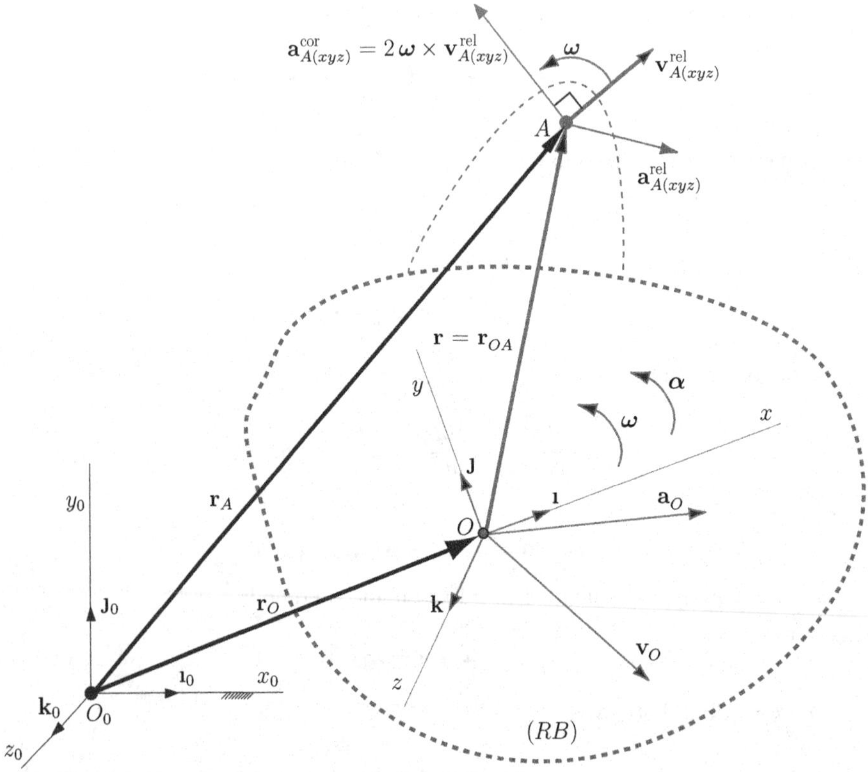

Fig. 3.2 Rigid body in motion; the point A is not assumed to be a point of the rigid body, $A \notin (RB)$

The body-fixed reference frame, xyz, has its origin at a point O of the rigid body ($O \in (RB)$), and is a moving reference frame relative to the primary reference. The unit vectors $\boldsymbol{\imath}, \boldsymbol{\jmath}$, and $\mathbf{k}$ of the body-fixed reference frame are not constant, because they rotate with the body-fixed reference frame.

The position vector of a point P of the rigid body ($P \in (RB)$) relative to the origin, O, of the body-fixed reference frame is the vector $\mathbf{r}_{OP}$. The velocity of P relative to O is

$$\frac{d\mathbf{r}_{OP}}{dt} = \mathbf{v}_{PO}^{\text{rel}} = \boldsymbol{\omega} \times \mathbf{r}_{OP},$$

where $\boldsymbol{\omega}$ is the angular velocity vector of the rigid body.

The position vector of a point A (the point A is not assumed to be a point of the rigid body, $A \notin (RB)$), relative to the origin O_0 of the primary reference frame is, Fig. 3.1,

$$\mathbf{r}_A = \mathbf{r}_O + \mathbf{r},$$

where

$$\mathbf{r} = \mathbf{r}_{OA} = x\boldsymbol{\imath} + y\boldsymbol{\jmath} + z\mathbf{k}$$

is the position vector of A relative to the origin O, of the body-fixed reference frame, and x, y, and z are the coordinates of A in terms of the body-fixed reference frame. The velocity of the point A is the time derivative of the position vector $\mathbf{r}_A$

$$\begin{aligned}\mathbf{v}_A &= \frac{d\mathbf{r}_O}{dt} + \frac{d\mathbf{r}}{dt} = \mathbf{v}_O + \mathbf{v}_{AO}^{\text{rel}} \\ &= \mathbf{v}_O + \frac{dx}{dt}\boldsymbol{\imath} + x\frac{d\boldsymbol{\imath}}{dt} + \frac{dy}{dt}\boldsymbol{\jmath} + y\frac{d\boldsymbol{\jmath}}{dt} + \frac{dz}{dt}\mathbf{k} + z\frac{d\mathbf{k}}{dt}.\end{aligned}$$

Using Poisson formulas, the total derivative of the position vector $\mathbf{r}$ is

$$\frac{d\mathbf{r}}{dt} = \dot{\mathbf{r}} = \dot{x}\boldsymbol{\imath} + \dot{y}\boldsymbol{\jmath} + \dot{z}\mathbf{k} + \boldsymbol{\omega} \times \mathbf{r}.$$

The velocity of A relative to the body-fixed reference frame is a derivative in the body-fixed reference frame

$$\mathbf{v}_{A(xyz)}^{\text{rel}} = \frac{^{(xyz)}d\,\mathbf{r}}{dt} = \frac{dx}{dt}\boldsymbol{\imath} + \frac{dy}{dt}\boldsymbol{\jmath} + \frac{dz}{dt}\mathbf{k} = \dot{x}\boldsymbol{\imath} + \dot{y}\boldsymbol{\jmath} + \dot{z}\mathbf{k}. \tag{3.33}$$

A general formula for the total derivative of a moving vector $\mathbf{r}$ may be written as

$$\frac{d\mathbf{r}}{dt} = \frac{^{(xyz)}d\,\mathbf{r}}{dt} + \boldsymbol{\omega} \times \mathbf{r}, \tag{3.34}$$

where $\dfrac{d\mathbf{r}}{dt} = \dfrac{^{(0)}d\mathbf{r}}{dt}$ is the derivative in the fixed (primary) reference frame (0) $(x_0y_0z_0)$, and $\dfrac{^{(xyz)}d\mathbf{r}}{dt}$ is the derivative in the rotating (mobile or body-fixed) reference frame (xyz).

The velocity of the point A relative to the primary reference frame is

$$\mathbf{v}_A = \mathbf{v}_O + \mathbf{v}^{\text{rel}}_{A(xyz)} + \boldsymbol{\omega} \times \mathbf{r}. \tag{3.35}$$

Equation 3.35 expresses the velocity of a point A as the sum of three terms:

- the velocity of a point O of the rigid body;
- the velocity $\mathbf{v}^{\text{rel}}_{A(xyz)}$ of A relative to the rigid body; and
- the velocity $\boldsymbol{\omega} \times \mathbf{r}$ of A relative to O due to the rotation of the rigid body.

The acceleration of the point A relative to the primary reference frame is obtained by taking the time derivative of Eq. 3.35

$$\begin{aligned}\mathbf{a}_A &= \mathbf{a}_O + \mathbf{a}_{AO}\\ &= \mathbf{a}_O + \mathbf{a}^{\text{rel}}_{A(xyz)} + 2\boldsymbol{\omega} \times \mathbf{v}^{\text{rel}}_{A(xyz)} + \boldsymbol{\alpha} \times \mathbf{r} + \boldsymbol{\omega} \times (\boldsymbol{\omega} \times \mathbf{r}),\end{aligned} \tag{3.36}$$

where

$$\mathbf{a}^{\text{rel}}_{A(xyz)} = \frac{^{(xyz)}d^2\mathbf{r}}{dt^2} = \frac{d^2x}{dt^2}\mathbf{\imath} + \frac{d^2y}{dt^2}\mathbf{\jmath} + \frac{d^2z}{dt^2}\mathbf{k} \tag{3.37}$$

is the acceleration of A relative to the body-fixed reference frame or relative to the rigid body. The term

$$\mathbf{a}^{\text{cor}}_{A(xyz)} = 2\boldsymbol{\omega} \times \mathbf{v}^{\text{rel}}_{A(xyz)}$$

is called the Coriolis acceleration. The direction of the Coriolis acceleration is obtained by rotating the linear relative velocity $\mathbf{v}^{\text{rel}}_{A(xyz)}$ through 90° in the direction of rotation given by $\boldsymbol{\omega}$.

In the case of planar motion, Eq. 3.36 becomes

$$\begin{aligned}\mathbf{a}_A &= \mathbf{a}_O + \mathbf{a}_{OA}\\ &= \mathbf{a}_O + \mathbf{a}^{\text{rel}}_{A(xyz)} + 2\boldsymbol{\omega} \times \mathbf{v}^{\text{rel}}_{A(xyz)} + \boldsymbol{\alpha} \times \mathbf{r} - \omega^2\mathbf{r}.\end{aligned} \tag{3.38}$$

The motion of the rigid body (RB) is described relative to the primary reference frame. The velocity $\mathbf{v}_A$ and the acceleration $\mathbf{a}_A$ of a point A are relative to the primary reference frame. The terms $\mathbf{v}^{\text{rel}}_{A(xyz)}$ and $\mathbf{a}^{\text{rel}}_{A(xyz)}$ are the velocity and acceleration of point A relative to the body-fixed reference frame, i.e., they are the velocity and acceleration measured by an observer moving with the rigid body, Fig. 3.2. If A is a point of the rigid body, $A \in (RB)$, $\mathbf{v}^{\text{rel}}_{A(xyz)} = \mathbf{0}$ and $\mathbf{a}^{\text{rel}}_{A(xyz)} = \mathbf{0}$.

Motion of a Point Relative to a Moving Reference Frame

The velocity and acceleration of an arbitrary point A relative to a point O of a rigid body, in terms of the body-fixed reference frame, are given by Eqs. 3.35 and 3.36

$$\mathbf{v}_A = \mathbf{v}_O + \mathbf{v}_{AO}^{\mathrm{rel}} + \boldsymbol{\omega} \times \mathbf{r}_{OA}, \tag{3.39}$$

$$\mathbf{a}_A = \mathbf{a}_O + \mathbf{a}_{AO}^{\mathrm{rel}} + 2\boldsymbol{\omega} \times \mathbf{v}_{AO}^{\mathrm{rel}} + \boldsymbol{\alpha} \times \mathbf{r}_{OA} + \boldsymbol{\omega} \times (\boldsymbol{\omega} \times \mathbf{r}_{OA}). \tag{3.40}$$

These results apply to any reference frame having a moving origin O and rotating with angular velocity $\boldsymbol{\omega}$ and angular acceleration $\boldsymbol{\alpha}$ relative to a primary reference frame (Fig. 3.2). The terms $\mathbf{v}_A$ and $\mathbf{a}_A$ are the velocity and acceleration of an arbitrary point A relative to the primary reference frame. The terms $\mathbf{v}_{AO}^{\mathrm{rel}}$ and $\mathbf{a}_{AO}^{\mathrm{rel}}$ are the velocity and acceleration of A relative to the secondary moving reference frame, i.e., they are the velocity and acceleration measured by an observer moving with the secondary reference frame. The Coriolis acceleration is $\mathbf{a}_{AO}^{\mathrm{cor}} = 2\boldsymbol{\omega} \times \mathbf{v}_{AO}^{\mathrm{rel}}$.

3.5 Slider-Crank (R-RRT) Mechanism

Exercise

The R-RRT (slider-crank) mechanism shown in Fig. 3.3 has the dimensions: $AB =$ 1 m and $BC = 1$ m. When the driver link 1 makes an angle $\phi = \phi_1 = \pi/6$ rad with the horizontal axis the instantaneous speed and the angular acceleration of the link 1 are ω=1 rad/s and α=–1 rad/s^2.

Find the velocities and the accelerations of the joints and the angular velocities and accelerations of the links for the given driver-link angle.

Solution

The point A is selected as the origin of the xyz reference frame. The position vectors of the joints B and C are:

$$\mathbf{r}_B = x_B\mathbf{\imath} + y_B\mathbf{\jmath} = \frac{\sqrt{3}}{2}\mathbf{\imath} + \frac{1}{2}\mathbf{\jmath}\ \text{m} \quad \text{and} \quad \mathbf{r}_C = x_C\mathbf{\imath} + y_C\mathbf{\jmath} = \sqrt{3}\mathbf{\imath} + 0\mathbf{\jmath}\ \text{m}.$$

The MATLAB® statements for the positions of the mechanism are:

```
AB=1; BC=1;
```

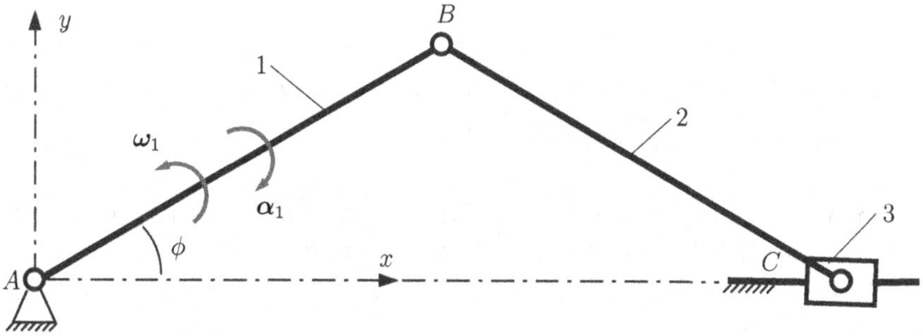

Fig. 3.3 Slider-crank (R-RRT) mechanism

```
phi = pi/6; % input angle
xA = 0; yA = 0; rA = [xA yA 0];
xB = AB*cos(phi); yB = AB*sin(phi);
rB = [xB yB 0];
yC = 0; xC = xB+sqrt(BC^2-(yC-yB)^2);
rC = [xC yC 0];
```

Velocity of Joint B

The velocity of the point $B = B_1$ on the link 1 is

$$\mathbf{v}_B = \mathbf{v}_{B_1} = \mathbf{v}_A + \mathbf{v}_{BA} = \mathbf{v}_A + \boldsymbol{\omega}_1 \times \mathbf{r}_{AB} = \boldsymbol{\omega}_1 \times \mathbf{r}_B,$$

where $\mathbf{v}_A \equiv \mathbf{0}$ is the velocity of the origin $A \equiv O$. The angular velocity of link 1 is

$$\boldsymbol{\omega}_1 = \boldsymbol{\omega} = \omega_1\,\mathbf{k} = 1\,\mathbf{k}\ \text{rad/s}.$$

The velocity of point B_2 on the link 2 is $\mathbf{v}_{B_2} = \mathbf{v}_{B_1}$ because the links 1 and 2 are connected at a rotational joint. The velocity of $B_1 = B_2$ is

$$\mathbf{v}_B = \mathbf{v}_{B_1} = \mathbf{v}_{B_2} = \begin{vmatrix} \mathbf{1} & \mathbf{J} & \mathbf{k} \\ 0 & 0 & \omega_1 \\ x_B & y_B & 0 \end{vmatrix} = \begin{vmatrix} \mathbf{1} & \mathbf{J} & \mathbf{k} \\ 0 & 0 & 1 \\ \frac{\sqrt{3}}{2} & \frac{1}{2} & 0 \end{vmatrix} = -\frac{1}{2}\mathbf{1} + \frac{\sqrt{3}}{2}\mathbf{J}\ \text{m/s}.$$

The magnitude of the velocity $\mathbf{v}_B$ is

$$|\mathbf{v}_B| = v_B = 1\ \text{m/s}.$$

The velocity $\mathbf{v}_B$ is perpendicular to the position vector $\mathbf{r}_B$ and has the direction given by the angular velocity $\boldsymbol{\omega}_1$ as shown in Fig. 3.4. The MATLAB commands for the velocity of the joint B are:

```
omega1 = [0 0  1 ]; % (rad/s)
vA = [0 0 0 ]; % (m/s) % velocity of A (fixed)
% A and B=B1 are two points on the rigid link 1
vB1 = vA + cross(omega1,rB); % velocity of B1
vB2 = vB1;
vB = norm(vB1); % norm() is the vector norm
fprintf('omega1 = [ %g, %g, %g ] (rad/s)\n', omega1)
fprintf('vB=vB1=vB2 = [ %g, %g, %g ] (m/s)\n', vB1)
fprintf('|vB|= %g (m/s)\n', vB)
```

The command `dot(u,v)` calculates the scalar product (or vector dot product) of the vectors `u` and `v`. The command `cross(u,v)` performs the cross product of the vectors `u` and `v`.

Velocity of Joint C

The points B_2 and C_2 are on the link 2 and

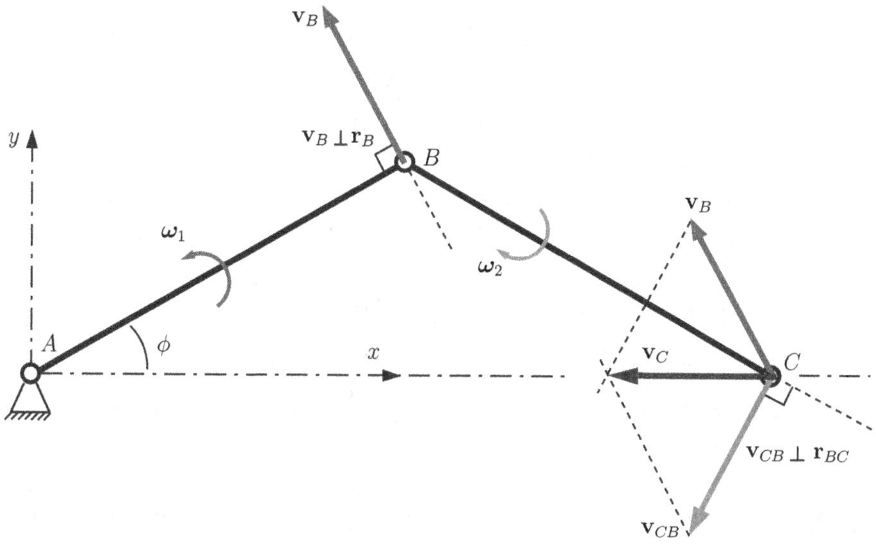

Fig. 3.4 Velocity field for the R-RRT mechanism

$$\mathbf{v}_C = \mathbf{v}_{C_2} = \mathbf{v}_B + \mathbf{v}_{CB} = \mathbf{v}_{B_2} + \boldsymbol{\omega}_2 \times \mathbf{r}_{BC} = \mathbf{v}_B + \boldsymbol{\omega}_2 \times (\mathbf{r}_C - \mathbf{r}_B), \qquad (3.41)$$

where the angular velocity of link 2 is $\boldsymbol{\omega}_2 = \omega_2 \mathbf{k}$ (ω_2 is unknown).

On the other hand, the velocity of C is along the vertical axis (x-axis) because the slider 2 translates along the x-axis

$$\mathbf{v}_C = \mathbf{v}_{C_3} = v_C \,\mathbf{\imath}. \qquad (3.42)$$

Equations 3.41 and 3.42 give

$$\mathbf{v}_B + \boldsymbol{\omega}_2 \times (\mathbf{r}_C - \mathbf{r}_B) = v_C \,\mathbf{\imath},$$

or

$$\mathbf{v}_B + \begin{vmatrix} \mathbf{\imath} & \mathbf{\jmath} & \mathbf{k} \\ 0 & 0 & \omega_2 \\ x_C - x_B & y_C - y_B & 0 \end{vmatrix} = v_C \,\mathbf{\imath}. \qquad (3.43)$$

Equation 3.43 represents a vectorial equation with two scalar components on x-axis and y-axis and with two unknowns ω_2 and v_C

$$v_{Bx} - \omega_2 (y_C - y_B) = v_C, \qquad (3.44)$$
$$v_{Bx} + \omega_2 (x_C - x_B) = 0, \qquad (3.45)$$

or

$$-\frac{1}{2} - \omega_2 \left(0 - \frac{1}{2}\right) = v_C,$$
$$\frac{\sqrt{3}}{2} + \omega_2 \left(\sqrt{3} - \frac{\sqrt{3}}{2}\right) = 0.$$

Thus,

$$\omega_2 = -1 \text{ rad/s} \text{ and } v_C = -1 \text{ m/s}.$$

The relative velocity of point C with respect to B is

$$\mathbf{v}_{CB} = \boldsymbol{\omega}_2 \times (\mathbf{r}_C - \mathbf{r}_B) = \begin{vmatrix} \mathbf{\imath} & \mathbf{\jmath} & \mathbf{k} \\ 0 & 0 & -1 \\ \sqrt{3} - \frac{\sqrt{3}}{2} & -\frac{1}{2} & 0 \end{vmatrix} = -\frac{1}{2}\mathbf{\imath} - \frac{\sqrt{3}}{2}\mathbf{\jmath} \text{ m/s}.$$

The relative velocity $\mathbf{v}_{CB}$ is perpendicular to $\mathbf{r}_{BC}$ and has the direction given by the angular velocity $\boldsymbol{\omega}_2$ as shown in Fig. 3.4.

In MATLAB the `sym` command constructs symbolic variables and expressions. The commands:

```
omega2z = sym('omega2z','real');
vCx = sym('vCx','real');
```

create a symbolic variables `omega2z` and `vCx` for the unknowns ω_2 and v_C. The commands `sym('omega2z','real')` and `sym('vCx','real')` also assume that `omega2z` and `vCx` are real numbers. The vectors $\boldsymbol{\omega}_2 = \omega_2 \mathbf{k}$ and $\mathbf{v}_C = v_C \mathbf{\imath}$ are expressed in MATLAB with:

```
omega2 = [ 0 0 omega2z ];
vC = [ vCx 0 0 ];
```

Equation 3.43 or $\mathbf{v}_C = \mathbf{v}_B + \boldsymbol{\omega}_2 \times (\mathbf{r}_C - \mathbf{r}_B)$ in MATLAB is

```
eqvC = vC - (vB2 + cross(omega2,rC-rB));
```

This vectorial equation has a component on:

- x-axis given by Eq. 3.44, or in MATLAB, `eqvC(1)`; and
- y-axis given by Eq. 3.45, or in MATLAB, `eqvC(2)`.

The two algebraic equations can be solved using the command `solve`:

```
eqvCx = eqvC(1);    % equation component on x-axis
eqvCy = eqvC(2);    % equation component on y-axis
solvC = solve(eqvCx,eqvCy);
```

with the solutions:

```
omega2zs = eval(solvC.omega2z);
vCxs = eval(solvC.vCx);
```

The angular velocity of the link 2 and the velocity of C in vectorial form are:

```
Omega2 = [0 0 omega2zs];
VC = [vCxs 0 0];
```

The relative velocity of point C with respect to B is:

```
vCB = cross(Omega2,rC-rB);
```

To display the correct expression for the equations `eqvCx` and `eqvCy` the following MATLAB statements can be used:

```
qvCx = vpa(eqvCx,6);
fprintf('x-axis: %s = 0 \n', char(qvCx))
qvCy = vpa(eqvCy,6);
fprintf('y-axis: %s = 0 \n', char(qvCy))
```

The command `vpa(S,D)` uses variable-precision arithmetic (vpa) to compute each element of `S` to `D` decimal digits of accuracy and the command `char()` creates a character array (string).

Acceleration of Joint B
The acceleration of the point $B = B_1$ on the link 1 is

$$\begin{aligned}
\mathbf{a}_B &= \mathbf{a}_{B_1} = \mathbf{a}_{B_2} = \mathbf{a}_A + \boldsymbol{\alpha}_1 \times \mathbf{r}_B + \boldsymbol{\omega}_1 \times (\boldsymbol{\omega}_1 \times \mathbf{r}_B) = \boldsymbol{\alpha}_1 \times \mathbf{r}_B - \omega_1^2 \mathbf{r}_B \\
&= \begin{vmatrix} \mathbf{1} & \mathbf{J} & \mathbf{k} \\ 0 & 0 & \alpha_1 \\ x_B & y_B & 0 \end{vmatrix} - \omega_1^2 \mathbf{r}_B = \begin{vmatrix} \mathbf{1} & \mathbf{J} & \mathbf{k} \\ 0 & 0 & -1 \\ \frac{\sqrt{3}}{2} & \frac{1}{2} & 0 \end{vmatrix} - 1^2 \left(\frac{\sqrt{3}}{2} \mathbf{1} + \frac{1}{2} \mathbf{J} \right) \\
&= \left(\frac{1}{2} - \frac{\sqrt{3}}{2} \right) \mathbf{1} - \left(\frac{1}{2} + \frac{\sqrt{3}}{2} \right) \mathbf{J} \ \mathrm{m/s^2}.
\end{aligned}$$

The angular acceleration of link 1 is $\boldsymbol{\alpha}_1 = -1\,\mathbf{k}$ rad/s^2. The normal acceleration of the point B is

$$\mathbf{a}_B^n = -\omega_1^2 \mathbf{r}_B = -1^2 \left(\frac{\sqrt{3}}{2} \mathbf{1} + \frac{1}{2} \mathbf{J} \right) = -\frac{\sqrt{3}}{2} \mathbf{1} - \frac{1}{2} \mathbf{J} \ \mathrm{m/s^2}.$$

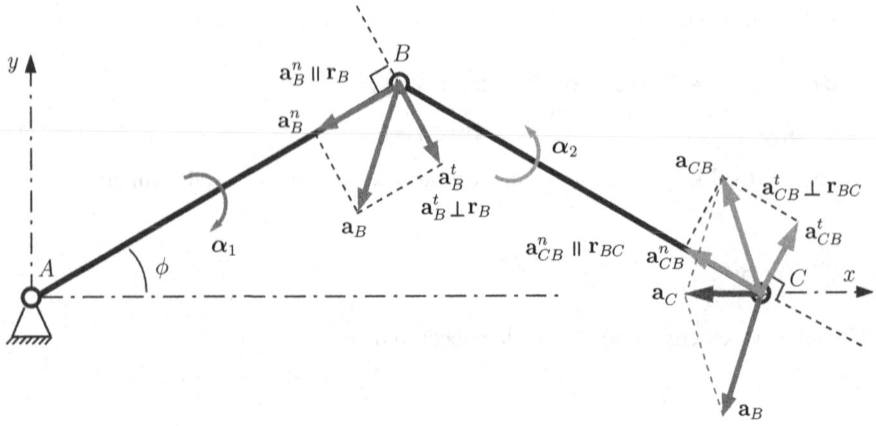

Fig. 3.5 Acceleration field for the R-RRT mechanism

The normal acceleration $\mathbf{a}_B^n$ is parallel to the vector $\mathbf{r}_B$ and the orientation is toward the center of rotation A (from B to A) as shown in Fig. 3.5. The tangential acceleration of the point B is

$$\mathbf{a}_B^t = \boldsymbol{\alpha}_1 \times \mathbf{r}_B = \begin{vmatrix} \mathbf{1} & \mathbf{J} & \mathbf{k} \\ 0 & 0 & \alpha_1 \\ x_B & y_B & 0 \end{vmatrix} = \begin{vmatrix} \mathbf{1} & \mathbf{J} & \mathbf{k} \\ 0 & 0 & -1 \\ \frac{\sqrt{3}}{2} & \frac{1}{2} & 0 \end{vmatrix} = \frac{1}{2}\mathbf{1} - \frac{\sqrt{3}}{2}\mathbf{J} \text{ m/s}^2.$$

The tangential acceleration $\mathbf{a}_B^t$ is perpendicular to the vector $\mathbf{r}_B$ and the orientation given by the vector $\boldsymbol{\alpha}_1$ as shown in Fig. 3.5. The MATLAB commands for the acceleration of the joint B are:

```
alpha1  =  [0 0 -1 ]; % (rad/s^2)
aA = [0 0 0 ]; % (m/s^2) acceleration of A
aB1 = aA + cross(alpha1,rB) - dot(omega1,omega1)*rB;
aB2 = aB1;
aBn = - dot(omega1,omega1)*rB;
aBt = cross(alpha1,rB);
```

Acceleration of Joint C
The points C_2 and B_2 are on the link 2 and

$$\mathbf{a}_C = \mathbf{a}_{C_2} = \mathbf{a}_{B_2} + \boldsymbol{\alpha}_2 \times \mathbf{r}_{BC} - \omega_2^2\,\mathbf{r}_{BC} = \mathbf{a}_B + \boldsymbol{\alpha}_2 \times (\mathbf{r}_C - \mathbf{r}_B) - \omega_2^2\,(\mathbf{r}_C - \mathbf{r}_B), \quad (3.46)$$

where the angular acceleration of link 2 is $\boldsymbol{\alpha}_2 = \alpha_2\mathbf{k}$ (α_2 is unknown). The slider C has a translational motion along x-axis and

$$\mathbf{a}_C = \mathbf{a}_{C_3} = a_C\,\mathbf{1}. \quad (3.47)$$

Equations 3.46 and 3.47 give

$$\mathbf{a}_B + \boldsymbol{\alpha}_2 \times (\mathbf{r}_C - \mathbf{r}_B) - \omega_2^2\,(\mathbf{r}_C - \mathbf{r}_B) = a_C\,\mathbf{ı},$$

or

$$\mathbf{a}_B + \begin{vmatrix} \mathbf{ı} & \mathbf{ȷ} & \mathbf{k} \\ 0 & 0 & \alpha_2 \\ x_C - x_B & y_C - y_B & 0 \end{vmatrix} - \omega_2^2\,[(x_C - x_B)\mathbf{ı} + (y_C - y_B)\mathbf{ȷ}] = a_C\,\mathbf{ı}. \qquad (3.48)$$

Equation 3.48 represents a vectorial equation with two scalar components on the x-axis and y-axis and with two unknowns α_2 and α_3

$$a_{Bx} - \alpha_2(y_C - y_B) - \omega_2^2(x_C - x_B) = a_C,$$
$$a_{By} + \alpha_2(x_C - x_B) - \omega_2^2(y_C - y_B) = 0,$$

or

$$\left(\frac{1}{2} - \frac{\sqrt{3}}{2}\right) - \alpha_2\left(0 - \frac{1}{2}\right) - (-1)^2\left(\sqrt{3} - \frac{\sqrt{3}}{2}\right) = a_C,$$
$$-\left(\frac{1}{2} + \frac{\sqrt{3}}{2}\right) + \alpha_2\left(\sqrt{3} - \frac{\sqrt{3}}{2}\right) - (-1)^2\left(0 - \frac{1}{2}\right) = 0.$$

Thus,

$$\alpha_2 = 1 \text{ rad/s}^2 \quad \text{and} \quad a_C = 1 - \sqrt{3} \text{ m/s}^2.$$

The normal relative acceleration of point C with respect to B is

$$\begin{aligned} \mathbf{a}_{CB}^n &= -\omega_2^2\mathbf{r}_{BC} = -\omega_2^2\,(\mathbf{r}_C - \mathbf{r}_B) \\ &= -\omega_2^2\,(\mathbf{r}_C - \mathbf{r}_B) = -\omega_2^2\,[(x_C - x_B)\mathbf{ı} + (y_C - y_B)\mathbf{ȷ}] \\ &= -(-1)^2\left[\left(\sqrt{3} - \frac{\sqrt{3}}{2}\right)\mathbf{ı} + \left(0 - \frac{1}{2}\right)\mathbf{ȷ}\right] \\ &= -\frac{\sqrt{3}}{2}\mathbf{ı} + \frac{1}{2}\mathbf{ȷ} \text{ m/s}^2. \end{aligned}$$

The normal relative acceleration of point C with respect to B, $\mathbf{a}_{CB}^n$, is parallel to the vector $\mathbf{r}_{BC}$ and the orientation is toward the center of rotation B (from C to B) as shown in Fig. 3.5. The tangential relative acceleration of the point C with respect to B is

$$\mathbf{a}_{CB}^t = \boldsymbol{\alpha}_2 \times \mathbf{r}_{BC} = \begin{vmatrix} \mathbf{ı} & \mathbf{ȷ} & \mathbf{k} \\ 0 & 0 & \alpha_2 \\ x_C - x_B & y_C - y_B & 0 \end{vmatrix}$$

$$= \begin{vmatrix} \mathbf{ı} & \mathbf{ȷ} & \mathbf{k} \\ 0 & 0 & 1 \\ \sqrt{3}-\frac{\sqrt{3}}{2} & 0-\frac{1}{2} & 0 \end{vmatrix} = \frac{1}{2}\mathbf{ı}+\frac{\sqrt{3}}{2}\mathbf{ȷ} \ \ \mathrm{m/s^2}.$$

The tangential relative acceleration $\mathbf{a}^t_{CB}$ is perpendicular to the vector $\mathbf{r}_{BC}$ and the orientation given by the vector $\boldsymbol{\alpha}_2$ as shown in Fig. 3.5.

To calculate $\boldsymbol{\alpha}_2$, $\mathbf{a}_C$, and $\mathbf{a}_{CB}$ the following commands are used with MATLAB:

```
alpha2z=sym('alpha2z','real');
aCx=sym('aCx','real');
alpha2 = [ 0 0 alpha2z ];      % alpha3z unknown
aC = [aCx 0 0 ];               % aCx unknown
eqaC=aC-(aB1+cross(alpha2,rC-rB)-...
     dot(Omega2,Omega2)*(rC-rB));
eqaCx = eqaC(1);     % equation component on x-axis
eqaCy = eqaC(2);     % equation component on y-axis
solaC = solve(eqaCx,eqaCy);
alpha2zs=eval(solaC.alpha2z);
aCxs=eval(solaC.aCx);
Alpha2 = [0 0 alpha2zs];
aCs = [aCxs 0 0];
aCB=cross(Alpha2,rC-rB)-dot(Omega2,Omega2)*(rC-rB);
aCBn=-dot(Omega2,Omega2)*(rC-rB);
aCBt=cross(Alpha2,rC-rB);
```

The MATLAB program for the velocities and accelerations is given in Appendix B.1. The results are shown at the end of the program.

3.6 Four-Bar (R-RRR) Mechanism

Exercise

The planar R-RRR mechanism is shown in Fig. 2.4. The following data are given: AB=0.150 m, BC=0.35 m, CD=0.30 m, CE=0.15 m. $x_A=y_A=0$, x_D=0.30 m, and y_D=0.30 m. For $\phi = \phi_1 = 45°$ the positions of B, C, and D are $x_B=y_B$=0.106066 m, x_C=0.0400698 m, y_C=0.449788 m, x_E=–0.0898952 m, and y_E=0.524681 m. The driver link 1 rotates with a constant angular speed $n=n_1$=60 rpm (revolutions per minute).

Find the velocities and the accelerations of the mechanism at the moment when the driver link 1 makes an angle $\phi = \phi_1 = 45°$ with the horizontal axis.

Solution

The angular velocity of link 1 is

$$\boldsymbol{\omega}_1 = \omega_1\,\mathbf{k} = \frac{\pi n}{30}\mathbf{k} = \frac{\pi(60)}{30}\mathbf{k} = 6.28319\mathbf{k} \text{ rad/s}.$$

The angular acceleration of link 1 is $\boldsymbol{\alpha}_1 = \dot{\boldsymbol{\omega}}_1 = \mathbf{0}$.

The MATLAB statements for the angular velocity and acceleration of link 1 are:

```
n = 60; omega1 = [ 0 0 pi*n/30 ]; alpha1 = [0 0 0 ];
```

Velocity and Acceleration of Joint B

The velocity of the point $B = B_1$ on the link 1 is

$$\mathbf{v}_B = \mathbf{v}_{B_1} = \mathbf{v}_A + \boldsymbol{\omega}_1 \times \mathbf{r}_{AB} = \boldsymbol{\omega}_1 \times \mathbf{r}_B,$$

where $\mathbf{v}_A \equiv \mathbf{0}$ is the velocity of the origin $A \equiv O$. The velocity of point B_2 on the link 2 is $\mathbf{v}_{B_2} = \mathbf{v}_{B_1}$ because between the links 1 and 2 there is a rotational joint. The velocity of $B = B_1 = B_2$ is

$$\mathbf{v}_B = \mathbf{v}_{B_1} = \mathbf{v}_{B_2} = \begin{vmatrix} \mathbf{\imath} & \mathbf{\jmath} & \mathbf{k} \\ 0 & 0 & \omega \\ x_B & y_B & 0 \end{vmatrix} = \begin{vmatrix} \mathbf{\imath} & \mathbf{\jmath} & \mathbf{k} \\ 0 & 0 & 6.28319 \\ 0.106066 & 0.106066 & 0 \end{vmatrix}$$
$$= -0.666432\mathbf{\imath} + 0.666432\mathbf{\jmath} \text{ m/s}.$$

The acceleration of the point $B = B_1 = B_2$ is

$$\mathbf{a}_B = \mathbf{a}_{B_1} = \mathbf{a}_{B_2} = \mathbf{a}_A + \boldsymbol{\alpha}_1 \times \mathbf{r}_B + \boldsymbol{\omega}_1 \times (\boldsymbol{\omega}_1 \times \mathbf{r}_B) = \boldsymbol{\alpha}_1 \times \mathbf{r}_B - \omega_1^2 \mathbf{r}_B$$
$$= -(6.28319)^2(0.106066\mathbf{\imath} + 0.106066\mathbf{\jmath}) = -4.18732\mathbf{\imath} - 4.18732\mathbf{\jmath} \text{ m/s}^2.$$

The MATLAB statements for the velocity and acceleration of the driver link 1 are:

```
vA = [0 0 0 ]; aA = [0 0 0 ];
vB1 = vA + cross(omega1,rB);
vB2 = vB1;
aB1 = aA + cross(alpha1,rB) - dot(omega1,omega1)*rB;
aB2 = aB1;
```

Velocity of Joint C

The points B_2 and C_2 are on the link 2 and

$$\mathbf{v}_{C_2} = \mathbf{v}_{B_2} + \boldsymbol{\omega}_2 \times \mathbf{r}_{BC} = \mathbf{v}_B + \boldsymbol{\omega}_2 \times (\mathbf{r}_C - \mathbf{r}_B), \tag{3.49}$$

where the angular velocity of link 2 is $\boldsymbol{\omega}_2 = \omega_2\,\mathbf{k}$ (ω_2 is unknown).

The points D_3 and C_3 are on the link 3 and

$$\mathbf{v}_{C_3} = \mathbf{v}_{D_3} + \boldsymbol{\omega}_3 \times \mathbf{r}_{DC} = \boldsymbol{\omega}_3 \times (\mathbf{r}_C - \mathbf{r}_D), \tag{3.50}$$

where $\mathbf{v}_D = \mathbf{v}_{D_3} \equiv \mathbf{0}$ and the angular velocity of link 3 is $\boldsymbol{\omega}_3 = \omega_3\,\mathbf{k}$. The numerical value of ω_3 is unknown.

Equations 3.49 and 3.50 give ($\mathbf{v}_{C_2} = \mathbf{v}_{C_3}$)

$$\mathbf{v}_B + \boldsymbol{\omega}_2 \times (\mathbf{r}_C - \mathbf{r}_B) = \boldsymbol{\omega}_3 \times (\mathbf{r}_C - \mathbf{r}_D),$$

or

$$\mathbf{v}_B + \begin{vmatrix} \mathbf{\imath} & \mathbf{\jmath} & \mathbf{k} \\ 0 & 0 & \omega_2 \\ x_C - x_B & y_C - y_B & 0 \end{vmatrix} = \begin{vmatrix} \mathbf{\imath} & \mathbf{\jmath} & \mathbf{k} \\ 0 & 0 & \omega_3 \\ x_C - x_D & y_C - y_D & 0 \end{vmatrix}. \tag{3.51}$$

Equation 3.51 represents a vectorial equation with two scalar components on the x-axis and y-axis and with two unknowns ω_2 and ω_3

$$v_{Bx} - \omega_2(y_C - y_B) = -\omega_3(y_C - y_D),$$
$$v_{By} + \omega_2(x_C - x_B) = \omega_3(x_C - x_D),$$

or

$$-0.666432 - \omega_2(0.449788 - 0.106066) = -\omega_3(0.449788 - 0.3),$$
$$0.666432 + \omega_2(0.0400698 - 0.106066) = \omega_3(0.0400698 - 0.3).$$

Thus,

$$\omega_2 = -3.43639 \text{ rad/s} \text{ and } \omega_3 = -3.43639 \text{ rad/s}.$$

The velocity of C is

$$\mathbf{v}_C = \mathbf{v}_D + \boldsymbol{\omega}_3 \times (\mathbf{r}_C - \mathbf{r}_D) = -\omega_3(y_C - y_D)\mathbf{\imath} + \omega_3(x_C - x_D)\mathbf{\jmath}$$
$$= -(-3.43639)(0.449788 - 0.3)\mathbf{\imath} + (-3.43639)(0.0400698 - 0.3)\mathbf{\jmath}$$
$$= 0.514728\mathbf{\imath} + 0.893221\mathbf{\jmath} \text{ m/s}.$$

The MATLAB commands for the angular velocities of links 2 and 3, and the velocity of C are:

```
omega2z = sym('omega2z','real');
omega3z = sym('omega3z','real');
omega2 = [ 0 0 omega2z ];
omega3 = [ 0 0 omega3z ];
eqvC=vB2+cross(omega2,rC-rB)-(vD+cross(omega3,rC-rD));
eqvCx = eqvC(1); eqvCy = eqvC(2);
solvC = solve(eqvCx,eqvCy);
omega2zs=eval(solvC.omega2z);
omega3zs=eval(solvC.omega3z);
Omega2 = [0 0 omega2zs];
Omega3 = [0 0 omega3zs];
vC = vB2 + cross(Omega2,rC-rB);
```

Velocity of Point E
The points E_3 and D_3 are on the link 3 and

$$\begin{aligned}
\mathbf{v}_E &= \mathbf{v}_{E_3} = \mathbf{v}_{D_3} + \boldsymbol{\omega}_3 \times \mathbf{r}_{DE} = \boldsymbol{\omega}_3 \times (\mathbf{r}_E - \mathbf{r}_D) \\
&= -\omega_3(y_E - y_D)\,\mathbf{i} + \omega_3(x_E - x_D)\,\mathbf{j} \\
&= -(-3.43639)(0.524681 - 0.3)\,\mathbf{i} + (-3.43639)(-0.0898952 - 0.3)\,\mathbf{j} \\
&= 0.772092\,\mathbf{i} + 1.33983\,\mathbf{j} \text{ m/s},
\end{aligned}$$

or in MATLAB:

```
vE = vD + cross(Omega3,rE-rD);
```

Acceleration of Joint C
The points C_2 and B_2 are on the link 2 and

$$\mathbf{a}_{C_2} = \mathbf{a}_{B_2} + \boldsymbol{\alpha}_2 \times \mathbf{r}_{BC} - \omega_2^2\,\mathbf{r}_{BC} = \mathbf{a}_B + \boldsymbol{\alpha}_2 \times (\mathbf{r}_C - \mathbf{r}_B) - \omega_2^2\,(\mathbf{r}_C - \mathbf{r}_B), \quad (3.52)$$

where the angular acceleration of link 2 is $\boldsymbol{\alpha}_2 = \alpha_2\,\mathbf{k}$ (α_2 is unknown).

The points C_3 and D_3 are on the link 3 and

$$\mathbf{a}_{C_3} = \mathbf{a}_{D_3} + \boldsymbol{\alpha}_3 \times \mathbf{r}_{DC} - \omega_3^2\,\mathbf{r}_{DC} = \boldsymbol{\alpha}_3 \times (\mathbf{r}_C - \mathbf{r}_D) - \omega_3^2\,(\mathbf{r}_C - \mathbf{r}_D), \quad (3.53)$$

where $\mathbf{a}_D = \mathbf{a}_{D_3} \equiv \mathbf{0}$ and the angular velocity of link 3 is $\boldsymbol{\alpha}_3 = \alpha_3\,\mathbf{k}$. The numerical value of α_3 is unknown.

Equations 3.52 and 3.53 give

$$\mathbf{a}_B + \boldsymbol{\alpha}_2 \times (\mathbf{r}_C - \mathbf{r}_B) - \omega_2^2\,(\mathbf{r}_C - \mathbf{r}_B) = \boldsymbol{\alpha}_3 \times (\mathbf{r}_C - \mathbf{r}_D) - \omega_3^2\,(\mathbf{r}_C - \mathbf{r}_D),$$

or

$$\begin{aligned}
&\mathbf{a}_B + \begin{vmatrix} \mathbf{i} & \mathbf{j} & \mathbf{k} \\ 0 & 0 & \alpha_2 \\ x_C - x_B & y_C - y_B & 0 \end{vmatrix} - \omega_2^2\,[(x_C - x_B)\,\mathbf{i} + (y_C - y_B)\,\mathbf{j}] \\
&= \begin{vmatrix} \mathbf{i} & \mathbf{j} & \mathbf{k} \\ 0 & 0 & \alpha_3 \\ x_C - x_D & y_C - y_D & 0 \end{vmatrix} - \omega_3^2\,[(x_C - x_D)\,\mathbf{i} + (y_C - y_D)\,\mathbf{j}]. \quad (3.54)
\end{aligned}$$

Equation 3.54 represents a vectorial equation with two scalar components on the x-axis and y-axis and with two unknowns α_2 and α_3

$$\begin{aligned}
a_{Bx} - \alpha_2(y_C - y_B) - \omega_2^2(x_C - x_B) &= -\alpha_3(y_C - y_D) - \omega_3^2(x_C - x_D), \\
a_{By} + \alpha_2(x_C - x_B) - \omega_2^2(y_C - y_B) &= \alpha_3(x_C - x_D) - \omega_3^2(y_C - y_D),
\end{aligned}$$

or

$$-4.18732 - \alpha_2(0.449788 - 0.106066) - (-3.43639)^2(0.0400698 - 0.106066)$$

$$= -\alpha_3(0.449788 - 0.3) - (-3.43639)^2(0.0400698 - 0.3),$$
$$-4.18732 + \alpha_2(0.0400698 - 0.106066) - (-3.43639)^2(0.449788 - 0.106066)$$
$$= \alpha_3(0.0400698 - 0.3) - (-3.43639)^2(0.449788 - 0.3).$$

Thus,

$$\alpha_2 = -8.97883 \text{ rad/s}^2 \text{ and } \alpha_3 = 22.6402 \text{ rad/s}^2.$$

The acceleration of C is

$$\begin{aligned}
\mathbf{a}_C &= \boldsymbol{\alpha}_3 \times (\mathbf{r}_C - \mathbf{r}_D) - \omega_3^2\,(\mathbf{r}_C - \mathbf{r}_D) \\
&= [-\alpha_3(y_C - y_D) - \omega_3^2(x_C - x_D)]\,\mathbf{ı} + [\alpha_3(x_C - x_D) - \omega_3^2(y_C - y_D)]\,\mathbf{j} \\
&= [-(22.6402)(0.449788 - 0.3) - (-3.43639)^2(0.0400698 - 0.3)]\,\mathbf{ı} \\
&+ [(22.6402)(0.0400698 - 0.3) - (-3.43639)^2(0.449788 - 0.3)]\,\mathbf{j} \\
&= -0.321767\,\mathbf{ı} - 7.65368\,\mathbf{j} \text{ m/s}^2.
\end{aligned}$$

The MATLAB commands for the angular accelerations of links 2 and 3, and the acceleration of C are:

```
alpha2z = sym('alpha2z','real');
alpha3z = sym('alpha3z','real');
alpha2 = [ 0 0 alpha2z ]; alpha3 = [ 0 0 alpha3z ];
eqaC2 = aB2+cross(alpha2,rC-rB)-...
         dot(Omega2,Omega2)*(rC-rB);
eqaC3 = aD+cross(alpha3,rC-rD)-...
         dot(Omega3,Omega3)*(rC-rD);
eqaC = eqaC2 - eqaC3;
eqaCx = eqaC(1);
eqaCy = eqaC(2);
solaC = solve(eqaCx,eqaCy);
alpha2zs = eval(solaC.alpha2z);
alpha3zs = eval(solaC.alpha3z);
Alpha2 = [0 0 alpha2zs];
Alpha3 = [0 0 alpha3zs];
aC=aB2+cross(Alpha2,rC-rB)-dot(Omega2,Omega2)*(rC-rB);
```

Acceleration of Point E

The points E and D are on the link 3 and the acceleration of E is

$$\begin{aligned}
\mathbf{a}_E &= \mathbf{a}_D + \boldsymbol{\alpha}_3 \times \mathbf{r}_{DE} - \omega_2^2\,\mathbf{r}_{DE} = \boldsymbol{\alpha}_3 \times (\mathbf{r}_E - \mathbf{r}_D) - \omega_3^2\,(\mathbf{r}_E - \mathbf{r}_D) \\
&= [-\alpha_3(y_E - y_D) - \omega_3^2(x_E - x_D)]\,\mathbf{ı} + [\alpha_3(x_E - x_D) - \omega_3^2(y_E - y_D)]\,\mathbf{j} \\
&= [-(22.6402)(0.524681 - 0.3) - (-3.43639)^2(-0.0898952 - 0.3)]\,\mathbf{ı} \\
&+ [(22.6402)(-0.0898952 - 0.3) - (-3.43639)^2(0.524681 - 0.3)]\,\mathbf{j} \\
&= -0.48265\,\mathbf{ı} - 11.4805\,\mathbf{j} \text{ m/s}^2.
\end{aligned}$$

The MATLAB command for the acceleration of E is

```
aE=aD+cross(Alpha3,rE-rD)-dot(Omega3,Omega3)*(rE-rD);
```

The MATLAB program with the results for the velocities and accelerations is given in Appendix B.2.

3.7 Inverted Slider-Crank Mechanism

Exercise
The following dimensions are given for the inverted slider-crank mechanism (shown in Fig. 3.6): AC=0.15 m and BC=0.2 m. The length AD is selected as 0.35 m ($AD = AC + BC$). The driver link 1 rotates with a constant speed of $n = n_1 = 30$ rpm.

Find the velocities and the accelerations of the mechanism when the angle of the driver link 1 with the horizontal axis is $\phi = \phi_1 = 60°$.

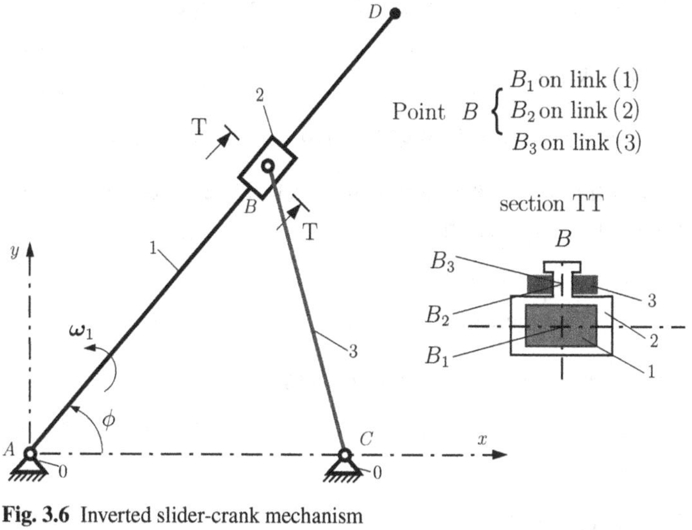

Fig. 3.6 Inverted slider-crank mechanism

Solution
A Cartesian reference frame with the origin at A is selected and the coordinates of joint A are $x_A = y_A = 0$. The coordinates of the joint C are $x_C = AC = 0.15$ m and $y_C = 0$. The coordinates of joint B for $\phi = \phi_1 = 60°$ are x_B=0.113535 m and y_B=0.196648 m. The position of joint B is calculated from the equations

$$\tan\phi = \frac{y_B}{x_B} \quad \text{and} \quad (x_B - x_C)^2 + (y_B - y_C)^2 = BC^2.$$

The MATLAB commands for the position vector of B are

```
eqB1 = 'xBsol*sin(phi) = yBsol*cos(phi)';
eqB2 = 'yBsol^2+(xC-xBsol)^2-BC^2 = 0';
solB = solve(eqB1, eqB2, 'xBsol, yBsol');
xBpositions = eval(solB.xBsol);
yBpositions = eval(solB.yBsol);
xB1 = xBpositions(1); xB2 = xBpositions(2);
yB1 = yBpositions(1); yB2 = yBpositions(2);
if (phi>=0 && phi<= pi)
 if yB1 >= 0 xB=xB1; yB=yB1; else xB=xB2; yB=yB2;
end end
if (phi>pi && phi<=2*pi)
 if yB1 < 0 xB=xB1; yB=yB1; else xB=xB2; yB=yB2;
end end
rB = [ xB, yB, 0 ];
```

The magnitude of the angular velocity of the driver link 1 is

$$\omega = \omega_1 = \dot{\phi}(t) = \frac{\pi n_1}{30} = \frac{\pi\,(30\text{ rpm})}{30} = 3.141 \text{ rad/s}.$$

The angular velocity of link 1 is

$$\boldsymbol{\omega} = \boldsymbol{\omega}_1 = \omega\,\mathbf{k} = 3.141\,\mathbf{k} \text{ rad/s}.$$

The link 2 and the driver link 1 have the same angular velocity $\boldsymbol{\omega}_1 = \boldsymbol{\omega}_2$. The angular acceleration of link 1 is $\boldsymbol{\alpha}_1 = \dot{\boldsymbol{\omega}}_1 = \mathbf{0}$.

Velocity and Acceleration of B_1
The velocity of the point B_1 on the link 1 is

$$\mathbf{v}_{B_1} = \mathbf{v}_A + \boldsymbol{\omega}_1 \times \mathbf{r}_B = \boldsymbol{\omega}_1 \times \mathbf{r}_B,$$

where $\mathbf{v}_A \equiv \mathbf{0}$ is the velocity of the origin $A \equiv O$.

The velocity of B_1 is

$$\mathbf{v}_{B_1} = \begin{vmatrix} \mathbf{\imath} & \mathbf{\jmath} & \mathbf{k} \\ 0 & 0 & \omega \\ x_B & y_B & 0 \end{vmatrix} = \begin{vmatrix} \mathbf{\imath} & \mathbf{\jmath} & \mathbf{k} \\ 0 & 0 & 3.141 \\ 0.113535 & 0.196648 & 0 \end{vmatrix} = -0.617787\mathbf{\imath} + 0.356679\mathbf{\jmath} \text{ m/s}.$$

The acceleration of the point B_1 on the link 1 is

$$\begin{aligned}\mathbf{a}_{B_1} &= \mathbf{a}_A + \boldsymbol{\alpha}_1 \times \mathbf{r}_B + \boldsymbol{\omega}_1 \times (\boldsymbol{\omega}_1 \times \mathbf{r}_B) = \boldsymbol{\alpha}_1 \times \mathbf{r}_B - \omega_1^2 \mathbf{r}_B \\ &= -\omega_1^2 \mathbf{r}_B = -3.141^2(0.113535\mathbf{\imath} + 0.196648\mathbf{\jmath}) = -1.12054\mathbf{\imath} - 1.94083\mathbf{\jmath} \text{ m/s}^2.\end{aligned}$$

Angular Velocity of Link 3

The velocity of the point B_2 on the link 2 is equal to the velocity of the point B_3 on the link 3 (link 2 and link 3 are connected with a rotational joint).

The points B_3 and C are on the link 3 and

$$\mathbf{v}_{B_2} = \mathbf{v}_{B_3} = \mathbf{v}_C + \boldsymbol{\omega}_3 \times \mathbf{r}_{CB} = \boldsymbol{\omega}_3 \times (\mathbf{r}_B - \mathbf{r}_C), \tag{3.55}$$

where $\mathbf{v}_C \equiv \mathbf{0}$ and the angular velocity of link 3 is $\boldsymbol{\omega}_3 = \omega_3\,\mathbf{k}$. The angular velocity of link 3 ω_3 is to be calculated.

The velocity of the point B_2 on the link 2 is calculated in terms of the velocity of the point B_1 on the link 1

$$\mathbf{v}_{B_2} = \mathbf{v}_{B_1} + \mathbf{v}^{rel}_{B_2B_1} = \mathbf{v}_{B_1} + \mathbf{v}_{B_{21}}, \tag{3.56}$$

where $\mathbf{v}^{rel}_{B_2B_1} = \mathbf{v}_{B_{21}}$ is the relative acceleration of B_2 with respect to B_1 on link 1. This relative velocity is parallel to the sliding direction AB, $\mathbf{v}_{B_{21}}||AB$, or

$$\mathbf{v}_{B_{21}} = v_{B_{21}} \cos\phi_1\,\mathbf{\imath} + v_{B_{21}} \sin\phi_1\,\mathbf{\jmath}, \tag{3.57}$$

where $\phi_1 = 45°$. Equations 3.55–3.57 give

$$\begin{vmatrix} \mathbf{\imath} & \mathbf{\jmath} & \mathbf{k} \\ 0 & 0 & \omega_3 \\ x_B - x_C & y_B - y_C & 0 \end{vmatrix} = \mathbf{v}_{B_1} + v_{B_{21}} \cos\phi_1\,\mathbf{\imath} + v_{B_{21}} \sin\phi_1\,\mathbf{\jmath}. \tag{3.58}$$

Equation 3.58 represents a vectorial equation with two scalar components on the x-axis and y-axis and with two unknowns ω_3 and $v_{B_{21}}$

$$-\omega_3(y_B - y_C) = v_{B_{1x}} + v_{B_{21}} \cos\phi_1,$$
$$\omega_3(x_B - x_C) = v_{B_{1y}} + v_{B_{21}} \sin\phi_1,$$

or

$$-\omega_3(0.196648 - 0) = -0.617787 + v_{B_{21}} \cos 60°,$$
$$\omega_3(0.113535 - 0.15) = 0.356679 + v_{B_{21}} \sin 60°.$$

Thus,

$$\omega_3 = 4.69102 \text{ rad/s and } v_{B_{21}} = -0.609381 \text{ m/s},$$

or in vectorial form

$$\boldsymbol{\omega}_3 = 4.69102\,\mathbf{k} \text{ rad/s and } \mathbf{v}_{B_{21}} = -0.304691\,\mathbf{\imath} - 0.527739\,\mathbf{\jmath} \text{ m/s}.$$

The velocity of B_3 (or B_2) is

$$\mathbf{v}_{B_2} = \mathbf{v}_{B_3} = \begin{vmatrix} \mathbf{\imath} & \mathbf{\jmath} & \mathbf{k} \\ 0 & 0 & 3.903 \\ 0.113535 - 0.15 & 0.196648 & \end{vmatrix} = -0.922477\,\mathbf{\imath} - 0.17106\,\mathbf{\jmath} \text{ m/s}.$$

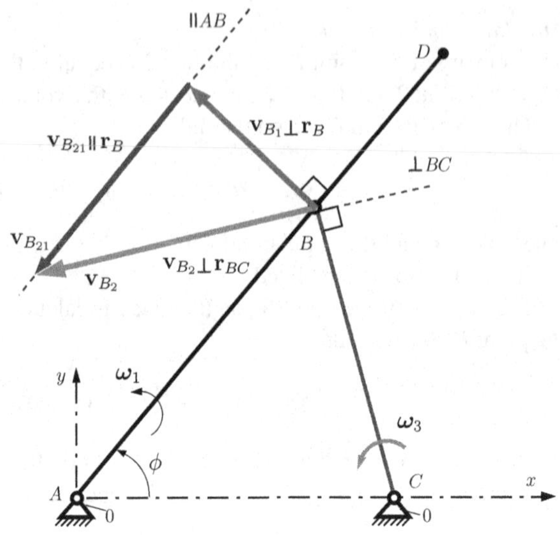

Fig. 3.7 Velocity field for the inverted slider-crank mechanism

The MATLAB commands for the ω_3, $\mathbf{v}_{B_{21}}$, and $\mathbf{v}_{B_3}$ are:

```
omega3z = sym('omega3z','real');    % omega3z unknown
omega3 = [ 0 0 omega3z ];
vB21 = sym('vB21','real');    % vB21 unknown
vB2B1 = [ vB21*cos(phi1) vB21*sin(phi1) 0 ];
vC = [0 0 0 ];
% vB2 = vB3 = vC + omega3 x (rB-rC)
vB3 = vC + cross(omega3,rB-rC);
vB2 = vB3;
% vB2 = vB1 + vB2B1
eqvB = vB2 - ( vB1 + vB2B1 );
eqvBx = eqvB(1); eqvBy = eqvB(2);
solvB = solve(eqvBx,eqvBy);
omega3zs = eval(solvB.omega3z);
vB21s = eval(solvB.vB21);
Omega3 = [0 0 omega3zs];
VB21 = vB21s*[cos(phi1) sin(phi1) 0];
VB3 = vC + cross(Omega3,rB-rC);
```

The velocity field for the inverted slider-crank mechanism is shown in Fig. 3.7.

Angular Acceleration of Link 3
The points B_3 and C are on the link 3 and

$$\mathbf{a}_{B_2} = \mathbf{a}_{B_3} = \mathbf{a}_C + \boldsymbol{\alpha}_3 \times \mathbf{r}_{CB} - \omega_3^2 \mathbf{r}_{CB} = \boldsymbol{\alpha}_3 \times \mathbf{r}_{CB} - \omega_3^2 \mathbf{r}_{CB}, \tag{3.59}$$

where $\mathbf{a}_C \equiv \mathbf{0}$ and the angular acceleration of link 3 is $\boldsymbol{\alpha}_3 = \alpha_3 \mathbf{k}$. The angular acceleration of link 3 α_3 is to be calculated.

The acceleration of the point B_2 on the link 2 is calculated in terms of the acceleration of the point B_1 on the link 1

$$\mathbf{a}_{B_2} = \mathbf{a}_{B_1} + \mathbf{a}^{rel}_{B_2B_1} + \mathbf{a}^{cor}_{B_2B_1} = \mathbf{a}_{B_1} + \mathbf{a}_{B_{21}} + \mathbf{a}^{cor}_{B_{21}}, \tag{3.60}$$

where $\mathbf{a}^{rel}_{B_2B_1} = \mathbf{a}_{B_{21}}$ is the relative acceleration of B_2 with respect to B_1 on link 1. This relative acceleration is parallel to the sliding direction AB, $\mathbf{a}_{B_{21}} || AB$, or

$$\mathbf{a}_{B_{21}} = a_{B_{21}} \cos\phi_1 \,\mathbf{\imath} + a_{B_{21}} \sin\phi_1 \,\mathbf{\jmath}. \tag{3.61}$$

The Coriolis acceleration of B_2 relative to B_1 is

$$\begin{aligned}
\mathbf{a}^{cor}_{B_{21}} &= 2\,\boldsymbol{\omega}_1 \times \mathbf{v}_{B_{21}} = 2\,\boldsymbol{\omega}_2 \times \mathbf{v}_{B_{21}} = 2\begin{vmatrix} \mathbf{\imath} & \mathbf{\jmath} & \mathbf{k} \\ 0 & 0 & \omega_1 \\ v_{B_{21}}\cos\phi_1 & v_{B_{21}}\sin\phi_1 & 0 \end{vmatrix} \\
&= 2(-\omega_1 v_{B_{21}} \sin\phi_1 \,\mathbf{\imath} + \omega_1 v_{B_{21}} \cos\phi_1 \,\mathbf{\jmath} \\
&= 2[-3.141(-0.609381)\sin 60^\circ \,\mathbf{\imath} + 3.141(-0.609381)\cos 60^\circ \,\mathbf{\jmath}] \\
&= 3.31588\mathbf{\imath} - 1.91443\mathbf{\jmath} \ \text{m/s}^2.
\end{aligned} \tag{3.62}$$

Equations 3.59–3.62 give

$$\begin{aligned}
&\begin{vmatrix} \mathbf{\imath} & \mathbf{\jmath} & \mathbf{k} \\ 0 & 0 & \alpha_3 \\ x_B - x_C & y_B - y_C & 0 \end{vmatrix} - \omega_3^2(\mathbf{r}_B - \mathbf{r}_C) \\
&= \mathbf{a}_{B_1} + a_{B_{21}}(\cos\phi_1 \mathbf{\imath} + \sin\phi_1 \mathbf{\jmath}) + 2\,\boldsymbol{\omega}_1 \times \mathbf{v}_{B_{21}}.
\end{aligned} \tag{3.63}$$

Equation 3.63 represents a vectorial equations with two scalar components on the x-axis and y-axis and with two unknowns α_3 and $a_{B_{21}}$

$$\begin{aligned}
-\alpha_3(y_B - y_C) - \omega_3^2(x_B - x_C) &= a_{B_{1x}} + a_{B_{21}} \cos\phi_1 - 2\omega_1 v_{B_{21}} \sin\phi_1, \\
\alpha_3(x_B - x_C) - \omega_3^2(y_B - y_C) &= a_{B_{1y}} + a_{B_{21}} \sin\phi_1 + 2\omega_1 v_{B_{21}} \cos\phi_1,
\end{aligned}$$

or

$$\begin{aligned}
&-\alpha_3(0.196648 - 0) - 3.903^2(0.113535 - 0.15) \\
&\quad = -1.12054 + a_{B_{21}} \cos 60^\circ + 3.31588, \\
&\alpha_3(0.113535 - 0.15) - 3.903^2(0.196648 - 0) \\
&\quad = -1.94083, + a_{B_{21}} \sin 60^\circ - 1.91443.
\end{aligned}$$

Thus,

$$\alpha_3 = -6.38024 \text{ rad/s}^2 \ \text{and} \ a_{B_{21}} = -0.276477 \ \text{m/s}^2.$$

The relative acceleration of B_2 with respect to B_1 is

$$\mathbf{a}_{B_{21}} = -0.276477\cos 60^\circ \imath - 0.276477\sin 60^\circ \jmath = -0.138239\imath - 0.239436\jmath \text{ m/s}^2,$$

and the acceleration of B_3 is

$$\mathbf{a}_{B_2} = \mathbf{a}_{B_3} = \boldsymbol{\alpha}_3 \times \mathbf{r}_{CB} - \omega_3^2 \mathbf{r}_{CB} =$$
$$-6.38024\mathbf{k} \times [(0.113535 - 0.15)\imath + (0.196648 - 0)\jmath]$$
$$4.69102^2[(0.113535 - 0.15)\imath + (0.196648 - 0)\jmath] =$$
$$2.0571\imath - 4.0947\jmath \text{ m/s}^2.$$

The MATLAB commands for the $\boldsymbol{\alpha}_3$, $\mathbf{a}_{B_{21}}$, and $\mathbf{a}_{B_3}$ are

```
alpha3z = sym('alpha3z','real');    % alpha3z unknown
alpha3 = [ 0 0 alpha3z ];
aB21 = sym('aB21','real');    % aB21 unknown
aB2B1 = [ aB21*cos(phi1) aB21*sin(phi1) 0 ];
aC = [ 0 0 0 ];

aB3=aC+cross(alpha3,rB-rC)-dot(Omega3,Omega3)*(rB-rC);
aB2 = aB3;

% aB2B1cor = 2 omega1 x vB2B1
aB2B1cor = 2*cross(omega1,VB21);
% aB2=aB1+aB2B1+aB2B1cor
eqaB = aB2 - ( aB1 + aB2B1 + aB2B1cor );
eqaBx = eqaB(1);
eqaBy = eqaB(2);

solaB = solve(eqaBx,eqaBy);
alpha3zs = eval(solaB.alpha3z);
aB21s = eval(solaB.aB21);

Alpha3 = [0 0 alpha3zs];
AB21 = aB21s*[cos(phi1) sin(phi1) 0];

AB3=aC+cross(Alpha3,rB-rC)-dot(Omega3,Omega3)*(rB-rC);
```

The relation between the angular velocities of link 2 and link 3 is

$$\boldsymbol{\omega}_2 = \boldsymbol{\omega}_3 + \boldsymbol{\omega}_{23},$$

and the relative angular velocity of link 2 with respect to link 3 is

$$\boldsymbol{\omega}_{23} = \boldsymbol{\omega}_2 - \boldsymbol{\omega}_3 = 3.141\mathbf{k} - 4.69102\mathbf{k} = -1.54942\mathbf{k} \text{ rad/s}.$$

The relative angular acceleration of link 2 with respect to link 3 is

$$\alpha_{23} = \alpha_2 - \alpha_3 = -\alpha_3 = 6.38024\mathbf{k} \ \ \text{rad/s}^2,$$

where $\alpha_2 = \alpha_1 = \mathbf{0}$.

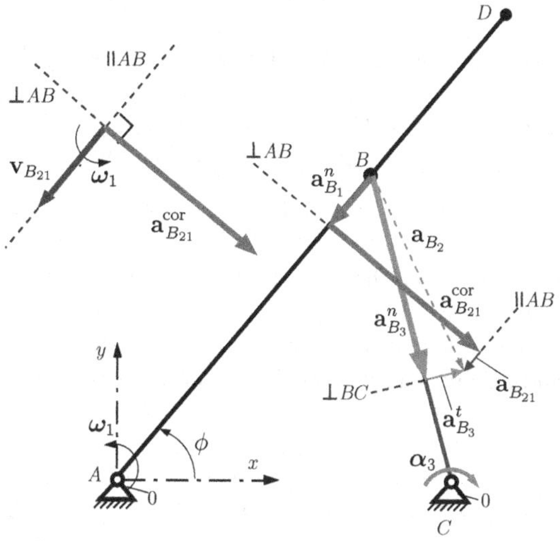

Fig. 3.8 Acceleration field for the inverted slider-crank mechanism

The MATLAB program and the results for the velocities and accelerations analysis are given in Appendix B.3.

The acceleration field for the inverted slider-crank mechanism is shown in Fig. 3.8.

3.8 R-RTR-RTR Mechanism

Exercise

The planar R-RTR-RTR mechanism is shown in Fig. 2.8. The following numerical data are given: $AB = 0.15$ m, $AC = 0.10$ m, $CD = 0.15$ m, $DF = 0.40$ m, and $AG = 0.30$ m. The constant angular speed of the driver link 1 is 50 rpm.

Find the velocities and accelerations of the mechanism when the angle of the driver link 1 with the horizontal axis is $\phi = \phi_1 = 30°$.

Solution

A Cartesian reference frame xOy is selected. The joint A is the origin of the reference frame, that is, $A \equiv O$, and $x_A = 0$, $y_A = 0$. The coordinates of the joint C are: $x_C = 0$, $y_C = AC = 0.1$ m. The coordinates of the joint B, for the given angle $\phi_1 = 30°$, are $x_B = AB\cos\phi_1 = 0.129904$ m and $y_B = AB\sin\phi_1 = 0.075$ m.

The coordinates of the joint D are $x_D = -0.147297$ m and $y_D = 0.128347$ m. The angle of links 2 (or link 3) and 5 (or link 4) with the horizontal axis are $\phi_2 = \phi_3 = -0.1901$ rad $= -10.8934°$ and $\phi_4 = \phi_5 = 2.4248$ rad $= 138.933°$.

The angular velocity of link 1 is constant and has the value

$$\boldsymbol{\omega}_1 = \omega_1\,\mathbf{k} = \frac{\pi n}{30}\mathbf{k} = \frac{\pi(50)}{30}\mathbf{k} = 5.23599\mathbf{k}\ \text{rad/s}.$$

The angular acceleration of link 1 is $\boldsymbol{\alpha}_1 = \dot{\boldsymbol{\omega}}_1 = \mathbf{0}$.

Velocity and Acceleration of $B_1 = B_2$

The velocity of the point B_1 on the link 1 is

$$\mathbf{v}_{B_1} = \mathbf{v}_A + \boldsymbol{\omega}_1 \times \mathbf{r}_{AB} = \boldsymbol{\omega}_1 \times \mathbf{r}_B,$$

where $\mathbf{v}_A \equiv \mathbf{0}$ is the velocity of the origin $A \equiv O$ and $\mathbf{r}_B = x_B\,\mathbf{\imath} + y_B\,\mathbf{\jmath} = 0.129904\,\mathbf{\imath} + 0.075\,\mathbf{\jmath}$ m.

The velocity of point B_2 on the link 2 is $\mathbf{v}_{B_2} = \mathbf{v}_{B_1}$ because between the links 1 and 2 there is a rotational joint. The velocity of $B_1 = B_2$ is

$$\mathbf{v}_{B_1} = \mathbf{v}_{B_2} = \begin{vmatrix} \mathbf{\imath} & \mathbf{\jmath} & \mathbf{k} \\ 0 & 0 & \omega \\ x_B & y_B & 0 \end{vmatrix} = \begin{vmatrix} \mathbf{\imath} & \mathbf{\jmath} & \mathbf{k} \\ 0 & 0 & 5.23599 \\ 0.129904 & 0.075 & 0 \end{vmatrix}$$
$$= -0.392699\,\mathbf{\imath} + 0.680175\,\mathbf{\jmath}\ \text{m/s}.$$

The acceleration of $B_1 = B_2$ is

$$\mathbf{a}_{B_1} = \mathbf{a}_{B_2} = \mathbf{a}_A + \boldsymbol{\alpha}_1 \times \mathbf{r}_B + \boldsymbol{\omega}_1 \times (\boldsymbol{\omega}_1 \times \mathbf{r}_B) = \boldsymbol{\alpha}_1 \times \mathbf{r}_B - \omega_1^2\mathbf{r}_B$$
$$= -\omega_1^2\mathbf{r}_B = -5.23599^2(0.129904\,\mathbf{\imath} + 0.075\,\mathbf{\jmath}) = -3.56139\,\mathbf{\imath} - 2.05617\,\mathbf{\jmath}\ \text{m/s}^2.$$

Angular Velocity of Link 3

The velocity of the point B_3 on the link 3 is calculated in terms of the velocity of the point B_2 on the link 2

$$\mathbf{v}_{B_3} = \mathbf{v}_{B_2} + \mathbf{v}^{\text{rel}}_{B_3B_2} = \mathbf{v}_{B_2} + \mathbf{v}_{B_{32}}, \tag{3.64}$$

where $\mathbf{v}^{\text{rel}}_{B_3B_2} = \mathbf{v}_{B_{32}}$ is the relative acceleration of B_3 with respect to B_2 on link 3. This relative velocity is parallel to the sliding direction BC, $\mathbf{v}_{B_{32}} || BC$, or

$$\mathbf{v}_{B_{32}} = v_{B_{32}} \cos\phi_2\,\mathbf{\imath} + v_{B_{32}} \sin\phi_2\,\mathbf{\jmath}, \tag{3.65}$$

where $\phi_2 = 4.715°$ is known from position analysis. The points B_3 and C are on the link 3 and

$$\mathbf{v}_{B_3} = \mathbf{v}_C + \boldsymbol{\omega}_3 \times \mathbf{r}_{CB} = \boldsymbol{\omega}_3 \times (\mathbf{r}_B - \mathbf{r}_C), \tag{3.66}$$

where $\mathbf{v}_C \equiv \mathbf{0}$ and the angular velocity of link 3 is $\boldsymbol{\omega}_3 = \omega_3\,\mathbf{k}$.

Equations 3.64–3.66 give

$$\begin{vmatrix} \mathbf{i} & \mathbf{j} & \mathbf{k} \\ 0 & 0 & \omega_3 \\ x_B - x_C & y_B - y_C & 0 \end{vmatrix} = \mathbf{v}_{B_2} + v_{B_{32}} \cos\phi_2 \mathbf{i} + v_{B_{32}} \sin\phi_2 \mathbf{j}. \tag{3.67}$$

Equation 3.67 represents a vectorial equations with two scalar components on the x-axis and y-axis and with two unknowns ω_3 and $v_{B_{32}}$

$$-\omega_3(y_B - y_C) = v_{B_2x} + v_{B_{32}} \cos\phi_2,$$
$$\omega_3(x_B - x_C) = v_{B_2y} + v_{B_{32}} \sin\phi_2,$$

or

$$-\omega_3(0.075 - 0.1) = -0.392699 + v_{B_{32}} \cos(-10.8934^\circ),$$
$$\omega_3(0.129904 - 0) = 0.680175 + v_{B_{32}} \sin(-10.8934^\circ).$$

Thus,

$$\omega_3 = \omega_2 = 4.48799 \text{ rad/s} \quad \text{and} \quad v_{B_{32}} = 0.514164 \text{ m/s},$$

or in vectorial form

$$\boldsymbol{\omega}_3 = \boldsymbol{\omega}_2 = 4.48799\mathbf{k} \text{ rad/s} \quad \text{and} \quad \mathbf{v}_{B_{32}} = 0.504899\mathbf{i} - 0.0971678\mathbf{j} \text{ m/s}.$$

The MATLAB commands for the $\boldsymbol{\omega}_3$ and $\mathbf{v}_{B_{32}}$ are:

```
omega3z=sym('omega3z','real');
vB32=sym('vB32','real');
omega3 = [ 0 0 omega3z ];
% omega3z unknown (to be calculated)
% vB32 unknown (to be calculated)
vC = [0 0 0 ];   % C is fixed
% vB3 = vC + omega3 x rCB
vB3 = vC + cross(omega3,rB-rC);
vB3B2 = vB32*[ cos(phi2) sin(phi2) 0];
% vB3 = vB2 + vB3B2 (vectorial equation)
eqvB = vB3 - vB2 - vB3B2;
eqvBx = eqvB(1); eqvBy = eqvB(2);
% two equations eqvBx & eqvBy with two unknowns
% solve for omega3z and vB32
solvB = solve(eqvBx,eqvBy);
omega3zs=eval(solvB.omega3z); vB32s=eval(solvB.vB32);
Omega3 = [0 0 omega3zs]; Omega2 = Omega3;
VB32 = vB32s*[cos(phi2) sin(phi2) 0];
```

Angular Acceleration of Link 3

The acceleration of the point B_3 on the link 3 is calculated in terms of the acceleration of the point B_2 on the link 2

$$\mathbf{a}_{B_3} = \mathbf{a}_{B_2} + \mathbf{a}^{rel}_{B_3B_2} + \mathbf{a}^{cor}_{B_3B_2} = \mathbf{a}_{B_2} + \mathbf{a}_{B_{32}} + \mathbf{a}^{cor}_{B_{32}}, \tag{3.68}$$

where $\mathbf{a}^{rel}_{B_3B_2} = \mathbf{a}_{B_{32}}$ is the relative acceleration of B_3 with respect to B_2 on link 3. This relative acceleration is parallel to the sliding direction BC, $\mathbf{a}_{B_{32}}||BC$, or

$$\mathbf{a}_{B_{32}} = a_{B_{32}} \cos\phi_2 \mathbf{\imath} + a_{B_{32}} \sin\phi_2 \mathbf{\jmath}. \tag{3.69}$$

The Coriolis acceleration of B_3 relative to B_2 is

$$\begin{aligned}\mathbf{a}^{cor}_{B_{32}} = 2\,\boldsymbol{\omega}_3 \times \mathbf{v}_{B_{32}} = 2\,\boldsymbol{\omega}_2 \times \mathbf{v}_{B_{32}} = 2\begin{vmatrix} \mathbf{\imath} & \mathbf{\jmath} & \mathbf{k} \\ 0 & 0 & \omega_3 \\ v_{B_{32}}\cos\phi_2 & v_{B_{32}}\sin\phi_2 & 0 \end{vmatrix}\\ = 2(-\omega_3 v_{B_{32}} \sin\phi_2 \mathbf{\imath} + \omega_3 v_{B_{32}} \cos\phi_2 \mathbf{\jmath})\\ = -2(4.48799)(0.514164)\sin(-10.8934°)\mathbf{\imath}\\ +2(4.48799)(0.514164)\cos(-10.8934°)\mathbf{\jmath}\\ = 0.872176\mathbf{\imath} + 4.53196\mathbf{\jmath}\ \text{m/s}^2.\end{aligned} \tag{3.70}$$

The points B_3 and C are on the link 3 and

$$\mathbf{a}_{B_3} = \mathbf{a}_C + \boldsymbol{\alpha}_3 \times \mathbf{r}_{CB} - \omega_3^2 \mathbf{r}_{CB}, \tag{3.71}$$

where $\mathbf{a}_C \equiv \mathbf{0}$ and the angular acceleration of link 3 is

$$\boldsymbol{\alpha}_3 = \alpha_3 \mathbf{k}.$$

Equations 3.68–3.71 give

$$\begin{aligned}\begin{vmatrix} \mathbf{\imath} & \mathbf{\jmath} & \mathbf{k} \\ 0 & 0 & \alpha_3 \\ x_B - x_C & y_B - y_C & 0 \end{vmatrix} - \omega_3^2(\mathbf{r}_B - \mathbf{r}_C)\\ = \mathbf{a}_{B_2} + a_{B_{32}}(\cos\phi_2 \mathbf{\imath} + \sin\phi_2 \mathbf{\jmath}) + 2\,\boldsymbol{\omega}_3 \times \mathbf{v}_{B_{32}}.\end{aligned} \tag{3.72}$$

Equation 3.72 represents a vectorial equations with two scalar components on the x-axis and y-axis and with two unknowns α_3 and $a_{B_{32}}$

$$-\alpha_3(y_B - y_C) - \omega_3^2(x_B - x_C) = a_{B_2x} + a_{B_{32}}\cos\phi_2 - 2\omega_3 v_{B_{32}}\sin\phi_2,$$
$$\alpha_3(x_B - x_C) - \omega_3^2(y_B - y_C) = a_{B_2y} + a_{B_{32}}\sin\phi_2 + 2\omega_3 v_{B_{32}}\cos\phi_2,$$

or

$$-\alpha_3(0.075 - 0.1) - 4.48799^2(0.121 - 0)$$

$$= -3.56139 + a_{B_{32}} \cos(-10.8934^\circ) - 2(4.48799)(0.514164)\sin(-10.8934^\circ),$$

$$\alpha_3(0.129904 - 0) - 4.48799^2(0.075 - 0.1)$$

$$= -2.05617 + a_{B_{32}} \sin(-10.8934^\circ) + 2(4.48799)(0.514164)\cos(-10.8934^\circ).$$

Thus,

$$\alpha_3 = \alpha_2 = 14.5363 \text{ rad/s}^2 \text{ and } a_{B_{32}} = 0.44409 \text{ m/s}^2.$$

The MATLAB commands for the $\boldsymbol{\alpha}_3$ and $\mathbf{a}_{B_{32}}$ are:

```
% Coriolis acceleration
aB3B2cor = 2*cross(Omega3,VB32);
alpha3z=sym('alpha3z','real');  % alpha3z unknown
aB32=sym('aB32','real');  % aB32 unknown
alpha3 = [ 0 0 alpha3z ];
aC = [0 0 0 ];    % C is fixed
aB3=aC+cross(alpha3,rB-rC)-dot(Omega3,Omega3)*(rB-rC);
aB3B2 = aB32*[ cos(phi2) sin(phi2) 0];
% aB3 = aB2 + aB3B2 + aB3B2cor
eqaB = aB3 - aB2 - aB3B2 - aB3B2cor;
eqaBx = eqaB(1); eqaBy = eqaB(2);
solaB = solve(eqaBx,eqaBy);
alpha3zs=eval(solaB.alpha3z);
aB32s=eval(solaB.aB32);
Alpha3 = [0 0 alpha3zs];
Alpha2 = Alpha3;
AB32 = aB32s*[cos(phi2) sin(phi2) 0];
```

Velocity and Acceleration of $D_3 = D_4$

The velocity of $D_3 = D_4$ is

$$\mathbf{v}_{D_3} = \mathbf{v}_{D_4} = \mathbf{v}_C + \boldsymbol{\omega}_3 \times \mathbf{r}_{CD} = \boldsymbol{\omega}_3 \times (\mathbf{r}_D - \mathbf{r}_C)$$

$$= \begin{vmatrix} \mathbf{1} & \mathbf{J} & \mathbf{k} \\ 0 & 0 & \omega_3 \\ x_D - x_C & y_D - y_C & 0 \end{vmatrix} = \begin{vmatrix} \mathbf{1} & \mathbf{J} & \mathbf{k} \\ 0 & 0 & 4.48799 \\ -0.147297 - 0 & 0.128347 - 0.1 & 0 \end{vmatrix}$$

$$= -0.127223\mathbf{1} - 0.661068\mathbf{J} \text{ m/s}.$$

The acceleration of $D_3 = D_4$ is

$$\mathbf{a}_{D_3} = \mathbf{a}_{D_4} = \mathbf{a}_C + \boldsymbol{\alpha}_3 \times \mathbf{r}_{CD} - \omega_3^2 \mathbf{r}_{CD} = \boldsymbol{\alpha}_3 \times (\mathbf{r}_D - \mathbf{r}_C) - \omega_3^2(\mathbf{r}_D - \mathbf{r}_C)$$

$$= \begin{vmatrix} \mathbf{1} & \mathbf{J} & \mathbf{k} \\ 0 & 0 & \alpha_3 \\ x_D - x_C & y_D - y_C & 0 \end{vmatrix} - \omega_3^2 \left[(x_D - x_C)\mathbf{1} + (y_D - y_C)\mathbf{J}\right]$$

$$= \begin{vmatrix} \mathbf{1} & \mathbf{J} & \mathbf{k} \\ 0 & 0 & 14.5363 \\ -0.147297 - 0 & 0.128347 - 0.1 & 0 \end{vmatrix}$$

$$-4.48799^2\,[(-0.147297-0)\,\imath+(0.128347-0.1)\,\jmath]$$
$$=2.55481\imath-2.71212\jmath\ \text{m/s}^2.$$

The MATLAB commands for the velocity and acceleration of $D_3 = D_4$ are:

```
% D3 & C points on link 3
vD3 = vC + cross(Omega3,rD-rC);
vD4 = vD3;
aD3=aC+cross(Alpha3,rD-rC)-dot(Omega3,Omega3)*(rD-rC);
aD4 = aD3;
```

Angular Velocity of Link 5
The velocity of the point D_5 on the link 5 is calculated in terms of the velocity of the point D_4 on the link 4

$$\mathbf{v}_{D_5}=\mathbf{v}_{D_4}+\mathbf{v}_{D_{54}}. \tag{3.73}$$

This relative velocity of D_5 with respect to D_4 is parallel to the sliding direction DA, $\mathbf{v}_{D_{54}}||DA$, or

$$\mathbf{v}_{D_{54}}=v_{D_{54}}\cos\phi_5\,\imath+v_{D_{54}}\sin\phi_5\,\jmath. \tag{3.74}$$

The points D_5 and A are on the link 5 and

$$\mathbf{v}_{D_5}=\mathbf{v}_A+\boldsymbol{\omega}_5\times\mathbf{r}_D, \tag{3.75}$$

where $\mathbf{v}_A\equiv\mathbf{0}$ and the angular velocity of link 5 is

$$\boldsymbol{\omega}_5=\omega_5\,\mathbf{k}.$$

Equations 3.73–3.75 give

$$\begin{vmatrix}\imath & \jmath & \mathbf{k}\\ 0 & 0 & \omega_5\\ x_D & y_D & 0\end{vmatrix}=\mathbf{v}_{D_4}+v_{D_{54}}(\cos\phi_5\,\imath+\sin\phi_5\,\jmath). \tag{3.76}$$

Equation 3.76 represents a vectorial equations with two scalar components on the x-axis and y-axis and with two unknowns ω_5 and $v_{D_{54}}$

$$-\omega_5 y_D=v_{D4x}+v_{D_{54}}\cos\phi_5,$$
$$\omega_5 x_D=v_{D4y}+v_{D_{54}}\sin\phi_5,$$

or

$$-\omega_5(0.128347)=-0.127223+v_{D_{54}}\cos(138.933^\circ),$$
$$\omega_5(-0.147297)=-0.661068+v_{D_{54}}\sin(138.933^\circ).$$

Thus,

$$\omega_5 = \omega_4 = 2.97887 \text{ rad/s and } v_{D_{54}} = 0.338367 \text{ m/s}.$$

The MATLAB commands for the $\boldsymbol{\omega}_5$ and $\mathbf{v}_{D_{54}}$ are:

```
omega5z=sym('omega5z','real');  % omega5z unknown
vD54=sym('vD54','real');  % vD54 unknown
omega5 = [ 0 0 omega5z ];
vD5 = vA + cross(omega5,rD);
vD5D4 = vD54*[ cos(phi5) sin(phi5) 0];
% vD5 = vD4 + vD5D4
eqvD = vD5 - vD4 - vD5D4;
eqvDx = eqvD(1); eqvDy = eqvD(2);
solvD = solve(eqvDx,eqvDy);
omega5zs=eval(solvD.omega5z);
vD54s=eval(solvD.vD54);
Omega5 = [0 0 omega5zs];
Omega4 = Omega5;
VD54 = vD54s*[cos(phi5) sin(phi5) 0];
```

Angular Acceleration of Link 5

The acceleration of the point D_5 on the link 5 is calculated in terms of the acceleration of the point D_4 on the link 4

$$\mathbf{a}_{D_5} = \mathbf{a}_{D_4} + \mathbf{a}_{D_{54}} + \mathbf{a}^{cor}_{D_{54}}, \tag{3.77}$$

This relative acceleration $\mathbf{a}_{B_{32}}$ is parallel to the sliding direction DA, $\mathbf{a}_{D_{54}}||DE$, or

$$\mathbf{a}_{D_{54}} = a_{D_{54}} \cos\phi_5 \mathbf{1} + a_{D_{54}} \sin\phi_5 \mathbf{J}. \tag{3.78}$$

The Coriolis acceleration of D_5 relative to D_4 is

$$\mathbf{a}^{cor}_{D_{54}} = 2\,\boldsymbol{\omega}_4 \times \mathbf{v}_{D_{54}} = 2\,\boldsymbol{\omega}_5 \times \mathbf{v}_{D_{54}} = 2\begin{vmatrix} \mathbf{1} & \mathbf{J} & \mathbf{k} \\ 0 & 0 & \omega_5 \\ v_{D_{54}}\cos\phi_5 & v_{D_{54}}\sin\phi_5 & 0 \end{vmatrix}$$
$$= 2(-\omega_5 v_{D_{54}} \sin\phi_5 \mathbf{1} + \omega_5 v_{D_{54}} \cos\phi_5 \mathbf{J})$$
$$= 2[-2.97887(0.338367)\sin(138.933^\circ)\mathbf{1} + 2.97887(0.338367)\cos(138.933^\circ)\mathbf{J}]$$
$$= -1.32434\mathbf{1} - 1.51987\mathbf{J} \text{ m/s}^2. \tag{3.79}$$

The points D_5 and A are on the link 5 and

$$\mathbf{a}_{D_5} = \mathbf{a}_A + \boldsymbol{\alpha}_5 \times \mathbf{r}_D - \omega_5^2 \mathbf{r}_D, \tag{3.80}$$

where $\mathbf{a}_A \equiv \mathbf{0}$ and the angular acceleration of link 5 is

$$\boldsymbol{\alpha}_5 = \alpha_5 \mathbf{k}.$$

Equations 3.77–3.80 give

$$\begin{vmatrix} \mathbf{\imath} & \mathbf{\jmath} & \mathbf{k} \\ 0 & 0 & \alpha_5 \\ x_D & y_D & 0 \end{vmatrix} - \omega_5^2 \mathbf{r}_D$$
$$= \mathbf{a}_{D_4} + a_{D_{54}}(\cos\phi_5\,\mathbf{\imath} + \sin\phi_5\,\mathbf{\jmath}) + 2\,\boldsymbol{\omega}_5 \times \mathbf{v}_{D_{54}}. \tag{3.81}$$

Equation 3.81 represents a vectorial equations with two scalar components on the x-axis and y-axis and with two unknowns α_5 and $a_{D_{54}}$

$$-\alpha_5 y_D - \omega_5^2 x_D = a_{D_{4x}} + a_{D_{54}} \cos\phi_5 - 2\omega_5 v_{D_{54}} \sin\phi_5,$$
$$\alpha_5 x_D - 2\omega_5^2 y_D = a_{D_{4y}} + a_{D_{54}} \sin\phi_5 + 2\omega_5 v_{D_{54}} \cos\phi_5,$$

or

$$-\alpha_5(0.128347) - 2.97887^2(-0.147297)$$
$$= 2.5548 + a_{D_{54}} \cos(138.933^\circ) - 2(2.97887)(0.338367)\sin(138.933^\circ),$$
$$\alpha_5(-0.147297) - 2.97887^2(0.128347)$$
$$= -2.71212 + a_{D_{54}} \sin(138.933^\circ) + 2(2.97887)(0.338367)\cos(138.933^\circ).$$

Thus,

$$\alpha_5 = \alpha_4 = 12.1939 \text{ rad/s}^2 \text{ and } a_{D_{54}} = 1.97423 \text{ m/s}^2.$$

The MATLAB commands for the $\boldsymbol{\alpha}_5$ and $\mathbf{a}_{D_{54}}$ are:

```
% Coriolis acceleration
aD5D4cor = 2*cross(Omega5,VD54);
alpha5z = sym('alpha5z','real');  % alpha5z unknown
aD54 = sym('aD54','real');  % aD54 unknown
alpha5 = [ 0 0 alpha5z ];
aD5 = aA + cross(alpha5,rD) - dot(Omega5,Omega5)*rD;
aD5D4 = aD54*[ cos(phi5) sin(phi5) 0];
eqaD = aD5 - aD4 - aD5D4 - aD5D4cor;
eqaDx = eqaD(1); eqaDy = eqaD(2);
solaD = solve(eqaDx,eqaDy);
alpha5zs = eval(solaD.alpha5z);
aD54s = eval(solaD.aD54);
Alpha5 = [0 0 alpha5zs];
Alpha4 = Alpha5;
AD54 = aD54s*[cos(phi5) sin(phi5) 0];
```

The MATLAB program and the results for the velocities and accelerations analysis are given in Appendix B.4.

3.9 Derivative Method

Another method for obtaining the velocities and/or accelerations of links and joints is to compute the derivatives of the positions and/or velocities with respect to time.

Exercise: R-RTR-RTR Mechanism
The derivative method will be explained using the planar R-RTR-RTR mechanism considered in Sect. 3.8 and shown in Fig. 2.8.

Solution
The angular velocity of link 1 is constant and has the value

$$n = 50 \text{ rpm and } \omega = \dot{\phi} = \frac{\pi n}{30} = \frac{5\pi}{3} \text{ rad/s},$$

or in MATLAB

```
n = 50 ;        % rpm of the driver link (constant)
omega = n*pi/30;     % rad/s
```

The velocity is obtained taking the derivative of the position with respect to time, t. The symbolic variable t is introduced in MATLAB with the statement `sym`:

```
t = sym('t','real');
```

The coordinates of the joint B are $x_B(t) = AB \cos\phi(t)$ and $y_B(t) = AB \sin\phi(t)$, and the position vector of B is $\mathbf{r}_B = x_B\mathbf{1} + x_B\mathbf{J}$. To calculate symbolically the position of the joint B, the following MATLAB commands are used:

```
xB = AB*cos(sym('phi(t)'));
yB = AB*sin(sym('phi(t)'));
% position vector of B in terms of phi(t)
rB = [ xB yB 0 ];
```

The statement `sym('phi(t)')` represents the mathematical function $\phi(t)$ and is introduced with the command `sym` that constructs symbolic numbers, variables and objects. The function `phi` has one argument, the time `t`. To calculate numerically the position of the joint B, the symbolic variables need to be substituted with the input data. To apply a transformation rule to a particular expression `expr`, type `subs(expr,lhs,rhs)`.

The statement `subs(expr,lhs,rhs)` replaces `lhs` with `rhs` in the symbolic expression `expr`. For the mechanism, the numerical values for the joint B are:

```
xBn = subs(xB, 'phi(t)', pi/6); % xB for phi(t)=pi/6
yBn = subs(yB, 'phi(t)', pi/6); % yB for phi(t)=pi/6
rBn = subs(rB, 'phi(t)', pi/6); % rB for phi(t)=pi/6
```

The numerical values of the vector rBn are printed with:

```
fprintf('rB = [ %g, %g, %g ] (m) \n', rBn)
```

The linear velocity vector of $B_1 = B_2$ is

$$\mathbf{v}_B = \mathbf{v}_{B_1} = \mathbf{v}_{B_2} = \dot{x}_B \mathbf{1} + \dot{y}_B \mathbf{J},$$

where

$$\dot{x}_B = \frac{dx_B}{dt} = -AB\,\dot{\phi}\sin\phi \quad \text{and} \quad \dot{y}_B = \frac{dy_B}{dt} = AB\,\dot{\phi}\cos\phi,$$

are the components of the velocity vector of $B_1 = B_2$. To calculate symbolically the components of the velocity vector using the MATLAB the command `diff(f,t)` is used, which gives the derivative of `f` with respect to `t`. The symbolical expression of the velocity vector of $B_1 = B_2$ is obtain with the statement

```
% vB=vB1=vB2 in terms of phi(t) and diff(phi(t),t)
vB = diff(rB,t);
```

The components, `vB(1)` and `vB(2)`, of the vector `vB` are symbolic expressions in terms of `phi(t)` and `diff(phi(t),t)`:

```
-3/20*sin(phi(t))*diff(phi(t),t)
3/20*cos(phi(t))*diff(phi(t),t)
```

The numerical values for the components of the velocity of $B_1 = B_2$ are

$$\dot{x}_B = -0.15\,(5\pi/3)\sin 30^\circ = -0.392699 \ \text{m/s},$$
$$\dot{y}_B = 0.15\,(5\pi/3)\cos 30^\circ = 0.680175 \ \text{m/s}.$$

To obtain the numerical values in MATLAB first `diff('phi(t)',t)` is replaced with `omega` and then `phi(t)` is replaced with `pi/6`

```
% replaces diff('phi(t)',t) with omega in vB
vBnn = subs(vB,diff('phi(t)',t),omega);
% replaces phi(t) with pi/6 in vBnn
vBn = subs(vBnn,'phi(t)',pi/6);
```

Instead of replacing `diff('phi(t)',t)` with `omega` and then replacing `'phi(t)'` with `pi/6`, a list with the symbolical variables `'phi(t)'`, `diff('phi(t)',t)`, and `diff('phi(t)',t,2)` is created:

```
slist={diff('phi(t)',t,2),diff('phi(t)',t),'phi(t)'};
```

Next, a list with the numerical values for `slist` is introduced:

```
nlist = {0, omega, pi/6};  % numbers for slist
% diff('phi(t)',t,2) -> 0
% diff('phi(t)',t) -> omega
% 'phi(t)' -> pi/6
```

The velocities and accelerations need to be calculated at the moment when the driver link makes an angle $\phi(t) = \pi/6$ with the horizontal and $\dot{\phi}(t) = \omega$ and $\ddot{\phi}(t) = \dot{\omega} = 0$. To obtain the numerical value for the symbolic vector `rB` the following statements are introduced:

```
% replaces slist with nlist in vB
vBn = subs(vB,slist,nlist);
%converts the symbolic vBn to a numeric object
VB = double(vBn);
fprintf('vB1 = vB2 = [ %g, %g, %g ] (m/s) \n', VB)
```

The statement `double(S)` converts the symbolic object `S` to a numeric object. The magnitude of the velocity vector $\mathbf{v}_{B_1} = \mathbf{v}_{B_2}$ is

$$v_{B_1} = v_{B_2} = |\mathbf{v}_{B_1}| = |\mathbf{v}_{B_2}| = \sqrt{\dot{x}_B^2 + \dot{y}_B^2}$$
$$= \sqrt{(-0.392699)^2 + 0.680175^2} = 0.785398 \text{ m/s}.$$

The MATLAB command `norm(v)` calculates the magnitude of a vector `v`. The magnitude of the velocity vector $\mathbf{v}_{B_1} = \mathbf{v}_{B_2}$ in MATLAB is:

```
VBn = norm(VB);
fprintf('|vB1| = |vB2| = %g (m/s) \n', VBn)
```

The linear acceleration vector of $B_1 = B_2$ is

$$\mathbf{a}_{B_1} = \mathbf{a}_{B_2} = \ddot{x}_B \mathbf{\imath} + \ddot{y}_B \mathbf{\jmath},$$

where

$$\ddot{x}_B = \frac{d\dot{x}_B}{dt} = -AB\,\dot{\phi}^2 \cos\phi - AB\,\ddot{\phi}\,\sin\phi,$$
$$\ddot{y}_B = \frac{d\dot{y}_B}{dt} = -AB\,\dot{\phi}^2 \sin\phi + AB\,\ddot{\phi}\,\cos\phi,$$

and

$$a_{B_1} = a_{B_2} = |\mathbf{a}_{B_1}| = |\mathbf{a}_{B_2}| = \sqrt{\ddot{x}_B^2 + \ddot{y}_B^2}.$$

For the considered mechanism the angular acceleration of the link 1 is $\ddot{\phi} = \dot{\omega} = 0$. The numerical values of the acceleration of B are

$$\ddot{x}_B = -0.15\,(5\pi/3)^2 \cos 30^\circ - 0.15\,(0) \sin 30^\circ = -3.56139 \text{ m/s}^2,$$
$$\ddot{y}_B = -0.15\,(5\pi/3)^2 \sin 30^\circ + 0.15\,(0) \cos 30^\circ = -2.05617 \text{ m/s}^2.$$

The MATLAB command used to calculate symbolically the acceleration vector is:

```
aB = diff(vB,t);   % acceleration of B1=B2
```

The numerical value for the vector aB is obtained with

```
% numerical value for aB
aBn = double(subs(aB,slist,nlist));
fprintf('aB1 = aB2 = [ %g, %g, %g ] (m/s^2) \n', aBn)
ABn = norm(aBn);
fprintf('|aB1| = |aB2| = %g (m/s^2) \n', ABn)
```

The coordinates of the joint D are x_D and y_D. The position of the joint D is calculated from the following equations

$$[x_D(t) - x_C]^2 + [y_D(t) - y_C]^2 = CD^2,$$
$$\frac{y_B(t) - y_C}{x_B(t) - x_C} = \frac{y_D(t) - y_C}{x_D(t) - x_C}.$$

The MATLAB commands used to calculate the position of D are:

```
eqnD1 = '( xDsol - xC )^2 + ( yDsol - yC )^2 = CD^2 ';
eqnD2 = '(yB-yC)/(xB-xC) = (yDsol-yC)/(xDsol-xC)';
solD = solve(eqnD1, eqnD2, 'xDsol, yDsol');
```

Two sets of solutions are found for the position of the joint D that are functions of the angle $\phi(t)$ (i.e., functions of time):

```
xDpositions = eval(solD.xDsol);
yDpositions = eval(solD.yDsol);
xD1 = xDpositions(1); xD2 = xDpositions(2);
yD1 = yDpositions(1); yD2 = yDpositions(2);
```

To determine the correct position of the joint D for the mechanism, an additional condition is needed. For the first quadrant, $0 \leq \phi \leq 90^\circ$, the condition is $x_D \leq x_C$. This condition using the MATLAB command is:

```
xD1n = subs(xD1,'phi(t)',pi/6); % xD1 for phi(t)=pi/6
if xD1n < xC
```

```
    xD = xD1; yD = yD1;
else
    xD = xD2; yD = yD2;
end
% position vector of D in term of phi(t)
rD = [ xD yD 0 ];
```

The numerical solutions are printed using MATLAB

```
xDn = subs(xD,'phi(t)',pi/6); % xD for phi(t)=pi/6
yDn = subs(yD,'phi(t)',pi/6); % yD for phi(t)=pi/6
rDn = [ xDn yDn 0 ]; % rD for phi(t)=pi/6
fprintf('rD = [ %g, %g, %g ] (m) \n', rDn)
```

The linear velocity vector of the joint $D_3 = D_4$ (on link 3 or link 4) is

$$\mathbf{v}_{D_3} = \mathbf{v}_{D_4} = \dot{x}_D \mathbf{\imath} + \dot{y}_D \mathbf{J},$$

where

$$\dot{x}_D = \frac{dx_D}{dt} \quad \text{and} \quad \dot{y}_D = \frac{dy_D}{dt},$$

are the components of the velocity vector of the joint D, respectively, on the x-axis and the y-axis. The magnitude of the velocity is

$$v_{D_3} = v_{D_4} = |\mathbf{v}_{D_3}| = |\mathbf{v}_{D_4}| = \sqrt{\dot{x}_D^2 + \dot{y}_D^2}.$$

To calculate symbolically the components of this velocity vector the following MATLAB commands are used:

```
% vD in terms of phi(t) and diff('phi(t)',t)
vD = diff(rD,t);
```

The numerical solutions are printed using MATLAB:

```
% numerical value for vD
vDn = double(subs(vD,slist,nlist));
fprintf('vD3 = vD4 = [ %g, %g, %g ] (m/s) \n', vDn)
fprintf('|vD3| = |vD4| = %g (m/s) \n', norm(vDn))
```

For the considered mechanism the numerical values are

$$\dot{x}_D = -0.127223 \text{ m/s} \quad \text{and} \quad \dot{y}_D = -0.661068 \text{ m/s}.$$

The linear acceleration vector of $D_3 = D_4$ is

$$\mathbf{a}_{D_3} = \mathbf{a}_{D_4} = \ddot{x}_D \mathbf{\imath} + \ddot{y}_D \mathbf{J},$$

where

$$\ddot{x}_D = \frac{d\dot{x}_D}{dt} \quad \text{and} \quad \ddot{y}_D = \frac{d\dot{y}_D}{dt}.$$

The magnitude of the acceleration is

$$a_{D_3} = a_{D_4} = |\mathbf{a}_{D_3}| = |\mathbf{a}_{D_4}| = \sqrt{\ddot{x}_D^2 + \ddot{y}_D^2}.$$

To calculate symbolically the components of the acceleration vector the following MATLAB commands are used:

```
aD = diff(vD,t);
```

The numerical values for the acceleration of $D_3 = D_4$ are

$$\ddot{x}_D = 2.5548 \text{ m/s}^2 \quad \text{and} \quad \ddot{y}_D = -2.71212 \text{ m/s}^2,$$

and can be printed using MATLAB:

```
% numerical value for aD
aDn = double(subs(aD,slist,nlist));
fprintf('aD3 = aD4 = [ %g, %g, %g ] (m/s^2) \n', aDn)
fprintf('|aD3| = |aD4| = %g (m/s^2) \n', norm(aDn))
```

The angle $\phi_2(t) = \phi_3(t)$ is determined as a function of time t from the equation of the slope of the line BC:

$$\tan\phi_2(t) = \tan\phi_3(t) = \frac{y_B(t) - y_C}{x_B(t) - x_C}.$$

The MATLAB function `atan(z)` gives the arc tangent of the number `z` and the angle ϕ_2 is calculated symbolically:

```
phi2 = atan((yB-yC)/(xB-xC));
```

The numerical value is given by:

```
phi2n = subs(phi2,'phi(t)',pi/6);
```

The numerical solution is printed using MATLAB:

```
fprintf('phi2 = phi3 = %g (degrees) \n', phi2n*180/pi)
```

The angular velocity $\omega_2(t) = \omega_3(t)$ is the derivative with respect to time of the angle $\phi_2(t)$

$$\omega_2 = \frac{d\phi_2(t)}{dt}.$$

Symbolically, the angular velocity $\omega_2 = \omega_3$ is calculated using MATLAB:

```
% omega2 in terms of phi(t) and diff('phi(t)',t)
dphi2 = diff(phi2,t);
```

and is the numerical value is printed using the MATLAB statements:

```
dphi2nn = subs(dphi2,diff('phi(t)',t),omega);
dphi2n = subs(dphi2nn,'phi(t)',pi/6);
fprintf('omega2 = omega3 = %g (rad/s) \n', dphi2n)
```

The angular acceleration $\alpha_2(t) = \alpha_3(t)$ is the derivative with respect to time of the angular velocity $\omega_2(t)$:

$$\alpha_2(t) = \frac{d\omega_2(t)}{dt}.$$

Symbolically, using MATLAB, the angular acceleration α_2 is:

```
ddphi2 = diff(dphi2,t);
```

The numerical solution is printed using MATLAB:

```
ddphi2n = double(subs(ddphi2,slist,nlist));

fprintf('alpha2 = alpha3 = %g (rad/s ^2) \n', ddphi2n)
```

The numerical values of the angles, angular velocities, and angular accelerations for the links 2 and 3 are:

$\phi_2 = \phi_3 = -10.8934$ rad, $\omega_2 = \omega_3 = 4.48799$ rad/s, $\alpha_2 = \alpha_3 = 14.5363$ rad/s^2.

The angle $\phi_4(t) = \phi_5(t)$ is determined as a function of time t from the following equation:

$$\tan\phi_4(t) = \tan\phi_5(t) = \frac{y_D(t) - y_E}{x_D(t) - x_E},$$

and symbolically using MATLAB:

```
ddphi4 = diff(dphi4,t);
```

The angular velocity $\omega_4(t) = \omega_5(t)$ is the derivative with respect to time of the angle $\phi_4(t)$

$$\omega_4 = \frac{d\phi_4(t)}{dt}.$$

To calculate symbolically the angular velocity ω_4 using MATLAB, the following command is used:

```
dphi4 = diff(phi4,t);
```

The angular acceleration $\alpha_4(t) = \alpha_5(t)$ is the derivative with respect to time of the angular velocity $\omega_4(t)$:

$$\alpha_4(t) = \frac{d\omega_4(t)}{dt},$$

and it is calculated symbolically with MATLAB:

```
ddphi4 = diff(dphi4,t);
```

The numerical values of the angles, angular velocities, and angular accelerations for the links 5 and 4 are:

$$\phi_5 = \phi_4 = 138.933 \text{ rad}, \ \ \omega_5 = \omega_4 = 2.97887 \text{ rad/s}, \ \ \alpha_5 = \alpha_4 = 12.1939 \text{ rad/s}^2.$$

The numerical solutions printed with MATLAB are:

```
dphi4n = double(subs(dphi4,slist,nlist));
fprintf('omega4 = omega5 = %g (rad/s) \n', dphi4n)
ddphi4n = double(subs(ddphi4,slist,nlist));
fprintf('alpha4 = alpha5 = %g (rad/s^2) \n', ddphi4n)
```

The MATLAB program for velocity and acceleration analysis and the results are given in Appendix B.5.

Exercise: Inverted Slider-Crank Mechanism
The mechanism considered in Sect. 3.7 (shown in Fig. 3.6) will be analyzed using the derivative method. The dimensions of the links are AC=0.15 m and BC=0.2 m. The driver link 1 rotates with a constant speed of $n = n_1 = 30$ rpm.

Find the velocities and the accelerations of the mechanism when the angle of the driver link 1 with the horizontal axis is $\phi = \phi_1 = 60°$.

Solution
A Cartesian reference frame with the origin at A is selected. The coordinates of joint A are $x_A = y_A = 0$. The coordinates of the joint C are $x_C = AC = 0.15$ m and $y_C = 0$. The position of joint B is calculated from the equations

$$\tan\phi(t) = \frac{y_B(t)}{x_B(t)} \quad \text{and} \quad [x_B(t) - x_C]^2 + [y_B(t) - y_C]^2 = BC^2,$$

or

$$x_B(t)\sin\phi(t) = y_B(t)\cos\phi(t),$$
$$[x_B(t) - x_C]^2 + [y_B(t) - y_C]^2 = BC^2. \tag{3.82}$$

The coordinates of joint B are $x_B = 0.113535$ m and $y_B = 0.196648$ m. The MATLAB statements for the positions are:

```
AC = 0.15 ; BC = 0.20 ; xA = 0 ; yA = 0 ;
xC = AC ; yC = 0 ;
n = 30 ; omega = n*pi/30;

t = sym('t','real') ;
phi = sym('phi(t)') ;
xB = sym('xB(t)') ;
yB = sym('yB(t)') ;

eqB1 = xB*sin(phi) - yB*cos(phi) ;
eqB2 = ( xB - xC )^2 + ( yB - yC )^2 - BC^2 ;

sp = {'phi(t)','xB(t)','yB(t)'} ;
np = {pi/3,'xBn','yBn'} ;
eqB1p = subs(eqB1,sp,np) ;
eqB2p = subs(eqB2,sp,np) ;
solBp = solve(eqB1p, eqB2p) ;
xBpositions = eval(solBp.xBn) ;
yBpositions = eval(solBp.yBn) ;
xB1 = xBpositions(1); xB2 = xBpositions(2) ;
yB1 = yBpositions(1); yB2 = yBpositions(2) ;
if yB1 > 0 xBp = xB1; yBp = yB1;
   else xBp = xB2; yBp = yB2; end
rB = [xBp yBp 0] ;
fp = {pi/3,xBp,yBp} ;
```

The linear velocity of point B on link 3 or 2 is

$$\mathbf{v}_{B_3} = \mathbf{v}_{B_2} = \dot{x}_B\mathbf{1} + \dot{y}_B\mathbf{J},$$

where

$$\dot{x}_B = \frac{dx_B}{dt} \quad \text{and} \quad \dot{y}_B = \frac{dy_B}{dt}.$$

The velocity analysis is carried out by differentiating Eq. 3.82:

$$\dot{x}_B\sin\phi + x_B\dot{\phi}\cos\phi = \dot{y}_B\cos\phi - y_B\dot{\phi}\sin\phi,$$
$$\dot{x}_B(x_B - x_C) + \dot{y}_B(y_B - y_C) = 0,$$

or

$$\dot{x}_B \sin\phi + x_B\,\omega \cos\phi = \dot{y}_B \cos\phi - y_B\,\omega \sin\phi,$$
$$\dot{x}_B(x_B - x_C) + \dot{y}_B(y_B - y_C) = 0. \quad (3.83)$$

The magnitude of the angular velocity of the driver link 1 is

$$\omega = \omega_1 = \dot{\phi} = \frac{\pi n_1}{30} = \frac{\pi\,(30\ \text{rpm})}{30} = 3.141\ \ \text{rad/s}.$$

The link 2 and the driver link 1 have the same angular velocity $\omega_1 = \omega_2$. For the given numerical data Eq. 3.83 becomes

$$\dot{x}_B \sin 60^\circ + 0.113535\,(3.141)\,\cos 60^\circ = \dot{y}_B \cos 60^\circ - 0.196648\,(3.141)\,\sin 60^\circ,$$
$$\dot{x}_B(0.113535 - 0.15) + \dot{y}_B(0.196648 - 0) = 0. \quad (3.84)$$

The solution of Eq. 3.84 gives

$$\dot{x}_B = -0.922477\ \ \text{m/s} \quad \text{and} \quad \dot{y}_B = -0.17106\ \ \text{m/s}.$$

The velocity of B is

$$\mathbf{v}_{B_3} = \mathbf{v}_{B_2} = -0.922477\mathbf{\imath} - 0.17106\mathbf{\jmath}\ \ \text{m/s},$$

$$|\mathbf{v}_{B_3}| = |\mathbf{v}_{B_2}| = \sqrt{(-0.922477)^2 + (-0.17106)^2} = 0.938203\ \ \text{m/s}.$$

The MATLAB statements for the velocity of $B_2 = B_3$ are:

```
deqB1 = diff(eqB1,t) ;
deqB2 = diff(eqB2,t) ;

sv = ...
 {diff('phi(t)',t),diff('xB(t)',t),diff('yB(t)',t)};
nv = {omega,'vxB','vyB'} ;

deqB1p=subs(deqB1,sv,nv) ;
deqB1n=subs(deqB1p,sp,fp) ;
deqB2p=subs(deqB2,sv,nv) ;
deqB2n=subs(deqB2p,sp,fp) ;

solvB = solve(deqB1n, deqB2n) ;
vBx = eval(solvB.vxB) ;
vBy = eval(solvB.vyB) ;

fv = {omega,vBx,vBy} ;
```

The acceleration analysis is obtained using the derivative of the velocities given by Eq. 3.83:

$$\ddot{x}_B \sin\phi + \dot{x}_B\,\omega\cos\phi + \dot{x}_B\,\omega\cos\phi - x_B\,\omega^2\sin\phi$$
$$= \ddot{y}_B\cos\phi - \dot{y}_B\,\omega\sin\phi - \dot{y}_B\,\omega\sin\phi + y_B\,\omega^2\cos\phi,$$
$$\ddot{x}_B(x_B - x_C) + \dot{x}_B^2 + \ddot{y}_B(y_B - y_C) + \dot{y}_B^2 = 0. \tag{3.85}$$

The magnitude of the angular acceleration of the driver link 1 is

$$\alpha = \dot{\omega} = \ddot{\phi} = 0.$$

Numerically, Eq. 3.85 gives

$$\ddot{x}_B \sin 60^\circ + 2\,(-0.922477)\,(3.141)\cos 60^\circ - 0.113535\,(3.141)^2 \sin 60^\circ$$
$$= \ddot{y}_B \cos 45^\circ - 2\,(-0.17106)\,(3.141)\sin 60^\circ + 0.196648\,(3.141)^2\cos 60^\circ,$$
$$\ddot{x}_B(0.113535 - 0.15)\ddot{y}_B(0.196648 - 0)$$
$$+(-0.922477)^2 + (-0.17106)^2 = 0. \tag{3.86}$$

The solution of Eq. 3.86 is

$$\ddot{x}_B = 2.0571 \ \text{m/s}^2 \ \text{ and } \ \ddot{y}_B = -4.0947 \ \text{m/s}^2.$$

The acceleration of B on link 3 or 2 is

$$\mathbf{a}_{B_3} = \mathbf{a}_{B_2} = \ddot{x}_B\,\mathbf{1} + \ddot{y}_B\,\mathbf{J} = 2.0571\,\mathbf{1} - 4.0947\,\mathbf{J} \ \text{m/s}^2,$$

$$|\mathbf{a}_{B_3}| = |\mathbf{a}_{B_2}| = \sqrt{(2.0571)^2 + (-4.0947)^2} = 4.58238 \ \text{m/s}^2.$$

The MATLAB statements for the acceleration of $B_2 = B_3$ are:

```
ddeqB1  =  diff(deqB1,t) ;
ddeqB2  =  diff(deqB2,t) ;

sa={diff('phi(t)',t,2),diff('xB(t)',t,2),...
    diff('yB(t)',t,2)};

na={0,'axB','ayB'} ;

ddeqB1p=subs(ddeqB1,sa,na) ;
ddeqB1n=subs(ddeqB1p,sv,fv) ;
ddeqB1f=subs(ddeqB1n,sp,fp) ;
ddeqB2p=subs(ddeqB2,sa,na) ;
ddeqB2n=subs(ddeqB2p,sv,fv) ;
ddeqB2f=subs(ddeqB2n,sp,fp) ;

solaB  =  solve(ddeqB1f, ddeqB2f) ;
aBx  =  eval(solaB.axB) ;
```

```
aBy = eval(solaB.ayB) ;

fa = {0,aBx,aBy};
```

The slope of the link 3 (the points *B* and *C* are on the straight line *BC*) is

$$\tan\phi_3(t) = \frac{y_B(t) - y_C}{x_B(t) - x_C},$$

or

$$[x_B(t) - x_C]\sin\phi_3(t) = [y_B(t) - y_C]\cos\phi_3(t). \tag{3.87}$$

The angle ϕ_3 is computed as follows:

$$\phi_3 = \arctan\frac{y_B - y_C}{x_B - x_C} = \arctan\frac{0.196648 - 0}{0.113535 - 0.15} = -79.4946^\circ.$$

The derivative of Eq. 3.87 yields

$$\dot{x}_B\sin\phi_3 + (x_B - x_C)\,\dot{\phi}_3\cos\phi_3 = \dot{y}_B\cos\phi_3 - (y_B - y_C)\,\dot{\phi}_3\sin\phi_3,$$

or

$$\dot{x}_B\sin\phi_3 + (x_B - x_C)\,\omega_3\cos\phi_3 = \dot{y}_B\cos\phi_3 - (y_B - y_C)\,\omega_3\sin\phi_3, \tag{3.88}$$

where $\omega_3 = \dot{\phi}_3$.

Numerically, Eq. 3.88 gives

$$-0.922477\sin(-79.4946^\circ) + (0.113535 - 0.15)\,\omega_3\cos(-79.4946^\circ)$$
$$= -0.17106\cos(-79.4946^\circ) - (0.196648 - 0)\,\omega_3\sin(-79.4946^\circ),$$

with the solution $\omega_3 = 4.69102$ rad/s.

The angular velocity of link 3 is

$$\boldsymbol{\omega}_3 = \omega_3\,\mathbf{k} = 4.69102\mathbf{k}\ \text{rad/s}.$$

The MATLAB statements for the angular velocity of link 3 are:

```
phi3 = atan((yB-yC)/(xB-xC)) ;
phi3n = subs(phi3,sp,fp) ;
dphi3 = diff(phi3,t) ;
dphi3nn = subs(dphi3,sv,fv) ;
dphi3n = subs(dphi3nn,sp,fp) ;
fprintf('omega3 = %g (rad/s) \n', double(dphi3n))
```

The angular acceleration of link 3, $\alpha_3 = \dot{\omega}_3 = \ddot{\phi}_3$, is obtained using the derivative of Eq. 3.88:

$$\ddot{x}_B \sin\phi_3 + \dot{x}_B\, \omega_3 \cos\phi_3$$
$$+\dot{x}_B\, \omega_3 \cos\phi_3 + (x_B - x_C)\, \dot{\omega}_3 \cos\phi_3 - (x_B - x_C)\, \omega_3^2 \sin\phi_3$$
$$= \ddot{y}_B \cos\phi_3 - \dot{y}_B\, \omega_3 \sin\phi_3$$
$$-\dot{y}_B\, \omega_3 \sin\phi_3 - (y_B - y_C)\, \dot{\omega}_3 \sin\phi_3 - (y_B - y_C)\, \omega_3^2 \cos\phi_3,$$

or

$$\ddot{x}_B \sin\phi_3 + 2\dot{x}_B\, \omega_3 \cos\phi_3 + (x_B - x_C)\, \alpha_3 \cos\phi_3 - (x_B - x_C)\, \omega_3^2 \sin\phi_3$$
$$= \ddot{y}_B \cos\phi_3 - 2\dot{y}_B\, \omega_3 \sin\phi_3 - (y_B - y_C)\, \alpha_3 \sin\phi_3 - (y_B - y_C)\, \omega_3^2 \cos\phi_3.$$

Numerically, the previous equation becomes

$$2.0571 \sin(-79.4946^\circ) + 2\,(-0.922477)\,(4.69102)\, \cos(-79.4946^\circ)$$
$$+(0.113535 - 0.15)\, \alpha_3 \cos(-79.4946^\circ)$$
$$-(0.113535 - 0.15)\,(4.69102)^2 \sin(-79.4946^\circ)$$
$$= -4.0947 \cos(-79.4946^\circ) - 2\,(-0.17106)\,(4.69102)\, \sin(-79.4946^\circ)$$
$$-(0.196648 - 0)\, \alpha_3 \sin(-79.4946^\circ)$$
$$-(0.196648 - 0)\,(4.69102)^2 \cos(-79.4946^\circ),$$

with the solution $\alpha_3 = -6.38024$ rad/s^2. The angular acceleration of link 3 is

$$\boldsymbol{\alpha}_3 = \alpha_3\, \mathbf{k} = -6.38024\mathbf{k}\ \text{rad/s}^2.$$

The MATLAB statements for the angular acceleration of link 3 are:

```
ddphi3 = diff(dphi3,t) ;
ddphi3nnn = subs(ddphi3,sa,fa) ;
ddphi3nn = subs(ddphi3nnn,sv,fv) ;
ddphi3n = subs(ddphi3nn,sp,fp) ;
fprintf('alpha3 = %g (rad/s^2 ) \n', double(ddphi3n))
```

The MATLAB program for velocity and acceleration analysis and the results using the derivative method are given in Appendix B.6.

Exercise: R-RTR Mechanism

The R-RTR mechanism shown in Fig. 3.9 has the dimensions: $AB = 0.1$ m, $AC = 0.1$ m, and $CD = 0.3$ m. The constant angular speed of the driver link 1 is $\omega = \omega_1 = \pi$ rad/s.

Find the velocities and the accelerations of the mechanism using the derivative method when the angle of the driver link 1 with the horizontal axis is $\phi = \phi_1 = \pi/4 = 45^\circ$.

Solution

The origin of the fixed reference frame is at $C \equiv 0$. The position of the fixed joint A

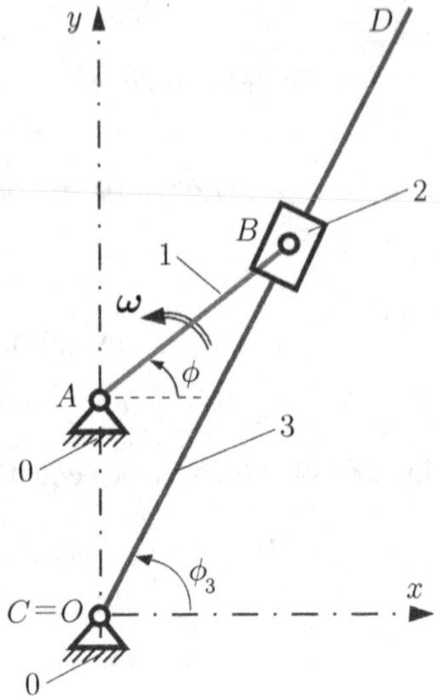

Fig. 3.9 R-RTR mechanism

is $x_A = 0$ and $y_A = AC = 0.1$ m. The position of joint B is

$$x_B(t) = x_A + AB\cos\phi(t), \quad y_B(t) = y_A + AB\sin\phi(t),$$

and for $\phi = 45°$, the position is

$$x_B = 0 + 0.1\cos 45° = 0.07071 \text{ m}, \ \ y_B = 0.1 + 0.1\sin 45° = 0.17071 \text{ m}.$$

The linear velocity vector of $B_1 = B_2$ is

$$\mathbf{v}_{B_1} = \mathbf{v}_{B_2} = \dot{x}_B\mathbf{\imath} + \dot{y}_B\mathbf{J},$$

where

$$\dot{x}_B = \frac{dx_B}{dt} = -AB\,\dot{\phi}\,\sin\phi, \ \ \dot{y}_B = \frac{dy_B}{dt} = AB\,\dot{\phi}\,\cos\phi.$$

With $\phi = 45°$ and $\dot{\phi} = \omega = \pi = 3.141$ rad/s:

$$\begin{aligned}
\dot{x}_B &= -0.1\,\pi\,\sin 45° = -0.222144 \text{ m/s},\\
\dot{y}_B &= 0.1\,\pi\,\cos 45° = -0.222144 \text{ m/s},\\
v_{B_1} &= v_{B_2} = |\mathbf{v}_{B_1}| = |\mathbf{v}_{B_2}| = \sqrt{\dot{x}_B^2 + \dot{y}_B^2} = 0.222144\sqrt{2} \text{ m/s}.
\end{aligned}$$

The linear acceleration vector of $B_1 = B_2$ is

$$\mathbf{a}_{B_1} = \mathbf{a}_{B_2} = \ddot{x}_B\mathbf{\imath} + \ddot{y}_B\mathbf{J},$$

where

$$\ddot{x}_B = \frac{d\dot{x}_B}{dt} = -AB\,\dot{\phi}^2 \cos\phi - AB\,\ddot{\phi}\,\sin\phi,$$
$$\ddot{y}_B = \frac{d\dot{y}_B}{dt} = -AB\,\dot{\phi}^2 \sin\phi + AB\,\ddot{\phi}\,\cos\phi.$$

The angular acceleration of link 1 is $\ddot{\phi} = \dot{\omega} = \alpha = 0$. The numerical values for the acceleration of B are

$$\ddot{x}_B = -0.1\,\pi^2 \cos 45^\circ - 0 = -0.697886 \text{ m/s}^2,$$
$$\ddot{y}_B = -0.1\,\pi^2 \sin 45^\circ + 0 = -0.697886 \text{ m/s}^2,$$
$$a_{B_1} = a_{B_2} = |\mathbf{a}_{B_1}| = |\mathbf{a}_{B_2}| = \sqrt{\ddot{x}_B^2 + \ddot{y}_B^2} = 0.697886\sqrt{2} \text{ m/s}^2.$$

The MATLAB statements for the velocity and acceleration of $B_1 = B_2$ are:

```
AB = 0.1; AC = 0.1; CD = 0.3; % (m)
phi1 = pi/4; omega = pi; alpha = 0;
xC = 0; yC = 0;
xA = 0; yA = AC;

t = sym('t','real');
xB1 = xA + AB*cos(sym('phi(t)'));
yB1 = yA + AB*sin(sym('phi(t)'));
rB = [ xB1 yB1 0 ]; %symbolic function of phi(t)
xBn = subs(xB1,'phi(t)',pi/4); % xB for phi(t)=pi/4
yBn = subs(yB1,'phi(t)',pi/4); % yB for phi(t)=pi/4
rBn = subs(rB,'phi(t)',pi/4); % rB for phi(t)=pi/4
fprintf('rB = [ %g, %g, %g ] (m)\n', rBn)

vB = diff(rB,t); %differentiates rB with respect to t
%list for symbolical variables phi'',phi',phi
slist={diff('phi(t)',t,2),diff('phi(t)',t),'phi(t)'};
%list for numerical values of phi''(t),phi'(t),phi(t)
nlist={alpha,omega,phi1}; %numerical values for slist
vBn = double(subs(vB,slist,nlist));
fprintf('vB1 = vB2 = [ %g, %g, %g ] (m/s)\n', vBn)
fprintf('|vB1| = |vB2| = %g (m/s)\n', norm(vBn))
%acceleration of B1=B2
aB = diff(vB,t); %differentiates vB with respect to t
aBn = double(subs(aB,slist,nlist));
fprintf('aB1 = aB2 = [ %g, %g, %g ] (m/s^2)\n', aBn)
fprintf('|aB1| = |aB2| = %g (m/s^2)\n', norm(aBn))
```

The points B and C are located on the same straight line BCD:

$$y_B(t) - y_C - [x_B(t) - x_C]\,\tan\phi_3(t) = 0. \tag{3.89}$$

The angle $\phi_3 = \phi_2$ is computed as follows:

$$\phi_3 = \phi_2 = \arctan\frac{y_B - y_C}{x_B - x_C},$$

and for $\phi = 45°$ is obtained as

$$\phi_3 = \arctan\frac{0.17071 - 0}{0.07071 - 0} = 67.5°.$$

The derivative of Eq. 3.89 yields

$$\dot{y}_B - \dot{y}_C - (\dot{x}_B - \dot{x}_C)\,\tan\phi_3 - (x_B - x_C)\frac{1}{\cos^2\phi_3}\,\dot{\phi}_3 = 0. \tag{3.90}$$

The angular velocity of link 3, $\omega_3 = \omega_2 = \dot{\phi}_3$, is computed as follows

$$\begin{aligned}\omega_3 = \omega_2 &= \frac{\cos^2\phi_3[\dot{y}_B - \dot{y}_C - (\dot{x}_B - \dot{x}_C)\,\tan\phi_3]}{x_B - x_C}\\ &= \frac{\cos^2\phi_3[\dot{y}_B - \dot{x}_B\,\tan\phi_3]}{x_B},\end{aligned}$$

and

$$\omega_3 = \frac{\cos^2 67.5°(0.222144 + 0.222144\tan 67.5°)}{0.07071} = 1.5708 \text{ rad/s}.$$

The angular acceleration of link 3, $\alpha_3 = \alpha_2 = \ddot{\phi}_3$, is computed from the time derivative of Eq. 3.90

$$\begin{aligned}&\ddot{y}_B - \ddot{y}_C - (\ddot{x}_B - \ddot{x}_C)\tan\phi_3 - 2(\dot{x}_B - \dot{x}_C)\frac{1}{\cos^2\phi_3}\dot{\phi}_3\\ &-2(x_B - x_C)\frac{\sin\phi_3}{\cos^3\phi_3}\,\dot{\phi}_3^2 - (x_B - x_C)\frac{1}{\cos^2\phi_3}\,\ddot{\phi}_3 = 0.\end{aligned}$$

The solution of the previous equation is

$$\begin{aligned}\alpha_3 = \alpha_2 = [&\ddot{y}_B - \ddot{y}_C - (\ddot{x}_B - \ddot{x}_C)\tan\phi_3 - 2(\dot{x}_B - \dot{x}_C)\frac{1}{\cos^2\phi_3}\dot{\phi}_3\\ &-2(x_B - x_C)\frac{\sin\phi_3}{\cos^3\phi_3}\,\dot{\phi}_3^2]\frac{\cos^2\phi_3}{x_B - x_C},\end{aligned}$$

and for the given numerical data:

$$\begin{aligned}\alpha_3 = \alpha_2 = [&-0.697886 + 0.697886\tan 67.5° + 2(-0.222144)\frac{1}{\cos^2 67.5°}1.5708\\ &-2(0.07071)\frac{\sin 67.5°}{\cos^3 67.5°}(1.5708)^2]\frac{\cos^2 67.5°}{0.07071} = 0 \text{ rad/s}^2.\end{aligned}$$

The MATLAB statements for the angular velocity and acceleration of links 2 and 3 are:

```
xB = sym('xB(t)'); % xB(t) symbolic
yB = sym('yB(t)'); % yB(t) symbolic
% list for the symbolical variables of B
% xB''(t), yB''(t), xB'(t), yB'(t), xB(t), yB(t)
sB={diff('xB(t)',t,2),diff('yB(t)',t,2),...
    diff('xB(t)',t),diff('yB(t)',t),'xB(t)','yB(t)'};
% list for the numerical values of the sB list
nB={aBn(1),aBn(2),vBn(1),vBn(2),xBn,yBn};
phi3 = atan((yB-yC)/(xB-xC));
phi3n = subs(phi3,sB,nB);
fprintf('phi2=phi3=%g(degrees)\n',...
        double(phi3n*180/pi))
dphi3 = diff(phi3,t);
dphi3n = subs(dphi3,sB,nB) ;
fprintf('omega2=omega3=%g (rad/s)\n',double(dphi3n))
ddphi3 = diff(dphi3,t);
ddphi3n = subs(ddphi3,sB,nB);
fprintf('alpha2=alpha3=%g(rad/s^2)\n',double(ddphi3n))
```

The MATLAB program for velocity and acceleration analysis, for the R-RTR mechanism using derivative method and the results are given in Appendix B.7.

For the R-RRR mechanism, shown in Fig. 2.4 and presented in Sect. 3.6, the MATLAB program for velocity and acceleration analysis using the derivative method is given in Appendix B.8.

3.10 Independent Contour Equations

This section provides an algebraic method to compute the velocities and accelerations of any closed kinematic chain. The classical method for obtaining the velocities and accelerations involves the computation of the derivative with respect to time of the position vectors. The method of contour equations avoids this task and uses only algebraic equations [Atanasiu (1973), Voinea et al. (1983)]. Using this approach, a numerical implementation is much more efficient. The method described here can be applied to planar and spatial mechanisms.

Figure 3.10 shows a monocontour closed kinematic chain with n rigid links. The joint A_i, $i = 0,1,2,...,n$ is the connection between the links (i) and $(i-1)$. The last link n is connected with the first link 0 of the chain. For the closed kinematic chain, a path is chosen from link 0 to link n. At the joint A_i there are two instantaneously coincident points: the point $A_{i,i}$ belonging to link (i), $A_{i,i} \in (i)$, and the point $A_{i,i-1}$ belonging to link $(i-1)$, $A_{i,i-1} \in (i-1)$.

The velocity equations for a simple closed kinematic chain are

$$\sum_{(i)} \boldsymbol{\omega}_{i,i-1} = \mathbf{0} \text{ and } \sum_{(i)} \mathbf{r}_{A_i} \times \boldsymbol{\omega}_{i,i-1} + \sum_{(i)} \mathbf{v}_{A_{i,i-1}} = \mathbf{0}, \tag{3.91}$$

where

$\boldsymbol{\omega}_{i,i-1}$ is the relative angular velocity of link (i) with respect to link $(i-1)$;

$\mathbf{r}_{A_i}$ is the position vector of the joint A_i;

$\mathbf{v}_{A_{i,i-1}} = \mathbf{v}^{\text{rel}}_{A_{i,i}A_{i,i-1}}$ is the relative velocity of $A_{i,i}$ on link (i) with respect to $A_{i,i-1}$ on link $(i-1)$. The acceleration equations for a simple closed kinematic chain are

$$\sum_{(i)} \boldsymbol{\alpha}_{i,i-1} + \sum_{(i)} \boldsymbol{\omega}_i \times \boldsymbol{\omega}_{i,i-1} = \mathbf{0} \text{ and}$$

$$\sum_{(i)} \mathbf{r}_{A_i} \times (\boldsymbol{\alpha}_{i,i-1} + \boldsymbol{\omega}_i \times \boldsymbol{\omega}_{i,i-1}) + \sum_{(i)} \mathbf{a}_{A_{i,i-1}} + \sum_{(i)} \mathbf{a}^{\text{cor}}_{A_{i,i-1}}$$

$$+ \sum_{(i)} \boldsymbol{\omega}_i \times (\boldsymbol{\omega}_i \times \mathbf{r}_{A_iA_{i+1}}) = \mathbf{0}, \tag{3.92}$$

where

$\boldsymbol{\alpha}_{i,i-1}$ is the relative angular acceleration of link (i) with respect to link $(i-1)$;

$\boldsymbol{\omega}_i$ is the absolute angular velocity of the link (i), or the angular velocity of link (i) with respect to the "fixed" reference frame $Oxyz$, $\boldsymbol{\omega}_i = \boldsymbol{\omega}_{i,0}$;

$\mathbf{a}_{A_{i,i-1}} = \mathbf{a}^{\text{rel}}_{A_{i,i}A_{i,i-1}}$ is the relative acceleration of $A_{i,i}$ on link (i) with respect to $A_{i,i-1}$ on link $(i-1)$;

$\mathbf{a}^{\text{cor}}_{A_{i,i-1}} = 2\boldsymbol{\omega}_{i-1} \times \mathbf{v}_{A_{i,i-1}}$ is the Coriolis acceleration;

$\mathbf{r}_{A_{i-1}A_i} = \mathbf{r}_{A_i} - \mathbf{r}_{A_{i-1}}$.

For a closed kinematic chain in planar motion the acceleration equations are

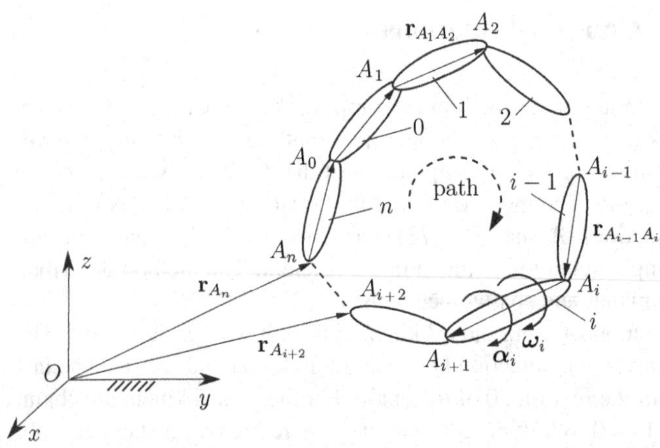

Fig. 3.10 Monocontour closed kinematic chain

$$\sum_{(i)} \boldsymbol{\alpha}_{i,i-1} = \mathbf{0} \text{ and}$$

$$\sum_{(i)} \mathbf{r}_{A_i} \times \boldsymbol{\alpha}_{i,i-1} + \sum_{(i)} \mathbf{a}_{A_{i,i-1}} + \sum_{(i)} \mathbf{a}^{\text{cor}}_{A_{i,i-1}} - \omega_i^2 \mathbf{r}_{A_iA_{i+1}} = \mathbf{0}. \tag{3.93}$$

For planar motion the following relations exist

$$\boldsymbol{\omega}_i \times (\boldsymbol{\omega}_i \times \mathbf{r}_{A_iA_{i+1}}) = -\omega_i^2 \mathbf{r}_{A_iA_{i+1}} \text{ and } \boldsymbol{\omega}_i \times \boldsymbol{\omega}_{i,i-1} = \mathbf{0}.$$

A systematic procedure, using the contour method, is presented below. The equations for velocities and accelerations are written for any closed contour of the mechanism. However, it is best to write the contour equations only for the independent loops of the diagram representing the mechanism.

1. Determine the position analysis of the mechanism.
2. Draw the contour diagram representing the mechanism and select the independent contours. For the contour diagram the numbered links are the nodes of the diagram and are represented by circles, and the joints are represented by lines that connect the nodes. Determine a path for each contour.
3. For each closed loop write the contour velocity relations, Eq. 3.91, and contour acceleration relations, Eq. 3.92. For a closed kinematic chain in planar motion Eq. 3.91 and Eq. 3.93 will be used.
4. Project on a Cartesian reference system the velocity and acceleration equations. Linear algebraic equations are obtained where the unknowns are:

   - the components of the relative angular velocities $\boldsymbol{\omega}_{j,j-1}$;
   - the components of the relative angular accelerations $\boldsymbol{\alpha}_{j,j-1}$;
   - the components of the relative linear velocities $\mathbf{v}_{Aj,j-1}$;
   - the components of the relative linear accelerations $\mathbf{a}_{Aj,j-1}$.

   Solve the algebraic system of equations and determine the unknown kinematic parameters.
5. Determine the absolute angular velocities $\boldsymbol{\omega}_j$ and the absolute angular accelerations $\boldsymbol{\alpha}_j$. Compute the velocities and accelerations of the characteristic points and joints.

Exercise: R-RTR-RTR Mechanism
For the planar R-RTR-RTR mechanism considered in Sect. 3.8 and shown in Fig. 3.11 the contour equations method will be applied and a MATLAB program for velocity and acceleration analysis will be presented.

Solution
The mechanism has five moving links and seven full joints. The number of independent contours is $n_c = c - n = 7 - 5 = 2$, where c is the number of joints and n is the number of moving links. The mechanism has two independent contours. The contour diagram of the mechanism is represented in Fig. 3.11. The first contour I contains the links 0, 1, 2, and 3, while the second contour II contains the links 0, 3,

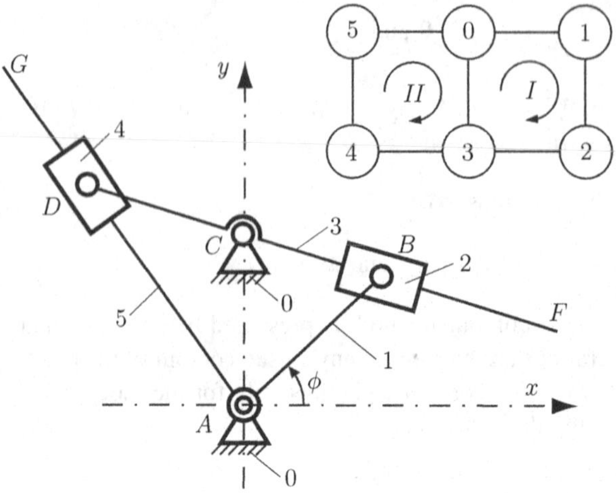

Fig. 3.11 R-RTR-RTR mechanism and contour diagram

4, and 5. Clockwise paths are chosen for each closed contours *I* and *II*.

Contour I: 0-1-2-3-0

Figure 3.12 shows the first independent contour *I* with:

- rotational joint R between the links 0 and 1 (joint A_R);
- rotational joint R between the links 1 and 2 (joint B_R);
- translational joint T between the links 2 and 3 (joint B_T);
- rotational joint R between the links 3 and 0 (joint C_R).

The angular velocity ω_{10} of the driver link is known:

$$\omega_{10} = \omega_1 = \omega = \frac{n\pi}{30} = \frac{50\pi}{30} \text{ rad/s} = 5\pi/3 \text{ rad/s}.$$

The origin of the reference frame is at the point $A(0,0)$. For the velocity analysis, using Eq. 3.91 the following equations are obtained

$$\begin{aligned} &\boldsymbol{\omega}_{10} + \boldsymbol{\omega}_{21} + \boldsymbol{\omega}_{03} = \mathbf{0}, \\ &\mathbf{r}_B \times \boldsymbol{\omega}_{21} + \mathbf{r}_C \times \boldsymbol{\omega}_{03} + \mathbf{v}^{\text{rel}}_{B_3B_2} = \mathbf{0}, \end{aligned} \tag{3.94}$$

where $\mathbf{r}_B = x_B\mathbf{\imath} + y_B\mathbf{\jmath}$, $\mathbf{r}_C = x_C\mathbf{\imath} + y_C\mathbf{\jmath}$, and

$$\begin{aligned} &\boldsymbol{\omega}_{10} = \omega_{10}\,\mathbf{k},\ \boldsymbol{\omega}_{21} = \omega_{21}\,\mathbf{k},\ \boldsymbol{\omega}_{03} = \omega_{03}\,\mathbf{k}, \\ &\mathbf{v}^{\text{rel}}_{B_3B_2} = \mathbf{v}_{B32} = v_{B_{32}} \cos\phi_2\mathbf{\imath} + v_{B_{32}} \sin\phi_2\mathbf{\jmath}. \end{aligned}$$

The unknowns are the relative angular and linear velocities: ω_{21}, ω_{03}, and $v_{B_{32}}$. The sign of the relative angular velocities is selected arbitrarily as positive. The numerical computation will then give the correct orientation (the correct sign) of

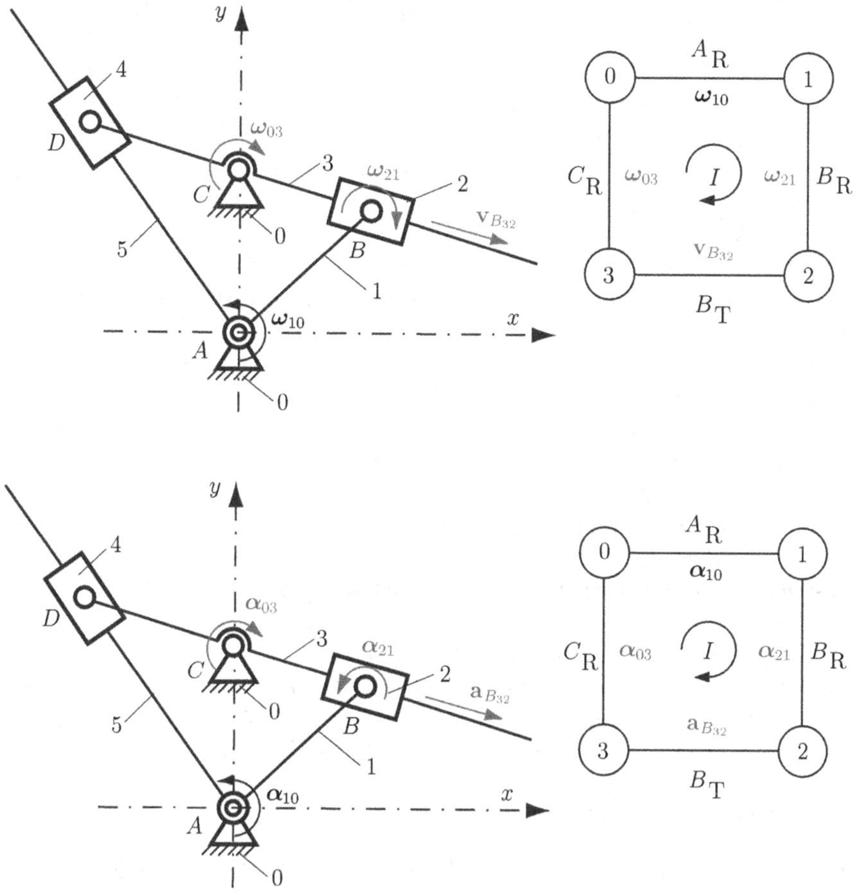

Fig. 3.12 First independent contour of R-RTR-RTR mechanism

the unknown vectors. The components of the vectors $\mathbf{r}_B$ and $\mathbf{r}_C$, and the angle ϕ_2 are already known from the position analysis of the mechanism. Equation 3.94 becomes

$$\omega_{10}\,\mathbf{k} + \omega_{21}\,\mathbf{k} + \omega_{03}\,\mathbf{k} = \mathbf{0},$$

$$\begin{vmatrix} \mathbf{1} & \mathbf{J} & \mathbf{k} \\ x_B & y_B & 0 \\ 0 & 0 & \omega_{21} \end{vmatrix} + \begin{vmatrix} \mathbf{1} & \mathbf{J} & \mathbf{k} \\ x_C & y_C & 0 \\ 0 & 0 & \omega_{03} \end{vmatrix} + v_{B_{32}} \cos\phi_2 \mathbf{1} + v_{B_{32}} \sin\phi_2 \mathbf{J} = \mathbf{0}. \qquad (3.95)$$

The unknown relative velocities are introduced with MATLAB as:

```
omega21v = [ 0 0 sym('omega21z','real') ];
omega03v = [ 0 0 sym('omega03z','real') ];
v32v = sym('vB32','real')*[ cos(phi2) sin(phi2) 0];
```

Equation 3.95 represents a system of three equations and with MATLAB commands gives:

```
eqIomega = omega10 + omega21v + omega03v;
eqIvz=eqIomega(3);
eqIv = cross(rB,omega21v) + cross(rC,omega03v) + v32v;
eqIvx=eqIv(1);
eqIvy=eqIv(2);
```

To display the equations the following MATLAB statements are used:

```
Ivz=vpa(eqIvz,6);
fprintf('%s = 0 \n', char(Ivz))
Ivx=vpa(eqIvx,6);
fprintf('%s = 0 \n', char(Ivx))
Ivy=vpa(eqIvy,6);
fprintf('%s = 0 \n', char(Ivy))
```

The system of equations can be solved using the MATLAB commands:

```
solIv=solve(eqIvz,eqIvx,eqIvy);
omega21 = [ 0 0 eval(solIv.omega21z) ];
omega03 = [ 0 0 eval(solIv.omega03z) ];
vB3B2 = eval(solIv.vB32)*[ cos(phi2) sin(phi2) 0];
```

and the following numerical solutions are obtained

$$\omega_{21} = -0.747998 \text{ rad/s}, \quad \omega_{03} = -4.48799 \text{ rad/s}, \quad \text{and} \quad v_{B_{32}} = 0.514164 \text{ m/s}.$$

To print the numerical values, the following MATLAB commands are used:

```
fprintf('omega21 = [ %g, %g, %g] (rad/s)\n', omega21)
fprintf('omega03 = [ %g, %g, %g] (rad/s)\n', omega03)
fprintf('vB32 = %g (m/s)\n', eval(solIv.vB32))
fprintf('vB3B2 = [ %g, %g, %d] (m/s)\n', vB3B2)
```

The absolute angular velocities of the links 2 and 3 are

$$\boldsymbol{\omega}_{20} = \boldsymbol{\omega}_{30} = -\boldsymbol{\omega}_{03} = 4.48799\,\mathbf{k} \text{ rad/s}.$$

The absolute linear velocity of $D_3 = D_4$ is

$$\mathbf{v}_{D_3} = \mathbf{v}_{D_4} = \mathbf{v}_C + \boldsymbol{\omega}_{30} \times \mathbf{r}_{CD} = -0.127223\,\mathbf{\imath} - 0.661068\,\mathbf{\jmath} \text{ m/s},$$

where $\mathbf{v}_C = \mathbf{0}$ and $\mathbf{r}_{CD} = \mathbf{r}_D - \mathbf{r}_C$. The MATLAB commands for the absolute velocities are:

```
omega30 = - omega03;
omega20 = omega30;
vC = [0 0 0 ];
vD3 = vC + cross(omega30,rD-rC);
fprintf('omega20=omega30=[%d,%d,%g](rad/s)\n',omega30)
fprintf('vD3 = vD4 = [ %g, %g, %g] (m/s)\n', vD3)
```

For the acceleration analysis, using Eq. 3.93 the following equations are obtained

$$\boldsymbol{\alpha}_{10} + \boldsymbol{\alpha}_{21} + \boldsymbol{\alpha}_{03} = \mathbf{0},$$
$$\mathbf{r}_B \times \boldsymbol{\alpha}_{21} + \mathbf{r}_C \times \boldsymbol{\alpha}_{03} + \mathbf{a}^{rel}_{B_3B_2} + \mathbf{a}^{cor}_{B_3B_2} - \omega_{10}^2 \mathbf{r}_{AB} - \omega_{20}^2 \mathbf{r}_{BC} = \mathbf{0}, \tag{3.96}$$

where

$$\boldsymbol{\alpha}_{10} = \alpha_{10}\,\mathbf{k},\ \boldsymbol{\alpha}_{21} = \alpha_{21}\,\mathbf{k},\ \boldsymbol{\alpha}_{03} = \alpha_{03}\,\mathbf{k},$$
$$\mathbf{a}^{rel}_{B_3B_2} = \mathbf{a}_{B_{32}} = a_{B_{32}} \cos\phi_2\,\mathbf{i} + a_{B_{32}} \sin\phi_2\,\mathbf{j},$$
$$\mathbf{a}^{cor}_{B_3B_2} = \mathbf{a}^c_{B_{32}} = 2\,\boldsymbol{\omega}_{20} \times \mathbf{v}_{B_{32}}.$$

The driver link has a constant angular velocity and $\alpha_{10} = \dot{\omega}_{10} = 0$. The unknown acceleration vectors using the MATLAB commands are:

```
alpha21v = [ 0 0 sym('alpha21z','real') ];
alpha03v = [ 0 0 sym('alpha03z','real') ];
a32v = sym('aB32','real')*[ cos(phi2) sin(phi2) 0];
```

Equation 3.96 represents a system of three equations and using MATLAB commands gives:

```
eqIalpha = alpha10 + alpha21v + alpha03v;
eqIaz=eqIalpha(3);
eqIa1=cross(rB,alpha21v)+cross(rC,alpha03v)+a32v+...
      2*cross(omega20,vB3B2);
eqIa2=-dot(omega1,omega1)*rB-...
      dot(omega20,omega20)*(rC-rB);
eqIa=eqIa1+eqIa2;
eqIax=eqIa(1); eqIay=eqIa(2);
```

The equations are displayed with the statements:

```
Iaz=vpa(eqIaz,6);
fprintf('%s = 0 \n', char(Iaz))
Iax=vpa(eqIax,6);
```

```
fprintf('%s = 0 \n', char(Iax))
Iay=vpa(eqIay,6);
fprintf('%s = 0 \n', char(Iay))
```

The unknowns are α_{21}, α_{03}, and $a_{B_{32}}$ or alpha21z, alpha03z, and aB32. The system of equations is solved using the MATLAB commands:

```
solIa=solve(eqIaz,eqIax,eqIay);
alpha21 = [ 0 0 eval(solIa.alpha21z) ];
alpha03 = [ 0 0 eval(solIa.alpha03z) ];
aB3B2 = eval(solIa.aB32)*[ cos(phi2) sin(phi2) 0];
```

The following numerical solutions are then obtained

$$\alpha_{21} = 14.5363 \text{ rad/s}^2,\ \alpha_{03} = -14.5363 \text{ rad/s}^2, \text{ and } a_{B_{32}} = 0.44409 \text{ m/s}^2.$$

To print the numerical values, the following MATLAB commands are used:

```
fprintf('alpha21=[ %g, %g, %g] (rad/s^2)\n', alpha21)
fprintf('alpha03=[ %g, %g, %g] (rad/s^2)\n', alpha03)
fprintf('aB32 = %g (m/s^2)\n', eval(solIa.aB32))
fprintf('aB3B2 = [ %g, %g, %d] (m/s^2)\n', aB3B2)
```

The absolute angular accelerations of the links 2 and 3 are

$$\boldsymbol{\alpha}_{20} = \boldsymbol{\alpha}_{30} = -\boldsymbol{\alpha}_{03} = 14.5363\,\mathbf{k}\ \text{rad/s}^2.$$

The absolute linear acceleration of $D_3 = D_4$ is obtained from the following equation

$$\mathbf{a}_{D_3} = \mathbf{a}_{D_4} = \mathbf{a}_C + \boldsymbol{\alpha}_{30} \times \mathbf{r}_{CD} - \omega_{30}^2 \mathbf{r}_{CD} = 2.55481\,\mathbf{\imath} - 2.71212\,\mathbf{\jmath}\ \text{m/s}^2,$$

where $\mathbf{a}_C = \mathbf{0}$. In MATLAB the absolute accelerations are:

```
alpha30 = - alpha03;
alpha20 = alpha30;
aC = [0 0 0 ];
aD3=aC+cross(alpha30,rD-rC)-...
    dot(omega20,omega20)*(rD-rC);
fprintf('alpha20=alpha30=[%d,%d,%g] (rad/s^2)\n',...
        alpha30)
fprintf('aD3=aD4= [ %g, %g, %g] (m/s^2)\n', aD3)
```

Contour II: 0-3-4-5-0
Figure 3.13 depicts the second independent contour *II*:

- rotational joint R between the links 0 and 3 (joint C_R);
- rotational joint R between the links 3 and 4 (joint D_R);

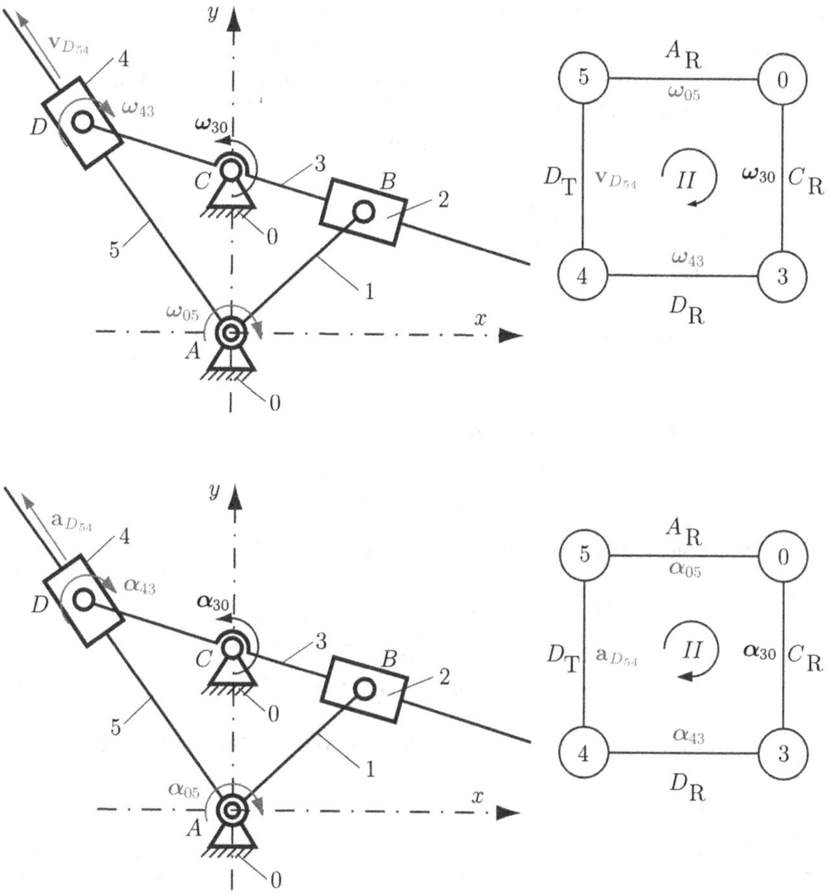

Fig. 3.13 Second independent contour of R-RTR-RTR mechanism

- translational joint T between the links 4 and 5 (joint D_T);
- rotational joint R between the links 5 and 0 (joint A_R).

For the velocity analysis, the following vectorial equations are used

$$\boldsymbol{\omega}_{30} + \boldsymbol{\omega}_{43} + \boldsymbol{\omega}_{05} = \mathbf{0},$$
$$\mathbf{r}_C \times \boldsymbol{\omega}_{30} + \mathbf{r}_D \times \boldsymbol{\omega}_{43} + \mathbf{r}_A \times \boldsymbol{\omega}_{05} + \mathbf{v}^{\text{rel}}_{D_5 D_4} = \mathbf{0}, \tag{3.97}$$

where $\mathbf{r}_D = x_D \mathbf{1} + y_D \mathbf{J}$, $\mathbf{r}_A = x_A \mathbf{1} + y_A \mathbf{J} = \mathbf{0}$, and

$$\boldsymbol{\omega}_{30} = \omega_{30}\,\mathbf{k},\ \boldsymbol{\omega}_{43} = \omega_{43}\,\mathbf{k},\ \boldsymbol{\omega}_{05} = \omega_{05}\,\mathbf{k},$$
$$\mathbf{v}^{\text{rel}}_{D_5 D_4} = \mathbf{v}_{D_{54}} = v_{D_{54}} \cos\phi_4\,\mathbf{1} + v_{D_{54}} \sin\phi_4\,\mathbf{J}.$$

The sign of the relative angular velocities is selected arbitrarily positive. The numerical computation will then give the correct orientation of the unknown vectors. The components of the vectors $\mathbf{r}_D$ and the angle ϕ_4 are already known from the position analysis of the mechanism.

The unknown vectors with MATLAB commands are:

```
omega43v = [ 0 0 sym('omega43z','real') ];
omega05v = [ 0 0 sym('omega05z','real') ];
v54v = sym('vD54','real')*[ cos(phi4) sin(phi4) 0];
```

Equation 3.97 becomes

$$\omega_{30}\,\mathbf{k} + \omega_{43}\,\mathbf{k} + \omega_{05}\,\mathbf{k} = \mathbf{0},$$

$$\begin{vmatrix} \mathbf{\imath} & \mathbf{\jmath} & \mathbf{k} \\ x_C & y_C & 0 \\ 0 & 0 & \omega_{30} \end{vmatrix} + \begin{vmatrix} \mathbf{\imath} & \mathbf{\jmath} & \mathbf{k} \\ x_D & y_D & 0 \\ 0 & 0 & \omega_{43} \end{vmatrix} + v_{D_{54}} \cos\phi_4\,\mathbf{\imath} + v_{D_{54}} \sin\phi_4\,\mathbf{\jmath} = \mathbf{0}. \quad (3.98)$$

Equation 3.98 projected onto the "fixed" reference frame $Oxyz$ gives

$$\begin{aligned} &\omega_{30} + \omega_{43} + \omega_{05} = 0, \\ &y_C\,\omega_{30} + y_D\,\omega_{43} + v_{D_{54}} \cos\phi_4 = 0, \\ &-x_C\,\omega_{30} - x_D\,\omega_{43} + v_{D_{54}} \sin\phi_4 = 0. \end{aligned} \quad (3.99)$$

The above system of equations using the following MATLAB commands becomes:

```
eqIIomega = omega30 + omega43v + omega05v;
eqIIvz=eqIIomega(3);
eqIIv=cross(rC,omega30)+cross(rD,omega43v)+v54v;
eqIIvx=eqIIv(1);
eqIIvy=eqIIv(2);
```

Equation 3.99 represents an algebraic system of three equations with three unknowns: ω_{43}, ω_{05}, and $v_{D_{54}}$. The system is solved using the MATLAB commands:

```
solIIv=solve(eqIIvz,eqIIvx,eqIIvy);
omega43 = [ 0 0 eval(solIIv.omega43z) ];
omega05 = [ 0 0 eval(solIIv.omega05z) ] ;
vD5D4 = eval(solIIv.vD54)*[ cos(phi4) sin(phi4) 0];
```

The following numerical solutions are obtained:

$$\omega_{43} = -1.50912 \text{ rad/s}, \ \omega_{05} = -2.97887 \text{ rad/s}, \text{ and } v_{D_{54}} = 0.338367 \text{ m/s}.$$

To print the numerical values with MATLAB, the following commands are used:

```
fprintf('omega43 = [ %g, %g, %g] (rad/s)\n', omega43)
fprintf('omega05 = [ %g, %g, %g] (rad/s)\n', omega05)
fprintf('vD54 = %g (m/s)\n', eval(solIIv.vD54))
fprintf('vD5D4 = [ %g, %g, %d] (m/s)\n', vD5D4)
```

The absolute angular velocities of the links 4 and 5 are

$$\boldsymbol{\omega}_{40} = \boldsymbol{\omega}_{50} = -\boldsymbol{\omega}_{05} = 2.97887\,\mathbf{k} \text{ rad/s},$$

and with MATLAB commands, they are:

```
omega50 = - omega05;
omega40 = omega50;
fprintf('omega40=omega50=[%d,%d,%g] (rad/s)\n',...
        omega50)
```

For the acceleration analysis, the following vectorial equations are used:

$$\begin{aligned}&\boldsymbol{\alpha}_{30} + \boldsymbol{\alpha}_{43} + \boldsymbol{\alpha}_{05} = \mathbf{0},\\&\mathbf{r}_C \times \boldsymbol{\alpha}_{30} + \mathbf{r}_D \times \boldsymbol{\alpha}_{43} + \mathbf{r}_A \times \boldsymbol{\alpha}_{05} + \mathbf{a}^{\text{rel}}_{D_5D_4} + \mathbf{a}^{\text{cor}}_{B_5B_4} - \omega^2_{30}\mathbf{r}_{CD} - \omega^2_{40}\mathbf{r}_{DA} = \mathbf{0},\end{aligned} \quad (3.100)$$

where

$$\begin{aligned}&\boldsymbol{\alpha}_{30} = \alpha_{30}\,\mathbf{k},\ \boldsymbol{\alpha}_{43} = \alpha_{43}\,\mathbf{k},\ \boldsymbol{\alpha}_{05} = \alpha_{05}\,\mathbf{k},\\&\mathbf{a}^{\text{rel}}_{D_5D_4} = \mathbf{a}_{D_{54}} = a_{D_{54}}\cos\phi_4\,\mathbf{ı} + a_{D_{54}}\sin\phi_4\,\mathbf{ȷ},\\&\mathbf{a}^{\text{cor}}_{D_{54}} = 2\,\boldsymbol{\omega}_{40} \times \mathbf{v}_{D_{54}}.\end{aligned}$$

The unknown acceleration vectors using the MATLAB commands are:

```
alpha43v = [ 0 0 sym('alpha43z','real') ];
alpha05v = [ 0 0 sym('alpha05z','real') ];
a54v = sym('aD54','real')*[ cos(phi4) sin(phi4) 0];
```

Equation 3.100 becomes

$$\begin{aligned}&\alpha_{30}\,\mathbf{k} + \alpha_{43}\,\mathbf{k} + \alpha_{05}\,\mathbf{k} = \mathbf{0},\\&\begin{vmatrix}\mathbf{ı} & \mathbf{ȷ} & \mathbf{k}\\ x_C & y_C & 0\\ 0 & 0 & \alpha_{30}\end{vmatrix} + \begin{vmatrix}\mathbf{ı} & \mathbf{ȷ} & \mathbf{k}\\ x_D & y_D & 0\\ 0 & 0 & \alpha_{43}\end{vmatrix} + a_{D_{54}}\cos\phi_4\,\mathbf{ı} + a_{D_{54}}\sin\phi_4\,\mathbf{ȷ}\\&+ \begin{vmatrix}\mathbf{ı} & \mathbf{ȷ} & \mathbf{k}\\ 0 & 0 & \omega_{40}\\ v_{D_{54}}\cos\phi_4 & v_{D_{54}}\sin\phi_4 & 0\end{vmatrix} - \omega^2_{30}[(x_D - x_C)\,\mathbf{ı} + (y_D - y_C)\,\mathbf{ȷ}]\\&-\omega^2_{40}[(x_A - x_D)\,\mathbf{ı} + (y_A - y_D)\,\mathbf{ȷ}] = \mathbf{0}.\end{aligned} \quad (3.101)$$

Equation 3.101 can be rewritten as

$$\begin{aligned}
&\alpha_{30} + \alpha_{43} + \alpha_{05} = 0,\\
&y_C\,\alpha_{30} + y_D\,\alpha_{43} + a_{54}\cos\phi_4 - 2\omega_{40}\,v_{54}\sin\phi_4\\
&-\omega_{30}^2(x_D - x_C) - \omega_{40}^2(0 - x_D) = 0,\\
&-x_C\,\alpha_{30} - x_D\,\alpha_{43} + a_{54}\sin\phi_4 + 2\omega_{40}\,v_{54}\cos\phi_4\\
&-\omega_{30}^2(y_D - y_C) - \omega_{40}^2(0 - y_D) = 0.
\end{aligned} \tag{3.102}$$

The contour acceleration equations using MATLAB commands are:

```
eqIIalpha = alpha30 + alpha43v + alpha05v;
eqIIaz=eqIIalpha(3);
eqIIa1=cross(rC,alpha30)+cross(rD,alpha43v)+a54v;
eqIIa2=2*cross(omega40,vD5D4);
eqIIa3=-dot(omega30,omega30)*(rD-rC)- ...
        dot(omega40,omega40)*(-rD);
eqIIa=eqIIa1+eqIIa2+eqIIa3;
eqIIax=eqIIa(1); eqIIay=eqIIa(2);
```

The unknowns in Eq. 3.102 are α_{43}, α_{05}, and $a_{D_{54}}$. To solve the system, the following MATLAB command is used:

```
solIIa=solve(eqIIaz,eqIIax,eqIIay);
alpha43 = [ 0 0 eval(solIIa.alpha43z) ];
alpha05 = [ 0 0 eval(solIIa.alpha05z) ] ;
aD5D4 = eval(solIIa.aD54)*[ cos(phi4) sin(phi4) 0];
```

The following numerical solutions are obtained:

$$\alpha_{43} = -2.3424\ \text{rad/s}^2,\ \alpha_{05} = -12.1939\ \text{rad/s}^2,\ \text{and}\ a_{D_{54}} = 1.97423\ \text{m/s}^2.$$

The MATLAB commands for displaying the solutions are:

```
fprintf('alpha43 = [ %g, %g, %g] (rad/s^2)\n',alpha43)
fprintf('alpha05 = [ %g, %g, %g] (rad/s^2)\n',alpha05)
fprintf('aD54 = %g (m/s^2)\n', eval(solIIa.aD54))
fprintf('aD5D4 = [ %g, %g, %d] (m/s^2)\n', aD5D4)
```

The absolute angular accelerations of the links 4 and 5 are

$$\boldsymbol{\alpha}_{40} = \boldsymbol{\alpha}_{50} = -\boldsymbol{\alpha}_{05} = 12.1939\mathbf{k}\ \text{rad/s}^2,$$

and with MATLAB they are:

```
alpha50 = - alpha05;
alpha40 = alpha50;
```

```
fprintf('alpha40=alpha50=[%d,%d,%g] (rad/s^2)\n',...
         alpha50)
```

The MATLAB program and results for the velocity and acceleration analysis using the contour method are given in Appendix B.9.

Chapter 4
Dynamic Force Analysis

4.1 Equation of Motion for General Planar Motion

The friction effects in the joints are assumed to be negligible. Figure 4.1 shows an arbitrary body with the total mass m. The body can be divided into n particles, the ith particle having mass, m_i, and the total mass is $m = \sum_{i=1}^{n} m_i$.

A rigid body can be considered as a collection of particles in which the number of particles approaches infinity and in which the distance between any two points remains constant. As N approaches infinity, each particle is treated as a differential mass element, and the summation is replaced by integration over the body $m = \int dm$. The position of the mass center of a collection of particles is defined by

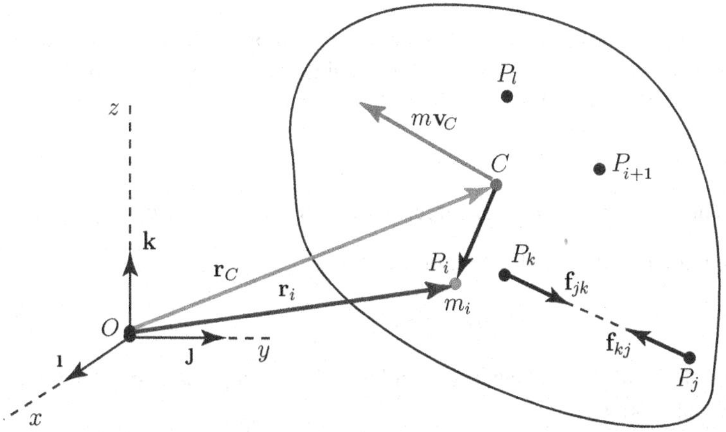

Fig. 4.1 Rigid body divided into particles

$$\mathbf{r}_C = \frac{1}{m}\sum_{i=1}^{n} m_i \mathbf{r}_i \quad \text{or} \quad \mathbf{r}_C = \frac{1}{m}\int \mathbf{r}\, dm, \tag{4.1}$$

where $\mathbf{r}_i = \mathbf{r}_{OP_i} = \mathbf{r}_{P_i}$ is the position vector from the origin O to the ith particle. The time derivative of Eq. 4.1 gives

$$\sum_{i=1}^{N} m_i \frac{d^2\mathbf{r}_i}{dt^2} = m\frac{d^2\mathbf{r}_C}{dt^2} = m\mathbf{a}_C, \tag{4.2}$$

where $\mathbf{a}_C$ is the acceleration of the mass center. Any particle of the system is acted on by two types of forces: internal forces (exerted by other particles that are also part of the system) and external forces (exerted by a particle or object not included in the system). Let $\mathbf{f}_{ij}$ be the internal force exerted on the jth particle by the ith particle. Newton's third law (action and reaction) states that the jth particle exerts a force on the ith particle of equal magnitude, and opposite direction, and collinear with the force exerted by the ith particle on the jth particle $\mathbf{f}_{ji} = -\mathbf{f}_{ij},\ j \neq i$. Newton's second law for the ith particle must include all of the internal forces exerted by all of the other particles in the system on the ith particle, plus the sum of any external forces exerted by particles, objects outside of the system on the ith particle

$$\sum_j \mathbf{f}_{ji} + \mathbf{F}_i^{\text{ext}} = m_i \frac{d^2\mathbf{r}_i}{dt^2}, \quad j \neq i, \tag{4.3}$$

where $\mathbf{F}_i^{\text{ext}}$ is the external force on the ith particle. Equation 4.3 is written for each particle in the collection of particles. Summing the resulting equations over all of the particles from $i = 1$ to N the following relation is obtained

$$\sum_i \sum_j \mathbf{f}_{ji} + \sum_i \mathbf{F}_i^{\text{ext}} = m\mathbf{a}_C, \quad j \neq i. \tag{4.4}$$

The sum of the internal forces includes pairs of equal and opposite forces. The sum of any such pair must be zero. The sum of all of the internal forces on the collection of particles is zero (Newton's third law) $\sum_i \sum_j \mathbf{f}_{ji} = \mathbf{0},\quad j \neq i.$

The term $\sum_i \mathbf{F}_i^{\text{ext}}$ is the sum of the external forces on the collection of particles $\sum_i \mathbf{F}_i^{\text{ext}} = \mathbf{F}$. The sum of the external forces acting on a closed system equals the product of the mass and the acceleration of the mass center

$$m\mathbf{a}_C = \mathbf{F}. \tag{4.5}$$

Equation 4.5 is Newton's second law for a rigid body and is applicable to planar and three-dimensional motions.

Resolving the sum of the external forces into Cartesian rectangular components

$$\mathbf{F} = F_x \mathbf{\imath} + F_y \mathbf{\jmath} + F_z \mathbf{k},$$

and the position vector of the mass center

$$\mathbf{r}_C = x_C(t)\,\boldsymbol{\imath} + y_C(t)\,\boldsymbol{\jmath} + z_C(t)\,\mathbf{k},$$

Newton's second law for the rigid body is

$$m\ddot{\mathbf{r}}_C = \mathbf{F}, \tag{4.6}$$

or

$$m\ddot{x}_C = F_x,\ m\ddot{y}_C = F_y,\ m\ddot{z}_C = F_z. \tag{4.7}$$

Figure 4.2 represents the rigid body moving with general planar motion in the (x,y) plane. The origin of the Cartesian reference frame is O. The mass center C of the rigid body is located in the plane of the motion. Let Oz be the axis through the fixed origin point O that is perpendicular to the plane of motion of the rigid body. Let Cz be the parallel axis through the mass center C. The rigid body has a general planar motion and the angular velocity vector is $\boldsymbol{\omega} = \omega\mathbf{k}$. The sum of the moments about O due to external forces and couples is

$$\sum \mathbf{M}_O = \frac{d\mathbf{H}_O}{dt} = \frac{d}{dt}[(\mathbf{r}_C \times m\mathbf{v}_C) + \mathbf{H}_C]. \tag{4.8}$$

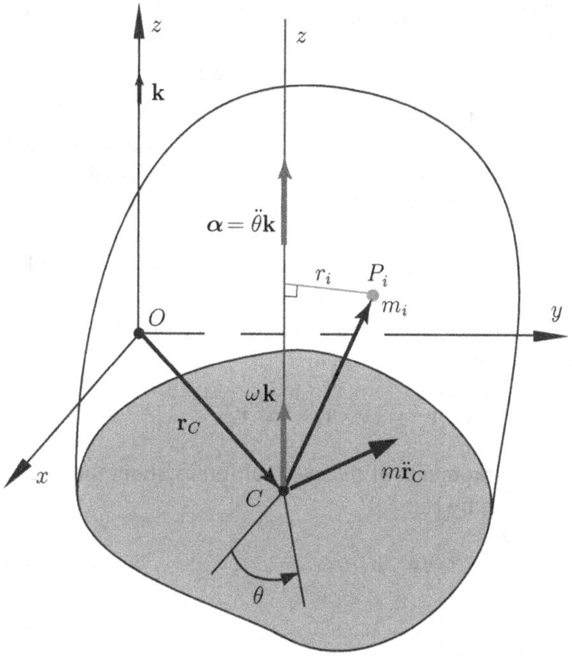

Fig. 4.2 Rigid body moving with general planar motion in the (x,y) plane

The total angular momentum of the system about O is $\mathbf{H}_O$, the total angular momentum of the system about C is $\mathbf{H}_C$, and $\mathbf{v}_C = \dot{\mathbf{r}}_C$ is the velocity of C. The magnitude of the angular momentum about Cz is $H_C = \sum_i m_i r_i^2\, \omega$. The summation $\sum_i m_i r_i^2$ or the integration over the body $\int r^2\, dm$ is defined as the mass moment of inertia I_{Cz} of the body about the z-axis through C

$$I_{Cz} = \sum_i m_i r_i^2.$$

The term r_i is the perpendicular distance from d_C to the P_i particle. The mass moment of inertia I_{Cz} is a constant property of the body and is a measure of the rotational inertia or resistance to change in angular velocity due to the radial distribution of the rigid body mass around the z-axis through C. The angular momentum of the rigid body about Cz (z-axis through C) is

$$H_C = I_{Cz}\, \omega \text{ or } \mathbf{H}_C = I_{Cz}\, \omega\, \mathbf{k} = I_{Cz}\, \boldsymbol{\omega}.$$

Substituting this expression into Eq. 4.8 gives

$$\sum \mathbf{M}_O = \frac{d}{dt}[(\mathbf{r}_C \times m\mathbf{v}_C) + I_{Cz}\boldsymbol{\omega}] = (\mathbf{r}_C \times m\,\mathbf{a}_C) + I_{Cz}\,\boldsymbol{\alpha}. \tag{4.9}$$

The rotational equation of motion for the rigid body is

$$I_{Cz}\,\boldsymbol{\alpha} = \sum \mathbf{M}_C \quad \text{or} \quad I_{Cz}\,\alpha\,\mathbf{k} = \sum M_C\,\mathbf{k}. \tag{4.10}$$

For general planar motion the angular acceleration is $\boldsymbol{\alpha} = \dot{\boldsymbol{\omega}} = \ddot{\theta}\,\mathbf{k}$, where the angle θ describes the position, or orientation, of the rigid body about a fixed axis. If the rigid body is a plate moving in the plane of motion, the mass moment of inertia of the rigid body about the z-axis through C becomes the polar mass moment of inertia of the rigid body about C, $I_{Cz} = I_C$. For this case the Eq. 4.10 gives

$$I_C\,\boldsymbol{\alpha} = \sum \mathbf{M}_C. \tag{4.11}$$

Consider the special case when the rigid body rotates about a fixed point O as shown in Fig. 4.3. The acceleration of the mass center is

$$\mathbf{a}_C = \mathbf{a}_O + \boldsymbol{\alpha} \times \mathbf{r}_C - \omega^2\,\mathbf{r}_C = \boldsymbol{\alpha} \times \mathbf{r}_C - \omega^2\,\mathbf{r}_C.$$

The relation between the sum of the moments of the external forces about the fixed point O and the product $I_{Cz}\,\boldsymbol{\alpha}$ is given by Eq. 4.9

$$\sum \mathbf{M}_O = \mathbf{r}_C \times m\,\mathbf{a}_C + I_{Cz}\,\boldsymbol{\alpha},$$

or

$$\sum \mathbf{M}_O = \mathbf{r}_C \times m(\boldsymbol{\alpha} \times \mathbf{r}_C - \omega^2\,\mathbf{r}_C) + I_{Cz}\,\boldsymbol{\alpha}$$

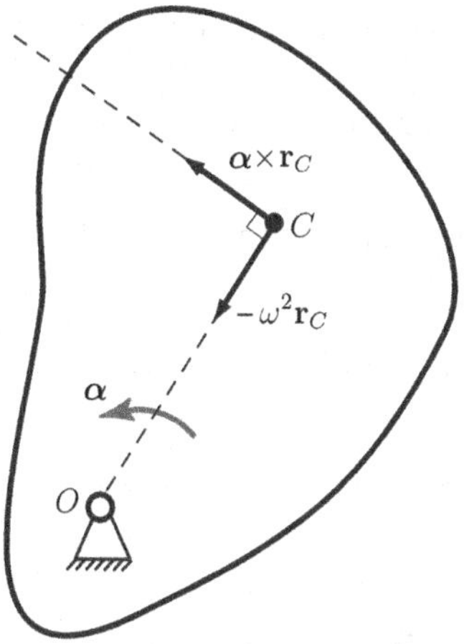

Fig. 4.3 Rotation about a fixed point O

$$
\begin{aligned}
&= m\mathbf{r}_C \times (\boldsymbol{\alpha} \times \mathbf{r}_C) + I_{Cz}\boldsymbol{\alpha} \\
&= m\left[(\mathbf{r}_C \cdot \mathbf{r}_C)\boldsymbol{\alpha} - (\mathbf{r}_C \cdot \boldsymbol{\alpha})\mathbf{r}_C\right] + I_{Cz}\boldsymbol{\alpha} \\
&= m\mathbf{r}_C^2\boldsymbol{\alpha} + I_{Cz}\boldsymbol{\alpha} = (mr_C^2 + I_{Cz})\boldsymbol{\alpha}.
\end{aligned}
$$

According to parallel-axis theorem

$$I_{Ozz} = I_{Cz} + mr_C^2,$$

where I_{Oz} denotes the mass moment of inertia of the rigid body about the z-axis through O. For the special case of rotation about a fixed point O one can use the formula

$$I_{Oz}\boldsymbol{\alpha} = \sum \mathbf{M}_O. \tag{4.12}$$

The general equations of motion for a rigid body in plane motion are (Fig. 4.3)

$$
\begin{aligned}
&\mathbf{F} = m\mathbf{a}_C \quad \text{or} \quad \mathbf{F} = m\ddot{\mathbf{r}}_C, \\
&\sum \mathbf{M}_C = I_{Cz}\boldsymbol{\alpha},
\end{aligned} \tag{4.13}
$$

or using the Cartesian components

$$
\begin{aligned}
m\ddot{x}_C &= \sum F_x, \\
m\ddot{y}_C &= \sum F_y, \\
I_{Cz}\ddot{\theta} &= \sum M_C.
\end{aligned} \tag{4.14}
$$

Equations 4.13 and 4.14, also known as the Newton–Euler equations of motion, are for plane motion, and are interpreted in two ways:

1. The forces and moments are known and the equations are solved for the motion of the rigid body (direct dynamics).
2. The motion of the rigid body is known and the equations are solved for the forces and moments (inverse dynamics).

The dynamic force analysis in this chapter is based on the known motion of the mechanism.

4.2 D'Alembert's Principle

Newton's second law can be written as

$$\mathbf{F} + (-m\mathbf{a}_C) = \mathbf{0}, \text{ or } \mathbf{F} + \mathbf{F}_{\text{in}} = \mathbf{0},$$

where the term $\mathbf{F}_{\text{in}} = -m\mathbf{a}_C$ is the *inertia force*. Newton's second law can be regarded as an "equilibrium" equation.

The total moment about a fixed point O is

$$\sum \mathbf{M}_O = (\mathbf{r}_C \times m\mathbf{a}_C) + I_{Cz}\boldsymbol{\alpha},$$

or

$$\sum \mathbf{M}_O + [\mathbf{r}_C \times (-m\mathbf{a}_C)] + (-I_{Cz}\boldsymbol{\alpha}) = \mathbf{0}. \tag{4.15}$$

The term $\mathbf{M}_{\text{in}} = -I_{Cz}\boldsymbol{\alpha}$ is the *inertia moment*. The sum of the moments about any point, including the moment due to the inertial force $-m\mathbf{a}$ acting at the mass center and the inertial moment, equals zero.

The equations of motion for a rigid body are analogous to the equations for static equilibrium:

> The sum of the forces equals zero and the sum of the moments about any point equals zero when the inertial forces and moments are taken into account.

This is called *D'Alembert's principle*. The dynamic force analysis is expressed in a form similar to static force analysis

$$\sum \mathbf{F} + \mathbf{F}_{\text{in}} = \mathbf{0}, \tag{4.16}$$

$$\sum \mathbf{M}_C + \mathbf{M}_{\text{in}} = \mathbf{0}, \tag{4.17}$$

where $\sum \mathbf{F}$ is the vector sum of all external forces (resultant of external force), and $\sum \mathbf{M}_C$ is the sum of all external moments about the center of mass C (resultant external moment).

For a rigid body in plane motion in the (x, y) plane,

$$\mathbf{a}_C = \ddot{x}_C \boldsymbol{\imath} + \ddot{y}_C \boldsymbol{\jmath}, \quad \boldsymbol{\alpha} = \alpha \mathbf{k},$$

with all external forces in that plane, Eqs. 4.16 and 4.17 become

$$\begin{aligned}
\sum F_x + F_{\text{in}x} &= \sum F_x + (-m\ddot{x}_C) = 0, \\
\sum F_y + F_{\text{in}y} &= \sum F_y + (-m\ddot{y}_C) = 0, \\
\sum M_C + M_{\text{in}} &= \sum M_C + (-I_C\,\alpha) = 0.
\end{aligned}$$

With d'Alembert's principle the moment summation can be about any arbitrary point P

$$\sum \mathbf{M}_P + \mathbf{M}_{\text{in}} + \mathbf{r}_{PC} \times \mathbf{F}_{\text{in}} = \mathbf{0},$$

where $\sum \mathbf{M}_P$ is the sum of all external moments about P, $\mathbf{M}_{\text{in}}$ is the inertia moment, $\mathbf{F}_{\text{in}}$ is the inertia force, and $\mathbf{r}_{PC}$ is a vector from P to C.

The dynamic analysis problem is reduced to a static force and moment balance problem where the inertia forces and moments are treated in the same way as external forces and moments.

4.3 Free-Body Diagrams

A free-body diagram is a drawing of a part of a complete system, isolated in order to determine the forces acting on that rigid body.

The following force convention is defined: $\mathbf{F}_{ij}$ represents the force exerted by link i on link j. Figure 4.4 shows the joint forces for one degree of freedom joints.

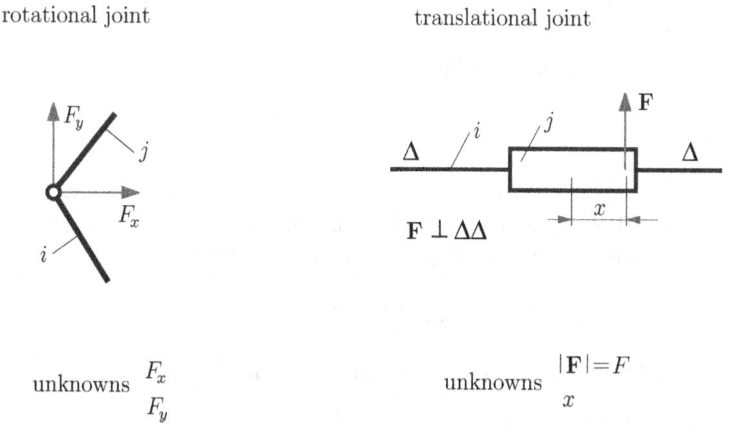

Fig. 4.4 Joint forces for one degree of freedom joints

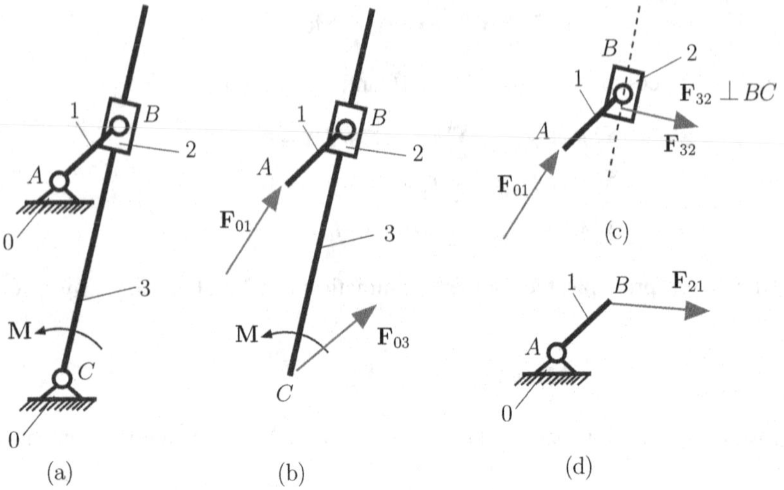

Fig. 4.5 Free-body diagrams

Figure 4.5 shows various free-body diagrams that are considered in the analysis of a R-RTR mechanism. In Fig. 4.5b, the free body consists of the three moving links isolated from the frame 0. The forces and moments acting on the system include an external driven moment **M**, and the forces transmitted from the frame at joint A, $\mathbf{F}_{01}$, and at joint C, $\mathbf{F}_{03}$. Figure 4.5c is a free-body diagram of the two links 1 and 2 and Fig. 4.5d is a free-body diagram of the two links 0 and 1.

The force analysis can be accomplished by examining individual links or a sub-system of links. In this way the joint forces between links as well as the required input force or moment for a given output load are computed.

4.4 Force Analysis Using Dyads

4.4.1 RRR Dyad

Figure 4.6 shows an RRR dyad with two links 2 and 3, and three pin joints, B, C, and D. First, the exterior unknown joint reaction forces are considered

$$\mathbf{F}_{12} = F_{12x}\mathbf{\imath} + F_{12y}\mathbf{\jmath} \quad \text{and} \quad \mathbf{F}_{43} = F_{43x}\mathbf{\imath} + F_{43y}\mathbf{\jmath}.$$

To determine $\mathbf{F}_{12}$ and $\mathbf{F}_{43}$, the following equations are written:

- sum of all forces on links 2 and 3 is zero

$$\sum \mathbf{F}^{(2\&3)} \Longrightarrow$$
$$m_2\,\mathbf{a}_{C_2} + m_3\,\mathbf{a}_{C_3} = \mathbf{F}_{12} + \mathbf{G}_2 + \mathbf{G}_2 + \mathbf{F}_{43},$$

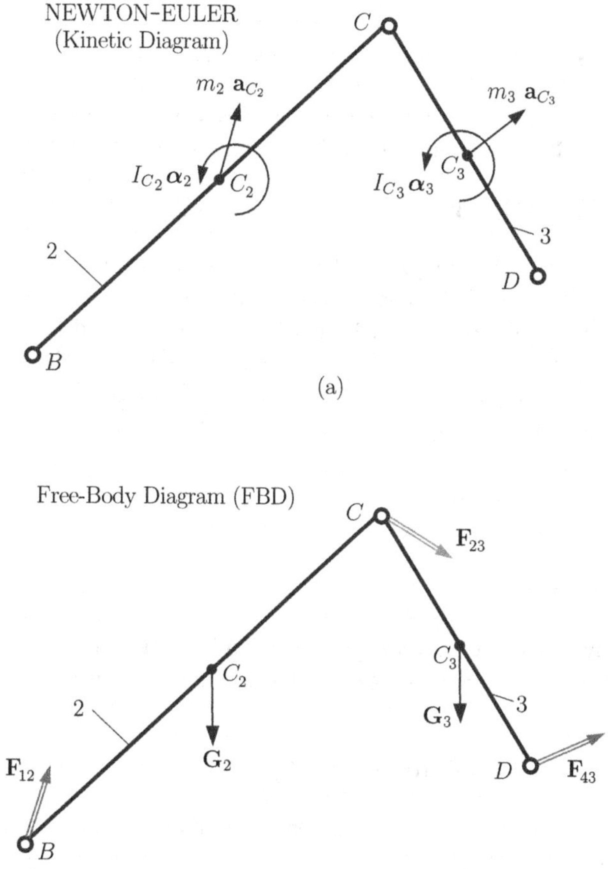

Fig. 4.6 RRR dyad: (a) kinetic diagram and (b) free-body diagram

or

$$\sum \mathbf{F}^{(2\&3)} \cdot \mathbf{\imath} \Longrightarrow$$
$$m_2\, a_{C_{2x}} + m_3\, a_{C_{3x}} = F_{12x} + F_{43x}, \tag{4.18}$$
$$\sum \mathbf{F}^{(2\&3)} \cdot \mathbf{\jmath} \Longrightarrow$$
$$m_2\, a_{C_{2y}} + m_3\, a_{C_{3y}} = F_{12y} - m_2\, g - m_3\, g + F_{43y}. \tag{4.19}$$

- sum of moments of all forces and moments on link 2 about C is zero

$$\sum \mathbf{M}_C^{(2)} \Longrightarrow$$
$$I_{C_2}\, \boldsymbol{\alpha}_2 + \mathbf{r}_{CC_2} \times m_2\, \mathbf{a}_{C_2} = \mathbf{r}_{CB} \times \mathbf{F}_{12} + \mathbf{r}_{CC_2} \times \mathbf{G}_2. \tag{4.20}$$

- sum of moments of all forces and moments on link 3 about C is zero

$$\sum \mathbf{M}_C^{(3)} \Longrightarrow$$
$$I_{C_3}\,\alpha_3 + \mathbf{r}_{CC_3} \times m_3\,\mathbf{a}_{C_3} = \mathbf{r}_{CD} \times \mathbf{F}_{43} + \mathbf{r}_{CC_3} \times \mathbf{G}_3. \tag{4.21}$$

The components F_{12x}, F_{12y}, F_{43x}, and F_{43y} are calculated from Eqs. 4.18–4.21.

The reaction force $\mathbf{F}_{32} = -\mathbf{F}_{23}$ is computed from the sum of all forces on link 2

$$\sum \mathbf{F}^{(2)} \Longrightarrow m_2\,\mathbf{a}_{C_2} = \mathbf{F}_{12} + \mathbf{G}_2 + \mathbf{F}_{32} \quad \text{or} \quad \mathbf{F}_{32} = m_2\,\mathbf{a}_{C_2} - \mathbf{F}_{12} - \mathbf{G}_2.$$

Figure 4.6a is a kinetic diagram (or Newton–Euler diagram) that represents the dynamic effects as specified by Newton–Euler equations of motion in terms of translational terms, $m\mathbf{a}_C$, and rotational terms, $I_C\,\alpha$. Representation of $m\mathbf{a}_C$ and $I_C\,\alpha$ from the kinetic diagram will guarantee that the force and moment sums determined from the free-body diagram are equated to their proper resultants. The equivalence between the kinetic diagram and the free-body diagram will be employed in the solution of dynamical problems.

4.4.2 RRT Dyad

Figure 4.7 shows an RRT dyad with the unknown joint reaction forces $\mathbf{F}_{12}$, $\mathbf{F}_{43}$, and $\mathbf{F}_{23} = -\mathbf{F}_{32}$. The joint reaction force $\mathbf{F}_{43}$ is perpendicular to the sliding direction $\mathbf{F}_{43} \perp \Delta$ or

$$\mathbf{F}_{43} \cdot \Delta = (F_{43x}\mathbf{1} + F_{43y}\mathbf{J}) \cdot (\cos\theta\mathbf{1} + \sin\theta\mathbf{J}) = 0. \tag{4.22}$$

In order to determine $\mathbf{F}_{12}$ and $\mathbf{F}_{43}$ the following equations are written:

- sum of all the forces on links 2 and 3 is zero

$$\sum \mathbf{F}^{(2\&3)} \Longrightarrow m_2\,\mathbf{a}_{C_2} + m_3\,\mathbf{a}_{C_3} = \mathbf{F}_{12} + \mathbf{G}_2 + \mathbf{G}_3 + \mathbf{F}_{43},$$

  or

$$\sum \mathbf{F}^{(2\&3)} \cdot \mathbf{1} \Longrightarrow m_2\,a_{C_{2x}} + m_3\,a_{C_{3x}} = F_{12x} + F_{43x}, \tag{4.23}$$
$$\sum \mathbf{F}^{(2\&3)} \cdot \mathbf{J} \Longrightarrow m_2\,a_{C_{2y}} + m_3\,a_{C_{3y}} = F_{12y} - m_2\,g - m_3\,g + F_{43y}. \tag{4.24}$$

- sum of moments of all the forces and the moments on link 2 about C is zero

$$\sum \mathbf{M}_C^{(2)} \Longrightarrow I_{C_2}\,\alpha_2 + \mathbf{r}_{CC_2} \times m_2\,\mathbf{a}_{C_2} = \mathbf{r}_{CB} \times \mathbf{F}_{12} + \mathbf{r}_{CC_2} \times \mathbf{G}_2. \tag{4.25}$$

The components F_{12x}, F_{12y}, F_{43x}, and F_{43y} are calculated from Eqs. 4.22–4.25.

The reaction force components F_{32x} and F_{32y} are computed from the sum of all the forces on link 2

$$\sum \mathbf{F}^{(2)} \Longrightarrow m_2\,\mathbf{a}_{C_2} = \mathbf{F}_{12} + \mathbf{G}_2 + \mathbf{F}_{32} \quad \text{or} \quad \mathbf{F}_{32} = m_2\,\mathbf{a}_{C_2} - \mathbf{F}_{12} - \mathbf{G}_2.$$

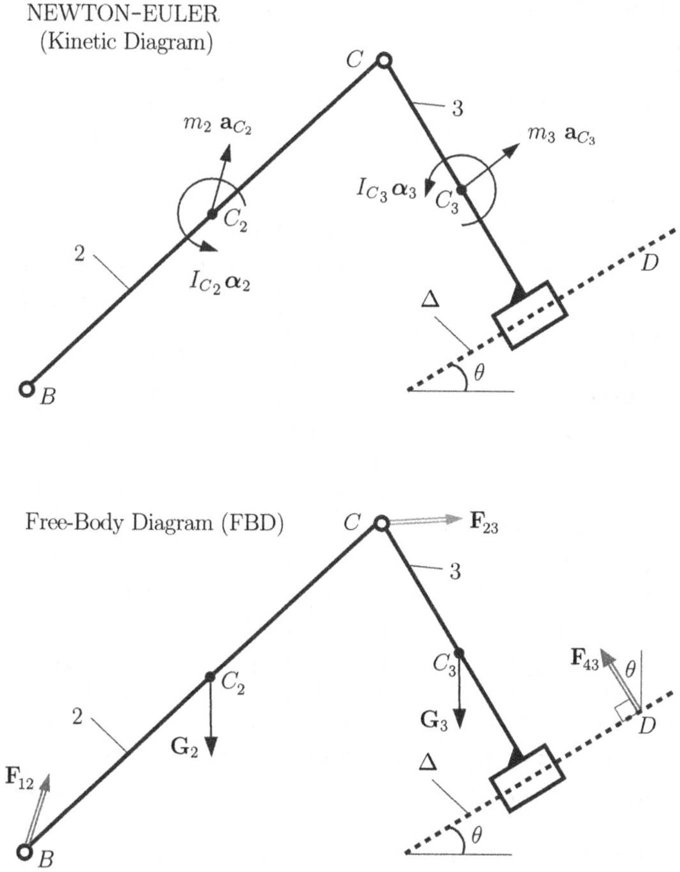

Fig. 4.7 RRT dyad

4.4.3 RTR Dyad

The unknown joint reaction forces $\mathbf{F}_{12}$ and $\mathbf{F}_{43}$, as shown in Fig. 4.8, are calculated from the relations:

- sum of all the forces on links 2 and 3 is zero

$$\sum \mathbf{F}^{(2\&3)} \Longrightarrow m_2\,\mathbf{a}_{C_2} + m_3\,\mathbf{a}_{C_3} = \mathbf{F}_{12} + \mathbf{G}_2 + \mathbf{G}_3 + \mathbf{F}_{43},$$

or

$$\sum \mathbf{F}^{(2\&3)} \cdot \mathbf{1} \Longrightarrow m_2\,a_{C_{2x}} + m_3\,a_{C_{3x}} = F_{12x} + F_{43x}, \tag{4.26}$$

$$\sum \mathbf{F}^{(2\&3)} \cdot \mathbf{J} \Longrightarrow m_2\,a_{C_{2y}} + m_3\,a_{C_{3y}} = F_{12y} - m_2\,g - m_3\,g + F_{43y}. \tag{4.27}$$

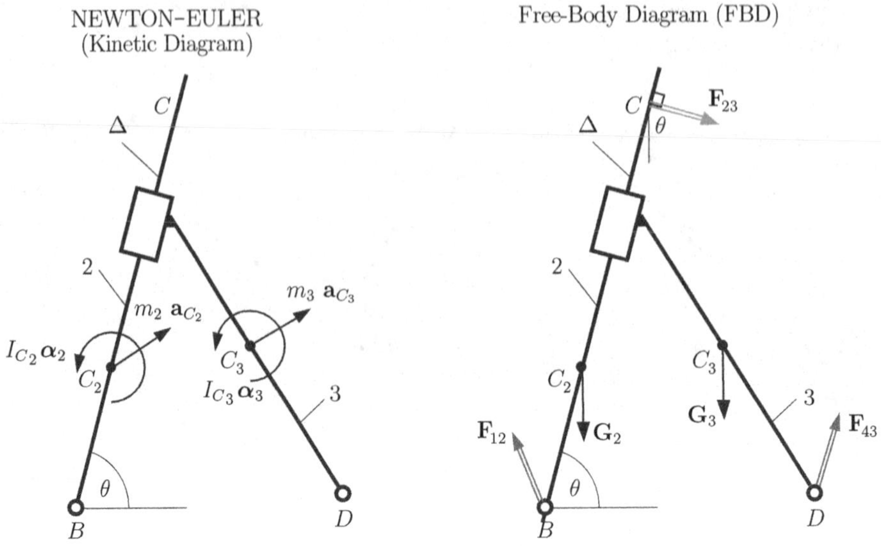

Fig. 4.8 RTR dyad

- sum of the moments of all the forces and moments on links 2 and 3 about B is zero

$$\sum \mathbf{M}_B^{(2\&3)} \Longrightarrow I_{C_2}\alpha_2 + I_{C_3}\alpha_3 + \mathbf{r}_{BC_2} \times m_2\,\mathbf{a}_{C_2} + \mathbf{r}_{BC_3} \times m_3\,\mathbf{a}_{C_3}$$
$$= \mathbf{r}_{BD} \times \mathbf{F}_{43} + \mathbf{r}_{BC_3} \times \mathbf{G}_3 + \mathbf{r}_{BC_2} \times \mathbf{G}_2. \tag{4.28}$$

- sum of all the forces on link 2 projected onto the sliding direction $\Delta = \cos\theta\mathbf{\imath} + \sin\theta\mathbf{\jmath}$ is zero

$$\sum \mathbf{F}^{(2)} \cdot \Delta = (\mathbf{F}_{12} + \mathbf{F}_2) \cdot (\cos\theta\mathbf{\imath} + \sin\theta\mathbf{\jmath}) = 0. \tag{4.29}$$

The components F_{12x}, F_{12y}, F_{43x}, and F_{43y} are calculated from Eqs. 4.26–4.29.

The force components F_{32x} and F_{32y} are computed from the sum of all the forces on link 2

$$\sum \mathbf{F}^{(2)} \Longrightarrow m_2\,\mathbf{a}_{C_2} = \mathbf{F}_{12} + \mathbf{G}_2 + \mathbf{F}_{32} \quad \text{or} \quad \mathbf{F}_{32} = m_2\,\mathbf{a}_{C_2} - (\mathbf{F}_{12} + \mathbf{G}_2).$$

4.5 Force Analysis Using Contour Method

An analytical method to compute joint forces that can be applied for both planar and spatial mechanisms will be presented. The method is based on the decoupling of a closed kinematic chain and writing the dynamic equilibrium equations. The kinematic links are loaded with external forces and inertia forces and moments.

A general monocontour closed kinematic chain is considered in Fig. 4.9. The joint force between the links $i-1$ and i (joint A_i) will be determined. When these two links $i-1$ and i are separated, the joint forces $\mathbf{F}_{i-1,i}$ and $\mathbf{F}_{i,i-1}$ are introduced and $\mathbf{F}_{i-1,i}+\mathbf{F}_{i,i-1}=\mathbf{0}$.

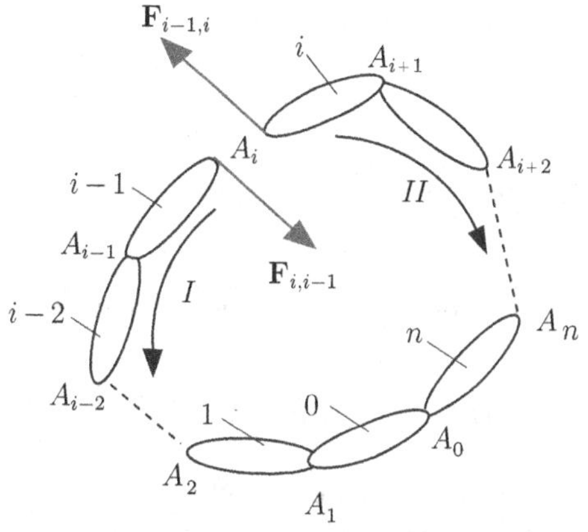

Fig. 4.9 Monocontour closed kinematic chain

It is helpful to "mentally disconnect" the two links $(i-1)$ and i, which create joint A_i, from the rest of the mechanism. The joint at A_i will be replaced by the joint forces $\mathbf{F}_{i-1,i}$ and $\mathbf{F}_{i,i-1}$. The closed kinematic chain has been transformed into two open kinematic chains, and two paths I and II are associated. The two paths start from A_i.

For the path I (counterclockwise), starting at A_i and following I the first joint encountered is A_{i-1}. For the link $i-1$ left behind, dynamic equilibrium equations are written according to the type of joint at A_{i-1}. Following the same path I, the next joint encountered is A_{i-2}. For the sub-system ($i-1$ and $i-2$) equilibrium conditions corresponding to the type of joint at A_{i-2} can be specified, and so on. A similar analysis is performed for the path II of the open kinematic chain. The number of equilibrium equations written is equal to the number of unknown scalars introduced by joint A_i (joint forces at this joint). For a joint, the number of equilibrium conditions is equal to the number of relative mobilities of the joint.

4.6 Slider-Crank (R-RRT) Mechanism

Figure 4.10 is a schematic diagram of a R-RRT (slider-crank) mechanism comprised of a crank 1, a connecting rod 2, and a slider 3. The mechanism shown in the figure has the dimensions: $AB = 1$ m and $BC = 1$ m. The driver link 1 rotates with a constant speed of $n = 30/\pi$ rpm. The point A is selected as the origin of the xyz reference frame. The moment when the driver link 1 makes an angle $\phi = \phi_1 = \pi/4$ rad with the horizontal axis will be considered for the dynamic force analysis.

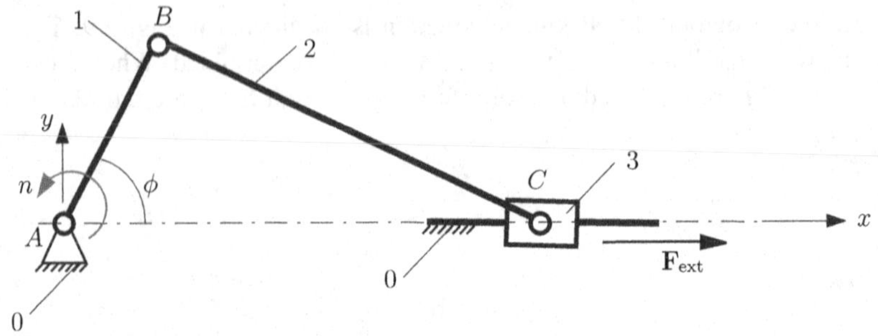

Fig. 4.10 Slider-crank (R-RRT) mechanism

The position vectors of the joints B and C, for $\phi = \phi_1 = \pi/4$ rad, are

$$\mathbf{r}_B = x_B\mathbf{\imath} + y_B\mathbf{\jmath} = \frac{\sqrt{2}}{2}\mathbf{\imath} + \frac{\sqrt{2}}{2}\mathbf{\jmath}\text{ m and } \mathbf{r}_C = x_C\mathbf{\imath} + y_C\mathbf{\jmath} = \sqrt{2}\mathbf{\imath} + 0\mathbf{\jmath}\text{ m.}$$

The angular velocities of links 1 and 2 are $\boldsymbol{\omega}_1 = \omega_1\,\mathbf{k} = 1\,\mathbf{k}$ rad/s and $\boldsymbol{\omega}_2 = \omega_2\,\mathbf{k} = -1\,\mathbf{k}$ rad/s. The angular accelerations of link 1 and 2 are $\boldsymbol{\alpha}_1$ and $\boldsymbol{\alpha}_2$. For this particular configuration of the mechanism $\boldsymbol{\alpha}_1 = \boldsymbol{\alpha}_2 = \mathbf{0}$. The velocity and acceleration of B are

$$\mathbf{v}_B = -\frac{\sqrt{2}}{2}\mathbf{\imath} + \frac{\sqrt{2}}{2}\mathbf{\jmath}\ \text{m/s} \quad\text{and}\quad \mathbf{a}_B = -\frac{\sqrt{2}}{2}\mathbf{\imath} - \frac{\sqrt{2}}{2}\mathbf{\jmath}\ \text{m/s}^2.$$

The velocity and acceleration of C are

$$\mathbf{v}_C = -\sqrt{2}\mathbf{\imath}\ \text{m/s} \quad\text{and}\quad \mathbf{a}_C = -\sqrt{2}\mathbf{\imath}\ \text{m/s}^2.$$

The center of mass of link 1 is C_1, the center of mass of link 2 is C_2, and the center of mass of slider 3 is C. The position vectors of the C_i, $i = 1,2,3$ are

$$\mathbf{r}_{C_1} = \mathbf{r}_B/2 = x_{C_1}\mathbf{\imath} + y_{C_1}\mathbf{\jmath} = \frac{\sqrt{2}}{4}\mathbf{\imath} + \frac{\sqrt{2}}{4}\mathbf{\jmath}\ \text{m},$$
$$\mathbf{r}_{C_2} = (\mathbf{r}_B + \mathbf{r}_C)/2 = x_{C_2}\mathbf{\imath} + y_{C_2}\mathbf{\jmath} = \frac{3\sqrt{2}}{4}\mathbf{\imath} + \frac{\sqrt{2}}{4}\mathbf{\jmath}\ \text{m},$$
$$\mathbf{r}_{C_3} = \mathbf{r}_C = x_{C_3}\mathbf{\imath} + y_{C_3}\mathbf{\jmath} = \sqrt{2}\mathbf{\imath}\ \text{m}.$$

The acceleration vectors of the C_i, $i = 1,2,3$ are

$$\mathbf{a}_{C_1} = \mathbf{a}_B/2 = a_{C_{1x}}\mathbf{\imath} + a_{C_{1y}}\mathbf{\jmath} = -\frac{\sqrt{2}}{4}\mathbf{\imath} - \frac{\sqrt{2}}{4}\mathbf{\jmath}\ \text{m/s}^2,$$
$$\mathbf{a}_{C_2} = (\mathbf{a}_B + \mathbf{a}_C)/2 = a_{C_{2x}}\mathbf{\imath} + a_{C_{2y}}\mathbf{\jmath} = -\frac{3\sqrt{2}}{4}\mathbf{\imath} - \frac{\sqrt{2}}{4}\mathbf{\jmath}\ \text{m/s}^2,$$
$$\mathbf{a}_{C_3} = \mathbf{a}_C = a_{C_x}\mathbf{\imath} + a_{C_y}\mathbf{\jmath} = -\sqrt{2}\mathbf{\imath}\ \text{m/s}^2.$$

The MATLAB® commands for the kinematics of the mechanism (positions, velocities, and accelerations) are:

```
AB = 1; BC = 1; phi = 45*(pi/180);
xA = 0; yA = 0;
rA = [xA yA 0];
xB = AB*cos(phi); yB = AB*sin(phi);
rB = [xB yB 0];
yC = 0; xC = xB+sqrt(BC^2-(yC-yB)^2);
rC = [xC yC 0];
n = 30/pi;
omega1 = [ 0 0 pi*n/30 ];
alpha1 = [0 0 0 ];
vA = [0 0 0 ]; aA = [0 0 0 ];
vB1 = vA + cross(omega1,rB); vB2 = vB1;
aB1=aA+cross(alpha1,rB)-dot(omega1,omega1)*rB;
aB2 = aB1;
omega2z = sym('omega2z','real');
vCx = sym('vCx','real');
omega2 = [ 0 0 omega2z ]; vC = [ vCx 0 0 ];
eqvC = vC - (vB2 + cross(omega2,rC-rB));
eqvCx = eqvC(1); eqvCy = eqvC(2);
solvC = solve(eqvCx,eqvCy);
omega2zs=eval(solvC.omega2z);
vCxs=eval(solvC.vCx); vCs = [vCxs 0 0];
Omega2 = [0 0 omega2zs];
alpha2z = sym('alpha2z','real');
aCx = sym('aCx','real');
alpha2 = [ 0 0 alpha2z ]; aC = [aCx 0 0 ];
eqaC = aC-(aB1+cross(alpha2,rC-rB)-...
       dot(Omega2,Omega2)*(rC-rB));
eqaCx = eqaC(1); eqaCy = eqaC(2);
solaC = solve(eqaCx,eqaCy);
alpha2zs = eval(solaC.alpha2z);
aCxs = eval(solaC.aCx);
alpha20 = [0 0 alpha2zs]; aCs = [aCxs 0 0];
alpha30 = [0 0 0];
rC1 = (rA+rB)/2;
fprintf('rC1 = [ %g, %g, %g ] (m)\n', rC1)
rC2 = (rB+rC)/2;
fprintf('rC2 = [ %g, %g, %g ] (m)\n', rC2)
rC3 = rC;
fprintf('rC3 = [ %g, %g, %g ] (m)\n', rC3)
aC1 = aB1/2;
fprintf('aC1 = [ %g, %g, %g ] (m/s^2\n', aC1)
```

```
aC2 = (aB1+aCs)/2;
fprintf('aC2 = [ %g, %g, %g ] (m/s^2)\n', aC2)
aC3 = aCs;
fprintf('aC3 = [ %g, %g, %g ] (m/s^2)\n', aC3)
```

The external driven force $\mathbf{F}_{ext}$ applied on link 3 is opposed to the motion of the link (opposed to $\mathbf{v}_C$). Because $\mathbf{v}_C = -\sqrt{2}\,\mathbf{\imath}$ m/s, the external force vector will be

$$\mathbf{F}_{ext} = [-\mathrm{Sign}(\mathbf{v}_C)]\ 100\mathbf{\imath} = 100\mathbf{\imath}\ \mathrm{N}.$$

The MATLAB commands for the external force on link 3 are:

```
fe = 100;
Fe = -sign(vCs(1))*[fe 0 0];
```

The signum function in MATLAB is `sign(x)`. If `x` is greater than zero `sign(x)` returns 1, if `x` is zero `sign(x)` returns zero, and if if `x` is less than zero `sign(x)` returns –1.

The height of the links 1 and 2 is $h = 0.01$ m. The width of the links 3 is $w_{slider} = 0.01$ m and the height is $h_{slider} = 0.01$ m (Fig. 4.11). All three moving links are rectangular prisms with the depth $d = 0.001$ m. The acceleration of gravity is $g = 10$ m/s^2. The MATLAB commands are:

```
h = 0.01; d = 0.001; hSlider = 0.01; wSlider = 0.01;
g = 10.;
```

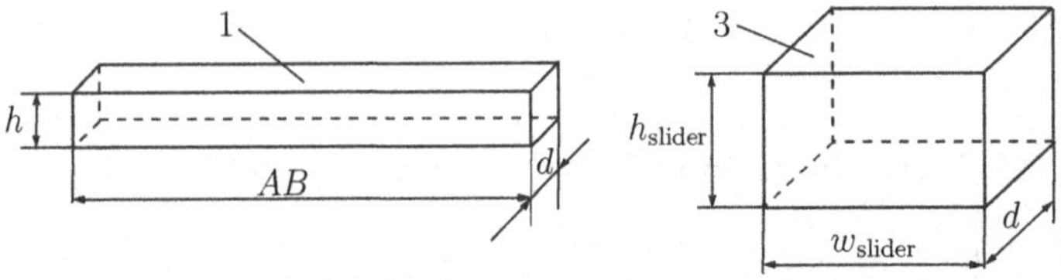

Fig. 4.11 Geometry of the links (not a scale drawing)

4.6.1 Inertia Forces and Moments

Link 1
The mass of the crank 1 is

$$m_1 = \rho_1 AB\,h\,d,$$

where the density of the material is ρ_1. For simplicity of calculation $m_1 = 1$ kg.

The inertia force on link 1 at C_1 is

$$\mathbf{F}_{\text{in}\,1} = -m_1\,\mathbf{a}_{C_1} = \frac{\sqrt{2}}{4}\,\mathbf{i} + \frac{\sqrt{2}}{4}\,\mathbf{j}\ \text{N}.$$

The gravitational force on crank 1 at C_1 is

$$\mathbf{G}_1 = -m_1\,g\,\mathbf{j} = -10\,\mathbf{j}\ \text{N}.$$

The mass moment of inertia of the link 1 about C_1 is

$$I_{C_1} = m_1\,(AB^2 + h^2)/12 = 0.0833417\ \text{kg m}^2.$$

The moment of inertia on link 1 is

$$\mathbf{M}_{\text{in}\,1} = -I_{C_1}\,\alpha_1 = \mathbf{0}.$$

Link 2
The mass of connecting rod 2 is

$$m_2 = \rho_2\,BC\,h\,d,$$

where the density of the material of link 2 is ρ_2. For simplicity of calculation $m_2 =$ 1 kg. The inertia force on link 2 at C_2 is

$$\mathbf{F}_{\text{in}\,2} = -m_2\,\mathbf{a}_{C_2} = \frac{3\sqrt{2}}{4}\,\mathbf{i} + \frac{\sqrt{2}}{4}\,\mathbf{j}\ \text{N}.$$

The gravitational force on link 2 at C_2 is

$$\mathbf{G}_2 = -m_2\,g\,\mathbf{j} = -10\,\mathbf{j}\ \text{N}.$$

The mass moment of inertia of link 2 about C_2 is

$$I_{C_2} = m_2\,(BC^2 + h^2)/12 = 0.0833417\ \text{kg m}^2.$$

The moment of inertia on link 2 is

$$\mathbf{M}_{\text{in}\,2} = -I_{C_2}\,\alpha_2 = \mathbf{0}.$$

Link 3
The mass of the link 3 is

$$m_3 = \rho_3\,h_{\text{slider}}\,w_{\text{slider}}\,d,$$

where the density of the material of link 3 is ρ_3. For simplicity of calculation $m_3 =$ 1 kg. The inertia force on link 3 at $C_3 = C$ is

$$\mathbf{F}_{\text{in}\,3} = -m_3\,\mathbf{a}_{C_3} = \sqrt{2}\,\mathbf{i}\ \text{N}.$$

The gravitational force on link 3 at $C_3 = C$ is

$$\mathbf{G}_3 = -m_3\, g\, \mathbf{J} = -10\mathbf{J} \text{ N}.$$

The mass moment of inertia of slider 3 about $C_3 = C$ is

$$I_{C_3} = m_3\,(h^2_{\text{slider}} + w^2_{\text{slider}})/12 = 0.0000166667 \text{ kg m}^2.$$

The moment of inertia on slider 3 is

$$\mathbf{M}_{\text{in}\,3} = -I_{C_3}\,\alpha_3 = \mathbf{0}.$$

The MATLAB commands for the forces and moments of inertia are:

```
m1 = 1;
IC1 = m1*(AB^2+h^2)/12;
G1 = [ 0 -m1*g 0 ];
Fin1 = - m1*aC1;
Min1 = - IC1*alpha1;
m2 = 1;
IC2 = m2*(BC^2+h^2)/12;
G2 = [ 0 -m2*g 0 ];
Fin2 = - m2*aC2;
Min2 = - IC2*alpha20;
m3 = 1 ;
IC3 = m3*(hSlider^2+wSlider^2)/12;
G3 = [ 0 -m3*g 0 ];
Fin3 = - m3*aC3;
Min3 = - IC3*alpha30;
```

For a given value of the crank angle ϕ ($\phi = \pi/4$) and a known driven force $\mathbf{F}_{\text{ext}}$ find the joint reactions and the drive (equilibrium) moment **M** on the crank.

4.6.2 Joint Forces and Drive Moment

4.6.2.1 Newton–Euler Equations of Motion

Figure 4.12 shows the free-body diagrams of the crank 1, the connecting rod 2, and the slider 3. For each moving link the dynamic equilibrium equations are applied (Newton–Euler equations of motion)

$$m\,\mathbf{a}_C = \sum \mathbf{F} \quad \text{and} \quad I_C\,\alpha = \sum \mathbf{M}_C,$$

where C is the center of mass of the link.

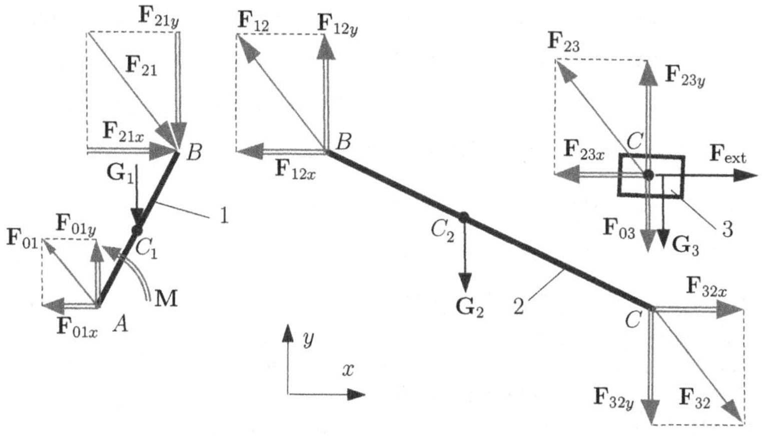

Fig. 4.12 Free-body diagrams

The force analysis starts with the link 3 because the external driven force $\mathbf{F}_{\text{ext}}$ on the slider is given.

The reaction joint force of the ground 0 on the slider 3, $\mathbf{F}_{03}$, is perpendicular to the sliding direction, x-axis: $\mathbf{F}_{03} \perp \mathbf{\imath}$ (Fig. 4.13). The application point Q of the reaction force $\mathbf{F}_{03}$ is determined using the Euler's moment equation

$$I_{C_3}\alpha_3 = \mathbf{r}_{CQ} \times \mathbf{F}_{03} \quad \text{or} \quad \mathbf{0} = \mathbf{r}_{CQ} \times \mathbf{F}_{03} \quad \Longrightarrow \quad \mathbf{r}_{CQ} = \mathbf{0} \quad \text{or} \quad C = Q.$$

It results that the reaction force $\mathbf{F}_{03}$ acts at C. For the slider 3 the vector sum of the net forces (external forces $\mathbf{F}_{\text{ext}}$, gravitational force $\mathbf{G}_3$, joint forces $\mathbf{F}_{23}$, $\mathbf{F}_{03}$) is equal to $m_3\mathbf{a}_{C_3}$ (Fig. 4.13)

$$m_3\,\mathbf{a}_{C_3} = \mathbf{F}_{23} + \mathbf{G}_3 + \mathbf{F}_{\text{ext}} + \mathbf{F}_{03},$$

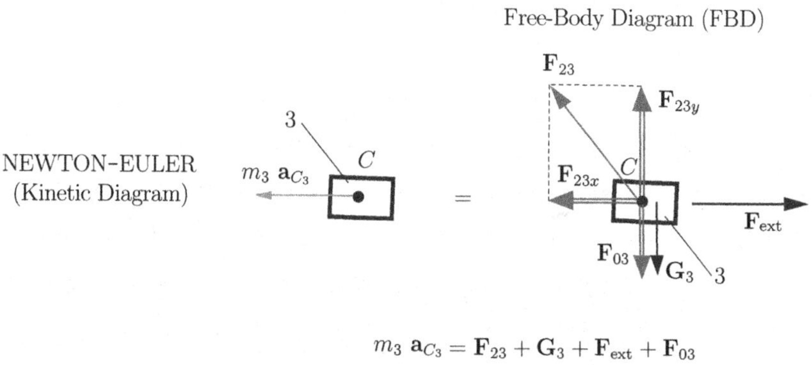

Fig. 4.13 Newton–Euler equations for slider 3

where $\mathbf{F}_{23} = F_{23x}\mathbf{\imath} + F_{23y}\mathbf{\jmath}$ and $\mathbf{F}_{03} = F_{03y}\mathbf{\jmath}$. Projecting the previous vectorial equation onto x and y axes gives

$$m_3\, a_{C_{3x}} = F_{23x} + F_{\text{ext}},$$
$$m_3\, a_{C_{3y}} = F_{23y} - m_3\, g + F_{03y},$$

or numerically

$$(1)(-\sqrt{2}) = F_{23x} + 100, \tag{4.30}$$
$$0 = F_{23y} - (1)(10) + F_{03y}. \tag{4.31}$$

There are two equations Eqs. 4.30 and 4.31 and three unknowns F_{03y}, F_{23x} and F_{23y} and that is why the analysis will continue with link 2.

The MATLAB commands for Newton-Euler equations for slider 3 are:

```
F03 = [ 0 sym('F03y','real') 0 ];
F23 = [ sym('F23x','real') sym('F23y','real') 0 ];
eqF3 = F03+F23+Fe+G3-m3*aC3;
eqF3x = eqF3(1);
eqF3y = eqF3(2);
```

For the connecting rod 2 (Fig. 4.14), Newton's equation gives

$$m_2\, \mathbf{a}_{C_2} = \mathbf{F}_{32} + \mathbf{G}_2 + \mathbf{F}_{12}.$$

The previous equation can be projected on x and y axes

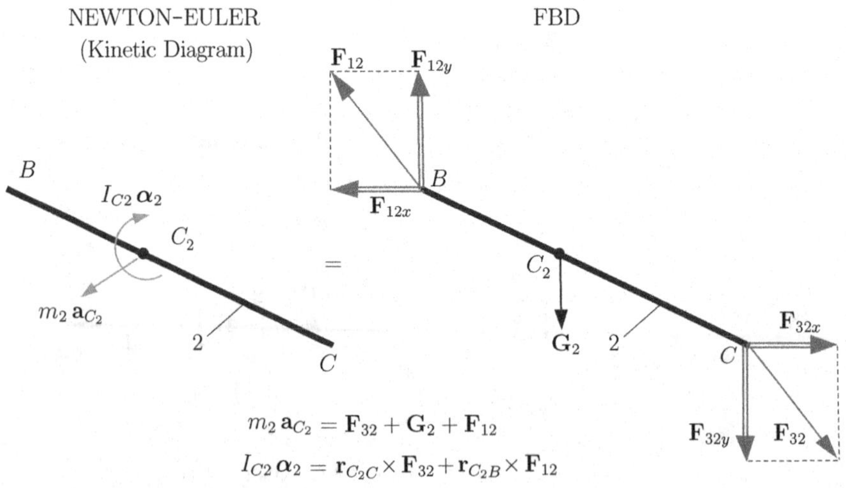

Fig. 4.14 Newton–Euler equations for link 2

$$m_2\, a_{C_{2x}} = F_{32x} + F_{12x},$$
$$m_2\, a_{C_{2y}} = F_{32y} - m_2\, g + F_{12y},$$

or numerically

$$(1)(-\frac{3\sqrt{2}}{4}) = F_{32x} + F_{12x}, \tag{4.32}$$

$$(1)(-\frac{\sqrt{2}}{4}) = F_{32y} - 1(10) + F_{12y}. \tag{4.33}$$

For the link 2 a moment equation can be written with respect to C_2

$$I_{C_2}\, \alpha_2 = \mathbf{r}_{C_2C} \times \mathbf{F}_{32} + \mathbf{r}_{C_2B} \times \mathbf{F}_{12},$$

or

$$I_{C_2}\, \alpha_2\, \mathbf{k} = \begin{vmatrix} \mathbf{i} & \mathbf{j} & \mathbf{k} \\ x_C - x_{C_2} & y_C - y_{C_2} & 0 \\ F_{32x} & F_{32y} & 0 \end{vmatrix} + \begin{vmatrix} \mathbf{i} & \mathbf{j} & \mathbf{k} \\ x_B - x_{C_2} & y_B - y_{C_2} & 0 \\ F_{12x} & F_{12y} & 0 \end{vmatrix},$$

or

$$I_{C_2}\, \alpha_2 = (x_C - x_{C_2})F_{32y} - (y_C - y_{C_2})F_{32x} + (x_B - x_{C_2})F_{12y} - (y_B - y_{C_2})F_{12x},$$

or numerically

$$0 = (\sqrt{2} - \frac{3\sqrt{2}}{4})F_{32y} - (-\frac{\sqrt{2}}{4})F_{32x}$$
$$+(\frac{\sqrt{2}}{2} - \frac{3\sqrt{2}}{4})F_{12y} - (\frac{\sqrt{2}}{2} - \frac{\sqrt{2}}{4})F_{12x}. \tag{4.34}$$

The MATLAB commands for the Newton–Euler equations for link 2 are:

```
F32 = -F23;
F12 = [ sym('F12x','real') sym('F12y','real') 0 ];
eqF2 = F32+F12+G2-m2*aC2;
eqF2x = eqF2(1);
eqF2y = eqF2(2);
eqM2 = cross(rB-rC2,F12)+cross(rC-rC2,F32)-...
       IC2*alpha20;
eqM2z = eqM2(3);
```

Equations 4.30–4.34 form a system of 5 equations with five scalar unknowns. The system can be solved using the `solve` statement:

```
sol32 = solve(eqF3x,eqF3y,eqF2x,eqF2y,eqM2z);
F03ys = eval(sol32.F03y);
```

```
F23xs = eval(sol32.F23x);
F23ys = eval(sol32.F23y);
F12xs = eval(sol32.F12x);
F12ys = eval(sol32.F12y);
F03s = [ 0, F03ys, 0 ];
F23s = [ F23xs, F23ys, 0 ];
F12s = [ F12xs, F12ys, 0 ];
```

The numerical values for the joint forces for the links 3 and 2 are

$$F_{03y} = -85 - \frac{3\sqrt{2}}{2} \text{ N},$$
$$F_{23x} = -100 - \sqrt{2} \text{ N}, \quad F_{23y} = 95 + \frac{3\sqrt{2}}{2} \text{ N},$$
$$F_{12x} = -\frac{1}{4}(400 + 7\sqrt{2}) \text{ N}, \quad F_{12y} = \frac{5}{4}(84 + \sqrt{2}) \text{ N},$$

or

$$F_{03} = |\mathbf{F}_{03}| = 85 + \frac{3\sqrt{2}}{2} \text{ N},$$
$$F_{23} = |\mathbf{F}_{23}| = \sqrt{F_{23x}^2 + F_{23y}^2} = \sqrt{\frac{38063}{2} + 485\sqrt{2}} \text{ N},$$
$$F_{12} = |\mathbf{F}_{12}| = \sqrt{F_{12x}^2 + F_{12y}^2} = \frac{1}{2}\sqrt{84137 + 2450\sqrt{2}} \text{ N}.$$

For the crank 1 (Fig. 4.15), there are two vectorial equations

$$m_1 \mathbf{a}_{C_1} = \mathbf{F}_{21} + \mathbf{G}_1 + \mathbf{F}_{01},$$
$$I_{C_1} \boldsymbol{\alpha}_1 = \mathbf{r}_{C_1 B} \times \mathbf{F}_{21} + \mathbf{r}_{C_1 A} \times \mathbf{F}_{01} + \mathbf{M},$$

where $\mathbf{M}$, is the input (motor) moment on the crank, $\mathbf{F}_{21} = -\mathbf{F}_{12}$, and $\mathbf{F}_{01} = F_{01x}\mathbf{ı} + F_{01y}\mathbf{ȷ}$. The above vectorial equations give three scalar equations on x, y, and z

$$m_1 \, a_{C_{1x}} = F_{21x} + F_{01x},$$
$$m_1 \, a_{C_{1y}} = F_{21y} - m_1 \, g + F_{01y},$$
$$I_{C1} \, \alpha_1 \, \mathbf{k} = \begin{vmatrix} \mathbf{ı} & \mathbf{ȷ} & \mathbf{k} \\ x_B - x_{C_1} & y_B - y_{C_1} & 0 \\ F_{21x} & F_{21y} & 0 \end{vmatrix} + \begin{vmatrix} \mathbf{ı} & \mathbf{ȷ} & \mathbf{k} \\ x_A - x_{C_1} & y_A - y_{C_1} & 0 \\ F_{01x} & F_{01y} & 0 \end{vmatrix}$$
$$+ M\,\mathbf{k} = \mathbf{0},$$

or

$$m_1 \, a_{C_{1x}} = F_{21x} + F_{01x},$$
$$m_1 \, a_{C_{1y}} = F_{21y} - m_1 \, g + F_{01y},$$

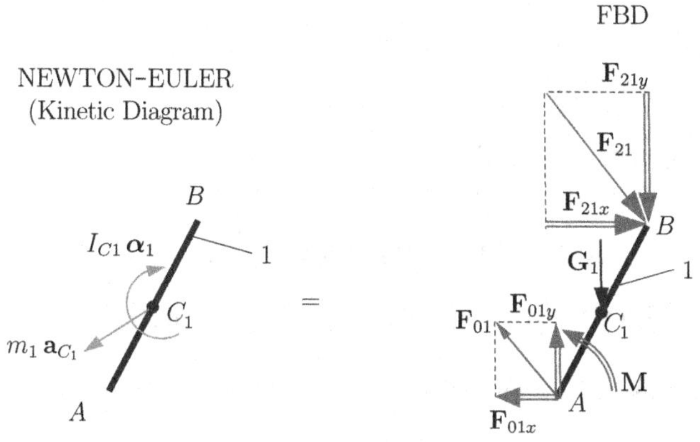

$$m_1\,\mathbf{a}_{C_1} = \mathbf{F}_{21} + \mathbf{F}_{in\,1} + \mathbf{G}_1 + \mathbf{F}_{01}$$
$$I_{C1}\,\boldsymbol{\alpha}_1 = \mathbf{r}_{C_1A} \times \mathbf{F}_{01} + \mathbf{r}_{C_1B} \times \mathbf{F}_{21} + \mathbf{M}$$

Fig. 4.15 Newton–Euler equations for link 1

$$I_{C_1}\,\alpha_1 = (x_B - x_{C_1})F_{21y} - (y_{C_2} - y_{C_1})F_{21x}$$
$$+(x_A - x_{C_1})F_{01y} - (y_A - y_{C_1})F_{01x} + M,$$

or numerically

$$1(-\frac{\sqrt{2}}{4}) = \frac{1}{4}(400 + 7\sqrt{2}) + F_{01x}, \tag{4.35}$$

$$1(-\frac{\sqrt{2}}{4}) = -\frac{5}{4}(84 + \sqrt{2}) - 1(10) + F_{01y}, \tag{4.36}$$

$$0 = (\frac{\sqrt{2}}{2} - \frac{\sqrt{2}}{4})\left[-\frac{5}{4}(84 + \sqrt{2})\right] - (\frac{\sqrt{2}}{2} - \frac{\sqrt{2}}{4})\left[\frac{1}{4}(400 + 7\sqrt{2})\right]$$
$$-\frac{\sqrt{2}}{4}F_{01y} + \frac{\sqrt{2}}{4}F_{01y} + M = 0. \tag{4.37}$$

The MATLAB commands for the Newton–Euler equations for the crank 1 are:

```
F01=m1*aC1-G1+F12s;
Mm=-cross(rB,-F12s)-cross(rC1,G1-m1*aC1)-IC1*alpha1;
```

Equations 4.35–4.37 give

$$F_{01x} = -2(50 + \sqrt{2})\ \text{N}, \quad F_{01y} = 115 + \sqrt{2}\ \text{N},$$
$$M = 3 + 105\sqrt{2}\ \text{N m},$$

or

$$F_{01} = |\mathbf{F}_{01}| = \sqrt{F_{01x}^2 + F_{01y}^2} = \sqrt{23235 + 630\sqrt{2}} \text{ N},$$
$$M = |\mathbf{M}| = 3 + 105\sqrt{2} \text{ N m}.$$

Another way of calculating the moment $\mathbf{M}$ required for dynamic equilibrium is to write the moment equation of motion for link 1 about the fixed point A

$$I_A\, \alpha_1\, \mathbf{k} = \mathbf{r}_{AC_1} \times \mathbf{G}_1 + \mathbf{r}_{AB} \times \mathbf{F}_{21} + \mathbf{M} \implies \mathbf{M} = \mathbf{r}_B \times \mathbf{F}_{12} - \mathbf{r}_{C_1} \times \mathbf{G}_1,$$

where $I_A = I_{C_1} + m_1\,(AB/2)^2$. The reaction force $\mathbf{F}_{01}$ does not appear in this moment equation.

The MATLAB program using Newton–Euler equations and the results are given in Appendix C.1.

4.6.2.2 D'Alembert's Principle

For each moving link the dynamic equilibrium equations are applied (d'Alembert's principle)

$$\sum \mathbf{F} + \mathbf{F}_{\text{in}} = \mathbf{0} \quad \text{and} \quad \sum \mathbf{M}_C + \mathbf{M}_{\text{in}} = \mathbf{0},$$

where C is the center of mass of the link. With d'Alembert's principle the moment summation can be about any arbitrary point P

$$\sum \mathbf{M}_P + \mathbf{M}_{\text{in}} + \mathbf{r}_{PC} \times \mathbf{F}_{\text{in}} = \mathbf{0}.$$

The force analysis starts with the link 3 because the external driven force $\mathbf{F}_{\text{ext}}$ is given. For the slider 3 the vector sum of all the forces (external forces $\mathbf{F}_{\text{ext}}$, gravitational force $\mathbf{G}_3$, inertia forces $\mathbf{F}_{\text{in}\,3}$, joint forces $\mathbf{F}_{23}$, $\mathbf{F}_{03}$) is zero (Fig. 4.16)

$$\sum \mathbf{F}^{(3)} = \mathbf{F}_{23} + \mathbf{F}_{\text{in}\,3} + \mathbf{G}_3 + \mathbf{F}_{\text{ext}} + \mathbf{F}_{03} = \mathbf{0},$$

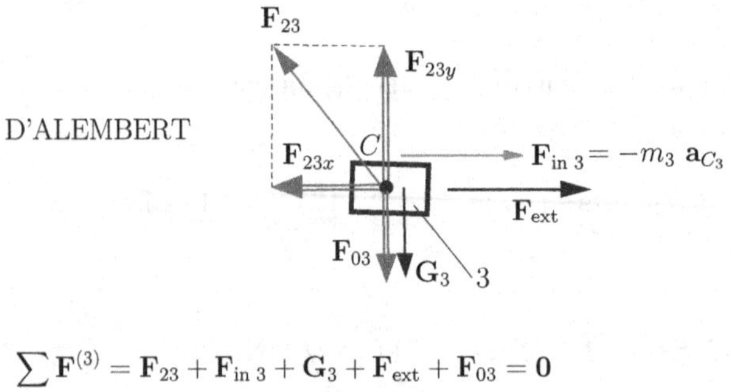

Fig. 4.16 D'Alembert's principle for slider 3

where $\mathbf{F}_{23} = F_{23x}\mathbf{I} + F_{23y}\mathbf{J}$ and $\mathbf{F}_{03} = F_{03y}\mathbf{J}$. Projecting this force onto x and y axes gives

$$\sum \mathbf{F}^{(3)} \cdot \mathbf{I} = F_{23x} + (-m_3\, a_{C_{3x}}) + F_{ext} = 0,$$
$$\sum \mathbf{F}^{(3)} \cdot \mathbf{J} = F_{23y} - m_3\, g + F_{03y} = 0,$$

or numerically

$$F_{23x} + (-1)(-\sqrt{2}) + 100 = 0, \tag{4.38}$$
$$F_{23y} - (1)(10) + F_{03y} = 0. \tag{4.39}$$

The MATLAB commands for slider 3 are:

```
F03 = [ 0 sym('F03y','real') 0 ];
F23 = [ sym('F23x','real') sym('F23y','real') 0 ];
eqF3 = F03+F23+Fe+G3+Fin3;
eqF3x = eqF3(1);
eqF3y = eqF3(2);
```

For the connecting rod 2 (Fig. 4.17), the sum of the forces is equal to zero

$$\sum \mathbf{F}^{(2)} = \mathbf{F}_{32} + \mathbf{F}_{in2} + \mathbf{G}_2 + \mathbf{F}_{12} = \mathbf{0},$$

The previous equation can be projected on x and y axes

$$\sum \mathbf{F}^{(2)} = \mathbf{F}_{32} + \mathbf{F}_{in\,2} + \mathbf{G}_2 + \mathbf{F}_{12} = 0$$
$$\sum \mathbf{M}_B^{(2)} = (\mathbf{r}_C - \mathbf{r}_B) \times \mathbf{F}_{32} + (\mathbf{r}_{C2} - \mathbf{r}_B) \times (\mathbf{F}_{in\,2} + \mathbf{G}_2) + \mathbf{M}_{in\,2} = 0$$

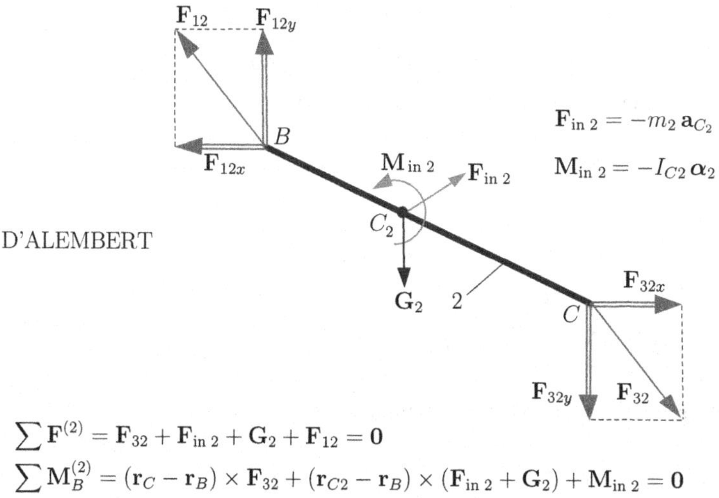

Fig. 4.17 D'Alembert's principle for link 2

$$\sum \mathbf{F}^{(2)} \cdot \imath = F_{32x} + (-m_2\, a_{C_{2x}}) + F_{12x} = 0,$$
$$\sum \mathbf{F}^{(2)} \cdot \jmath = F_{32y} + (-m_2\, a_{C_{2y}}) - m_2\, g + F_{12y} = 0,$$

or numerically

$$F_{32x} + (-1)(-\frac{3\sqrt{2}}{4}) + F_{12x} = 0, \tag{4.40}$$

$$F_{32y} + (-1)(-\frac{\sqrt{2}}{4}) - 1(10) + F_{12y} = 0. \tag{4.41}$$

For the link 2 a moment equation can be written with respect to C_2

$$\sum \mathbf{M}^{(2)}_{C_2} = \mathbf{r}_{C_2C} \times \mathbf{F}_{32} + \mathbf{r}_{C_2B} \times \mathbf{F}_{12} + \mathbf{M}_{\text{in}2} = \mathbf{0}.$$

Instead of the previous equation the sum of the moments with respect to B can be used

$$\sum \mathbf{M}^{(2)}_{B} = \mathbf{r}_{BC} \times \mathbf{F}_{32} + \mathbf{r}_{BC_2} \times (\mathbf{F}_{\text{in}2} + \mathbf{G}_2) + \mathbf{M}_{\text{in}2} = \mathbf{0},$$

or

$$\begin{vmatrix} \imath & \jmath & \mathbf{k} \\ x_C - x_B & y_C - y_B & 0 \\ F_{32x} & F_{32y} & 0 \end{vmatrix} + \begin{vmatrix} \imath & \jmath & \mathbf{k} \\ x_{C_2} - x_B & y_{C_2} - y_B & 0 \\ -m_2\, a_{C_{2x}} & -m_2\, a_{C_{2y}} - m_2\, g & 0 \end{vmatrix}$$
$$-I_{C_2}\, \alpha_2\, \mathbf{k} = \mathbf{0},$$

or

$$(x_C - x_B)F_{32y} - (y_C - y_B)F_{32x} + (x_{C_2} - x_B)(-m_2\, a_{C_{2y}} - m_2\, g)$$
$$- (y_{C_2} - y_B)(-m_2\, a_{C_{2x}}) - I_{C_2}\, \alpha_2 = 0,$$

or numerically

$$(\sqrt{2} - \frac{\sqrt{2}}{2})F_{32y} - (-\frac{\sqrt{2}}{2})F_{32x} + (\frac{3\sqrt{2}}{4} - \frac{\sqrt{2}}{2})\left[-1(-\frac{\sqrt{2}}{4}) - 1(10)\right]$$
$$-(\frac{\sqrt{2}}{4} - \frac{\sqrt{2}}{2})\left[-1(-\frac{3\sqrt{2}}{4})\right] - 0 = 0. \tag{4.42}$$

For the connecting rod 2 the MATLAB commands are:

```
F32 = -F23;
F12 = [ sym('F12x','real') sym('F12y','real') 0 ];
eqF2 = F32+F12+G2+Fin2;
eqF2x = eqF2(1);
eqF2y = eqF2(2);
```

```
eqM2 = cross(rB-rC2,F12)+cross(rC-rC2,F32)+Min2;
eqM2z = eqM2(3);
```

For the crank 1 (Fig. 4.18), there are two vectorial equations

$$\sum \mathbf{F}^{(1)} = \mathbf{F}_{21} + \mathbf{F}_{\text{in}1} + \mathbf{G}_1 + \mathbf{F}_{01} = 0,$$

$$\sum \mathbf{M}_A^{(1)} = \mathbf{r}_B \times \mathbf{F}_{21} + \mathbf{r}_{C_1} \times (\mathbf{F}_{\text{in}1} + \mathbf{G}_1) + \mathbf{M}_{\text{in}1} + \mathbf{M} = 0,$$

where $M = |\mathbf{M}|$ is the magnitude of the input moment on the crank, $\mathbf{F}_{21} = -\mathbf{F}_{12}$, and $\mathbf{F}_{01} = F_{01x}\mathbf{I} + F_{01y}\mathbf{J}$.

The above vectorial equations give three scalar equations on x, y, and z

$$\sum \mathbf{F}^{(1)} \cdot \mathbf{I} = F_{21x} + (-m_1 a_{C_{1x}}) + F_{01x} = 0,$$

$$\sum \mathbf{F}^{(1)} \cdot \mathbf{J} = F_{21y} + (-m_1 a_{C_{1y}}) - m_1 g + F_{01y} = 0,$$

$$\begin{vmatrix} \mathbf{I} & \mathbf{J} & \mathbf{k} \\ x_B & y_B & 0 \\ F_{21x} & F_{21y} & 0 \end{vmatrix} + \begin{vmatrix} \mathbf{I} & \mathbf{J} & \mathbf{k} \\ x_{C_1} & y_{C_1} & 0 \\ -m_1 a_{C_{1x}} & -m_1 a_{C_{1y}} - m_1 g & 0 \end{vmatrix}$$

$$-I_{C_1} \alpha_1 \mathbf{k} + M \mathbf{k} = 0,$$

or

$$F_{21x} + (-m_1 a_{C_{1x}}) + F_{01x} = 0,$$

$$F_{21y} + (-m_1 a_{C_{1y}}) - m_1 g + F_{01y} = 0,$$

$$x_B F_{21y} - y_B F_{21x} + x_{C_1}(-m_1 a_{C_{1y}} - m_1 g) - y_{C_1}(-m_1 a_{C_{1x}}) - I_{C_1} \alpha_1 + M = 0,$$

$$\sum \mathbf{F}^{(1)} = \mathbf{F}_{21} + \mathbf{F}_{\text{in}1} + \mathbf{G}_1 + \mathbf{F}_{01} = 0$$

$$\sum \mathbf{M}_A^{(1)} = \mathbf{r}_B \times \mathbf{F}_{21} + \mathbf{r}_{C1} \times (\mathbf{F}_{\text{in}1} + \mathbf{G}_1) + \mathbf{M}_{\text{in}1} + \mathbf{M} = 0$$

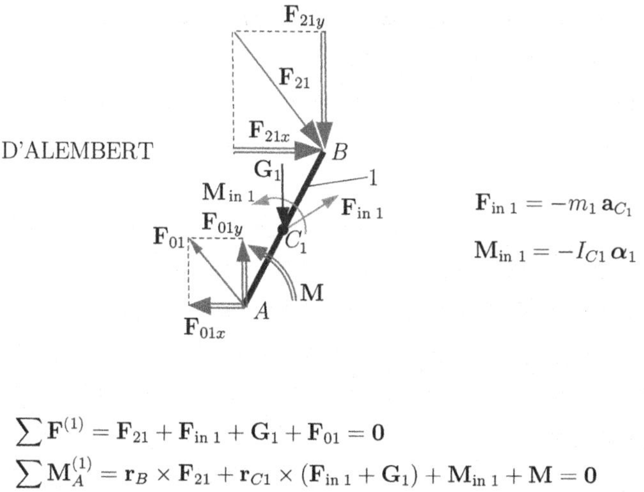

Fig. 4.18 D'Alembert's principle for link 1

or numerically

$$\frac{1}{4}(400+7\sqrt{2})+\left[-1(-\frac{\sqrt{2}}{4})\right]+F_{01x}=0, \tag{4.43}$$

$$-\frac{5}{4}(84+\sqrt{2})+\left[-1(-\frac{\sqrt{2}}{4})\right]-1(10)+F_{01y}=0, \tag{4.44}$$

$$\begin{aligned}&\frac{\sqrt{2}}{2}\left[-\frac{5}{4}(84+\sqrt{2})\right]-\frac{\sqrt{2}}{2}\left[\frac{1}{4}(400+7\sqrt{2})\right]\\&+\frac{\sqrt{2}}{4}\left[-1(-\frac{\sqrt{2}}{4})-1(10)\right]-\frac{\sqrt{2}}{4}\left[-1(-\frac{\sqrt{2}}{4})\right]\\&-0+M=0.\end{aligned} \tag{4.45}$$

For the crank 1 the MATLAB commands are:

```
F01 = [ sym('F01x','real') sym('F01y','real') 0 ];
Mm = [ 0 0 sym('Mmz','real') ];
eqF1 = F01+Fin1+G1-F12;
eqF1x = eqF1(1);
eqF1y = eqF1(2);
eqM1 = cross(rB-rC1,-F12)+cross(rA-rC1,F01)+Min1+Mm;
eqM1z = eqM1(3);
```

Equations 4.38–4.45 form a system of 8 equations with eight scalar unknowns.

The MATLAB commands for solving the system of equations are:

```
sol321 = solve(eqF3x,eqF3y,eqF2x,eqF2y,eqM2z,...
               eqF1x,eqF1y,eqM1z);
F03ys = eval(sol321.F03y);
F23xs = eval(sol321.F23x);
F23ys = eval(sol321.F23y);
F12xs = eval(sol321.F12x);
F12ys = eval(sol321.F12y);
F01xs = eval(sol321.F01x);
F01ys = eval(sol321.F01y);
Mmzs = eval(sol321.Mmz);
```

The following numerical solutions are obtained

$$F_{03y}=-85-\frac{3\sqrt{2}}{2}\ \text{N},$$

$$F_{23x}=-100-\sqrt{2}\ \text{N},\quad F_{23y}=95+\frac{3\sqrt{2}}{2}\ \text{N},$$

$$F_{12x} = -\frac{1}{4}(400+7\sqrt{2})\ \text{N},\quad F_{12y} = \frac{5}{4}(84+\sqrt{2})\ \text{N},$$
$$F_{01x} = -2(50+\sqrt{2})\ \text{N},\quad F_{01y} = 115+\sqrt{2}\ \text{N},$$
$$M = 3+105\sqrt{2}\ \text{N m}.$$

The MATLAB program using D'Alembert principle and the results are given in Appendix C.2.

4.6.2.3 Dyad Method

$B_R C_R C_T$ Dyad

Figure 4.19 shows the dyad $B_R C_R C_T$ with the unknown joint reactions $\mathbf{F}_{12}$ and $\mathbf{F}_{03}$. The joint reaction $\mathbf{F}_{03}$ is perpendicular to the sliding direction $\mathbf{F}_{03} \perp \Delta = \mathbf{1}$ or $\mathbf{F}_{03} = F_{03y}\mathbf{J}$. The following equations are written to determine $\mathbf{F}_{12}$ and $\mathbf{F}_{03}$

- Newton's equation for links 2 and 3, $\sum \mathbf{F}^{(2\&3)} \Longrightarrow$

$$m_2\,\mathbf{a}_{C_2} + m_3\,\mathbf{a}_{C_3} = \mathbf{F}_{12} + \mathbf{G}_2 + \mathbf{G}_3 + \mathbf{F}_{03} + \mathbf{F}_{\text{ext}},$$

or

$$m_2\,a_{C_{2x}} + m_3\,a_{C_{3x}} = F_{12x} + F_{\text{ext}},$$
$$m_2\,a_{C_{2y}} + m_3\,a_{C_{3y}} = F_{12y} - m_2\,g - m_3\,g + F_{03y} - m_2\,a_{C_{2y}},$$

or

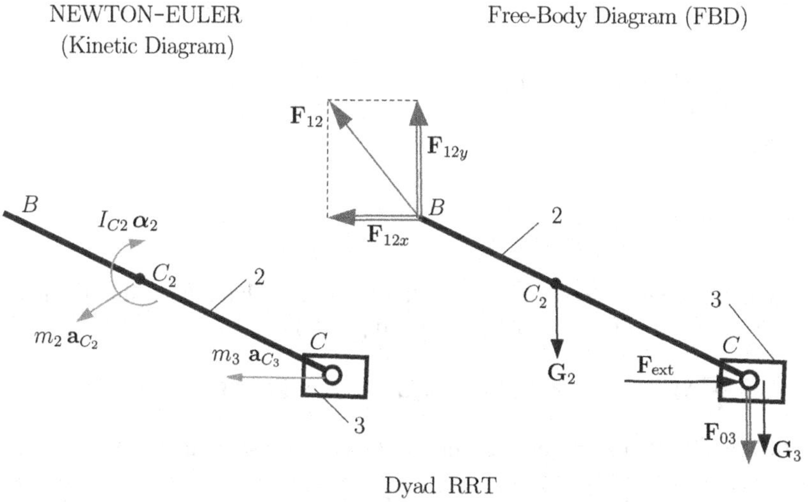

Fig. 4.19 Newton–Euler diagrams for dyad $B_R C_R C_T$

$$F_{12x} + 102.475 = 0, \tag{4.46}$$
$$F_{12y} + F_{03y} - 19.6464 = 0. \tag{4.47}$$

- Euler equation of moments for links 2 about C_R, $\sum \mathbf{M}_C^{(2)} \Longrightarrow$

$$I_{C_2}\alpha_2 + \mathbf{r}_{CC_2} \times m_2\,\mathbf{a}_{C_2} = \mathbf{r}_{CB} \times \mathbf{F}_{12} + \mathbf{r}_{CC_2} \times \mathbf{G}_2,$$

or

$$-0.707105\,F_{12y} - 0.707105\,F_{12x} + 3.03552 = 0. \tag{4.48}$$

Equations 4.46–4.48 give

$$F_{12x} = -102.475 \text{ N}, \; F_{12y} = 106.768 \text{ N}, \text{ and } F_{03y} = -87.1213 \text{ N}.$$

The joint reaction force $\mathbf{F}_{32}$ is calculated from

$$\mathbf{F}_{32} = m_2\mathbf{a}_{C_2} - (\mathbf{G}_2 + \mathbf{F}_{12}) = 101.414\mathbf{\imath} - 97.1213\mathbf{\jmath} \;\; \text{N}.$$

The MATLAB commands for finding the unknowns using the dyad method are:

```
F03 = [ 0 sym('F03y','real') 0 ];
F12 = [ sym('F12x','real') sym('F12y','real') 0 ];
eqF23 = F03+Fe+G3+F12+G2-m3*aC3-m2*aC2;
eqF32x = eqF32(1);
eqF32y = eqF32(2);
eqM2C = cross(rB-rC,F12)+cross(rC2-rC,G2)-...
        IC2*alpha20-cross(rC2-rC,m2*aC2);
eqM2Cz = eqM2C(3);
sol32=solve(eqF32x,eqF32y,eqM2Cz);
F03ys=eval(sol32.F03y);
F12xs=eval(sol32.F12x);
F12ys=eval(sol32.F12y);
F03s = [ 0, F03ys, 0 ];
F12s = [ F12xs, F12ys, 0 ];
F32 = m2*aC2-(F12s+G2);
```

The moment $\mathbf{M}$ required for dynamic equilibrium is calculated from the moment equation of for link 1 (Fig. 4.20) about the fixed point A

$$\sum \mathbf{M}_A^{(1)} \Longrightarrow I_{C_1}\alpha_1 + \mathbf{r}_{C_1} \times m_1\,\mathbf{a}_{C_1} = \mathbf{r}_B \times \mathbf{F}_{21} + \mathbf{G}_1 + \mathbf{M}.$$

Thus, $M = 151.492$ N m. The joint reaction force $\mathbf{F}_{01}$ is calculated from

$$\sum \mathbf{F}^{(1)} \Longrightarrow m_1\,\mathbf{a}_{C_1} = -\mathbf{F}_{12} + \mathbf{G}_1 + \mathbf{F}_{01},$$

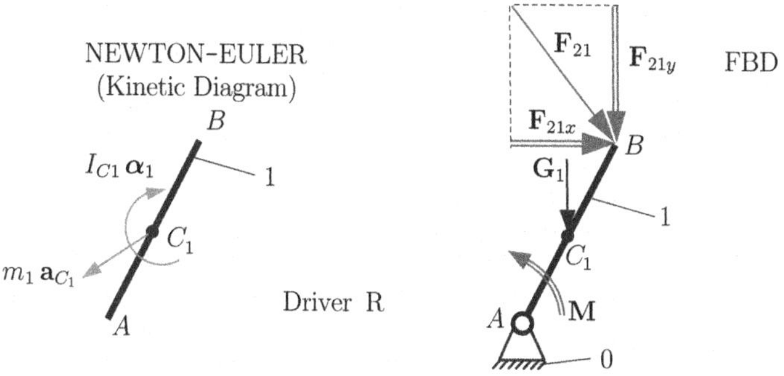

Fig. 4.20 Newton–Euler diagram for driver link

and $\mathbf{F}_{01} = -102.828\mathbf{\imath} + 116.414\mathbf{\jmath}$ N. The MATLAB program for the dyad method and the results are given in Appendix C.3.

D'Alembert's principle can be applied for the dyad method using the diagrams shown in Fig. 4.21.

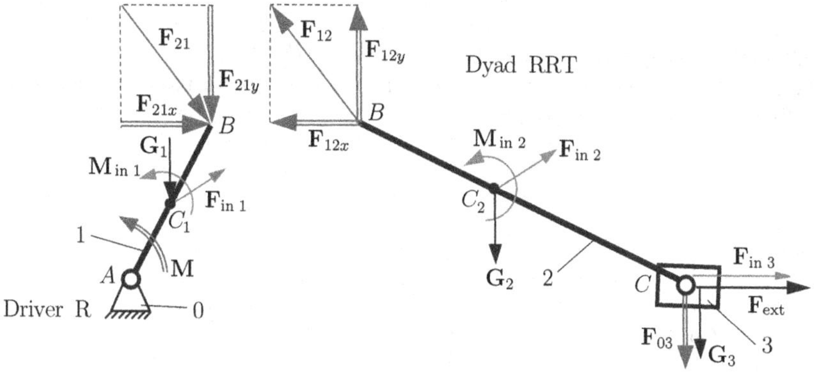

Fig. 4.21 D'Alembert's principle for the dyad method

4.6.2.4 Contour Method

The diagram representing the mechanism is shown in Fig. 4.22 and has one contour, 0-1-2-3-0. The dynamic force analysis can start with any joint.

Reaction $\mathbf{F}_{03}$

The reaction force $\mathbf{F}_{03}$ is perpendicular to the sliding direction of joint C_T ($C_{\text{Translation}}$) as shown in Fig. 4.23

$$\mathbf{F}_{03} = F_{03y}\,\mathbf{J}.$$

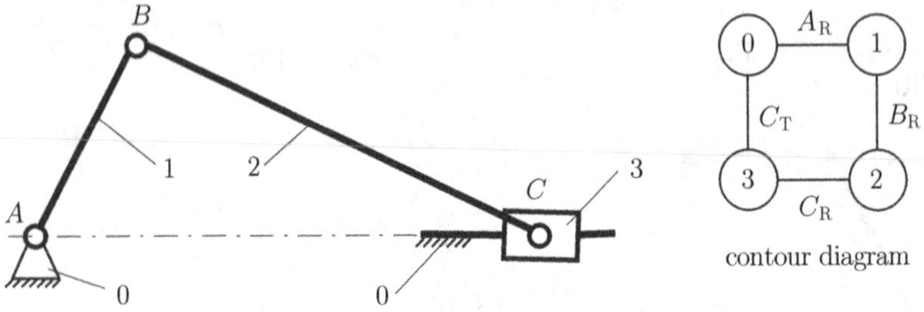

Fig. 4.22 Contour diagram representing the mechanism

The application point of the unknown reaction force $\mathbf{F}_{03}$ is computed from a moment equation about C_R (C_{Rotation}) for link 3 (path I), as shown in Fig. 4.23

$$\sum \mathbf{M}_C^{(3)} = \mathbf{r}_{CP} \times \mathbf{F}_{03} = (\mathbf{r}_P - \mathbf{r}_C) \times \mathbf{F}_{03} = \mathbf{0},$$

or

$$x F_{05y} = 0 \ \Rightarrow x = 0.$$

The application point of the reaction force $\mathbf{F}_{03}$ is at C ($P \equiv C$).

The magnitude of the reaction force F_{03y} is obtained from a moment equation about B_R for the links 3 and 2 (path I)

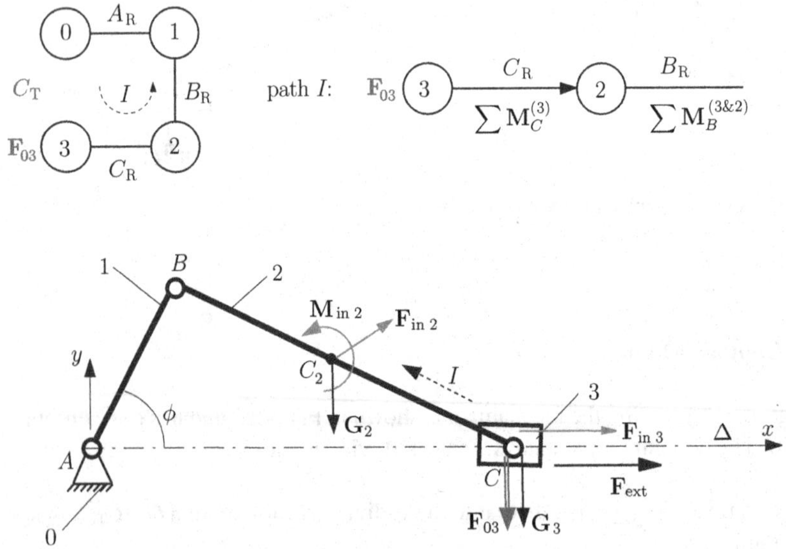

Fig. 4.23 Diagram for calculating the reaction force $\mathbf{F}_{03}$

$$\sum \mathbf{M}_B^{(3\&2)} = \mathbf{r}_{BC} \times (\mathbf{F}_{03} + \mathbf{F}_{\text{in}3} + \mathbf{G}_3 + \mathbf{F}_{\text{ext}})$$
$$+\mathbf{r}_{BC_2} \times (\mathbf{F}_{\text{in}2} + \mathbf{G}_2) + \mathbf{M}_{\text{in}2} = \mathbf{0},$$

or

$$\begin{vmatrix} \mathbf{\imath} & \mathbf{\jmath} & \mathbf{k} \\ x_C - x_B & y_C - y_B & 0 \\ F_{\text{in}3x} + F_{\text{ext}} & F_{03y} + F_{\text{in}3y} - m_3 g & 0 \end{vmatrix} + \begin{vmatrix} \mathbf{\imath} & \mathbf{\jmath} & \mathbf{k} \\ x_{C_2} - x_B & y_{C_2} - y_B & 0 \\ F_{\text{in}2x} & F_{\text{in}2y} - m_2 g & 0 \end{vmatrix}$$
$$+M_{\text{in}2}\,\mathbf{k} = \mathbf{0},$$

or numerically

$$\frac{3}{2} - \frac{5}{\sqrt{2}} + 45\sqrt{2} + \frac{F_{03y}}{\sqrt{2}} = 0.$$

The reaction F_{03y} is

$$F_{03y} = -85 - \frac{3\sqrt{2}}{2}\ \text{N}.$$

The MATLAB statements for finding $\mathbf{F}_{03}$ are:

```
% Joint C_T
F03=[ 0 sym('F03y','real') 0 ];
eqM32B=cross(rC-rB,F03+G3+Fin3+Fe)+...
       cross(rC2-rB,Fin2+G2)+Min2;
eqM32Bz=eqM32B(3);
fprintf('%s = 0 (1)\n', char(vpa(eqM32Bz,6)))
fprintf('Eq(1) => F03y \n')
solF03=solve(eqM32Bz);
F03ys=eval(solF03);
F03s=[ 0, F03ys, 0 ];
fprintf('F03 = [ %g, %g, %g ] (N)\n', F03s)
```

Reaction $\mathbf{F}_{23}$

The pin joint at C_R, between 2 and 3, is replaced with the reaction force (Fig. 4.24)

$$\mathbf{F}_{23} = -\mathbf{F}_{32} = F_{23x}\mathbf{\imath} + F_{23y}\mathbf{\jmath}.$$

For the path I, an equation for the forces projected onto the sliding direction of the joint C_T is written for link 3

$$\sum \mathbf{F}^{(3)} \cdot \mathbf{\imath} = (\mathbf{F}_{23} + \mathbf{F}_{\text{in}3} + \mathbf{G}_3 + \mathbf{F}_{\text{ext}}) \cdot \mathbf{\imath}$$
$$= F_{23x} + F_{\text{in}3x} + F_{\text{ext}} = F_{23x} + 100 + \sqrt{2} = 0. \quad (4.49)$$

For the path II, shown Fig. 4.24, a moment equation about B_R is written for link 2

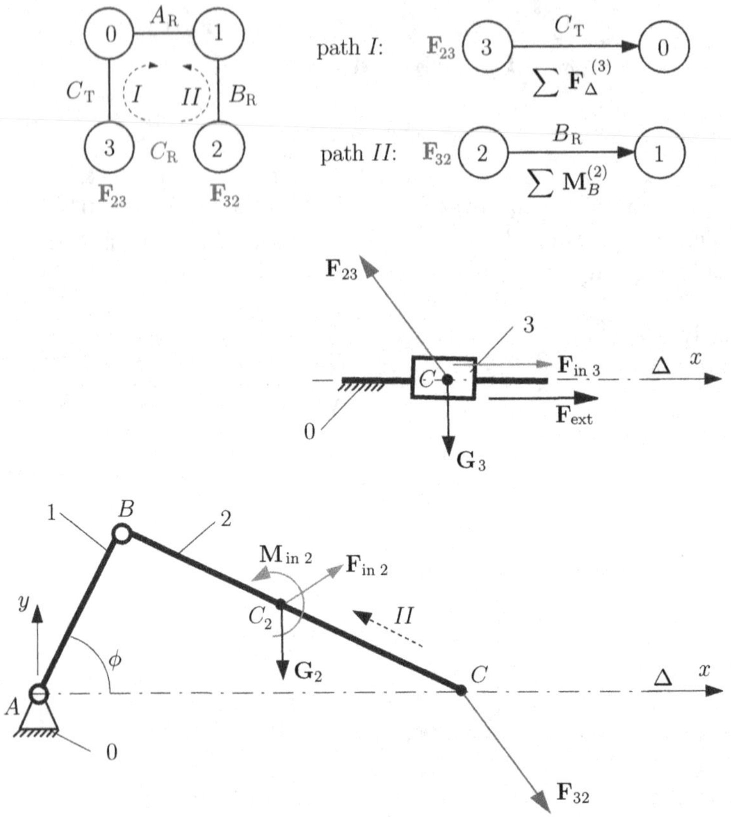

Fig. 4.24 Diagram for calculating the reaction force $\mathbf{F}_{23}$

$$\sum \mathbf{M}_B^{(2)} = \mathbf{r}_{BC} \times (-\mathbf{F}_{23}) + \mathbf{r}_{BC_2} \times (\mathbf{F}_{\text{in}2} + \mathbf{G}_2) + \mathbf{M}_{\text{in}2} = \mathbf{0},$$

$$\begin{vmatrix} \mathbf{\imath} & \mathbf{\jmath} & \mathbf{k} \\ x_C - x_B & y_C - y_B & 0 \\ -F_{23x} & -F_{23y} & 0 \end{vmatrix} + \begin{vmatrix} \mathbf{\imath} & \mathbf{\jmath} & \mathbf{k} \\ x_{C_2} - x_B & y_{C_2} - y_B & 0 \\ F_{\text{in}2x} & F_{\text{in}2y} - m_2 g & 0 \end{vmatrix} + M_{\text{in}2}\,\mathbf{k} = \mathbf{0},$$

or numerically

$$\frac{1}{2} - \frac{5\sqrt{2}}{2} - \frac{F_{23x}\sqrt{2}}{2} - \frac{F_{23y}\sqrt{2}}{2} = 0. \tag{4.50}$$

The joint force $\mathbf{F}_{23}$ is obtained from the system of Eqs. 4.49 and 4.50

$$F_{23x} = -100 - \sqrt{2}\ \text{N} \quad \text{and} \quad F_{23y} = 95 + \frac{3\sqrt{2}}{2}\ \text{N}.$$

The MATLAB statements for finding $\mathbf{F}_{23}$ are:

```
% Joint C_R
F23 = [ sym('F23x','real') sym('F23y','real') 0 ];
eqF3 = F23+Fe+G3+Fin3;
eqF3x = eqF3(1);
eqM2B = cross(rC-rB,-F23)+cross(rC2-rB,Fin2+G2)+Min2;
eqM2Bz = eqM2B(3);
fprintf('%s = 0 (2)\n', char(vpa(eqF3x,6)))
fprintf('%s = 0 (3)\n', char(vpa(eqM2Bz,6)))
fprintf('Eqs(2)-(3) => F23x, F23y \n')
solF23=solve(eqF3x,eqM2Bz);
F23xs=eval(solF23.F23x);
F23ys=eval(solF23.F23y);
F23s = [ F23xs, F23ys, 0 ];
fprintf('F23 = [ %g, %g, %g ] (N)\n', F23s)
```

Reaction $\mathbf{F}_{12}$
The pin joint at B_R, between 1 and 2, is replaced with the reaction force (Fig. 4.25)

$$\mathbf{F}_{12} = -\mathbf{F}_{21} = F_{12x}\mathbf{\imath} + F_{12y}\mathbf{\jmath}.$$

For the path I, shown in Fig. 4.25, a moment equation about C_R is written for link 2

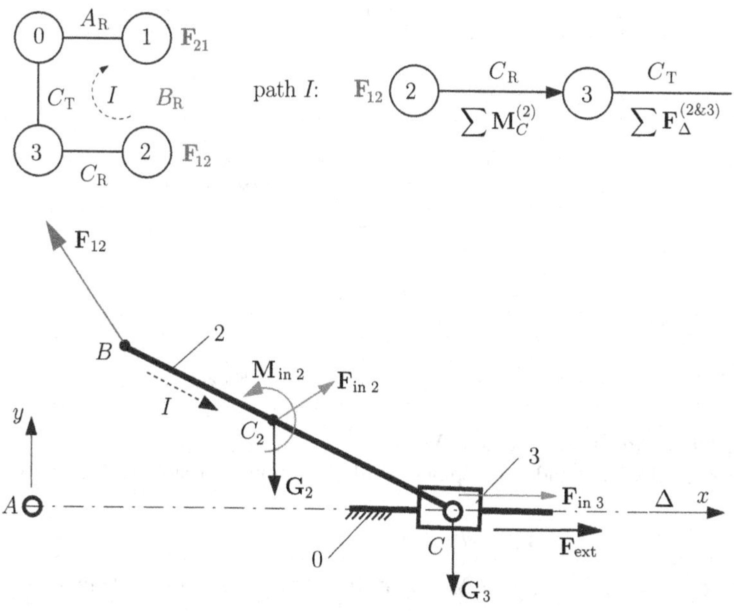

Fig. 4.25 Diagram for calculating the reaction force $\mathbf{F}_{12}$

$$\sum \mathbf{M}_C^{(2)} = \mathbf{r}_{CB} \times \mathbf{F}_{12} + \mathbf{r}_{CC_2} \times (\mathbf{F}_{\text{in}2} + \mathbf{G}_2) + \mathbf{M}_{\text{in}2} = \mathbf{0},$$

$$\begin{vmatrix} \mathbf{\imath} & \mathbf{\jmath} & \mathbf{k} \\ x_B - x_C & y_B - y_C & 0 \\ F_{12x} & F_{12y} & 0 \end{vmatrix} + \begin{vmatrix} \mathbf{\imath} & \mathbf{\jmath} & \mathbf{k} \\ x_{C_2} - x_C & y_{C_2} - y_C & 0 \\ F_{\text{in}2x} & F_{\text{in}2y} - m_2 g & 0 \end{vmatrix} + M_{\text{in}2}\,\mathbf{k} = \mathbf{0},$$

or numerically

$$-\frac{1}{2} + \frac{5\sqrt{2}}{2} - \frac{F_{12x}\sqrt{2}}{2} - \frac{F_{12y}\sqrt{2}}{2} = 0. \tag{4.51}$$

Continuing on path I, an equation for the forces projected onto the sliding direction of the joint C_T is written for links 2 and 3

$$\sum \mathbf{F}^{(2\&3)} \cdot \mathbf{\imath} = (\mathbf{F}_{12} + \mathbf{F}_{\text{in}2} + \mathbf{G}_2 + \mathbf{F}_{\text{in}3} + \mathbf{G}_3 + \mathbf{F}_{\text{ext}}) \cdot \mathbf{\imath}$$
$$= F_{12x} + F_{\text{in}2x} + F_{\text{in}3x} + F_{\text{ext}} = F_{23x} + 100 + \sqrt{2} + \frac{3\sqrt{2}}{2} = 0. \tag{4.52}$$

The joint force $\mathbf{F}_{12}$ is obtained from the system of Eqs. 4.52 and 4.51

$$F_{12x} = -\frac{1}{4}(400 + 7\sqrt{2}) \text{ N and } F_{12y} = \frac{5}{4}(84 + \sqrt{2}) \text{ N}.$$

The MATLAB statements for finding $\mathbf{F}_{12}$ are:

```
% Joint B_R
F12 = [ sym('F12x','real') sym('F12y','real') 0 ];
eqM2C = cross(rB-rC,F12)+cross(rC2-rC,Fin2+G2)+Min2;
eqM2Cz = eqM2C(3);
eqF23 = (F12+Fin2+G2+G3+Fin3+Fe);
eqF23x = eqF23(1);
fprintf('%s = 0 (4)\n', char(vpa(eqM2Cz,6)))
fprintf('%s = 0 (5)\n', char(vpa(eqF23x,6)))
fprintf('Eqs(4)-(5) => F12x, F12y \n')
solF12 = solve(eqM2Cz,eqF23x);
F12xs = eval(solF12.F12x);
F12ys = eval(solF12.F12y);
F12s = [ F12xs, F12ys, 0 ];
fprintf('F12 = [ %g, %g, %g ] (N)\n', F12s)
```

Reaction $\mathbf{F}_{01}$ *and Equilibrium Moment* $\mathbf{M}$
The pin joint A_R, between 0 and 1, is replaced with the unknown reaction (Fig. 4.26)

$$\mathbf{F}_{01} = F_{01x}\mathbf{\imath} + F_{01y}\mathbf{\jmath}.$$

The unknown equilibrium moment is $\mathbf{M} = M\mathbf{k}$. If the path I is followed (Fig. 4.26) for the pin joint B_R, a moment equation is written for link 1

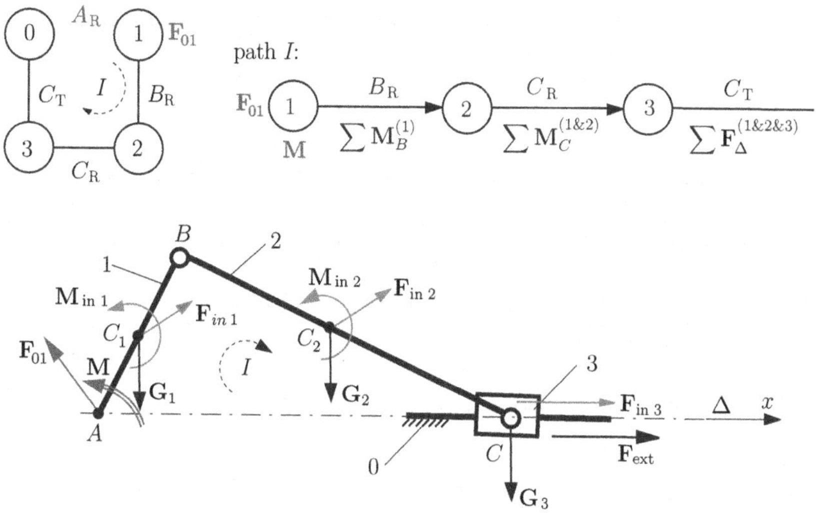

Fig. 4.26 Diagram for calculating the reaction force $\mathbf{F}_{01}$

$$\sum \mathbf{M}_B^{(1)} = \mathbf{r}_{BA} \times \mathbf{F}_{01} + \mathbf{r}_{BC_1} \times (\mathbf{F}_{\text{in}\,1} + \mathbf{G}_1) + \mathbf{M}_{\text{in}\,1} + \mathbf{M} = \mathbf{0},$$

$$\begin{vmatrix} \mathbf{i} & \mathbf{j} & \mathbf{k} \\ -x_B & -y_B & 0 \\ F_{01x} & F_{01y} & 0 \end{vmatrix} + \begin{vmatrix} \mathbf{i} & \mathbf{j} & \mathbf{k} \\ x_{C_1} - x_B & y_{C_1} - y_B & 0 \\ F_{\text{in}\,1x} & F_{\text{in}\,1y} - m_1 g & 0 \end{vmatrix} + M\mathbf{k} = \mathbf{0},$$

or numerically

$$\frac{5\sqrt{2}}{2} + \frac{F_{01x}\sqrt{2}}{2} + \frac{F_{01y}\sqrt{2}}{2} + M = 0. \tag{4.53}$$

Continuing on path *I* the next joint encountered is the pin joint C_R, and a moment equation is written for links 1 and 2

$$\sum \mathbf{M}_C^{(1\&2)} = \mathbf{r}_{CA} \times \mathbf{F}_{01} + \mathbf{r}_{CC_1} \times (\mathbf{F}_{\text{in}\,1} + \mathbf{G}_1) + \mathbf{M}_{\text{in}\,1} + \mathbf{M}$$
$$+ \mathbf{r}_{CC_2} \times (\mathbf{F}_{\text{in}\,2} + \mathbf{G}_2) + \mathbf{M}_{\text{in}\,2} = \mathbf{0},$$

$$\begin{vmatrix} \mathbf{i} & \mathbf{j} & \mathbf{k} \\ -x_C & -y_C & 0 \\ F_{01x} & F_{01y} & 0 \end{vmatrix} + \begin{vmatrix} \mathbf{i} & \mathbf{j} & \mathbf{k} \\ x_{C_1} - x_C & y_{C_1} - y_C & 0 \\ F_{\text{in}\,1x} & F_{\text{in}\,1y} - m_1 g & 0 \end{vmatrix} + M\mathbf{k}$$
$$+ \begin{vmatrix} \mathbf{i} & \mathbf{j} & \mathbf{k} \\ x_{C_2} - x_C & y_{C_2} - y_C & 0 \\ F_{\text{in}\,2x} & F_{\text{in}\,2y} - m_2 g & 0 \end{vmatrix} + M_{\text{in}\,2}\mathbf{k} = \mathbf{0},$$

or numerically

$$-\sqrt{2}F_{01y} + M - 1 + 10\sqrt{2} = 0. \tag{4.54}$$

Continuing on path I the next joint encountered is the slider joint C_T, and a force equation is written for links 1, 2, and 3

$$\sum \mathbf{F}^{(1\&2\&3)} \cdot \mathbf{\imath} = (\mathbf{F}_{01} + \mathbf{F}_{\text{in}\,1} + \mathbf{G}_1 + \mathbf{F}_{\text{in}\,2} + \mathbf{G}_2 + \mathbf{F}_{\text{in}\,3} + \mathbf{G}_3 + \mathbf{F}_{\text{ext}}) \cdot \mathbf{\imath}$$

$$= F_{01x} + F_{\text{in}\,1x} + F_{\text{in}\,2x} + F_{\text{in}\,3x} + F_{\text{ext}} = F_{12x} + 100 + \sqrt{2} + \frac{3\sqrt{2}}{2} = 0. \quad (4.55)$$

From Eqs. 4.53–4.55 the components F_{01x}, F_{01y} and M are computed

$$F_{01x} = -2(50 + \sqrt{2})\text{ N} \quad \text{and} \quad F_{01y} = 115 + \sqrt{2}\text{ N},$$
$$M = 3 + 105\sqrt{2}\text{ N m}.$$

The MATLAB statements for finding $\mathbf{F}_{01}$ and $\mathbf{M}$ are:

```
% Joint A_R
F01 = [ sym('F01x','real') sym('F01y','real') 0 ];
Mm = [ 0 0 sym('Mmz','real') ];
eqM1B = cross(-rB,F01)+cross(rC1-rB,Fin1+G1)+Min1+Mm;
eqM1Bz = eqM1B(3);
eqM12C=cross(-rC,F01)+cross(rC1-rC,Fin1+G1)+Min1+...
Mm+cross(rC2-rC,Fin2+G2)+Min2;
eqM12Cz = eqM12C(3);
eqF123 = (F01+Fin1+G1+Fin2+G2+Fin3+G3+Fe);
eqF123x = eqF123(1);
fprintf('%s = 0 (6\n', char(vpa(eqM1Bz,6)))
fprintf('%s = 0 (7)\n', char(vpa(eqM12Cz,6)))
fprintf('%s = 0 (8)\n', char(vpa(eqF123x,6)))
fprintf('Eqs(6)-(8) => F01x, F01y, Mmz \n')
solF01 = solve(eqM1Bz,eqM12Cz,eqF123x);
F01xs = eval(solF01.F01x);
F01ys = eval(solF01.F01y);
Mmzs = eval(solF01.Mmz);
F01s = [ F01xs, F01ys, 0 ];
Mms = [ 0, 0, Mmzs ];
fprintf('F01 = [ %g, %g, %g ] (N)\n', F01s)
fprintf('Mm = [ %g, %g, %g ] (N m)\n', Mms)
```

The MATLAB program using the contour method and the results are given in Appendix C.4.

4.7 R-RTR-RTR Mechanism

Exercise

The planar R-RTR-RTR mechanism considered is shown in Fig. 4.27. The following numerical data are given: $AB = 0.15$ m, $AC = 0.10$ m, $CD = 0.15$ m, $DF = 0.40$ m, and $AG = 0.30$ m. The height of the links 1, 3, and 5 is $h = 0.010$ m. The width of the links 2 and 4 is $w_{\text{slider}} = 0.050$ m, and the height is $h_{\text{slider}} = 0.020$ m. All five moving links are rectangular prisms with the depth $d = 0.001$ m. The angular velocity of the driver link 1 is $n = 50$ rpm. The density of the material is $\rho_{Steel} = \rho = 8000$ kg/m^3. The gravitational acceleration is $g = 9.807$ m/s^2. The center of mass locations of the links $i = 1, 2, ..., 5$ are designated by $C_i(x_{C_i}, y_{C_i}, 0)$.

The external moment applied on link 5 is opposed to the motion of the link: $\mathbf{M}_{5\text{ext}} = -\text{Sign}(\omega_5)\,|M_{\text{ext}}|\,\mathbf{k}$ where $|M_{\text{ext}}| = 100$ N m and ω_5 is the angular velocity of link 5.

Find the motor moment $\mathbf{M}_m$ required for the dynamic equilibrium and the joint reaction forces when the driver link 1 makes an angle $\phi = \dfrac{\pi}{6}$ rad with the horizontal axis.

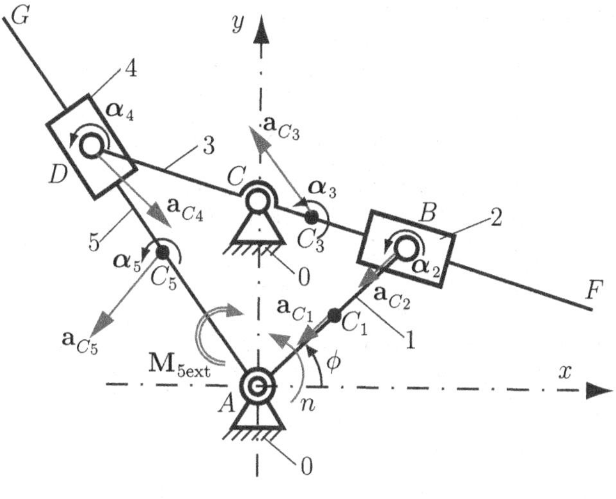

Fig. 4.27 R-RTR-RTR mechanism

Solution

The position vectors (in meters) of the joints were calculated in Sect. 2.4:

- position of joint A: $\mathbf{r}_A = \mathbf{0}$;
- position of joint B: $\mathbf{r}_B = x_B\,\mathbf{\imath} + y_B\,\mathbf{\jmath} = 0.129904\,\mathbf{\imath} + 0.075\,\mathbf{\jmath}$;
- position of joint C: $\mathbf{r}_C = y_C\,\mathbf{\jmath} = 0.1\,\mathbf{\jmath}$;
- position of joint D: $\mathbf{r}_D = x_D\,\mathbf{\imath} + y_D\,\mathbf{\jmath} = -0.147297\,\mathbf{\imath} + 0.128347\,\mathbf{\jmath}$;

- position of F: $\mathbf{r}_F = x_F\,\imath + y_F\,\jmath = 0.245495\,\imath + 0.0527544\,\jmath$; and
- position of G: $\mathbf{r}_G = x_G\,\imath + y_G\,\jmath = -0.226182\,\imath + 0.197083\,\jmath$.

The angles of the links with the horizontal are $\phi_2 = \phi_3 = -10.8934^\circ$ and $\phi_4 = \phi_5 = 138.933^\circ$. The position vector of the center of mass of link 1 is

$$\mathbf{r}_{C_1} = x_{C_1}\,\imath + y_{C_1}\,\jmath = \frac{x_B}{2}\,\imath + \frac{y_B}{2}\,\jmath = 0.0649519\,\imath + 0.0375\,\jmath \ \text{m}.$$

The position vector of the center of mass of slider 2 is

$$\mathbf{r}_{C_2} = x_{C_2}\,\imath + y_{C_2}\,\jmath = \mathbf{r}_B.$$

The position vector of the center of mass of link 3 is

$$\mathbf{r}_{C_3} = x_{C_3}\,\imath + y_{C_3}\,\jmath = \frac{x_D + x_F}{2}\,\imath + \frac{y_D + y_F}{2}\,\jmath = 0.049099\,\imath + 0.0905509\,\jmath \ \text{m}.$$

The position vector of the center of mass of slider 4 is

$$\mathbf{r}_{C_4} = x_{C_4}\,\imath + y_{C_4}\,\jmath = \mathbf{r}_D.$$

The position vector of the center of mass of link 5 is

$$\mathbf{r}_{C_5} = x_{C_5}\,\imath + y_{C_5}\,\jmath = \frac{x_G}{2}\,\imath + \frac{y_G}{2}\,\jmath = -0.113091\,\imath + 0.0985417\,\jmath \ \text{m}.$$

The velocity and acceleration analysis was carried out in Sect. 3.8:

- acceleration of joint B: $\mathbf{a}_{B_1} = \mathbf{a}_{B_2} = -3.56139\,\imath - 2.05617\,\jmath \ \text{m/s}^2$;
- acceleration of joint D: $\mathbf{a}_{D_3} = \mathbf{a}_{D_4} = 2.5548\,\imath - 2.71212\,\jmath \ \text{m/s}^2$;
- acceleration of joint F: $\mathbf{a}_F = -4.258\,\imath + 4.52021\,\jmath \ \text{m/s}^2$;
- acceleration of joint G: $\mathbf{a}_G = -0.396144\,\imath - 4.50689\,\jmath \ \text{m/s}^2$;
- angular velocity of link 5: $\boldsymbol{\omega}_5 = 2.97887\,\mathbf{k} \ \text{rad/s}$;
- angular acceleration of link 1: $\boldsymbol{\alpha}_1 = 0\,\mathbf{k} \ \text{rad/s}^2$;
- angular acceleration of links 2 and 3: $\boldsymbol{\alpha}_2 = \boldsymbol{\alpha}_3 = 14.5363\,\mathbf{k} \ \text{rad/s}^2$; and
- angular acceleration of links 4 and 5: $\boldsymbol{\alpha}_4 = \boldsymbol{\alpha}_5 = 12.1939\,\mathbf{k} \ \text{rad/s}^2$.

The acceleration vector of the center of mass of link 1 is

$$\mathbf{a}_{C_1} = \frac{\mathbf{a}_{B_1}}{2} = -1.78069\,\imath - 1.02808\,\jmath \ \text{m/s}^2.$$

The acceleration vector of the center of mass of slider 2 is

$$\mathbf{a}_{C_2} = \mathbf{a}_{B_2} = -3.56139\,\imath - 2.05617\,\jmath \ \text{m/s}^2.$$

The acceleration vector of the center of mass of link 3 is

$$\mathbf{a}_{C_3} = \frac{\mathbf{a}_{D_3} + \mathbf{a}_F}{2} = -0.8516\,\imath + 0.904041\,\jmath \ \text{m/s}^2.$$

The acceleration vector of the center of mass of slider 4 is

$$\mathbf{a}_{C_4} = \mathbf{a}_{D_4} = 2.5548\,\mathbf{\imath} - 2.71212\,\mathbf{\jmath} \ \text{m/s}^2.$$

The acceleration vector of the center of mass of link 5 is

$$\mathbf{a}_{C_5} = \frac{\mathbf{a}_G}{2} = -0.198072\,\mathbf{\imath} - 2.25344\,\mathbf{\jmath} \ \text{m/s}^2.$$

The MATLAB program for positions, velocities, and accelerations is:

```
AB=0.15; AC=0.10; CD=0.15; DF=0.40; AG=0.30;
phi = pi/6;
xA = 0; yA = 0; rA = [xA yA 0];
xC = 0; yC = AC; rC = [xC yC 0];
xB = AB*cos(phi); yB = AB*sin(phi); rB = [xB yB 0];
eqnD1='(xDsol-xC)^2+(yDsol-yC)^2=CD^2';
eqnD2='(yB-yC)/(xB-xC)=(yDsol-yC)/(xDsol-xC)';
solD = solve(eqnD1, eqnD2, 'xDsol, yDsol');
xDpositions = eval(solD.xDsol);
yDpositions = eval(solD.yDsol);
xD1 = xDpositions(1); xD2 = xDpositions(2);
yD1 = yDpositions(1); yD2 = yDpositions(2);
if (phi>=0&&phi<=pi/2)||(phi >= 3*pi/2&&phi<=2*pi)
if xD1 <= xC xD=xD1;yD=yD1; else xD=xD2;yD=yD2;end
else
if xD1 >= xC xD=xD1;yD=yD1; else xD=xD2;yD=yD2;end
end
rD = [xD yD 0];
phi2 = atan((yB-yC)/(xB-xC)); phi3 = phi2;
phi4 = atan((yD-yA)/(xD-xA))+pi; phi5 = phi4;
xF = xD+DF*cos(phi3); yF = yD+DF*sin(phi3);
rF = [xF yF 0];
xG = xA+AG*cos(phi5); yG = yA+AG*sin(phi5);
rG = [xG yG 0];
xC1 = xB/2; yC1 = yB/2; rC1 = [xC1 yC1 0];
rC2 = rB;
xC3 = (xD+xF)/2; yC3 = (yD+yF)/2; rC3 = [xC3 yC3 0];
rC4 = rD;
xC5 = (xA+xG)/2; yC5 = (yA+yG)/2; rC5 = [xC5 yC5 0];
n = 50.;
omega1 = [ 0 0 pi*n/30 ]; alpha1 = [0 0 0 ];
vA = [0 0 0 ]; aA = [0 0 0 ];
vB1 = vA + cross(omega1,rB); vB2 = vB1;
aB1 = aA + cross(alpha1,rB) - dot(omega1,omega1)*rB;
aB2 = aB1;
omega3z = sym('omega3z','real');
alpha3z = sym('alpha3z','real');
```

```
vB32 = sym('vB32','real');
aB32 = sym('aB32','real');
omega3 = [ 0 0 omega3z ];
vC = [0 0 0 ];
vB3 = vC + cross(omega3,rB-rC);
vB3B2 = vB32*[ cos(phi2) sin(phi2) 0];
eqvB = vB3 - vB2 - vB3B2;
eqvBx = eqvB(1); eqvBy = eqvB(2);
solvB = solve(eqvBx,eqvBy);
omega3zs = eval(solvB.omega3z);
vB32s = eval(solvB.vB32);
Omega3 = [0 0 omega3zs]; Omega2 = Omega3;
v32 = vB32s*[cos(phi2) sin(phi2) 0];
vD3 = vC + cross(Omega3,rD-rC); vD4 = vD3;
aB3B2cor = 2*cross(Omega3,v32);
alpha3 = [ 0 0 alpha3z ];
aC = [0 0 0 ];
aB3 = aC + cross(alpha3,rB-rC) - ...
      dot(Omega3,Omega3)*(rB-rC);
aB3B2 = aB32*[ cos(phi2) sin(phi2) 0];
eqaB = aB3 - aB2 - aB3B2 - aB3B2cor;
eqaBx = eqaB(1); eqaBy = eqaB(2);
solaB = solve(eqaBx,eqaBy);
alpha3zs = eval(solaB.alpha3z);
aB32s = eval(solaB.aB32);
Alpha3 = [0 0 alpha3zs]; Alpha2 = Alpha3;
aD3 = aC + cross(Alpha3,rD-rC) - ...
      dot(Omega3,Omega3)*(rD-rC);
aD4=aD3;
omega5z = sym('omega5z','real');
alpha5z = sym('alpha5z','real');
vD54 = sym('vD54','real');
aD54 = sym('aD54','real');
omega5 = [ 0 0 omega5z ];
vD5 = vA + cross(omega5,rD-rA);
vD5D4 = vD54*[ cos(phi5) sin(phi5) 0];
eqvD = vD5 - vD4 - vD5D4;
eqvDx = eqvD(1); eqvDy = eqvD(2);
solvD = solve(eqvDx,eqvDy);
omega5zs = eval(solvD.omega5z);
vD54s = eval(solvD.vD54);
Omega5 = [0 0 omega5zs];
v54 = vD54s*[cos(phi5) sin(phi5) 0];
Omega4 = Omega5;
aD5D4cor = 2*cross(Omega5,v54);
```

```
alpha5 = [ 0 0 alpha5z ];
aD5 = aA + cross(alpha5,rD-rE) - ...
      dot(Omega5,Omega5)*(rD-rA);
aD5D4 = aD54*[ cos(phi5) sin(phi5) 0];
eqaD = aD5 - aD4 - aD5D4 - aD5D4cor;
eqaDx = eqaD(1);
eqaDy = eqaD(2);
solaD = solve(eqaDx,eqaDy);
alpha5zs = eval(solaD.alpha5z);
aD54s = eval(solaD.aD54);
Alpha5 = [0 0 alpha5zs]; Alpha4 = Alpha5;
aF = aC + cross(Alpha3,rF-rC) - ...
     dot(Omega3,Omega3)*(rF-rC);
aG = aA + cross(Alpha5,rG-rA) - ...
     dot(Omega5,Omega5)*(rG-rA);
aC1 = aB1/2;
aC2 = aB2;
aC3 = (aD3+aF)/2;
aC4 = aD3;
aC5 = (aA+aG)/2;
```

The external moment applied on link 5 is opposed to the motion of the link

$$\mathbf{M}_{5\text{ext}} = -\text{Sign}(\omega_5)\,|M_{\text{ext}}|\,\mathbf{k} = -\text{Sign}(2.97887)\,(100)\,\mathbf{k} = -100\,\mathbf{k} \text{ Nm}.$$

4.7.1 Inertia Forces and Moments

Link 1
The mass of the link is

$$m_1 = \rho\, AB\, h\, d = 8000(0.15)(0.01)(0.001) = 0.012 \text{ kg}.$$

The inertia force of driver 1 at C_1 is

$$\mathbf{F}_{\text{in}\,1} = -m_1\,\mathbf{a}_{C_1} = -0.012(-1.78069\,\mathbf{\imath} - 1.02808\,\mathbf{\jmath}) = 0.0213683\,\mathbf{\imath} + 0.012337\,\mathbf{\jmath} \text{ N}.$$

The gravitational force on link 1 at C_1 is

$$\mathbf{G}_1 = -m_1\, g\,\mathbf{\jmath} = -0.012(9.807)\,\mathbf{\jmath} = -0.117684\,\mathbf{\jmath} \text{ N}.$$

The mass moment of inertia of link 1 with respect to C_1 is

$$I_{C_1} = m_1\,(AB^2 + h^2)/12 = 0.012(0.15^2 + 0.01^2)/12 = 2.26 \times 10^{-5} \text{ kg m}^2.$$

The moment of inertia of driver 1 is

$$\mathbf{M}_{\text{in}\,1} = -I_{C_1}\,\alpha_1 = \mathbf{0}.$$

To calculate the inertia force and the moment the following MATLAB commands are used:

```
m1 = rho*AB*h*d;
Fin1 = -m1*aC1;
G1 = [0,-m1*g,0];
IC1 = m1*(AB^2+h^2)/12;
Min1 = -IC1*alpha1;
```

Link 2
The mass of the slider 2 is

$$m_2 = \rho\, h_{\text{slider}}\, w_{\text{slider}}\, d = 8000(0.02)(0.05)(0.001) = 0.008 \text{ kg}.$$

The inertia force of slider 2 at C_2 is

$$\mathbf{F}_{\text{in}\,2} = -m_2\,\mathbf{a}_{C_2} = -0.008(-3.56139\,\mathbf{\imath} - 2.05617\,\mathbf{\jmath}) = 0.0284911\,\mathbf{\imath} + 0.0164493\,\mathbf{\jmath} \text{ N}.$$

The gravitational force of slider 2 at C_2 is

$$\mathbf{G}_2 = -m_2\,g\,\mathbf{\jmath} = -0.008(9.807)\,\mathbf{\jmath} = -0.078456\,\mathbf{\jmath} \text{ N}.$$

The mass moment of inertia of slider 2 with respect to C_2 is

$$I_{C_2} = m_2\,(h_{\text{slider}}^2 + w_{\text{slider}}^2)/12 = 0.008(0.02^2 + 0.05^2)/12 = 1.93333 \times 10^{-6} \text{ kg m}^2.$$

The moment of inertia of slider 2 is

$$\mathbf{M}_{\text{in}\,2} = -I_{C_2}\alpha_2 = -1.93333 \times 10^{-6}\,(14.5363)\,\mathbf{k} = -2.81035 \times 10^{-5}\,\mathbf{k} \text{ Nm}.$$

The MATLAB commands to calculate the inertia force and the moment are:

```
m2 = rho*hSlider*wSlider*d;
Fin2 = -m2*aC2;
G2 = [0,-m2*g,0];
IC2 = m2*(hSlider^2+wSlider^2)/12;
Min2 = -IC2*Alpha2;
```

Link 3
The mass of the link is

$$m_3 = \rho\; DF\; h\; d = 8000(0.4)(0.01)(0.001) = 0.032 \text{ kg}.$$

The inertia force of link 3 is

$$\mathbf{F}_{\text{in}3} = -m_3\,\mathbf{a}_{C_3} = -0.032(-0.8516\,\mathbf{ı}+0.904041\,\mathbf{ȷ}) = 0.0272512\,\mathbf{ı} - 0.0289293\,\mathbf{ȷ}\ \text{N}.$$

The gravitational force of link 3 is

$$\mathbf{G}_3 = -m_3\,g\,\mathbf{ȷ} = -0.032(9.807)\,\mathbf{ȷ} = -0.313824\,\mathbf{ȷ}\ \text{N}.$$

The mass moment of inertia with respect to C_3 is

$$I_{C_3} = m_3\,(DF^2+h^2)/12 = 0.032(0.4^2+0.01^2)/12 = 0.000426933\ \text{kg m}^2.$$

The inertia moment on link 3 is

$$\mathbf{M}_{\text{in}3} = -I_{C_3}\alpha_3 = -0.000426933(14.5363)\,\mathbf{k} = -0.00620602\,\mathbf{k}\ \text{N m}.$$

Link 4
The mass of the link is

$$m_4 = \rho\,h_{\text{slider}}\,w_{\text{slider}}\,d = 8000(0.02)(0.05)(0.001) = 0.008\ \text{kg}.$$

The inertia force is

$$\mathbf{F}_{\text{in}4} = -m_4\,\mathbf{a}_{C_4} = -0.008(2.5548\,\mathbf{ı} - 2.71212\,\mathbf{ȷ}) = -0.0204384\,\mathbf{ı} + 0.021697\,\mathbf{ȷ}\ \text{N}.$$

The gravitational force is

$$\mathbf{G}_4 = -m_4\,g\,\mathbf{ȷ} = -0.008(9.807)\,\mathbf{ȷ} = -0.078456\,\mathbf{ȷ}\ \text{N}.$$

The mass moment of inertia with respect to C_4 is

$$I_{C_4} = m_4(h^2_{\text{slider}} + w^2_{\text{slider}})/12 = 0.008(0.02^2+0.05^2)/12 = 1.93333\times10^{-6}\ \ \text{kg m}^2.$$

The moment of inertia is

$$\mathbf{M}_{\text{in}4} = -I_{C_4}\,\alpha_4 = -1.93333\times10^{-6}(12.1939)\,\mathbf{k} = -2.35748\times10^{-5}\,\mathbf{k}\ \text{N m}.$$

Link 5
The mass of the link is

$$m_5 = \rho\,AG\,h\,d = 8000(0.3)(0.01)(0.001) = 0.024\ \text{kg}.$$

The inertia force is

$$\mathbf{F}_{\text{in}5} = -m_5\,\mathbf{a}_{C_5} = -0.024(-0.198072\,\mathbf{ı} - 2.25344\,\mathbf{ȷ}) = 0.004754\,\mathbf{ı} + 0.054083\,\mathbf{ȷ}\ \text{N}.$$

The gravitational force is

$$\mathbf{G}_5 = -m_5\,g\,\mathbf{ȷ} = -0.024(9.807)\,\mathbf{ȷ} = -0.235368,\mathbf{ȷ}\ \text{N}.$$

The mass moment of inertia with respect to C_5 is

$$I_{C_5} = m_5\,(AG^2 + h^2)/12 = 0.024(0.3^2 + 0.01^2)/12 = 0.0001802\ \text{kg m}^2.$$

The moment of inertia is

$$\mathbf{M}_{\text{in}\,5} = -I_{C_5}\,\alpha_5 = -0.0001802\,(12.1939)\,\mathbf{k} = -0.00219734\mathbf{k}\ \text{N m}.$$

The MATLAB commands to calculate the inertia force and the moment for links 3, 4, and 5 are:

```
m3 = rho*DF*h*d;

Fin3 = -m3*aC3;
G3 = [0,-m3*g,0];
IC3 = m3*(DF^2+h^2)/12;
Min3 = -IC3*Alpha3;

m4 = rho*hSlider*wSlider*d;

Fin4 = -m4*aC4;
G4 = [0,-m4*g,0];
IC4 = m4*(hSlider^2+wSlider^2)/12;
Min4 = -IC4*Alpha4;

m5 = rho*AG*h*d;

Fin5 = -m5*aC5;
G5 = [0,-m5*g,0];
IC5 = m5*(AG^2+h^2)/12;
Min5 = -IC5*Alpha5;
```

4.7.2 Joint Forces and Drive Moment

4.7.2.1 Newton–Euler Equations of Motion

The force analysis starts with the link 5 because the external moment $\mathbf{M}_{5\text{ext}}$ is given. Figure 4.28 shows the free-body diagram of the link 5. The joint reaction force of the ground 0 on the link 5 at the joint F is $\mathbf{F}_{05} = F_{05x}\mathbf{ı} + F_{05y}\mathbf{ȷ}$. The joint reaction force of the link 4 on the link 5 is $\mathbf{F}_{45} = F_{45x}\mathbf{ı} + F_{45y}\mathbf{ȷ}$. The application point of the force $\mathbf{F}_{45}$ is $P(x_P, y_P)$ and the position vector of P is $\mathbf{r}_P = x_P\mathbf{ı} + y_P\mathbf{ȷ}$.

The symbolical six unknowns F_{05x}, F_{05y}, F_{45x}, F_{45y}, x_P, and y_P are introduced in MATLAB using the commands:

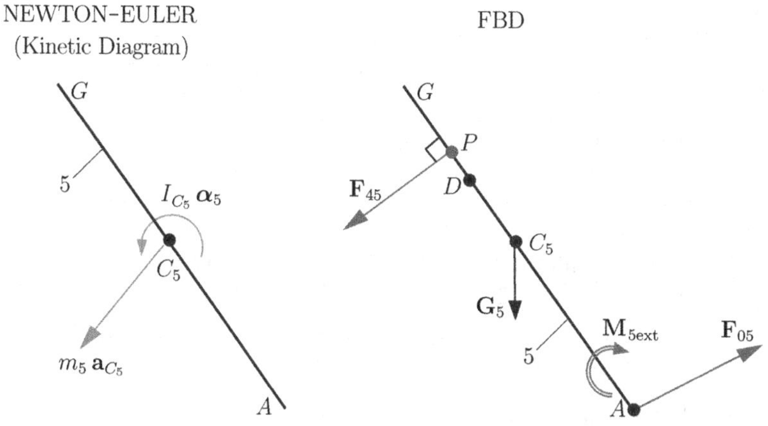

$$m_5\, \mathbf{a}_{C_5} = \mathbf{F}_{05} + \mathbf{G}_5 + \mathbf{F}_{45}$$

$$I_{C_5}\, \alpha_5 = \mathbf{r}_{C_5 A} \times \mathbf{F}_{05} + \mathbf{r}_{C_5 P} \times \mathbf{F}_{45} + \mathbf{M}_{5ext}$$

Fig. 4.28 Link 5 of the R-RTR-RTR mechanism

```
F05x=sym('F05x','real');
F05y=sym('F05y','real');
F45x=sym('F45x','real');
F45y=sym('F45y','real');
xP=sym('xP','real');
yP=sym('yP','real');
F05=[F05x,F05y,0]; % unknown reaction of 0 on 5
F45=[F45x,F45y,0]; % unknown reaction of 4 on 5
rP=[xP,yP,0]; % unknown application point of F45
```

The point P, the application of the force $\mathbf{F}_{45}$, is located on the direction DE, that is

$$(\mathbf{r}_D - \mathbf{r}_A) \times (\mathbf{r}_P - \mathbf{r}_A) = \mathbf{0} \quad \text{or} \quad \mathbf{r}_D \times \mathbf{r}_P = \mathbf{0}, \tag{4.56}$$

or

$$-0.128347\, x_P - 0.147297\, y_P = 0.$$

Equation 4.56 is written in MATLAB as:

```
eqP=cross(rD-rA,rP-rA);
eqPz=eqP(3);
```

The direction of the unknown joint force $\mathbf{F}_{45}$ is perpendicular to the sliding direction $\mathbf{r}_{DA}$

$$\mathbf{F}_{45} \cdot \mathbf{r}_{DA} = 0 \quad \text{or} \quad \mathbf{F}_{45} \cdot \mathbf{r}_D = 0. \tag{4.57}$$

Numerically Eq. 4.57 is

$$-0.147297\,F_{45x} + 0.128347\,F_{45y} = 0.$$

Equation 4.57 in MATLAB is:

```
eqF45DE=dot(F45,rD-rA);
```

For the link 5 the vector sum of the net forces, gravitational force $\mathbf{G}_5$, joint forces $\mathbf{F}_{05}$, $\mathbf{F}_{45}$, is equal to $m_5\,\mathbf{a}_{C_5}$ (Fig. 4.28)

$$m_5\,\mathbf{a}_{C_5} = \mathbf{F}_{05} + \mathbf{F}_{45} + \mathbf{G}_5,$$

or using MATLAB commands:

```
eqF5=F05+F45+G5-m5*aC5;
```

Projecting the previous vectorial onto x and y axes gives

$$m_5\,a_{C_{5x}} = F_{05x} + F_{45x}, \tag{4.58}$$
$$m_5\,a_{C_{5y}} = F_{05y} + F_{45y} - m_5\,g, \tag{4.59}$$

or

$$F_{05x} + F_{45x} + (0.475373)\ 10^{-2} = 0 \text{ and } F_{05y} + F_{45y} - 0.181285 = 0.$$

Equations 4.58 and 4.59 in MATLAB are:

```
eqF5x=eqF5(1); % projection on x-axis
eqF5y=eqF5(2); % projection on y-axis
```

The vector sum of the moments that act on link 5 with respect to the center of mass C_5 is equal to $I_{C_5}\,\alpha_5$ (Fig. 4.28)

$$I_{C_5}\,\boldsymbol{\alpha}_5 = \mathbf{r}_{C_5A} \times \mathbf{F}_{05} + \mathbf{r}_{C_5P} \times \mathbf{F}_{45} + \mathbf{M}_{5\text{ext}}, \tag{4.60}$$

or

$$0.113091\,F_{05y} + 0.0985417\,F_{05x} + (x_P + 0.113091)\,F_{45y}$$
$$-(y_P - 0.0985417)\,F_{45x} - 100.002 = 0.$$

Equation 4.60 in MATLAB is:

```
eqMC5=cross(rE-rC5,F05)+cross(rP-rC5,F45)+Me-...
      IC5*Alpha5;
eqMC5z=eqMC5(3); % projection on z-axis
```

There are five equations (Eqs. 4.56–4.60) and six unknowns, and that is why the analysis will continue with the slider 4. The diagrams of the slider 4 are shown in Fig. 4.29. The joint reaction force of the link 3 on the slider 4 at $D = C_4$ is $\mathbf{F}_{34} = F_{34x}\mathbf{I} + F_{34y}\mathbf{J}$ and the joint reaction force of the link 5 on the slider 4 is $\mathbf{F}_{54} = -\mathbf{F}_{45} = -F_{45x}\mathbf{I} - F_{45y}\mathbf{J}$.

NEWTON-EULER (Kinetic Diagram)

FBD

$$m_4\,\mathbf{a}_{C_4} = \mathbf{F}_{34} + \mathbf{G}_4 + \mathbf{F}_{54}$$
$$I_{C_4}\,\boldsymbol{\alpha}_4 = \mathbf{r}_{C_4 P} \times \mathbf{F}_{54}$$

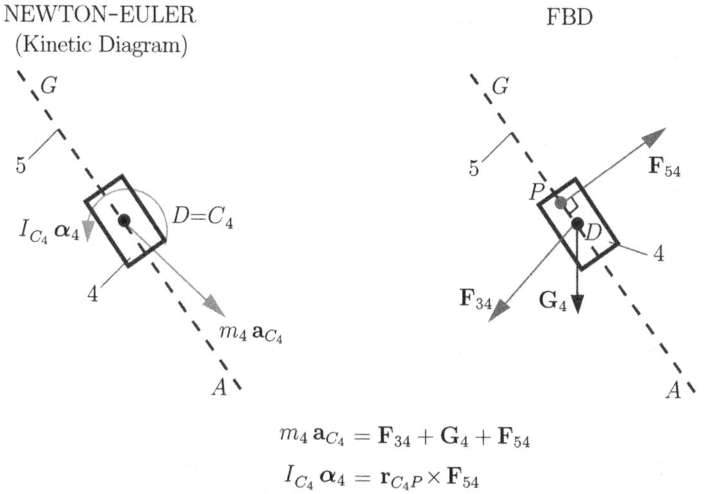

Fig. 4.29 Slider 4 of the R-RTR-RTR mechanism

The MATLAB commands are:

```
F34x=sym('F34x','real');
F34y=sym('F34y','real');
F34=[F34x,F34y,0]; % unknown joint force of 3 on 4
F54=-F45; % joint force of 5 on 4
```

For the slider 4, according to Newton's equations of motion, the vector sum of the net forces, gravitational force $\mathbf{G}_4$, joint forces $\mathbf{F}_{34}$, $\mathbf{F}_{54}$), is equal to $m_4\,\mathbf{a}_{C_4}$

$$m_4\,\mathbf{a}_{C_4} = \mathbf{F}_{34} + \mathbf{F}_{54} + \mathbf{G}_4,$$

or using MATLAB commands:

```
eqF4=F34-F45+G4-m4*aC4;
```

Projecting the previous vectorial onto x and y axes gives

$$m_4\,a_{C_{4x}} = F_{34x} + F_{54x}, \tag{4.61}$$
$$m_4\,a_{C_{4y}} = F_{34y} + F_{54y} - m_4\,g, \tag{4.62}$$

Using MATLAB the previous equations are:

```
eqF4x=eqF4(1);
eqF4y=eqF4(2);
```

Equations 4.61 and 4.62 can be written numerically as

$$F_{34x} - F_{45x} - 0.0204384 = 0 \text{ and } F_{34y} - F_{45y} - 0.0567590 = 0.$$

The vector sum of the moments that act on slider 4 with respect to the center of mass $D = C_4$ is equal to $I_{C_4}\alpha_4$

$$I_{C_4}\alpha_4 = \mathbf{r}_{C_4P} \times \mathbf{F}_{54}, \tag{4.63}$$

or in MATLAB:

```
eqMC4=cross(rP-rC4,F54)-IC4*Alpha4;
eqMC4z=eqMC4(3);
```

The numerical expression of Eq. 4.63 is

$$-(x_P + 0.147297)\,F_{45y} + (y_P - 128347)\,F_{45x} - (0.235748)\,10^{-4} = 0.$$

There are eight equations (Eqs. 4.56–4.63) with eight unknowns F_{05x}, F_{05y}, F_{45x}, F_{45y}, x_P, y_P, F_{34x}, and F_{34y}. The system is solved using MATLAB:

```
sol45=solve(eqF5x,eqF5y,eqMC5z,eqF45DE,eqPz,...
            eqF4x,eqF4y,eqMC4z);
F05xs=eval(sol45.F05x);
F05ys=eval(sol45.F05y);
F05s=[ F05xs, F05ys, 0 ];
F45xs=eval(sol45.F45x);
F45ys=eval(sol45.F45y);
F45s=[ F45xs, F45ys, 0 ];
F34xs=eval(sol45.F34x);
F34ys=eval(sol45.F34y);
F34s=[ F34xs, F34ys, 0 ];
yPs=eval(sol45.yP);
rPs=[xPs, yPs, 0];
```

The following numerical solutions are obtained

$$\mathbf{F}_{05} = 336.192\,\imath + 386.015\,\jmath \text{ N},$$
$$\mathbf{F}_{45} = -336.197\,\imath - 385.834\,\jmath \text{ N},$$
$$\mathbf{F}_{34} = -336.176\,\imath - 385.777\,\jmath \text{ N, and}$$
$$\mathbf{r}_P = -0.147297\,\imath + 0.128347\,\jmath \text{ m}.$$

The force analysis continues with the link 3. Figure 4.30 shows the diagrams of the link 3. The joint reaction force of the link 4 on the link 3 is $\mathbf{F}_{43} = -\mathbf{F}_{34} = 336.176\mathbf{\imath} + 385.777\mathbf{\jmath}$ N. The joint reaction force of the ground 0 on the link 3 at the joint C is $\mathbf{F}_{03} = F_{03x}\mathbf{\imath} + F_{03y}\mathbf{\jmath}$. The joint reaction force of the link 2 on the link 3 is $\mathbf{F}_{423} = F_{23x}\mathbf{\imath} + F_{23y}\mathbf{\jmath}$. The application point of the force $\mathbf{F}_{23}$ is $Q(x_Q, y_Q)$ and the position vector of Q is $\mathbf{r}_Q = x_Q\mathbf{\imath} + y_Q\mathbf{\jmath}$.

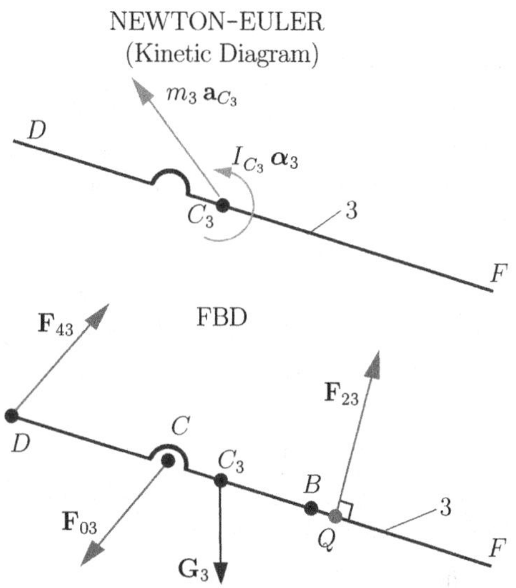

$$m_3\,\mathbf{a}_{C_3} = \mathbf{F}_{43} + \mathbf{F}_{03} + \mathbf{G}_3 + \mathbf{F}_{23}$$

$$I_{C_3}\,\alpha_3 = \mathbf{r}_{C_3D} \times \mathbf{F}_{43} + \mathbf{r}_{C_3C} \times \mathbf{F}_{03} + \mathbf{r}_{C_3Q} \times \mathbf{F}_{23}$$

Fig. 4.30 Link 3 of the R-RTR-RTR mechanism

The symbolical six unknowns F_{03x}, F_{03y}, F_{23x}, F_{23y}, x_Q, and y_Q are introduced in MATLAB using the commands:

```
F03x=sym('F03x','real');
F03y=sym('F03y','real');
F23x=sym('F23x','real');
F23y=sym('F23y','real');
xQ=sym('xQ','real');
yQ=sym('yQ','real');

F03=[F03x,F03y,0]; % unknown joint force of 0 on 3
F23=[F23x,F23y,0]; % unknown joint force of 2 on 3
% unknown application point of force F23
rQ=[xQ,yQ,0];
```

The point Q, the application of the force $\mathbf{F}_{23}$, is located on the direction BC, that is

$$(\mathbf{r}_B - \mathbf{r}_C) \times (\mathbf{r}_Q - \mathbf{r}_C) = \mathbf{0}. \tag{4.64}$$

Equation 4.64 is written in MATLAB as:

```
eqQ=cross(rB-rC,rQ-rC);
eqQz=eqQ(3);
```

and numerically is

$$0.129904\, y_Q + 0.025\, x_Q - 0.0129904 = 0.$$

The direction of the unknown joint force $\mathbf{F}_{23}$ is perpendicular to the sliding direction $\mathbf{r}_{BC}$

$$\mathbf{F}_{23} \cdot \mathbf{r}_{BC} = 0, \tag{4.65}$$

or in MATLAB:

```
eqF23BC = dot(F23,rB-rC);
```

Equation 4.65 can be written numerically as

$$0.129904\, F_{23x} + 0.025\, F_{23y} = 0.$$

For the link 3 the vector sum of the net forces, gravitational force $\mathbf{G}_3$, joint forces $\mathbf{F}_{43}, \mathbf{F}_{03}, \mathbf{F}_{23}$, is equal to $m_3\, \mathbf{a}_{C_3}$ (Fig. 4.30)

$$m_3\, \mathbf{a}_{C_3} = \mathbf{F}_{43} + \mathbf{F}_{03} + \mathbf{F}_{23} + \mathbf{G}_3,$$

or using MATLAB commands:

```
eqF3=F43+F03+F23+G3-m3*aC3;
```

Projecting the previous vectorial onto x and y axes gives

$$m_3\, a_{C_{3x}} = F_{43x} + F_{03x} + F_{23x}, \tag{4.66}$$

$$m_3\, a_{C_{3y}} = F_{43x} + F_{03y} + F_{23y} - m_3\, g, \tag{4.67}$$

or using MATLAB:

```
eqF3x=eqF3(1); % projection on x-axis
eqF3y=eqF3(2); % projection on y-axis
```

Equations 4.66 and 4.67 can be written numerically as

$$F_{03x} + F_{23x} + 336.203 = 0 \text{ and } F_{03y} + F_{23y} + 385.435 = 0.$$

The vector sum of the moments that act on link 3 with respect to the center of mass C_3 is equal to $I_{C_3}\,\alpha_3$ (Fig. 4.30)

$$I_{C_3}\,\alpha_3 = \mathbf{r}_{C_3D} \times \mathbf{F}_{43} + \mathbf{r}_{C_3C} \times \mathbf{F}_{03} + \mathbf{r}_{C_3Q} \times \mathbf{F}_{23}, \qquad (4.68)$$

or in MATLAB:

```
eqMC3=cross(rD-rC3,F43)+cross(rC-rC3,F03)+...
      cross(rQ-rC3,F23)-IC3*Alpha3;
eqMC3z=eqMC3(3); % projection on z-axis
```

The numerical expression of Eq. 4.68 is

$$-88.4776 - 0.0490990\,F_{03y} + 0.00944911\,F_{03x}$$
$$+(xQ - 0.049099)\,F_{23y} - (yQ - 0.0905509)\,F_{23x} = 0.$$

There are five equations (Eqs. 4.64–4.68) and six unknowns, and that is why the analysis will continue with the slider 2. The diagrams of the slider 2 are shown in Fig. 4.31. The joint reaction force of the link 1 on the slider 2 at B is $\mathbf{F}_{12} = F_{12x}\mathbf{I} + F_{12y}\mathbf{J}$ and the joint reaction force of the link 3 on the slider 2 is $\mathbf{F}_{32} = -\mathbf{F}_{23} = -F_{23x}\mathbf{I} - F_{23y}\mathbf{J}$. The MATLAB commands are:

```
F12x=sym('F12x','real');
F12y=sym('F12y','real');
F12=[ F12x, F12y, 0 ]; % unknown joint force of 1 on 2
F32=-F23; % joint force of 3 on 2
```

$$m_2\,\mathbf{a}_{C_2} = \mathbf{F}_{32} + \mathbf{G}_2 + \mathbf{F}_{12}$$
$$I_{C_2}\,\alpha_2 = \mathbf{r}_{C_2Q} \times \mathbf{F}_{32}$$

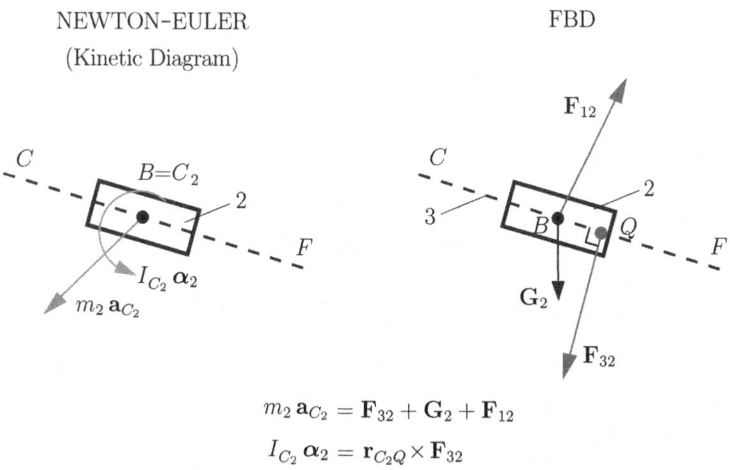

Fig. 4.31 Slider 2 of the R-RTR-RTR mechanism

For the slider 2 the vector sum of the net forces, gravitational force $\mathbf{G}_2$, joint forces $\mathbf{F}_{32}$, $\mathbf{F}_{12}$, is equal to $m_2\,\mathbf{a}_{C_2}$

$$m_2\,\mathbf{a}_{C_2} = \mathbf{F}_{32} + \mathbf{F}_{12} + \mathbf{G}_2,$$

or using MATLAB commands:

```
eqF2=F32+F12+G2-m2*aC2;
```

Projecting the previous vectorial onto x and y axes gives

$$m_2\,a_{C_{2x}} = F_{32x} + F_{12x}, \tag{4.69}$$

$$m_2\,a_{C_{2y}} = F_{32y} + F_{12y} - m_2\,g, \tag{4.70}$$

or using MATLAB:

```
eqF2x=eqF2(1);
eqF2y=eqF2(2);
```

Equations 4.69 and 4.70 can be written numerically as

$$-F_{23x} + F_{12x} + 0.0284911 = 0 \text{ and } -F_{23y} + F_{12y} - 0.0620067 = 0.$$

The vector sum of the moments that act on slider 2 with respect to the center of mass $B = C_2$ is equal to $I_{C_2}\,\alpha_2$

$$I_{C_2}\,\alpha_2 = \mathbf{r}_{C_2Q} \times \mathbf{F}_{32}, \tag{4.71}$$

or in MATLAB:

```
eqMC2=cross(rQ-rC2,F32)-IC2*Alpha2;
eqMC2z=eqMC2(3); % projection on z-axis
```

The numerical expression of Eq. 4.71 is given by

$$-(x_Q - 0.129904)\,F_{23y} + (y_Q - 0.075)\,F_{23x} - (0.281035)\,10^{-4} = 0.$$

There are eight equations (Eqs. 4.64–4.71) with eight unknowns F_{03x}, F_{03y}, F_{23x}, F_{23y}, x_Q, y_Q, F_{12x}, and F_{12y}. The system is solved using MATLAB:

```
sol23=solve(eqF3x,eqF3y,eqMC3z,eqF23BC,eqQz,...
            eqF2x,eqF2y,eqMC2z);
F03xs=eval(sol23.F03x);
F03ys=eval(sol23.F03y);
F03s=[ F03xs, F03ys, 0 ];
F23xs=eval(sol23.F23x);
F23ys=eval(sol23.F23y);
F23s=[ F23xs, F23ys, 0 ];
```

```
F12xs=eval(sol23.F12x);
F12ys=eval(sol23.F12y);
F12s=[ F12xs, F12ys, 0 ];
xQs=eval(sol23.xQ);
yQs=eval(sol23.yQ);
rQs=[xQs, yQs, 0];
```

The following numerical solutions are obtained

$$\mathbf{F}_{03} = -431.027\mathbf{\imath} - 878.152\mathbf{\jmath} \text{ N},$$
$$\mathbf{F}_{23} = 94.8234\mathbf{\imath} + 492.717\mathbf{\jmath} \text{ N},$$
$$\mathbf{F}_{12} = 94.7949\mathbf{\imath} + 492.779\mathbf{\jmath} \text{ N, and}$$
$$\mathbf{r}_Q = 0.129904\mathbf{\imath} + 0.075\mathbf{\jmath} \text{ m}.$$

The force analysis ends with the driver link 1. Figure 4.32 shows the diagrams of the link 1. The joint reaction force of the link 2 on the link 1 is $\mathbf{F}_{21} = -\mathbf{F}_{12} = -94.7949\mathbf{\imath} - 492.779\mathbf{\jmath}$ N. The joint reaction force of the ground 0 on the link 1 at the joint A is $\mathbf{F}_{01} = F_{01x}\mathbf{\imath} + F_{01y}\mathbf{\jmath}$. For the link 1 the vector sum of the net forces, gravitational force $\mathbf{G}_1$, joint forces $\mathbf{F}_{01}$, $\mathbf{F}_{21}$, is equal to $m_1\mathbf{a}_{C_1}$ (Fig. 4.32)

$$m_1\mathbf{a}_{C_1} = \mathbf{F}_{01} - \mathbf{F}_{12} + \mathbf{G}_1 \implies \mathbf{F}_{01} = m_1\mathbf{a}_{C_1} + \mathbf{F}_{12} - \mathbf{G}_1,$$

or with MATLAB:

```
F01=m1*aC1+F12s-G1;
```

The vector sum of the moments that act on link 1 with respect to the center of mass C_1 is equal to $I_{C_1}\alpha_1$

NEWTON–EULER   FBD
(Kinetic Diagram)

$$m_1\mathbf{a}_{C_1} = \mathbf{F}_{01} + \mathbf{G}_1 + \mathbf{F}_{21}$$
$$I_{C_1}\alpha_1 = \mathbf{r}_{C_1A} \times \mathbf{F}_{01} + \mathbf{r}_{C_1B} \times \mathbf{F}_{21} + \mathbf{M}_{\text{mot}}$$

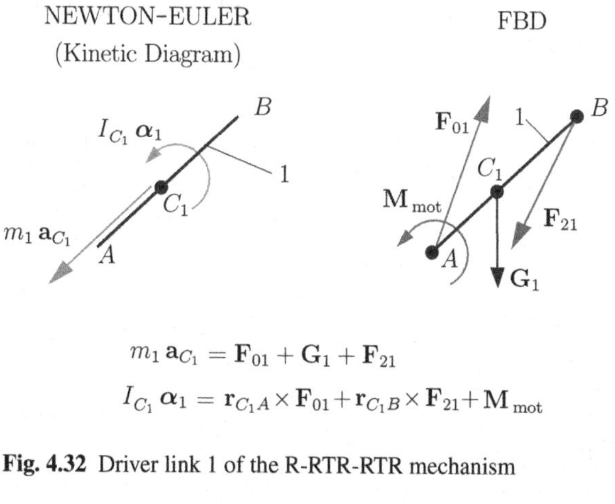

Fig. 4.32 Driver link 1 of the R-RTR-RTR mechanism

$$I_{C_1}\,\boldsymbol{\alpha}_1 = \mathbf{r}_{C_1A}\times\mathbf{F}_{01} - \mathbf{r}_{C_1B}\times\mathbf{F}_{12} + \mathbf{M}_{\text{mot}},$$

and the equilibrium moment (motor moment) is

$$\mathbf{M}_{\text{mot}} = I_{C_1}\,\boldsymbol{\alpha}_1 - \mathbf{r}_{C_1A}\times\mathbf{F}_{01} + \mathbf{r}_{C_1B}\times\mathbf{F}_{12}.$$

In MATLAB the equilibrium moment is:

```
Mm=IC1*alpha1-cross(rA-rC1,F01)+cross(rB-rC1,F12s);
```

Another way of calculating the equilibrium moment is to take the sum of the moments that act on link 1 with respect A

$$I_{C_1}\,\boldsymbol{\alpha}_1 + \mathbf{r}_{C_1}\times m_1\,\mathbf{a}_{C_1} = \mathbf{r}_{C_1}\times\mathbf{G}_1 + \mathbf{r}_B\times(-\mathbf{F}_{12}) + \mathbf{M}_{\text{mot}},$$

and the equilibrium moment is

$$\mathbf{M}_{\text{mot}} = \mathbf{r}_B\times\mathbf{F}_{12} + \mathbf{r}_{C_1}\times(m_1\,\mathbf{a}_{C_1} - \mathbf{G}_1) + I_{C_1}\,\boldsymbol{\alpha}_1,$$

or in MATLAB:

```
Mm=cross(rB,F12s)+cross(rC1,m1*aC1-G1)+IC1*alpha1;
```

The joint reaction force of the ground 0 on the link 1 is $\mathbf{F}_{01} = 94.7736\,\mathbf{\imath} + 492.884\,\mathbf{\jmath}$ N, and the equilibrium moment is $\mathbf{M}_{\text{mot}} = 56.9119\,\mathbf{k}$ N m.

The MATLAB program for the R-RTR-RTR mechanism using Newton–Euler equations of motion and the results are given in Appendix C.5.

4.7.2.2 Dyad Method

The dynamic force analysis starts with the last dyad (links 5 and 4) because the external moment $\mathbf{M}_{5\text{ext}}$ on link 5 is known.

$E_R\,D_T\,D_R$ Dyad

Figure 4.33 shows the forces and the moments that act on the dyad $E_R\,D_T\,D_R$. The unknown joint reaction forces are $\mathbf{F}_{05} = F_{05x}\,\mathbf{\imath} + F_{05y}\,\mathbf{\jmath}$, $\mathbf{F}_{34} = F_{34x}\,\mathbf{\imath} + F_{34y}\,\mathbf{\jmath}$, or in MATLAB

```
F05x=sym('F05x','real');
F05y=sym('F05y','real');
F34x=sym('F34x','real');
F34y=sym('F34y','real');
F05=[ F05x, F05y, 0 ] ;
F34=[ F34x, F34y, 0 ] ;
```

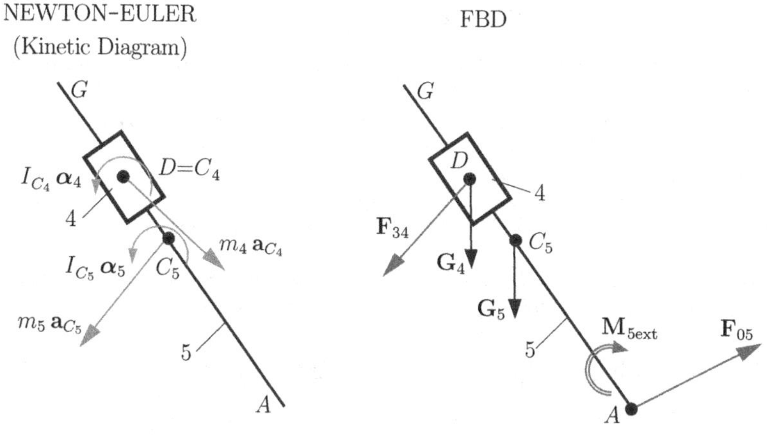

Fig. 4.33 $E_R D_T D_R$ dyad of the R-RTR-RTR mechanism

Newton's equation for links 5 and 4

$$m_5\,\mathbf{a}_{C_5} + m_4\,\mathbf{a}_{C_4} = \mathbf{F}_{05} + \mathbf{G}_5 + \mathbf{G}_4 + \mathbf{F}_{34} \implies$$
$$\sum \mathbf{F}^{(5\&4)} = \mathbf{F}_{05} + \mathbf{G}_5 + \mathbf{G}_4 + \mathbf{F}_{34} - m_5\,\mathbf{a}_{C_5} - m_4\,\mathbf{a}_{C_4} = \mathbf{0}. \tag{4.72}$$

Equation 4.72 has a component on the x-axis, $\sum \mathbf{F}^{(5\&4)} \cdot \mathbf{\imath}$, a component on the y-axis, $\sum \mathbf{F}^{(5\&4)} \cdot \mathbf{\jmath}$, and the MATLAB commands are:

```
eqF45=F05+G5+G4+F34-m5*aC5-m4*aC4;
% projection on x-axis
eqF45x=eqF45(1);
% projection on y-axis
eqF45y=eqF45(2);
```

Euler's equation of moments for links 5 and 4 about D_R gives

$$I_{C_5}\,\alpha_5 + \mathbf{r}_{DC_5} \times m_5\,\mathbf{a}_{C_5} + I_{C_4}\,\alpha_4 = \mathbf{r}_{DA} \times \mathbf{F}_{05} + \mathbf{r}_{DC_5} \times \mathbf{G}_5 + \mathbf{M}_{5\text{ext}} \implies$$
$$\sum \mathbf{M}_D^{(5\&4)} = (\mathbf{r}_A - \mathbf{r}_D) \times \mathbf{F}_{05} + (\mathbf{r}_{C_5} - \mathbf{r}_D) \times (\mathbf{G}_5 - m_5\,\mathbf{a}_{C_5}) + \mathbf{M}_{5\text{ext}}$$
$$-I_{C_5}\,\alpha_5 - I_{C_4}\,\alpha_4 = \mathbf{0}. \tag{4.73}$$

The MATLAB commands for Eq. 4.73 are:

```
eqMD45=cross(rA-rD,F05)+cross(rC5-rD,G5-m5*aC5)+....
       Me-IC5*Alpha5-IC4*Alpha4;
% projection on z-axis
eqMD45z=eqMD45(3);
```

Newton's equation for link 4 projected on the sliding direction AD is

$$(m_4\,\mathbf{a}_{C_4})\cdot\mathbf{r}_{AD} = (\mathbf{F}_{34}+\mathbf{G}_4+\mathbf{F}_{54})\cdot\mathbf{r}_{AD} \implies$$
$$\sum\mathbf{F}^{(4)}\cdot\mathbf{r}_{AD} = (\mathbf{F}_{34}+\mathbf{G}_4-m_4\,\mathbf{a}_{C_4})\cdot(\mathbf{r}_D-\mathbf{r}_A) = 0. \tag{4.74}$$

The force of the link 5 on link 4 is $\mathbf{F}_{54}$ and $\mathbf{F}_{54}\cdot\mathbf{r}_{ED}=0$. The MATLAB command for Eq. 4.74 is:

```
eqF4DA=dot(F34+G4-m4*aC4,rD-rA);
```

There are four equations (Eqs. 4.72–4.74) with four unknowns F_{05x}, F_{05y}, F_{34x}, F_{34y}. The system is solved using MATLAB:

```
solDI=solve(eqF45x, eqF45y , eqMD45z, eqF4DA);
F05xs=eval(solDI.F05x);
F05ys=eval(solDI.F05y);
F34xs=eval(solDI.F34x);
F34ys=eval(solDI.F34y);
F05s=[ F05xs, F05ys, 0 ];
F34s=[ F34xs, F34ys, 0 ];
```

The force of the link 4 on link 5, $\mathbf{F}_{45}$, is calculated from Newton's equation for link 5

$$m_5\,\mathbf{a}_{C_5} = \mathbf{F}_{05}+\mathbf{G}_5+\mathbf{F}_{45} \implies$$
$$\mathbf{F}_{45} = m_5\,\mathbf{a}_{C_5}-\mathbf{G}_5-\mathbf{F}_{05},$$

and the MATLAB command is:

```
F45=m5*aC5-G5-F05s;
```

The application point of the joint force $\mathbf{F}_{45}$ is $P(x_P, y_P)$. The point P is on the line AD or

$$\mathbf{r}_{AD}\times\mathbf{r}_{AP} = \mathbf{0} \quad \text{or} \quad (\mathbf{r}_D-\mathbf{r}_A)\times(\mathbf{r}_P-\mathbf{r}_A) = \mathbf{0},$$

and with MATLAB:

```
eqP=cross(rD-rA,rP-rA);
eqPz=eqP(3);
```

The second equation needed to calculate x_P and y_P is the moment equation on link 4 about $D = C_4$

$$I_{C_4}\,\alpha_4 = \mathbf{r}_{C_4P}\times(-\mathbf{F}_{45}).$$

The previous equation with MATLAB is:

```
eqM4=cross(rP-rC4,-F45)-IC4*Alpha4;
eqM4z=eqM4(3);
```

The coordinates x_P and y_P are calculated using the MATLAB commands:

```
solP=solve(eqPz,eqM4z);
xPs=eval(solP.xP);
yPs=eval(solP.yP);
rPs=[xPs, yPs, 0];
```

$C_R B_T B_R$ Dyad

Figure 4.34 shows the forces and the moments that act on the dyad $C_R B_T B_R$ (links 3 and 2). The unknown joint reaction forces are $\mathbf{F}_{03} = F_{03x}\mathbf{ı} + F_{03y}\mathbf{J}$, $\mathbf{F}_{12} = F_{12x}\mathbf{ı} + F_{12y}\mathbf{J}$, or in MATLAB:

```
F03x=sym('F03x','real');
F03y=sym('F03y','real');
F12x=sym('F12x','real');
F12y=sym('F12y','real');
F03=[ F03x, F03y, 0 ]; F12=[ F12x, F12y, 0 ];
```

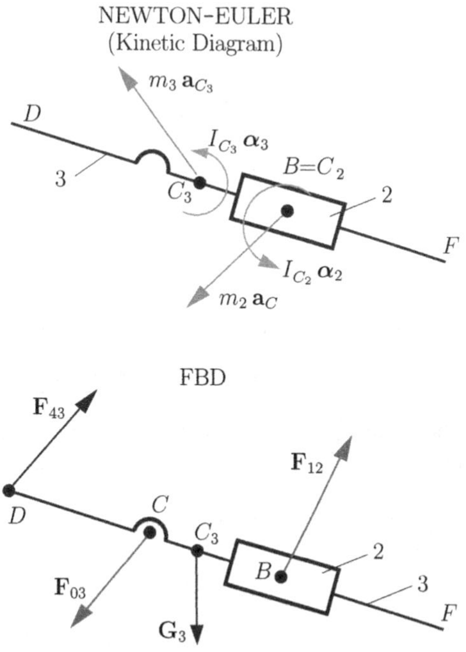

Fig. 4.34 $C_R B_T B_R$ dyad of the R-RTR-RTR mechanism

The joint force $\mathbf{F}_{43} = -\mathbf{F}_{34}$ was calculated from the previous dyad *EDD*

```
F43=-F34s;
```

The sum of all the forces that act on links 3 and 2 is

$$m_3\,\mathbf{a}_{C_3} + m_2\,\mathbf{a}_{C_2} = \mathbf{F}_{43} + \mathbf{F}_{03} + \mathbf{G}_3 + \mathbf{G}_2 + \mathbf{F}_{12} \quad \Longrightarrow$$
$$\sum \mathbf{F}^{(3\&2)} = \mathbf{F}_{43} + \mathbf{F}_{03} + \mathbf{G}_3 + \mathbf{G}_2 + \mathbf{F}_{12} - m_3\,\mathbf{a}_{C_3} - m_2\,\mathbf{a}_{C_2} = \mathbf{0}. \tag{4.75}$$

Equation 4.75 has a component on the x-axis, $\sum \mathbf{F}^{(3\&2)} \cdot \mathbf{\imath}$, a component on the y-axis, $\sum \mathbf{F}^{(3\&2)} \cdot \mathbf{\jmath}$, and the MATLAB commands are:

```
eqF23=F43+F03+G3-m3*aC3+G2-m2*aC2+F12;

eqF23x=eqF23(1); % projection on x-axis
eqF23y=eqF23(2); % projection on y-axis
```

The sum of moments of all the forces and moments on links 3 and 2 about B_R is zero

$$I_{C_3}\,\alpha_3 + \mathbf{r}_{BC_3} \times m_3\,\mathbf{a}_{C_3} + I_{C_2}\,\alpha_2 = \mathbf{r}_{BD} \times \mathbf{F}_{43} + \mathbf{r}_{BC} \times \mathbf{F}_{03} + \mathbf{r}_{BC_3} \times \mathbf{G}_3 \quad \Longrightarrow$$
$$\sum \mathbf{M}_B^{(3\&2)} = (\mathbf{r}_D - \mathbf{r}_B) \times \mathbf{F}_{43} + (\mathbf{r}_C - \mathbf{r}_B) \times \mathbf{F}_{03} + (\mathbf{r}_{C_3} - \mathbf{r}_B) \times (\mathbf{G}_3 - m_3\,\mathbf{a}_{C_3})$$
$$-I_{C_3}\,\alpha_3 - I_{C_2}\,\alpha_2 = \mathbf{0}. \tag{4.76}$$

The MATLAB commands for Eq. 4.76 are:

```
eqMB3=cross(rD-rB,F43)+cross(rC-rB,F03)+...
      cross(rC3-rB,G3-m3*aC3);
eqMB2=-IC3*Alpha3-IC2*Alpha2;
eqMB23=eqMB3+eqMB2;
eqMB23z=eqMB23(3);
```

The sum of all the forces on link 2 projected on the sliding direction *BC* is

$$(m_2\,\mathbf{a}_{C_2}) \cdot \mathbf{r}_{BC} = (\mathbf{F}_{12} + \mathbf{G}_2 + \mathbf{F}_{32}) \cdot \mathbf{r}_{BC} \quad \Longrightarrow$$
$$\sum \mathbf{F}^{(2)} \cdot \mathbf{r}_{BC} = (\mathbf{F}_{12} + \mathbf{G}_2 - m_2\,\mathbf{a}_{C_2}) \cdot (\mathbf{r}_C - \mathbf{r}_B) = 0. \tag{4.77}$$

The force of the link 3 on link 2 is $\mathbf{F}_{32}$ and $\mathbf{F}_{32} \cdot \mathbf{r}_{BC} = 0$. The MATLAB command for Eq. 4.77 is:

```
eqF2BC=dot(F12+G2-m2*aC2, rC-rB);
```

There are four equations (Eqs. 4.75–4.77) with four unknowns F_{03x}, F_{03y}, F_{12x}, F_{12y}. The system is solved using MATLAB:

```
solDII = solve(eqF23x, eqF23y , eqMB23z, eqF2BC);

F03xs=eval(solDII.F03x);
F03ys=eval(solDII.F03y);

F12xs=eval(solDII.F12x);
F12ys=eval(solDII.F12y);

F03s=[ F03xs, F03ys, 0 ];
F12s=[ F12xs, F12ys, 0 ];
```

The force of the link 3 on link 2, $\mathbf{F}_{32}$, is calculated from the sum of all the forces on link 2

$$m_2\,\mathbf{a}_{C_2} = \mathbf{F}_{32} + \mathbf{G}_2 + \mathbf{F}_{12} \quad \Longrightarrow$$
$$\mathbf{F}_{32} = m_2\,\mathbf{a}_{C_2} - \mathbf{G}_2 - \mathbf{F}_{12},$$

and the MATLAB command is:

```
F32=m2*aC2-G2-F12s;
```

The application point of the joint force $\mathbf{F}_{32}$ is $Q(x_Q, y_Q)$. The point Q is on the line BC or

$$\mathbf{r}_{BC} \times \mathbf{r}_{QC} = \mathbf{0} \quad \text{or} \quad (\mathbf{r}_C - \mathbf{r}_B) \times (\mathbf{r}_Q - \mathbf{r}_C) = \mathbf{0},$$

and with MATLAB:

```
eqQ=cross(rC-rB,rQ-rC);
eqQz=eqQ(3);
```

The second equation needed to calculate x_Q and y_Q is the sum of all the moments on link 2 about $B = C_2$

$$I_{C_2}\,\alpha_2 = \mathbf{r}_{C_2Q} \times \mathbf{F}_{32},$$

and with MATLAB:

```
eqM2=cross(rQ-rC2,F32)-IC2*Alpha2;
eqM2z=eqM2(3);
```

The coordinates x_Q and y_Q are calculated using the MATLAB commands:

```
solQ=solve(eqQz,eqM2z);

xQs=eval(solQ.xQ);
yQs=eval(solQ.yQ);

rQs=[xQs, yQs, 0];
```

The joint reaction force of the ground on the link 1 and the equilibrium moment (drive moment) shown in Fig. 4.32 are calculated using the procedure presented in the previous sub-section. The MATLAB program using the dyad method and the results are given in Program C.6.

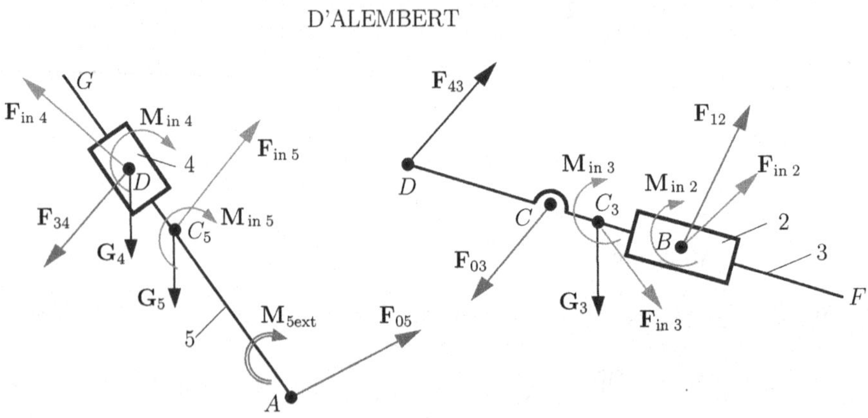

Fig. 4.35 D'Alembert's principle for $A_R D_T D_R$ and $C_R B_T B_R$ dyads

D'Alembert's principle can be applied for the dyad method using the diagrams shown in Fig. 4.35.

4.7.2.3 Contour Method

The contour diagram representing the mechanism is shown in Fig. 4.36. It has two contours 0-1-2-3-0 and 0-3-4-5-0.

Reaction Force $\mathbf{F}_{05}$
The rotation joint A_R between the links 0 and 5 is replaced with the unknown reaction force $\mathbf{F}_{05}$ (Fig. 4.37)

$$\mathbf{F}_{05} = F_{05x}\mathbf{\imath} + F_{05y}\mathbf{\jmath}.$$

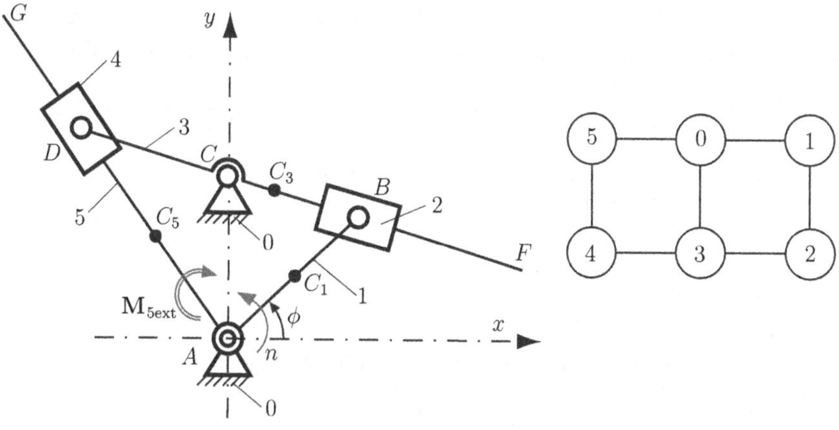

Fig. 4.36 Contour diagram representing the R-RTR-RTR mechanism

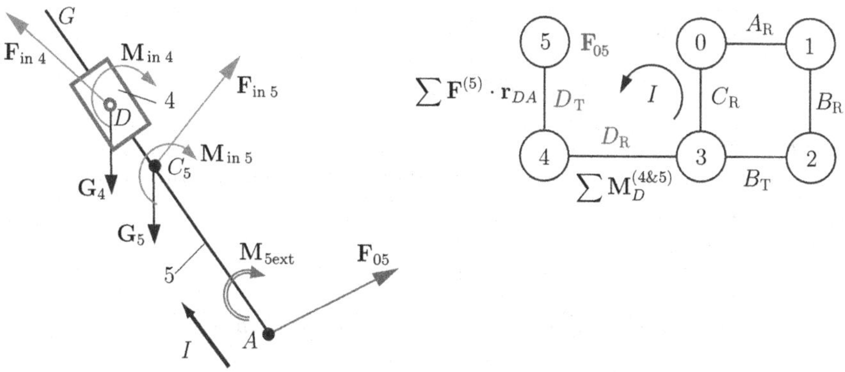

Fig. 4.37 Diagram for calculating the reaction force $\mathbf{F}_{05}$

With MATLAB, the force $\mathbf{F}_{05}$ is written as:

```
F05=[ sym('F05x','real'), sym('F05y','real'), 0 ];
```

Following the path I, as shown in Fig. 4.37, a force equation is written for the translation joint D_{T}. The projection of all forces, that act on the link 5, onto the sliding direction $\mathbf{r}_{DA}$ is zero

$$\sum \mathbf{F}^{(5)} \cdot \mathbf{r}_{DA} = (\mathbf{F}_{05} + \mathbf{G}_5 + \mathbf{F}_{\text{in}5}) \cdot \mathbf{r}_{DA} = 0, \tag{4.78}$$

where $\mathbf{r}_{DA} = \mathbf{r}_A - \mathbf{r}_D$. Equation 4.78 with MATLAB becomes:

```
eqAR1=dot(F05+G5+Fin5,rA-rD);
```

where the command `dot(a,b)` gives the scalar product of the vectors `a` and `b`. Continuing on the path I, a moment equation is written for the rotation joint D_R

$$\sum \mathbf{M}_D^{(4\&5)} = \mathbf{r}_{DA} \times \mathbf{F}_{05} + \mathbf{r}_{DC_5} \times (\mathbf{G}_5 + \mathbf{F}_{\text{in}5}) + \mathbf{M}_{\text{in}4} + \mathbf{M}_{\text{in}5} + \mathbf{M}_{\text{ext}} = \mathbf{0}, \quad (4.79)$$

where $\mathbf{r}_{DC_5} = \mathbf{r}_{C_5} - \mathbf{r}_D$. Equation 4.79 with MATLAB gives:

```
eqAR2=cross(rA-rD,F05)+cross(rC5-rD,G5+Fin5)+...
          Me+Min4+Min5;
eqAR2z=eqAR2(3);
```

The system of two equations is solved using MATLAB commands:

```
solF05=solve(eqAR1,eqAR2z);
F05s=[ eval(solF05.F05x), eval(solF05.F05y), 0 ];
```

The following numerical solution is obtained

$$\mathbf{F}_{05} = 336.192\mathbf{\imath} + 386.015\mathbf{\jmath}\ \text{N}.$$

Reaction Force $\mathbf{F}_{45}$

The translation joint D_T between the links 4 and 5 is replaced with the unknown reaction force $\mathbf{F}_{45}$ (Fig. 4.38)

$$\mathbf{F}_{45} = -\mathbf{F}_{54} = F_{45x}\mathbf{\imath} + F_{45y}\mathbf{\jmath}.$$

The position of the application point P of the force $\mathbf{F}_{45}$ is unknown

$$\mathbf{r}_P = x_P\mathbf{\imath} + y_P\mathbf{\jmath},$$

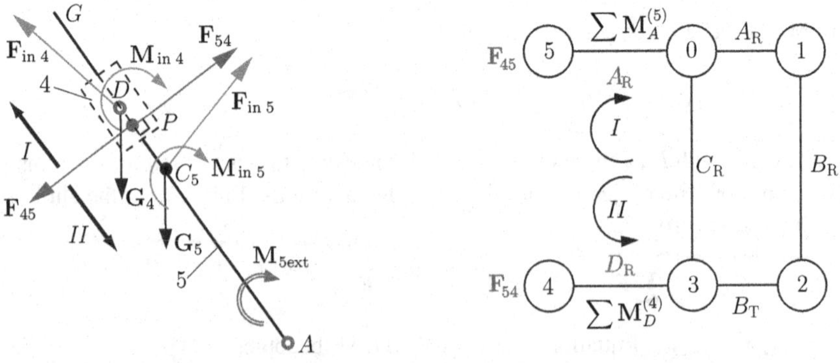

Fig. 4.38 Diagram for calculating the reaction force $\mathbf{F}_{45}$

where x_P and y_P are the plane coordinates of the point P. The force $\mathbf{F}_{45}$ and its point of application P with MATLAB is written as:

```
F45=[ sym('F45x','real'), sym('F45y','real'), 0 ];
F54=-F45;
rP=[ sym('xP','real'), sym('yP','real'), 0 ];
```

Following the path I (Fig. 4.38), a moment equation is written for the rotation joint E_R

$$\sum \mathbf{M}_E^{(5)} = \mathbf{r}_{AP} \times \mathbf{F}_{45} + \mathbf{r}_{AC_5} \times (\mathbf{G}_5 + \mathbf{F}_{\text{in}5}) + \mathbf{M}_{\text{in}5} + \mathbf{M}_{5\text{ext}} = \mathbf{0}, \tag{4.80}$$

where $\mathbf{r}_{AP} = \mathbf{r}_P - \mathbf{r}_A$ and $\mathbf{r}_{AC_5} = \mathbf{r}_{C_5} - \mathbf{r}_A$. One can write Eq. 4.80 using the MATLAB commands:

```
eqDT1=cross(rP-rA,F45)+cross(rC5-rA,G5+Fin5)+Me+Min5;
eqDT1z=eqDT1(3);
```

Following the path II (Fig. 4.38), a moment equation is written for the rotation joint D_R

$$\sum \mathbf{M}_D^{(4)} = \mathbf{r}_{DP} \times \mathbf{F}_{54} + \mathbf{M}_{\text{in}4} = \mathbf{0}, \tag{4.81}$$

where $\mathbf{r}_{DP} = \mathbf{r}_P - \mathbf{r}_D$ and $\mathbf{F}_{54} = -\mathbf{F}_{45}$. Equation 4.81 with MATLAB is:

```
eqDT2=cross(rP-rD,F54)+Min4;
eqDT2z=eqDT2(3);
```

The direction of the unknown joint force $\mathbf{F}_{45}$ is perpendicular to the sliding direction $\mathbf{r}_{AD}$

$$\mathbf{F}_{45} \cdot \mathbf{r}_{AD} = 0, \tag{4.82}$$

and using the MATLAB command:

```
eqF45DA=dot(F45,rD-rA);
```

The application point P of the force $\mathbf{F}_{45}$ is located on the direction AD, that is

$$(\mathbf{r}_D - \mathbf{r}_A) \times (\mathbf{r}_P - \mathbf{r}_A) = \mathbf{0}. \tag{4.83}$$

One can write Eq. 4.83 using the MATLAB commands:

```
eqP=cross(rD-rA,rP-rA);
eqPz=eqP(3);
```

The system of four equations is solved using the MATLAB command:

```
solF45=solve(eqDT1z,eqDT2z,F45DA,eqPz);
F45s=[ eval(solF45.F45x), eval(solF45.F45y), 0 ];
rPs=[ eval(solF45.xP), eval(solF45.yP), 0 ];
```

The following numerical solutions are obtained

$$\mathbf{F}_{45} = -336.197\mathbf{\imath} - 385.834\mathbf{\jmath} \text{ N and } \mathbf{r}_P = -0.147297\mathbf{\imath} + 0.128347\mathbf{\jmath} \text{ m}.$$

Reaction Force $\mathbf{F}_{34}$

The rotation joint D_R between the links 3 and 4 is replaced with the unknown reaction force $\mathbf{F}_{34}$ (Fig. 4.39)

$$\mathbf{F}_{34} = -\mathbf{F}_{34} = F_{34x}\mathbf{\imath} + F_{34y}\mathbf{\jmath},$$

and with MATLAB:

```
F34=[ sym('F34x','real'), sym('F34y','real'), 0 ];
F43=-F34;
```

Following the path I, a force equation can be written for the translation joint D_T. The projection of all forces, that act on the link 4, onto the sliding direction AD is zero

$$\sum \mathbf{F}^{(4)} \cdot \mathbf{r}_{AD} = (\mathbf{F}_{34} + \mathbf{G}_4 + \mathbf{F}_{\text{in}4}) \cdot \mathbf{r}_{AD} = 0, \tag{4.84}$$

where $\mathbf{r}_{AD} = \mathbf{r}_D - \mathbf{r}_A$. Equation 4.84 using MATLAB gives:

```
eqDR1=dot(F34+G4+Fin4,rD-rA);
```

Continuing on the path I (Fig. 4.39), a moment equation is written for the rotation joint A_R

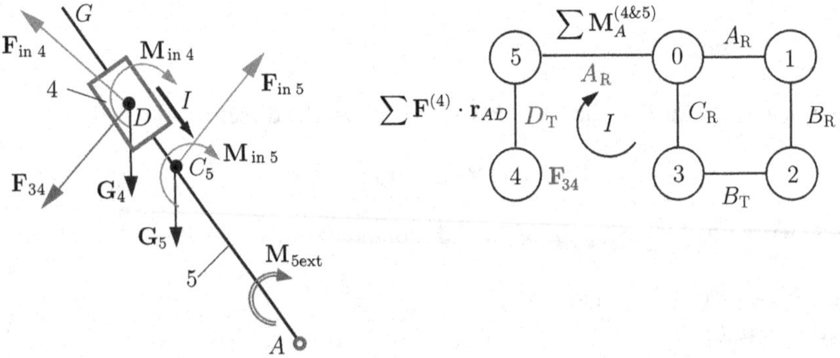

Fig. 4.39 Diagram for calculating the reaction force $\mathbf{F}_{34}$

$$\sum \mathbf{M}_A^{(4\&5)} = \mathbf{r}_{AC_4} \times (\mathbf{G}_4 + \mathbf{F}_{\text{in}4}) + \mathbf{r}_{AD} \times \mathbf{F}_{34} + \mathbf{M}_{\text{in}4}$$
$$+ \mathbf{r}_{AC_5} \times (\mathbf{G}_5 + \mathbf{F}_{\text{in}5}) + \mathbf{M}_{\text{in}5} + \mathbf{M}_{5\text{ext}} = \mathbf{0}, \quad (4.85)$$

where $\mathbf{r}_{AC_5} = \mathbf{r}_{C_5} - \mathbf{r}_A$, and $\mathbf{r}_{AC_4} = \mathbf{r}_{C_4} - \mathbf{r}_A$. Equation 4.85 with MATLAB becomes:

```
eqDR24=cross(rC4-rA,G4+Fin4)+cross(rD-rA,F34)+Min4;
eqDR25=cross(rC5-rA,G5+Fin5)+Me+Min5;
eqDR2=eqDR24+eqDR25;
eqDR2z=eqDR2(3);
```

The system of two equations is solved using the MATLAB commands:

```
solF34=solve(eqDR1,eqDR2z);
F34s=[ eval(solF34.F34x), eval(solF34.F34y), 0 ];
```

The following numerical solution is obtained

$$\mathbf{F}_{34} = -336.176\mathbf{\imath} - 385.777\mathbf{\jmath} \text{ N}.$$

Reaction Force $\mathbf{F}_{03}$

The rotation joint C_R between the links 0 and 3 is replaced with the unknown reaction force $\mathbf{F}_{03}$ (Fig. 4.40)

$$\mathbf{F}_{03} = F_{03x}\mathbf{\imath} + F_{03y}\mathbf{\jmath}.$$

With MATLAB the force $\mathbf{F}_{03}$ is written as:

```
F03=[ sym('F03x','real'), sym('F03y','real'), 0 ];
```

Following the path I (Fig. 4.40), a force equation is written for the translation joint B_T. The projection of all forces, that act on the link 3, onto the sliding direction CD

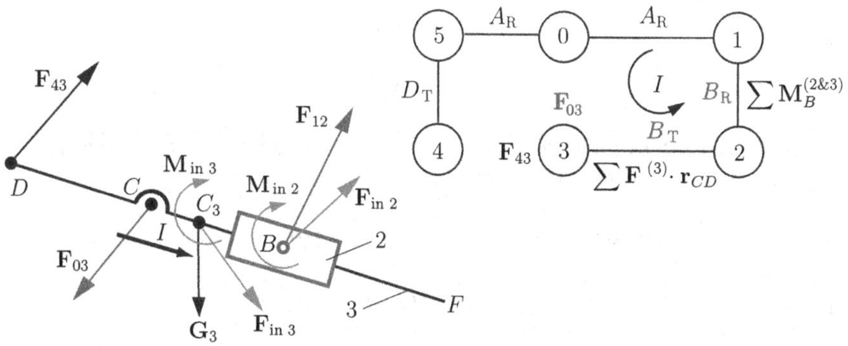

Fig. 4.40 Diagram for calculating the reaction force $\mathbf{F}_{03}$

is zero

$$\sum \mathbf{F}^{(3)} \cdot \mathbf{r}_{CD} = (\mathbf{F}_{03} + \mathbf{F}_{43} + \mathbf{G}_3 + \mathbf{F}_{\text{in}3}) \cdot \mathbf{r}_{CD} = 0, \tag{4.86}$$

where $\mathbf{r}_{CD} = \mathbf{r}_D - \mathbf{r}_C$. Equation 4.86 with the MATLAB command is

```
eqCR1=dot(F03-F34s+G3+Fin3,rD-rC);
```

Continuing on the path *II* (Fig. 4.40), a moment equation is written for the rotation joint B_R

$$\sum \mathbf{M}_B^{(3\&2)} = \mathbf{r}_{BC_3} \times (\mathbf{G}_3 + \mathbf{F}_{\text{in}3}) + \mathbf{r}_{BC} \times \mathbf{F}_{03} + \mathbf{r}_{BD} \times \mathbf{F}_{43} + \mathbf{M}_{\text{in}2} + \mathbf{M}_{\text{in}3} = \mathbf{0}, \tag{4.87}$$

where $\mathbf{r}_{BC_3} = \mathbf{r}_{C_3} - \mathbf{r}_B$, $\mathbf{r}_{BC} = \mathbf{r}_C - \mathbf{r}_B$, and $\mathbf{r}_{BD} = \mathbf{r}_D - \mathbf{r}_B$. With MATLAB Eq. 4.87 gives:

```
eqCR2=cross(rC3-rB,G3+Fin3)+cross(rC-rB,F03)+...
      cross(rD-rB,-F34s)+Min2+Min3;
eqCR2z=eqCR2(3);
```

To solve the system of two equations the MATLAB commands are used:

```
solF03=solve(eqCR1,eqCR2z);
F03s=[ eval(solF03.F03x), eval(solF03.F03y), 0 ];
```

The following numerical solution is obtained

$$\mathbf{F}_{03} = -431.027\,\mathbf{\imath} - 878.152\,\mathbf{\jmath} \text{ N}.$$

Reaction Force $\mathbf{F}_{23}$

The translation joint B_T between the links 2 and 3 is replaced with the unknown reaction force $\mathbf{F}_{23}$ (Fig. 4.41)

$$\mathbf{F}_{23} = -\mathbf{F}_{32} = F_{23x}\mathbf{\imath} + F_{23y}\mathbf{J}.$$

The position of the application point Q of the force $\mathbf{F}_{23}$ is unknown

$$\mathbf{r}_Q = x_Q\mathbf{\imath} + y_Q\mathbf{J},$$

where x_Q and y_Q are the plane coordinates of the point Q. The force $\mathbf{F}_{23}$ and its point of application Q are written in MATLAB as:

```
F23=[ sym('F23x','real'), sym('F23y','real'), 0 ];
F32=-F23;
rQ=[ sym('xQ','real'), sym('yQ','real'), 0 ];
```

Following the path *I* (Fig. 4.41), a moment equation is written for the rotation joint C_R

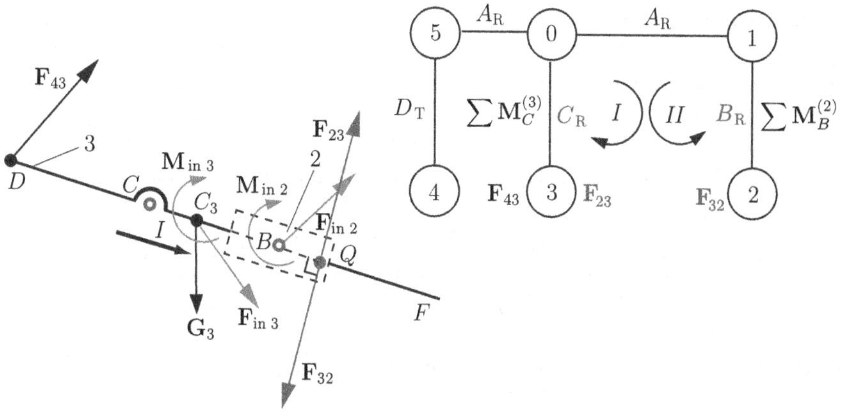

Fig. 4.41 Diagram for calculating the reaction force $\mathbf{F}_{23}$

$$\sum \mathbf{M}_C^{(3)} = \mathbf{r}_{CQ} \times \mathbf{F}_{23} + \mathbf{r}_{CC_3} \times (\mathbf{G}_3 + \mathbf{F}_{\text{in}\,3}) + \mathbf{r}_{CD} \times \mathbf{F}_{43} + \mathbf{M}_{\text{in}\,3} = \mathbf{0}, \qquad (4.88)$$

where $\mathbf{r}_{CQ} = \mathbf{r}_Q - \mathbf{r}_C$, $\mathbf{r}_{CC_3} = \mathbf{r}_{C_3} - \mathbf{r}_C$, and $\mathbf{r}_{CD} = \mathbf{r}_D - \mathbf{r}_C$. Using MATLAB, Eq. 4.88 is written as:

```
eqBT1=cross(rQ-rC,F23)+cross(rC3-rC,G3+Fin3)+...
      cross(rD-rC,-F34s)+Min3;
eqBT1z=eqBT1(3);
```

Following the path *II* (Fig. 4.41), a moment equation is written for the rotation joint B_R

$$\sum \mathbf{M}_B^{(2)} = \mathbf{r}_{BQ} \times \mathbf{F}_{32} + \mathbf{M}_{\text{in}\,2} = \mathbf{0}, \qquad (4.89)$$

where $\mathbf{r}_{BQ} = \mathbf{r}_Q - \mathbf{r}_B$. Equation 4.89 with MATLAB becomes:

```
eqBT2=cross(rQ-rB,F32)+Min2;
eqBT2z=eqBT2(3);
```

The direction of the unknown joint force $\mathbf{F}_{23}$ is perpendicular to the sliding direction BC. The following relation is written

$$\mathbf{F}_{23} \cdot \mathbf{r}_{BC} = 0,$$

or with MATLAB, it is:

```
eqF23BC=dot(F23,rC-rB);
```

The application point Q of the force $\mathbf{F}_{23}$ is located in the direction BC, that is

$$(\mathbf{r}_B - \mathbf{r}_C) \times (\mathbf{r}_Q - \mathbf{r}_C) = \mathbf{0}. \qquad (4.90)$$

Equation 4.90 with MATLAB gives:

```
eqQ=cross(rB-rC,rQ-rC);
eqQz=eqQ(3);
```

The system of four equations is solved using the MATLAB command:

```
solF23=solve(eqBT1z,eqBT2z,F23BC,eqQz);
F23s=[ eval(solF23.F23x), eval(solF23.F23y), 0 ];
rQs=[ eval(solF23.xQ), eval(solF23.yQ), 0 ];
```

The following numerical solutions are obtained

$$\mathbf{F}_{23} = 94.8233\,\mathbf{\imath} + 492.717\,\mathbf{\jmath}\ \text{N} \ \text{ and } \ \mathbf{r}_Q = 0.129904\,\mathbf{\imath} + 0.075\,\mathbf{\jmath}\ \text{m}.$$

Reaction Force $\mathbf{F}_{12}$

The rotation joint B_R between the links 1 and 2 is replaced with the unknown reaction force $\mathbf{F}_{12}$ (Fig. 4.42)

$$\mathbf{F}_{12} = -\mathbf{F}_{21} = F_{12x}\mathbf{\imath} + F_{12y}\mathbf{\jmath}.$$

With MATLAB it is written as:

```
F12=[ sym('F12x','real'), sym('F12y','real'), 0 ];
F21=-F12;
```

Following the path I (Fig. 4.42), a force equation is written for the translation joint B_T. The projection of all forces, that act on the link 2, onto the sliding direction BC is zero

$$\sum \mathbf{F}^{(2)} \cdot \mathbf{r}_{BC} = (\mathbf{F}_{12} + \mathbf{G}_2 + \mathbf{F}_{\text{in}2}) \cdot \mathbf{r}_{BC} = 0. \tag{4.91}$$

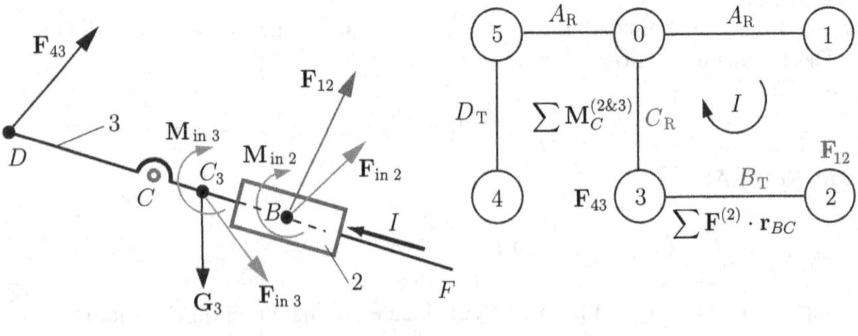

Fig. 4.42 Diagram for calculating the reaction force $\mathbf{F}_{12}$

Using MATLAB it is written as:

```
eqBR1=dot(F12+G2+Fin2,rC-rB);
```

Continuing on the path I, a moment equation is written for the rotation joint C_R

$$\sum \mathbf{M}_C^{(2\&3)} = \mathbf{r}_{CB} \times \mathbf{F}_{12} + \mathbf{r}_{CC_2} \times (\mathbf{G}_2 + \mathbf{F}_{\text{in}\,2}) + \mathbf{M}_{\text{in}\,2} + \mathbf{r}_{CC_3} \times (\mathbf{G}_3 + \mathbf{F}_{\text{in}\,3}) + \mathbf{r}_{CD} \times \mathbf{F}_{43} + \mathbf{M}_{\text{in}\,3} = \mathbf{0}, \quad (4.92)$$

where $\mathbf{r}_{CB} = \mathbf{r}_B - \mathbf{r}_C$, $\mathbf{r}_{CC_2} = \mathbf{r}_{C_2} - \mathbf{r}_C$, $\mathbf{r}_{CC_3} = \mathbf{r}_{C_3} - \mathbf{r}_C$, and $\mathbf{r}_{CD} = \mathbf{r}_D - \mathbf{r}_C$. Using the MATLAB, commands Eq. 4.92 gives:

```
eqBR2=cross(rB-rC,F12)+cross(rC2-rC,G2+Fin2)+Min2...
+cross(rC3-rC,G3+Fin3)+cross(rD-rC,-F34s)+Min3;
eqBR2z=eqBR2(3);
```

The system of two equations is solved using the MATLAB commands:

```
solF12=solve(eqBR1,eqBR2z);
F12s=[ eval(solF12.F12x), eval(solF12.F12y), 0 ];
```

and the following numerical solution is obtained

$$\mathbf{F}_{12} = 94.7949\,\mathbf{\imath} + 492.779\,\mathbf{\jmath}\ \text{N}.$$

Motor Moment $\mathbf{M}_{\text{mot}}$

The motor moment needed for the dynamic equilibrium of the mechanism is $\mathbf{M}_{\text{mot}} = M_{\text{mot}}\,\mathbf{k}$ (Fig. 4.43). Following the path I (Fig. 4.43), a moment equation is written for the rotation joint A_R

$$\sum \mathbf{M}_A^{(1)} = \mathbf{r}_{AB} \times \mathbf{F}_{21} + \mathbf{r}_{AC_1} \times (\mathbf{G}_1 + \mathbf{F}_{\text{in}\,1}) + \mathbf{M}_{\text{in}\,1} + \mathbf{M}_{\text{mot}} = \mathbf{0}. \quad (4.93)$$

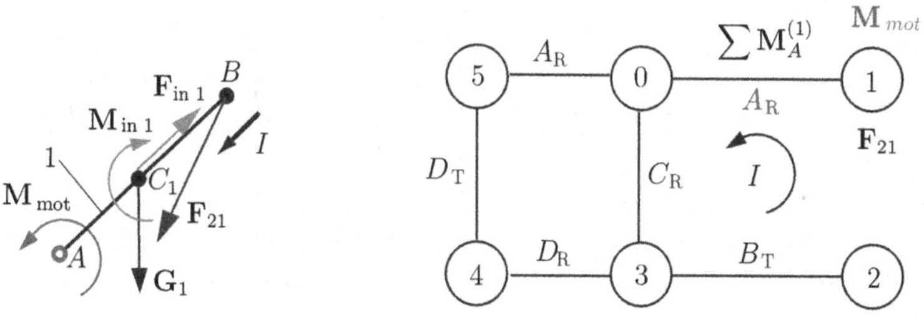

Fig. 4.43 Diagram for calculating the motor moment $\mathbf{M}_{\text{mot}}$

Equation 4.93 is solved using the MATLAB command:

```
M1m=-(cross(rB,-F12s)+cross(rC1,G1+Fin1)+Min1);
```

The numerical solution is

$$\mathbf{M}_{mot} = 56.9119\,\mathbf{k}\ \text{N m}.$$

Reaction Force $\mathbf{F}_{01}$

The rotation joint A_R between the links 0 and 1 is replaced with the unknown reaction force $\mathbf{F}_{01}$ (Fig. 4.44)

$$\mathbf{F}_{01} = -\mathbf{F}_{10} = F_{01x}\mathbf{\imath} + F_{01y}\mathbf{J},$$

With MATLAB it is written as:

```
F01=[ sym('F01x','real'), sym('F01y','real'), 0 ];
```

Following the path I (Fig. 4.44), a moment equation is written for the rotation joint B_R

$$\sum \mathbf{M}_B^{(1)} = \mathbf{r}_{BA} \times \mathbf{F}_{01} + \mathbf{r}_{BC_1} \times (\mathbf{G}_1 + \mathbf{F}_{in\,1}) + \mathbf{M}_{in\,1} + \mathbf{M}_{mot} = \mathbf{0}, \quad (4.94)$$

where $\mathbf{r}_{BA} = -\mathbf{r}_B$, and $\mathbf{r}_{BC_1} = \mathbf{r}_{C_1} - \mathbf{r}_B$. Equation 4.94 using the MATLAB commands gives:

```
eqAAR1=cross(-rB,F01)+cross(rC1-rB,G1+Fin1)+Min1+M1m;
eqAAR1z=eqAAR1(3);
```

Continuing on the path I (Fig. 4.44), a force equation is written for the translation joint B_T.

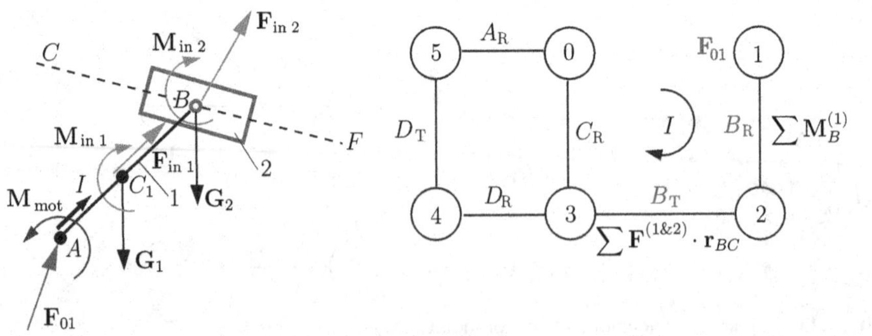

Fig. 4.44 Diagram for calculating the reaction force $\mathbf{F}_{01}$

The projection of all forces, that act on the links 1 and 2, onto the sliding direction BC is zero

$$\sum \mathbf{F}^{(1\&2)} \cdot \mathbf{r}_{BC} = (\mathbf{F}_{01} + \mathbf{G}_1 + \mathbf{F}_{\text{in}\,1} + \mathbf{G}_2 + \mathbf{F}_{\text{in}\,2}) \cdot \mathbf{r}_{BC} = 0, \tag{4.95}$$

or with MATLAB it is:

```
eqAAR2=dot(F01+G1+Fin1+G2+Fin2,rC-rB);
```

The system of two equations is solved using the MATLAB commands:

```
solF01=solve(eqAAR1z, eqAAR2);
F01s=[ eval(solF01.F01x), eval(solF01.F01y), 0 ];
```

The following numerical solution is obtained

$$\mathbf{F}_{01} = 94.7736\mathbf{\imath} + 492.884\mathbf{\jmath} \;\; \text{N}.$$

The MATLAB program for the dynamic force analysis is presented in Appendix C.7.

Chapter 5
Direct Dynamics: Newton–Euler Equations of Motion

The Newton–Euler equations of motion for a rigid body in plane motion are

$$m\ddot{\mathbf{r}}_C = \sum \mathbf{F} \quad \text{and} \quad I_{Czz}\alpha = \sum \mathbf{M}_C,$$

or using the Cartesian components

$$m\ddot{x}_C = \sum F_x, \; m\ddot{y}_C = \sum F_y, \; \text{and} \; I_{Czz}\ddot{\theta} = \sum M_C.$$

The forces and moments are known and the differential equations are solved for the motion of the rigid body (direct dynamics).

5.1 Compound Pendulum

Exercise
Figure 5.1a depicts a compound pendulum of mass m and length L. The pendulum is connnected to the ground by a pin joint and is free to swing in a vertical plane. The link is moving and makes an instant angle $\theta(t)$ with the horizontal. The local acceleration of gravity is g. Numerical application: $L = 3$ ft, $g = 32.2$ ft/s^2, $G = mg = 12$ lb. Find and solve the Newton–Euler equations of motion.

Solution
The system of interest is the link during the interval of its motion. The link in rotational motion is constrained to move in a vertical plane. First, a reference frame will be introduced. The plane of motion will be designated the (x, y) plane. The y-axis is vertical, with the positive sense directed vertically upward. The x-axis is horizontal and is contained in the plane of motion. The z-axis is also horizontal and is perpendicular to the plane of motion. These axes define an inertial reference frame. The unit vectors for the inertial reference frame are $\mathbf{\imath}$, $\mathbf{\jmath}$, and $\mathbf{k}$. The angle between the x and the link axis is denoted by θ. The link is moving and hence the angle is changing with time at the instant of interest. In the static equilibrium position of the link,

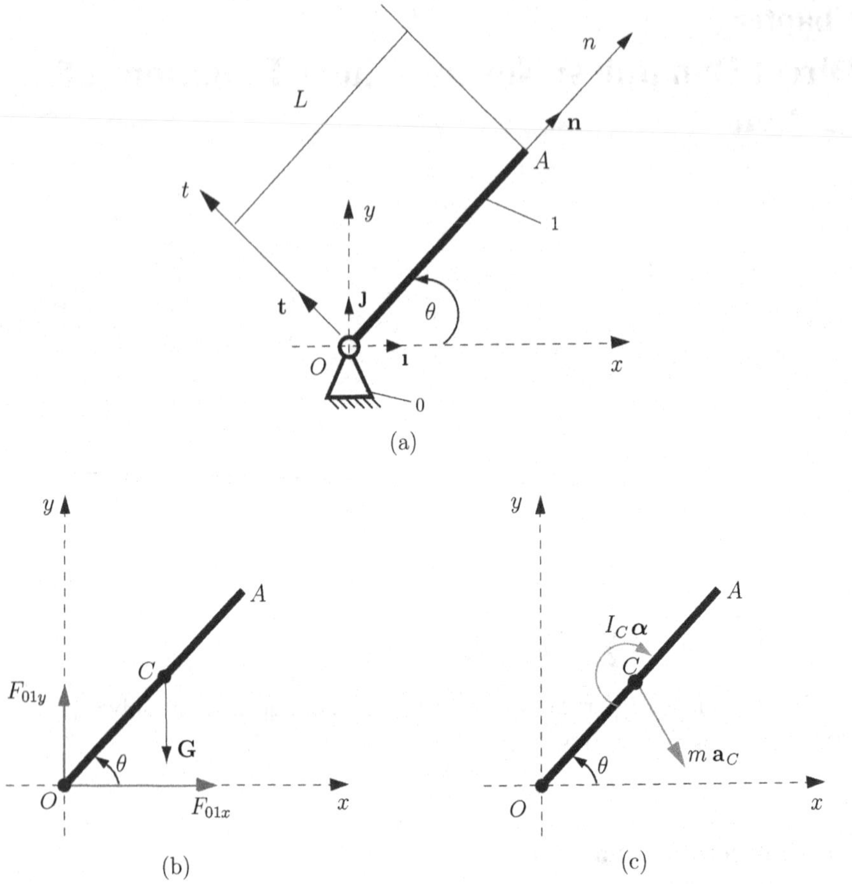

Fig. 5.1 Compound pendulum

the angle, θ, is equal to $-\pi/2$. The system has one degree of freedom. The angle, θ, is an appropriate generalized coordinate describing this degree of freedom. The system has a single moving body. The only motion permitted that body is rotation about a fixed horizontal axis (z-axis). The body is connected to the ground with the rotating pin joint (R) at O. The mass center of the link is at the point C. As the link is uniform, its mass center is coincident with its geometric center.

Kinematics
The mass center, C, is at a distance $L/2$ from the pivot point O and the position vector is

$$\mathbf{r}_{OC} = \mathbf{r}_C = x_C\mathbf{\imath} + y_C\mathbf{J}, \tag{5.1}$$

where x_C and y_C are the coordinates of C

$$x_C = \frac{L}{2}\cos\theta \quad \text{and} \quad y_C = \frac{L}{2}\sin\theta. \tag{5.2}$$

The link is constrained to move in a vertical plane, with its pinned location, O, serving as a pivot point. The motion of the link is planar, consisting of pure rotation about the pivot point. The directions of the angular velocity and angular acceleration vectors will be perpendicular to this plane, in the z-direction.The angular velocity of the link can be expressed as

$$\boldsymbol{\omega} = \omega\mathbf{k} = \frac{d\theta}{dt}\mathbf{k} = \dot{\theta}\mathbf{k}, \tag{5.3}$$

ω is the rate of rotation of the link. The positive sense is clockwise (consistent with the x and y directions defined above). This problem involves only a single moving rigid body and the angular velocity vector refers to that body. For this reason, no explicit indication of the body, 1, is included in the specification of the angular velocity vector $\boldsymbol{\omega} = \boldsymbol{\omega}_1$. The angular acceleration of the link can be expressed as

$$\boldsymbol{\alpha} = \dot{\boldsymbol{\omega}} = \alpha\mathbf{k} = \frac{d^2\theta}{dt^2}\mathbf{k} = \ddot{\theta}\mathbf{k}, \tag{5.4}$$

α is the angular acceleration of the link. The positive sense is clockwise.

The velocity of the mass center can be related to the velocity of the pivot point using the relationship between the velocities of two points attached to the same rigid body

$$\mathbf{v}_C = \mathbf{v}_O + \boldsymbol{\omega} \times \mathbf{r}_{OC} = \begin{vmatrix} \mathbf{ı} & \mathbf{ȷ} & \mathbf{k} \\ 0 & 0 & \omega \\ x_C & y_C & 0 \end{vmatrix} = \omega(-y_C\mathbf{ı} + x_C\mathbf{ȷ})$$
$$= \frac{L\omega}{2}(-\sin\theta\mathbf{ı} + \cos\theta\mathbf{ȷ}) = \frac{L\dot{\theta}}{2}(-\sin\theta\mathbf{ı} + \cos\theta\mathbf{ȷ}). \tag{5.5}$$

The velocity of the pivot point, O, is zero. The acceleration of the mass center can be related to the acceleration of the pivot point ($\mathbf{a}_O = \mathbf{0}$) using the relationship between the accelerations of two points attached to the same rigid body

$$\mathbf{a}_C = \mathbf{a}_O + \boldsymbol{\alpha} \times \mathbf{r}_{OC} + \boldsymbol{\omega} \times (\boldsymbol{\omega} \times \mathbf{r}_{OC}) = \mathbf{a}_O + \boldsymbol{\alpha} \times \mathbf{r}_{OC} - \omega^2\mathbf{r}_{OC}$$
$$= \begin{vmatrix} \mathbf{ı} & \mathbf{ȷ} & \mathbf{k} \\ 0 & 0 & \alpha \\ x_C & y_C & 0 \end{vmatrix} - \omega^2(x_C\mathbf{ı} + y_C\mathbf{ȷ}) = \alpha(-y_C\mathbf{ı} + x_C\mathbf{ȷ}) - \omega^2(x_C\mathbf{ı} + y_C\mathbf{ȷ})$$
$$= -(\alpha y_C + \omega^2 x_C)\mathbf{ı} + (\alpha x_C - \omega^2 y_C)\mathbf{ȷ}$$
$$= -\frac{L}{2}(\alpha\sin\theta + \omega^2\cos\theta)\mathbf{ı} + \frac{L}{2}(\alpha\cos\theta - \omega^2\sin\theta)\mathbf{ȷ}$$
$$= -\frac{L}{2}(\ddot{\theta}\sin\theta + \dot{\theta}^2\cos\theta)\mathbf{ı} + \frac{L}{2}(\ddot{\theta}\cos\theta - \dot{\theta}^2\sin\theta)\mathbf{ȷ}. \tag{5.6}$$

It is also useful to define a set of body-fixed coordinate axes. These are axes that move with the link (body-fixed axes). The n-axis is along the length of the link, the positive direction running from the origin O toward the mass center C. The unit vector of the n-axis is $\mathbf{n}$. The t-axis will be perpendicular to the link and be contained in the plane of motion as shown in Fig. 5.1a.The unit vector of the t-axis is $\mathbf{t}$ and $\mathbf{n}\times\mathbf{t}=\mathbf{k}$. The velocity of the mass center C in the body-fixed reference frame is

$$\mathbf{v}_C=\mathbf{v}_O+\boldsymbol{\omega}\times\mathbf{r}_{OC}=\begin{vmatrix}\mathbf{n} & \mathbf{t} & \mathbf{k}\\ 0 & 0 & \omega\\ \frac{L}{2} & 0 & 0\end{vmatrix}=\frac{L\omega}{2}\mathbf{t}=\frac{L\dot{\theta}}{2}\mathbf{t}, \tag{5.7}$$

where $\mathbf{r}_{OC}=(L/2)\mathbf{n}$. The acceleration of the mass center C in the body-fixed reference frame is

$$\mathbf{a}_C=\mathbf{a}_O+\boldsymbol{\alpha}\times\mathbf{r}_{OC}-\omega^2\mathbf{r}_{OC}=\frac{L\alpha}{2}\mathbf{t}-\omega^2\frac{L}{2}\mathbf{n}=\frac{L\ddot{\theta}}{2}\mathbf{t}-\dot{\theta}^2\frac{L}{2}\mathbf{n}, \tag{5.8}$$

or

$$\mathbf{a}_C=\mathbf{a}_C^t+\mathbf{a}_C^n,$$

with the components

$$\mathbf{a}_C^t=\frac{L\ddot{\theta}}{2}\mathbf{t}\ \text{and}\ \mathbf{a}_C^n=-\frac{L\dot{\theta}^2}{2}\mathbf{n}.$$

Newton–Euler Equation of Motion

The link is rotating about a fixed axis. The mass moment of inertia of the link about the fixed pivot point O can be evaluated from the mass moment of inertia about the mass center C using the transfer theorem. Thus

$$I_O=I_C+m\left(\frac{L}{2}\right)^2=\frac{mL^2}{12}+\frac{mL^2}{4}=\frac{mL^2}{3}. \tag{5.9}$$

The pin is frictionless and is capable of exerting horizontal and vertical forces on the link at O

$$\mathbf{F}_{01}=F_{01x}\mathbf{\imath}+F_{01y}\mathbf{\jmath}, \tag{5.10}$$

where F_{01x} and F_{01y} are the components of the pin force on the link in the fixed-axes system.

The force driving the motion of the link is gravity. The weight of the link is acting through its mass center and will cause a moment about the pivot point. This moment will give the link a tendency to rotate about the pivot point. This moment will be given by the cross product of the vector from the pivot point, O, to the mass center, C, crossed into the weight force $\mathbf{G}=-mg\mathbf{\jmath}$.

As the pivot point, O, of the link is fixed, the appropriate moment summation point will be about that pivot point. The sum of the moments about this point will be equal to the mass moment of inertia about the pivot point multiplied by the angular acceleration of the link. The only contributor to the moment is the weight of the link. Thus we should be able to directly determine the angular acceleration from the moment equation. The sum of the forces acting on the link should be equal to the product of the link mass and the acceleration of its mass center. This should be useful in determining the forces exerted by the pin on the link.

The free-body diagram shows the link at the instant of interest, Fig. 5.1b. The link is acted upon by its weight acting vertically downward through the mass center of the link. The link is acted upon by the pin force at its pivot point. The motion diagram shows the link at the instant of interest, Fig. 5.1c. The motion diagram shows the relevant acceleration information. The Newton–Euler equations of motion for the link are

$$m\mathbf{a}_C = \Sigma\mathbf{F} = \mathbf{G} + \mathbf{F}_{01}, \tag{5.11}$$

$$I_C\,\alpha = \Sigma\mathbf{M}_C = \mathbf{r}_{CO} \times \mathbf{F}_{01}. \tag{5.12}$$

Since the rigid body has a fixed point at O the equations of motion state that the moment sum about the fixed point must be equal to to the product of the link mass moment of inertia about that point and the link angular acceleration. Thus,

$$I_O\,\alpha = \Sigma\mathbf{M}_O = \mathbf{r}_{OC} \times \mathbf{G}. \tag{5.13}$$

Using Eqs. 5.4, 5.9, and 5.13 the equation of motion is

$$\frac{mL^2}{3}\ddot{\theta}\mathbf{k} = \begin{vmatrix} \mathbf{\imath} & \mathbf{\jmath} & \mathbf{k} \\ \frac{L}{2}\cos\theta & \frac{L}{2}\sin\theta & 0 \\ 0 & -mg & 0 \end{vmatrix}, \tag{5.14}$$

or

$$\ddot{\theta} = -\frac{3g}{2L}\cos\theta. \tag{5.15}$$

The equation of motion, Eq. 5.15, is a non-linear, second-order, differential equation relating the second time derivative of the angle, θ, to the value of that angle and various problem parameters g and L. The equation is non-linear due to the presence of the $\cos\theta$, where $\theta(t)$ is the unknown function of interest.

The force exerted by the pin on the link is obtained from Eq. 5.11

$$\mathbf{F}_{01} = m\mathbf{a}_C - \mathbf{G},$$

and the components of the force are

$$F_{01x} = m\ddot{x}_C = -\frac{mL}{2}(\ddot{\theta}\sin\theta + \dot{\theta}^2\cos\theta),$$

$$F_{01y} = m\ddot{y}_C + mg = \frac{mL}{2}(\ddot{\theta}\cos\theta - \dot{\theta}^2\sin\theta) + mg. \tag{5.16}$$

Using the moving reference frame (body-fixed) the components of the reaction force on n and t axes are

$$F_{01n} = m a_C^n - mg\sin\theta = -\frac{mL\dot{\theta}^2}{2} + mg\sin\theta,$$
$$F_{01t} = m a_C^t - mg\cos\theta = \frac{mL\ddot{\theta}}{2} + mg\cos\theta. \tag{5.17}$$

If the link is released from rest, then the initial value of the angular velocity is zero $\omega(t=0) = \omega(0) = \dot{\theta}(0) = 0$ rad/s. If the initial angle is $\theta(0) = 0$ radians, then the cosine of that initial angle is unity and the sine is zero. The initial angular acceleration can be determined from Eq. 5.15

$$\ddot{\theta}(0) = \alpha(0) = -\frac{3g}{2L}\cos\theta(0) = -\frac{3g}{2L} = 16.1 \text{ rad/s}^2. \tag{5.18}$$

The negative sign indicates that the initial angular acceleration of the link is counterclockwise, as one would expect.

The initial reaction force components can be evaluated from Eq. 5.16

$$F_{01x}(0) = 0 \text{ lb},$$
$$F_{01y}(0) = \frac{mL}{2}\ddot{\theta}(0) + mg = \frac{mg}{4} = 3 \text{ lb}.$$

The equation of motion, Eq. 5.15, is obtained symbolically using the MATLAB® commands:

```
syms L m g t
omega = [0 0 diff('theta(t)',t)];
alpha = diff(omega,t);
c = cos(sym('theta(t)'));
s = sin(sym('theta(t)'));
xC = (L/2)*c;
yC = (L/2)*s;
rC = [xC yC 0];
G = [0 -m*g 0];
IC = m*L^2/12;
IO = IC + m*(L/2)^2;
MO = cross(rC,G);
eq = -IO*alpha+MO;
eqz = eq(3);
```

The MATLAB statement `diff(X,'t')` or `diff(X,sym('t'))` differentiates a symbolic expression `X` with respect to `t`, and the statement `diff(X,'t',n)` and `diff(X,n,'t')` differentiates `X` `n` times, where `n` is a positive integer.

An analytical solution to the differential equation is difficult to obtain. Numerical approaches have the advantage of being simple to apply even for complex mechanical systems. The MATLAB function `ode45` is used to solve the differential equation.

The differential equation $\ddot{\theta} = -\frac{3g}{2L}\cos\theta$ is of order 2. The equation has to be re-written as a first-order system. Let $x_1 = \theta$ and $x_2 = \dot{\theta}$, this gives the first-order system

$$\dot{x}_1 = x_2,$$
$$\dot{x}_2 = -\frac{3g}{2L}\cos x_1.$$

The MATLAB commands for the right-hand side of the first-order system are:

```
eqI  = subs(eqz,{L,m,g},{3,12/32.2,32.2});
eqI1  = subs(eqI,diff('theta(t)',t,2),'ddtheta');
eqI2  = subs(eqI1,diff('theta(t)',t),sym('x(2)'));
eqI3  = subs(eqI2,'theta(t)',sym('x(1)'));
dx1  = sym('x(2)');
dx2  = solve(eqI3,'ddtheta');
dx1dt  = char(dx1);
dx2dt  = char(dx2);
```

An `inline` function `g` is defined for the right-hand side of the first-order system. Note that `g` must be specified as a column vector using `[...;...]` (not a row vector using `[...,...]`). In the definition of `g`, `x(1)` was used for x_1 and `x(2)` was used for x_2. The definition of `g` should have the form:

```
g = inline(sprintf('[%s;%s]',dx1dt,dx2dt),'t','x');
```

The statement has to have the form `inline(...,'t','y')`, even if `t` does not occur in your formula. The first component of `g` is `x(2)`. The statement `sprintf` writes formatted data to string. The time `t` is going from an initial value `t0` to a final value `f`:

```
t0 = 0;
tf = 10;
time = [0 tf];
```

The initial conditions at $t_0 = 0$ are $\theta(0) = \pi/4$ rad and $\dot{\theta}(0) = 0$ rad/s or in MATLAB:

```
x0 = [pi/4; 0]; % define initial conditions
```

The numerical solution of all the components of the solution for `t` going from `t0` to `f` is obtained using the command:

```
[t,xs] = ode45(g, time, x0);
```

where `x0` is the initial value vector at the starting point `t0`.

One can obtain a vector `t` and a matrix `xs` with the coordinates of these points using `ode45` command.

The vector of `x1` values in the first column of `xs` is obtained by using `xs(:,1)` and the vector of `x2` values in the second column of `xs` is obtained by using `xs(:,2)`:

```
x1 = xs(:,1);
x2 = xs(:,2);
```

The plot of the solution curves are obtained using the commands:

```
subplot(3,1,1),plot(t,x1,'r'),...
xlabel('t'),ylabel('\theta'),grid,...
subplot(3,1,2),plot(t,x2,'g'),...
xlabel('t'),ylabel('\omega'),grid,...
subplot(3,1,3),plot(x1,x2),...
xlabel('\theta'),ylabel('\omega'),grid
```

The plots using MATLAB are shown in Fig. 5.2. In general, the error tends to grow as one goes further from the initial conditions. To obtain numerical values at specific `t` values one can specify a vector `tp` of `t` values and use `[ts,xs] = ode45(g, tp, x0)`. The first element of the vector `tp` is the initial value and the vector `tp` must have at least 3 elements. To obtain the solution with the initial values at `t = 0, 0.5, 1.0, 1.5, ... , 10` one can use:

```
[ts,xs] = ode45(g, 0:0.5:10, x0);
[ts,xs]
```

and the results are displayed as a table with 3 columns `ts, x1 = xs(:,1), x2 = xs(:,1)`.

A MATLAB computer program to solve the governing differential equation is given in Appendix D.1.

The differential equation can be solved numerically by m-file functions. First create a function file, `R.m` as shown below:

```
function dx = R(t,x);
dx = zeros(2,1); % a column vector
W = 12; L = 3; g = 32.2; m = W/g;
dx(1) = x(2);
dx(2) = -3*g*cos(x(1))/(2*L);
```

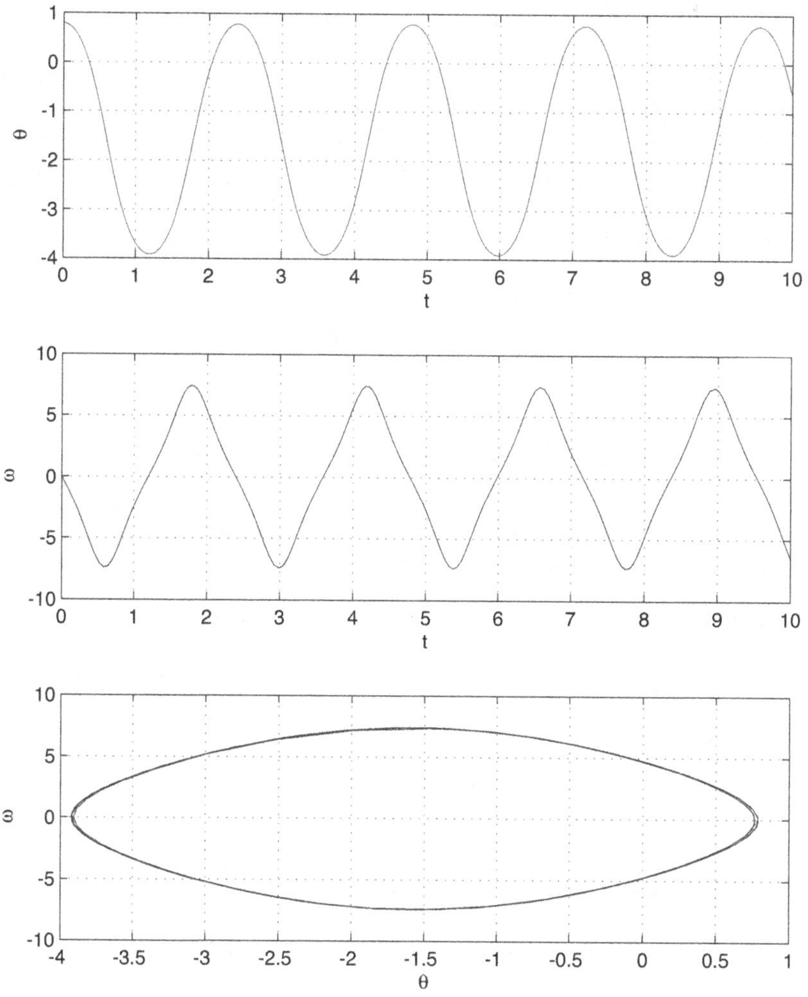

Fig. 5.2 Plot of solution curves θ and $\omega = \dot{\theta}$

The ode solver provided by MATLAB (`ode45`) is used to solve the differential equation:

```
tfinal=10;
time=[0 tfinal];
x0=[pi/4 0]; % x(1)(0)=pi/4; x(2)(0)=0
[t,x]=ode45(@R, time, x0);
```

The MATLAB program is given in Appendix D.2.

5.2 Double Pendulum

Exercise

A two-link planar chain (double pendulum) is considered, Fig. 5.3a. The links 1 and 2 have the masses m_1 and m_2 and the lengths $AB = L_1$ and $BD = L_2$. The system is free to move in a vertical plane. The local acceleration of gravity is g. Numerical application: $m_1 = m_2 = 1$ kg, $L_1 = 1$ m, $L_2 = 0.5$ m, and $g = 10$ m/s^2. Find and solve the equations of motion.

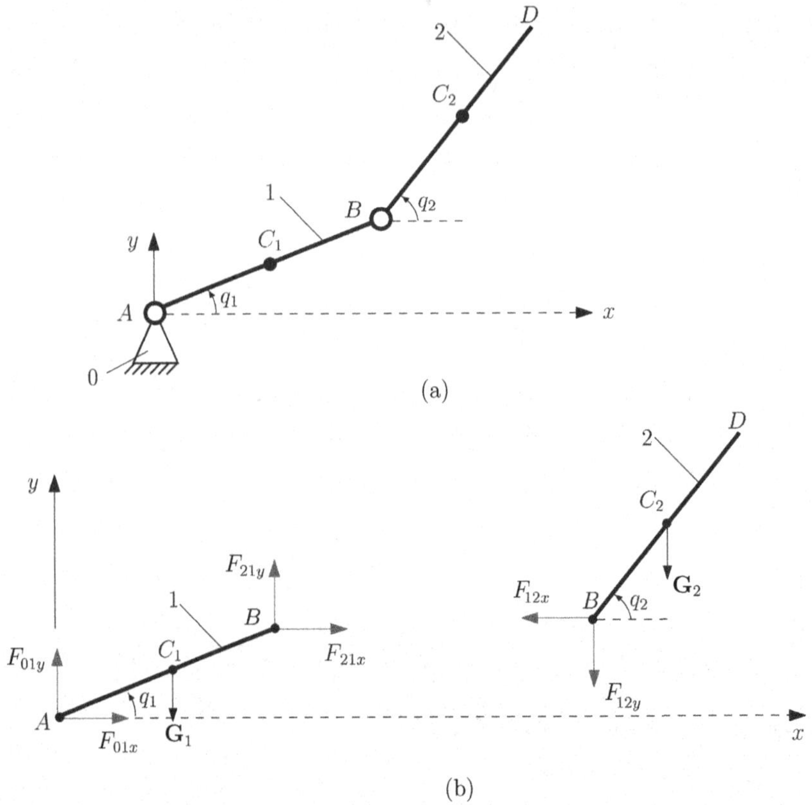

Fig. 5.3 Double pendulum

Solution

The plane of motion is the (x, y) plane with the y-axis vertical, with the positive sense directed upward. The origin of the reference frame is at A. The mass centers of the links are designated by $C_1(x_{C_1}, y_{C_1}, 0)$ and $C_2(x_{C_2}, y_{C_2}, 0)$.

The number of degrees of freedom are computed using the relation

$$M = 3n - 2c_5 - c_4,$$

where n is the number of moving links, c_5 is the number of one degree of freedom joints, and c_4 is the number of two degrees of freedom joints. For the double pendulum $n = 2$, $c_5 = 2$, $c_4 = 0$, and the system has two degrees of freedom, $M = 2$, and two generalized coordinates. The angles $q_1(t)$ and $q_2(t)$ are selected as the generalized coordinates as shown in Fig. 5.3a.

Kinematics

The position vector of the center of the mass C_1 of the link 1 is

$$\mathbf{r}_{C_1} = x_{C_1}\mathbf{\imath} + y_{C_1}\mathbf{J},$$

where x_{C_1} and y_{C_1} are the coordinates of C_1

$$x_{C_1} = \frac{L_1}{2}\cos q_1 \quad \text{and} \quad y_{C_1} = \frac{L_1}{2}\sin q_1.$$

The position vector of the center of the mass C_2 of the link 2 is

$$\mathbf{r}_{C_2} = x_{C_2}\mathbf{\imath} + y_{C_2}\mathbf{J},$$

where x_{C_2} and y_{C_2} are the coordinates of C_2

$$x_{C_2} = L_1\cos q_1 + \frac{L_2}{2}\cos q_2 \quad \text{and} \quad y_{C_2} = L_1\sin q_1 + \frac{L_2}{2}\sin q_2.$$

The velocity vector of C_1 is the derivative with respect to time of the position vector of C_1

$$\mathbf{v}_{C_1} = \dot{\mathbf{r}}_{C_1} = \dot{x}_{C_1}\mathbf{\imath} + \dot{y}_{C_1}\mathbf{J},$$

where

$$\dot{x}_{C_1} = -\frac{L_1}{2}\dot{q}_1\sin q_1 \quad \text{and} \quad \dot{y}_{C_1} = \frac{L_1}{2}\dot{q}_1\cos q_1.$$

The velocity vector of C_2 is the derivative with respect to time of the position vector of C_2

$$\mathbf{v}_{C_2} = \dot{\mathbf{r}}_{C_2} = \dot{x}_{C_2}\mathbf{\imath} + \dot{y}_{C_2}\mathbf{J},$$

where

$$\dot{x}_{C_2} = -L_1\dot{q}_1\sin q_1 - \frac{L_2}{2}\dot{q}_2\sin q_2 \quad \text{and} \quad \dot{y}_{C_2} = L_1\dot{q}_1\cos q_1 + \frac{L_2}{2}\dot{q}_2\cos q_2.$$

The acceleration vector of C_1 is the double derivative with respect to time of the position vector of C_1

$$\mathbf{a}_{C_1} = \ddot{\mathbf{r}}_{C_1} = \ddot{x}_{C_1}\mathbf{\imath} + \ddot{y}_{C_1}\mathbf{J},$$

where

$$\ddot{x}_{C_1} = -\frac{L_1}{2}\ddot{q}_1 \sin q_1 - \frac{L_1}{2}\dot{q}_1^2 \cos q_1 \quad \text{and} \quad \ddot{y}_{C_1} = \frac{L_1}{2}\ddot{q}_1 \cos q_1 - \frac{L_1}{2}\dot{q}_1^2 \sin q_1.$$

The acceleration vector of C_2 is the double derivative with respect to time of the position vector of C_2

$$\mathbf{a}_{C_2} = \ddot{\mathbf{r}}_{C_2} = \ddot{x}_{C_2}\mathbf{\imath} + \ddot{y}_{C_2}\mathbf{J},$$

where

$$\ddot{x}_{C_2} = -L_1\ddot{q}_1 \sin q_1 - L_1\dot{q}_1^2 \cos q_1 - \frac{L_2}{2}\ddot{q}_2 \sin q_2 - \frac{L_2}{2}\dot{q}_2^2 \cos q_2,$$
$$\ddot{y}_{C_2} = L_1\ddot{q}_1 \cos q_1 - L_1\dot{q}_1^2 \sin q_1 + \frac{L_2}{2}\ddot{q}_2 \cos q_2 - \frac{L_2}{2}\dot{q}_2^2 \sin q_2.$$

The MATLAB commands for the linear accelerations of the mass centers C_1 and C_2 are:

```
L1 = 1; L2 = 0.5; m1 = 1; m2 = 1; g = 10;
t = sym('t','real');
xB = L1*cos(sym('q1(t)'));
yB = L1*sin(sym('q1(t)'));
rB = [xB yB 0];
rC1 = rB/2;
vC1 = diff(rC1,t);
aC1 = diff(vC1,t);
xD = xB + L2*cos(sym('q2(t)'));
yD = yB + L2*sin(sym('q2(t)'));
rD = [xD yD 0];
rC2 = (rB + rD)/2;
vC2 = diff(rC2,t);
aC2 = diff(vC2,t);
```

The angular velocity vectors of the links 1 and 2 are

$$\boldsymbol{\omega}_1 = \dot{q}_1\mathbf{k} \quad \text{and} \quad \boldsymbol{\omega}_2 = \dot{q}_2\mathbf{k}.$$

The angular acceleration vectors of the links 1 and 2 are

$$\boldsymbol{\alpha}_1 = \ddot{q}_1\mathbf{k} \quad \text{and} \quad \boldsymbol{\alpha}_2 = \ddot{q}_2\mathbf{k}.$$

The MATLAB commands for the angular accelerations of the links 1 and 2 are

```
omega1 = [0 0 diff('q1(t)',t)];
alpha1 = diff(omega1,t);
omega2 = [0 0 diff('q2(t)',t)];
alpha2 = diff(omega2,t);
```

Newton–Euler Equations of Motion
The weight forces on the links 1 and 2 are

$$\mathbf{G}_1 = -m_1 g \mathbf{J} \quad \text{and} \quad \mathbf{G}_2 = -m_2 g \mathbf{J},$$

and in MATLAB:

```
G1 = [0 -m1*g 0];
G2 = [0 -m2*g 0];
```

The mass moment of inertia of the link 1 with respect to the center of mass C_1 is

$$I_{C_1} = \frac{m_1 L_1^2}{12}.$$

The mass moment of inertia of the link 1 with respect to the fixed point of rotation A is

$$I_A = I_{C_1} + m_1 \left(\frac{L_1}{2}\right)^2 = \frac{m_1 L_1^2}{3}.$$

The mass moment of inertia of the link 2 with respect to the center of mass C_2 is

$$I_{C_2} = \frac{m_2 L_2^2}{12}.$$

The MATLAB commands for the mass moments of inertia are:

```
IC1 = m1*L1^2/12;
IA = IC1 + m1*(L1/2)^2;
IC2 = m2*L2^2/12;
```

The equations of motion of the pendulum are inferred using the Newton–Euler method. There are two rigid bodies in the system and the Newton–Euler equations are written for each link using the free-body diagrams shown in Fig. 5.3b.

Link 1
The Newton–Euler equations for the link 1 are

$$m_1 \mathbf{a}_{C_1} = \mathbf{F}_{01} + \mathbf{F}_{21} + \mathbf{G}_1,$$
$$I_{C_1} \boldsymbol{\alpha}_1 = \mathbf{r}_{C_1A} \times \mathbf{F}_{01} + \mathbf{r}_{C_1B} \times \mathbf{F}_{21},$$

where $\mathbf{F}_{01}$ is the joint reaction of the ground 0 on the link 1 at point A, and $\mathbf{F}_{21}$ is the joint reaction of the link 2 on the link 1 at point B

$$\mathbf{F}_{01} = F_{01x}\mathbf{I} + F_{01y}\mathbf{J} \quad \text{and} \quad \mathbf{F}_{21} = F_{21x}\mathbf{I} + F_{21y}\mathbf{J}.$$

Since the link 1 has a fixed point of rotation at A the moment sum about the fixed point must be equal to to the product of the link mass moment of inertia about that point and the link angular acceleration. Thus,

$$I_A\alpha_1 = \mathbf{r}_{AC_1} \times \mathbf{G}_1 + \mathbf{r}_{AB} \times \mathbf{F}_{21}, \tag{5.19}$$

or

$$\frac{m_1 L_1^2}{3}\ddot{q}_1\mathbf{k} = \begin{vmatrix} \mathbf{\imath} & \mathbf{\jmath} & \mathbf{k} \\ x_{C_1} & y_{C_1} & 0 \\ 0 & -m_1 g & 0 \end{vmatrix} + \begin{vmatrix} \mathbf{\imath} & \mathbf{\jmath} & \mathbf{k} \\ x_B & y_B & 0 \\ F_{21x} & F_{21y} & 0 \end{vmatrix},$$

or

$$\frac{m_1 L_1^2}{3}\ddot{q}_1\mathbf{k} = (-m_1 g x_{C_1} + F_{21y}x_B - F_{21x}y_B)\mathbf{k}.$$

The equation of motion for link 1 is

$$\frac{m_1 L_1^2}{3}\ddot{q}_1 = \left(-m_1 g\frac{L_1}{2}\cos q_1 + F_{21y}L_1\cos q_1 - F_{21x}L_1\sin q_1\right). \tag{5.20}$$

Link 2
The Newton–Euler equations for the link 2 are

$$m_2\mathbf{a}_{C_2} = \mathbf{F}_{12} + \mathbf{G}_2, \tag{5.21}$$

$$I_{C_2}\alpha_2 = \mathbf{r}_{C_2B} \times \mathbf{F}_{12}, \tag{5.22}$$

where $\mathbf{F}_{12} = -\mathbf{F}_{21}$ is the joint reaction of the link 1 on the link 2 at B. Equation 5.22 becomes

$$\begin{aligned} m_2\ddot{x}_{C_2} &= -F_{21x}, \\ m_2\ddot{y}_{C_2} &= -F_{21y} - m_2 g, \\ \frac{mL_2^2}{12}\ddot{q}_2\mathbf{k} &= \begin{vmatrix} \mathbf{\imath} & \mathbf{\jmath} & \mathbf{k} \\ x_B - x_{C_2} & y_B - y_{C_2} & 0 \\ -F_{21x} & -F_{21y} & 0 \end{vmatrix}, \end{aligned} \tag{5.23}$$

or

$$m_2\left(-L_1\ddot{q}_1\sin q_1 - L_1\dot{q}_1^2\cos q_1 - \frac{L_2}{2}\ddot{q}_2\sin q_2 - \frac{L_2}{2}\dot{q}_2^2\cos q_2\right)$$
$$= -F_{21x}, \tag{5.24}$$

$$m_2\left(L_1\ddot{q}_1\cos q_1 - L_1\dot{q}_1^2\sin q_1 + \frac{L_2}{2}\ddot{q}_2\cos q_2 - \frac{L_2}{2}\dot{q}_2^2\sin q_2\right)$$
$$= -F_{21y} - m_2 g, \tag{5.25}$$

$$\frac{m_2 L_2^2}{12}\ddot{q}_2 = \frac{L_2}{2}(-F_{21y}\cos q_2 + F_{21x}\sin q_2). \tag{5.26}$$

The reaction components F_{21x} and F_{21y} are obtained from Eqs. 5.24 and 5.25

$$F_{21x} = m_2\left(L_1\ddot{q}_1\sin q_1 + L_1\dot{q}_1^2\cos q_1 + \frac{L_2}{2}\ddot{q}_2\sin q_2 + \frac{L_2}{2}\dot{q}_2^2\cos q_2\right),$$
$$F_{21y} = -m_2\left(L_1\ddot{q}_1\cos q_1 - L_1\dot{q}_1^2\sin q_1 + \frac{L_2}{2}\ddot{q}_2\cos q_2 - \frac{L_2}{2}\dot{q}_2^2\sin q_2\right)$$
$$+m_2 g. \tag{5.27}$$

The equations of motion are obtained substituting F_{21x} and F_{21y} in Eqs. 5.20 and 5.26

$$\frac{m_2L_1^2}{3}\ddot{q}_1 = -m_1g\frac{L_1}{2}\cos q_1$$
$$-m_2\left(L_1\ddot{q}_1\cos q_1 - L_1\dot{q}_1^2\sin q_1 + \frac{L_2}{2}\ddot{q}_2\cos q_2 - \frac{L_2}{2}\dot{q}_2^2\sin q_2 - g\right)L_1\cos q_1$$
$$-m_2\left(L_1\ddot{q}_1\sin q_1 + L_1\dot{q}_1^2\cos q_1 + \frac{L_2}{2}\ddot{q}_2\sin q_2 + \frac{L_2}{2}\dot{q}_2^2\cos q_2\right)L_1\sin q_1, \tag{5.28}$$

$$\frac{m_2L_2^2}{12}\ddot{q}_2$$
$$= \frac{m_2L_2}{2}\left(L_1\ddot{q}_1\cos q_1 - L_1\dot{q}_1^2\sin q_1 + \frac{L_2}{2}\ddot{q}_2\cos q_2 - \frac{L_2}{2}\dot{q}_2^2\sin q_2 - g\right)\cos q_2$$
$$+\frac{m_2L_2}{2}\left(L_1\ddot{q}_1\sin q_1 + L_1\dot{q}_1^2\cos q_1 + \frac{L_2}{2}\ddot{q}_2\sin q_2 + \frac{L_2}{2}\dot{q}_2^2\cos q_2\right)\sin q_2. \tag{5.29}$$

The equations of motion represent two non-linear differential equations. The initial conditions (Cauchy problem) are necessary to solve the equations. At $t = 0$ the initial conditions are

$$q_1(0) = q_{10},\ \dot{q}_1(0) = \omega_{10},$$
$$q_2(0) = q_{20},\ \dot{q}_2(0) = \omega_{20}.$$

The equations of motion for the mechanical system will be solved using MATLAB. First the reaction joint force $\mathbf{F}_{21}$ is calculated from Eq. 5.21:

```
F21 = -m2*aC2 + G2;
```

The moment equations for each link, Eqs. 5.19 and 5.22, using MATLAB are:

```
EqA = -IA*alpha1 + cross(rB, F21) + cross(rC1, G1);
Eq2 = -IC2*alpha2 + cross(rB - rC2, -F21);
```

Two lists `slist` and `nlist` are created:

```
slist={diff('q1(t)',t,2),diff('q2(t)',t,2),...
    diff('q1(t)',t),diff('q2(t)',t),'q1(t)','q2(t)'};
nlist={'ddq1', 'ddq2', 'x(2)', 'x(4)', 'x(1)','x(3)';
% diff('q1(t)',t,2) will be replaced by 'ddq1'
% diff('q2(t)',t,2) will be replaced by 'ddq2'
% diff('q1(t)',t) will be replaced by 'x(2)'
% diff('q2(t)',t) will be replaced by 'x(4)'
% 'q1(t)' will be replaced by 'x(1)'
% 'q2(t)' will be replaced by 'x(3)'
```

In the equations of motion `EqA` and `Eq2` the symbolical variables in `slist` are replaced with the symbolical variables in `nlist`:

```
eq1 = subs(EqA(3),slist,nlist);
eq2 = subs(Eq2(3),slist,nlist);
```

The previous equations are solved in terms of `'ddq1'` and `'ddq2'`

```
sol = solve(eq1,eq2,'ddq1, ddq2');
```

The second-order ODE system of two equations has to be re-written as a first-order system.

Let `x(1)`$=q_1(t)$, `x(2)`$=\dot{q}_1(t)$, `x(3)`$=q_2(t)$, and `x(4)`$=\dot{q}_2(t)$, this gives the first-order system:

```
d[x(1)]/dt = x(2), d[x(2)]/dt = ddq1,
d[x(3)]/dt = x(4), d[x(4)]/dt = ddq2.
```

The MATLAB commands for the first-order ODE system are:

```
dx1 = sym('x(2)');
dx2 = sol.ddq1;
dx3 = sym('x(4)');
dx4 = sol.ddq2;
dx1dt = char(dx1);
dx2dt = char(dx2);
dx3dt = char(dx3);
dx4dt = char(dx4);
```

The `inline` function `g` is defined for the right-hand side of the first-order system:

```
g = inline(sprintf('[%s;%s;%s;%s]',...
            dx1dt,dx2dt,dx3dt,dx4dt),'t','x');
```

The time t is going from an initial value t0 to a final value f:

```
t0 = 0; tf = 5; time = [0 tf];
```

The initial conditions at $t_0 = 0$ are $q_1(0) = -\pi/4$ rad, $\dot{q}_1(0) = 0$ rad/s, $q_2(0) = -\pi/3$ rad, $\dot{q}_2(0) = 0$ rad/s, or in MATLAB:

```
x0 = [-pi/4; 0; -pi/3; 0]; % define initial conditions
```

The numerical solution of all the components of the solution for t going from t0 to f is obtained using the command:

```
[t,xs] = ode45(g, time, x0);
```

where x0 is the initial value vector at the starting point t0. The plot of the solution curves q_1 and q_2 are obtained using the commands:

```
x1 = xs(:,1);
x3 = xs(:,3);
subplot(2,1,1),plot(t,x1*180/pi,'r'),...
xlabel('t (s)'),ylabel('q1 (deg)'),grid,...
subplot(2,1,2),plot(t,x3*180/pi,'b'),...
xlabel('t (s)'),ylabel('q2 (deg)'),grid
```

The plots using MATLAB are shown in Fig. 5.4 and the MATLAB program is given in Appendix D.3.

Instead of using the inline function g the system of differential equations can be solved numerically by m-file functions. The function file, RR.m is created using the statements:

```
sol = solve(eq1,eq2,'ddq1, ddq2');
dx2 = sol.ddq1;
dx4 = sol.ddq2;
dx2dt = char(dx2);
dx4dt = char(dx4);
% create the function file RR.m
fid = fopen('RR.m','w+');
fprintf(fid,'function dx = RR(t,x)\n');
fprintf(fid,'dx = zeros(4,1);\n');
fprintf(fid,'dx(1) = x(2);\n');
fprintf(fid,'dx(2) = ');
fprintf(fid,dx2dt);
fprintf(fid,';\n');
fprintf(fid,'dx(3) = x(4);\n');
fprintf(fid,'dx(4) = ');
```

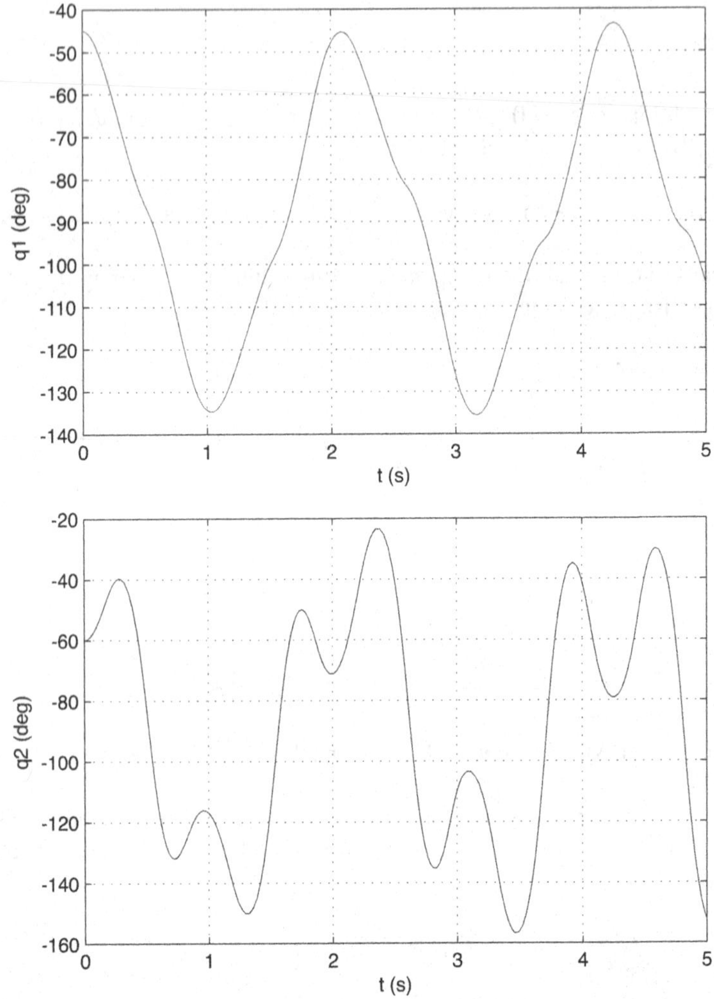

Fig. 5.4 Plot of solution curves q_1 and q_2

```
fprintf(fid,dx4dt);
fprintf(fid,';');
fclose(fid);
cd(pwd);
```

The terms `dx2dt` and `dx4dt` are calculated symbolically from the previous program (Appendix D.3). The MATLAB command `fid = fopen(file,perm)` opens the file `file` in the mode specified by `perm`. The mode `'w+'` deletes the contents of an existing file, or creates a new file, and opens it for reading and writing. The statement `fclose(fid)` closes the file associated with file identifier `fid`, the statement `cd` changes the current working directory, and `pwd` displays the current working directory. The `ode45` solver is used for the system of differential equations:

```
t0 = 0; tf = 5; time = [0 tf];
x0 = [-pi/4 0 -pi/3 0];
[t,xs] = ode45(@RR, time, x0);
```

The computing time for solving the system of differential equations is shorter using the function file `RR.m`. The MATLAB program is given in Appendix D.4.

5.3 One-Link Planar Robot Arm

Exercise

The robot arm shown in Fig. 5.5 is characterized by the length $L = 1$ m. The mass of the rigid body is $m = 1$ kg and the gravitational acceleration is $g = 9.81$ m/s^2. The initial conditions, at $t = 0$ s, are $\theta(0) = \pi/18$ rad and $\dot{\theta}(0) = 0$. The robot arm can be brought from an initial state of rest to a final state of rest in such a way that θ has the specified value $\theta_f = \pi/3$ rad. In the case of the robot arm the set of contact forces transmitted from 0 to 1 in order to drive the link 1 can be replaced with a couple of torque $\mathbf{T}_{01}$. The expression of $\mathbf{T}_{01}$ is

$$\mathbf{T}_{01} = T_{01x}\mathbf{\imath} + T_{01y}\mathbf{\jmath} + T_{01z}\mathbf{k} = T_{01z}\mathbf{k}.$$

The following feedback control law is used

$$T_{01z} = -\beta\dot{\theta} - \gamma(\theta - \theta_f) + 0.5\,gLm\cos\theta.$$

The constant gains are: $\beta = 45$ N m s/rad and $\gamma = 30$ N m/rad. Write a MATLAB program for solving the equations of motion.

Solution

The equation of motion for the robot arm is obtained symbolically using the MATLAB commands:

```
syms t
L = 1; m = 1; g = 9.81;
c = cos(sym('theta(t)')); s = sin(sym('theta(t)'));
xC = (L/2)*c; yC = (L/2)*s;
rC = [xC yC 0];
omega = [0 0 diff('theta(t)',t)];
alpha = diff(omega,t);
G = [0 -m*g 0];
```

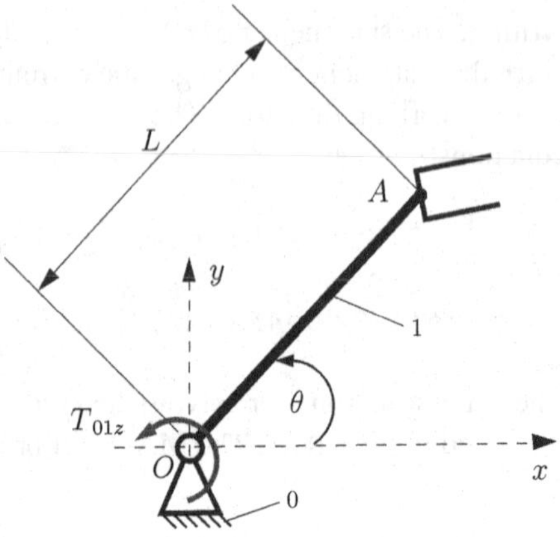

Fig. 5.5 One-link robot arm

```
IO = m*L^2/3;
beta = 45;
gamma = 30;
qf = pi/3;
T01z = -beta*diff('theta(t)',t)-...
        gamma*(sym('theta(t)')-qf)+0.5*g*L*m*c;
T01 = [0 0 T01z];
eq = -IO*alpha + cross(rC,G) + T01;
eqz = eq(3);
```

The equation has to be rewritten as a first-order system ($x_1 = \theta$ and $x_2 = \dot{\theta}$):

```
slist = {diff('theta(t)',t,2),diff('theta(t)',t),...
         'theta(t)'};
nlist = {'ddtheta', 'x(2)' , 'x(1)'};
eqI = subs(eqz,slist,nlist);
dx1 = sym('x(2)');
dx2 = solve(eqI,'ddtheta');
dx1dt = char(dx1);
dx2dt = char(dx2);
```

An `inline` function `g` is defined for the right-hand side of the first-order system:

```
g = inline(sprintf('[%s;%s]',dx1dt,dx2dt),'t','x');
```

and the solution is obtained using the commands:

```
time = [0 10];
x0 = [pi/18; 0]; % define initial conditions
[ts,xs] = ode45(g, 0:1:10, x0);
```

```
plot(ts,xs(:,1)*180/pi,'LineWidth',1.5),...
xlabel('t (s)'),ylabel('\theta (deg)'),...
grid,axis([0, 10, 0, 70])
fprintf('Results \n \n')
fprintf(' t(s) theta(rad) omega(rad/s) \n')
[ts,xs]
```

The plot of θ for the considered time interval, using MATLAB, is shown in Fig. 5.6. The MATLAB program and the results are given in Appendix D.5.

The system of differential equations can be solved numerically by m-file functions. The m-file function `Rrobot.m` is created:

```
function dx = Rrobot(t,x);
dx = zeros(2,1);
dx(1) = x(2);
dx(2) = -135*x(2)-90*x(1)+30*pi;
```

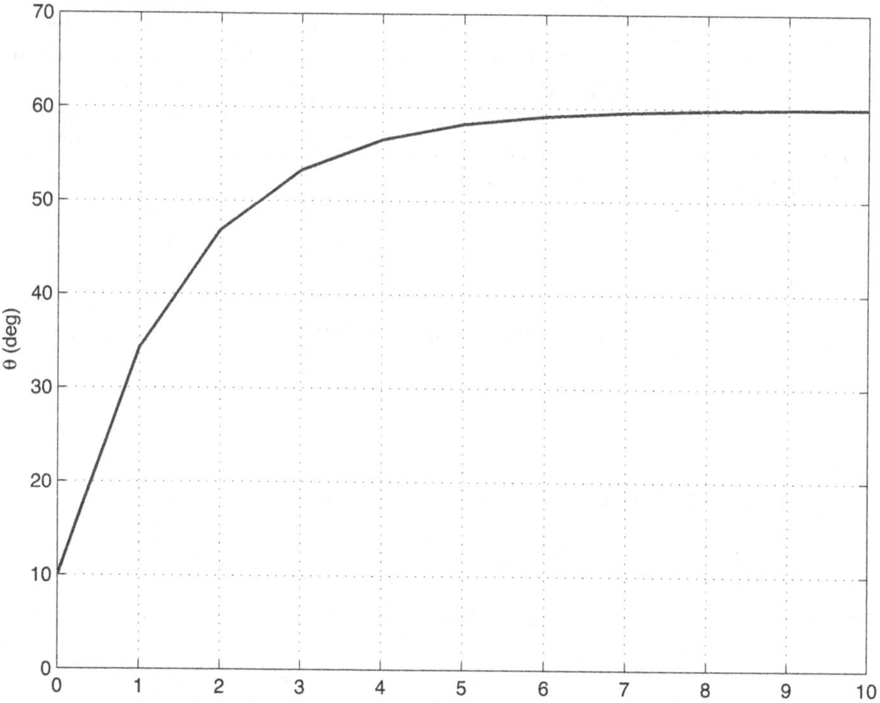

Fig. 5.6 Solution plot of θ

The `ode45` solver is used to solve the differential equations:

```
time = [0 10]; x0 = [pi/18 0];
[ts,xs] = ode45(@Rrobot, 0:1:10, x0);
```

and the MATLAB program is given in Appendix D.6.

5.4 Two-Link Planar Robot Arm

Exercise
A two-link planar robot arm is shown in Fig. 5.7. The lengths of the links are $L_1 =$ 1 m and $L_2 = 1$ m. The masses of the rigid links are $m_1 = 1$ kg and $m_2 = 1$ kg. The gravitational acceleration is $g = 9.81$ m/s^2. The generalized coordinates are $q_1(t)$ and $q_2(t)$ as shown in Fig. 5.7.

The initial conditions, at $t = 0$ s, are $q_1(0) = -\pi/18$ rad, $\dot{q}_1(0) = 0$ rad/s, $q_2(0) =$ $\pi/6$ rad, and $\dot{q}_2(0) = 0$ rad/s.

The robot arm can be brought from an initial state of rest to a final state of rest in such a way that q_1 and q_2 have the specified values $q_{1f} = \pi/6$ rad and $q_{2f} = \pi/3$ rad.

The set of contact forces transmitted from 0 to 1 can be replaced with a couple of torque $\mathbf{T}_{01} = T_{01z}\,\mathbf{k}$ applied to 1 at A. Similarly, the set of contact forces transmitted from 1 to 2 can be replaced with a couple of torque $\mathbf{T}_{12} = T_{12z}\,\mathbf{k}$ applied to 2 at B. The law of action and reaction then guarantees that the set of contact forces transmitted from 1 to 2 is equivalent to a couple of torque $-\mathbf{T}_{12}$ to 1 at B. The following feedback control laws are given

$$
\begin{aligned}
T_{01z} &= -\beta_{01}\,\dot{q}_1 - \gamma_{01}\,(q_1 - q_{1f}) + 0.5\,g\,L_1\,m_1\cos(q_1) + g\,L_1\,m_2\cos(q_1),\\
T_{12z} &= -\beta_{12}\,\dot{q}_2 - \gamma_{12}\,(q_2 - q_{2f}) + 0.5\,g\,L_2\,m_2\cos(q_2).
\end{aligned}
$$

The constant gains are: $\beta_{01} = 450$ N m s/rad, $\gamma_{01} = 300$ N m/rad, $\beta_{12} = 200$ N m s/rad, and $\gamma_{12} = 300$ N m/rad.

Write a MATLAB program for solving the equations of motion.

Solution
The MATLAB commands for the kinematics of the robot arm are:

```
L1 = 1; L2 = 1; m1 = 1; m2 = 1; g = 9.81;
t = sym('t','real');
xB = L1*cos(sym('q1(t)')); yB = L1*sin(sym('q1(t)'));
rB = [xB yB 0];
rC1 = rB/2; vC1 = diff(rC1,t); aC1 = diff(vC1,t);
xD = xB + L2*cos(sym('q2(t)'));
yD = yB + L2*sin(sym('q2(t)'));
rD = [xD yD 0];
rC2 = (rB + rD)/2; vC2 = diff(rC2,t);
aC2 = diff(vC2,t);
```

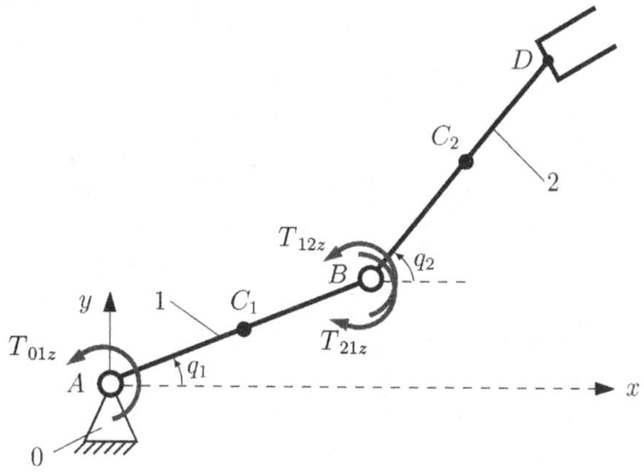

Fig. 5.7 Two-link robot arm

```
omega1 = [0 0 diff('q1(t)',t)];
alpha1 = diff(omega1,t);
omega2 = [0 0 diff('q2(t)',t)];
alpha2 = diff(omega2,t);
```

The weight forces on the links and the mass moment of inertia of the links are:

```
G1 = [0 -m1*g 0]; G2 = [0 -m2*g 0];
IC1=m1*L1^2/12; IA=IC1 + m1*(L1/2)^2;
IC2=m2*L2^2/12;
```

The joint reaction force $\mathbf{F}_{21}$ is calculated with:

```
F21 = -m2*aC2 + G2;
```

The control torques are given by:

```
b01 = 450; g01 = 300;
b12 = 200; g12 = 300;
q1f = pi/6;
q2f = pi/3;
T01z = -b01*diff('q1(t)',t)-g01*(sym('q1(t)')-q1f)...
       +0.5*g*L1*m1*cos(sym('q1(t)'))...
       +g*L1*m2*cos(sym('q1(t)'));
T01 = [0 0 T01z];
T12z = -b12*diff('q2(t)',t)-g12*(sym('q2(t)')-q2f)...
       +0.5*g*L2*m2*cos(sym('q2(t)'));
T12 = [0 0 T12z];
```

The moment equations for each link, Eqs. 5.19 and 5.22, using MATLAB are:

```
EqA=-IA*alpha1+cross(rB,F21)+cross(rC1,G1)+T01-T12;
Eq2 = -IC2*alpha2 + cross(rB - rC2, -F21) + T12;
slist = {diff('q1(t)',t,2),diff('q2(t)',t,2),...
  diff('q1(t)',t),diff('q2(t)',t),'q1(t)','q2(t)'};
nlist = {'ddq1','ddq2','x(2)','x(4)','x(1)''x(3)'};
eq1 = subs(EqA(3),slist,nlist);
eq2 = subs(Eq2(3),slist,nlist);
sol = solve(eq1,eq2,'ddq1, ddq2');
dx2 = sol.ddq1;
dx4 = sol.ddq2;
dx2dt = char(dx2);
dx4dt = char(dx4);
```

The equations of motion are complex and a m-file function `RRrobot.m` is constructed with the commands:

```
fid = fopen('RRrobot.m','w+');
fprintf(fid,'function dx = RRrobot(t,x)\n');
fprintf(fid,'dx = zeros(4,1);\n');
fprintf(fid,'dx(1) = x(2);\n');
fprintf(fid,'dx(2) = ');
fprintf(fid,dx2dt);
fprintf(fid,';\n');
fprintf(fid,'dx(3) = x(4);\n');
fprintf(fid,'dx(4) = ');
fprintf(fid,dx4dt);
fprintf(fid,';');
fclose(fid); cd(pwd);
```

The system of differential equations is solved using `ode45`:

```
t0 = 0; tf = 15;
time = [0 tf];
x0 = [-pi/18 0 pi/6 0];
[t,xs] = ode45(@RRrobot, time, x0);

x1 = xs(:,1);
x2 = xs(:,2);
x3 = xs(:,3);
x4 = xs(:,4);

subplot(2,1,1),plot(t,x1*180/pi,'r'),...
xlabel('t (s)'),ylabel('q1 (deg)'),grid,...
```

```
subplot(2,1,2),plot(t,x3*180/pi,'b'),...
xlabel('t (s)'),ylabel('q2 (deg)'),grid
[ts,xs] = ode45(@RRrobot,0:1:5,x0);
```

The plots of q_1 and q_2 for the considered time interval, using MATLAB, are shown in Fig. 5.8 and the MATLAB program is given in Appendix D.7.

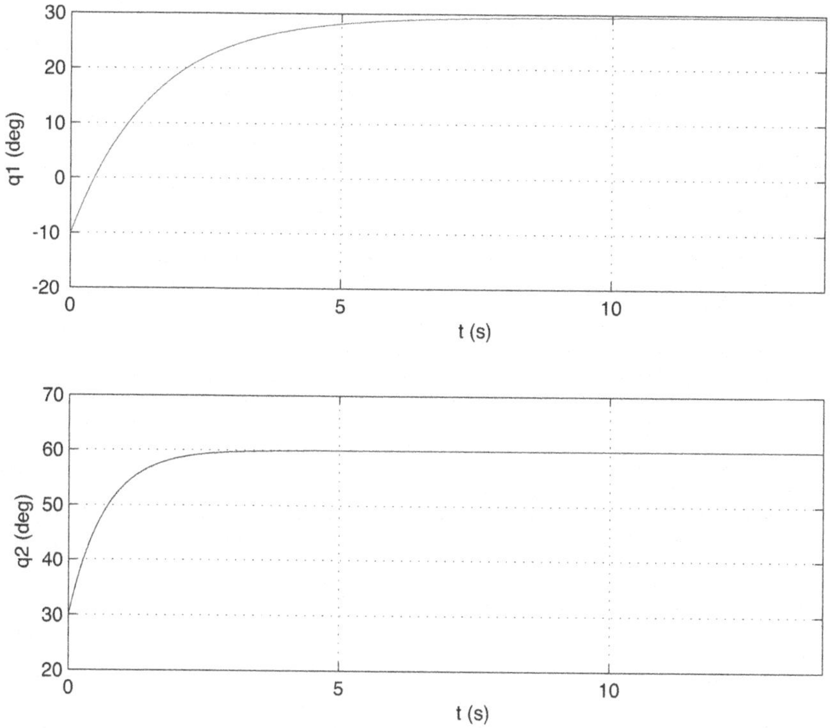

Fig. 5.8 Solution plots of q_1 and q_2

Chapter 6
Analytical Dynamics of Open Kinematic Chains

6.1 Generalized Coordinates and Constraints

Consider a system of N particles: $\{S\} = \{P_1, P_2, \ldots P_i \ldots P_N\}$. The position vector of the ith particle in the Cartesian reference frame is $\mathbf{r}_i = \mathbf{r}_i(x_i, y_i, z_i)$ and can be expressed as

$$\mathbf{r}_i = x_i\mathbf{\imath} + y_i\mathbf{\jmath} + z_i\mathbf{k}, \quad i = 1, 2, \ldots, N.$$

The system of N particles requires $n = 3N$ physical coordinates to specify its position. To analyze the motion of the system in many cases, it is more convenient to use a set of variables different from the physical coordinates. Let us consider a set of variables $q_1, q_2, \ldots, q_{3N}$ related to the physical coordinates by

$$\begin{aligned}
x_1 &= x_1(q_1, q_2, \ldots, q_{3N}),\\
y_1 &= y_1(q_1, q_2, \ldots, q_{3N}),\\
z_1 &= z_1(q_1, q_2, \ldots, q_{3N}),\\
&\ \ \vdots\\
x_{3N} &= x_{3N}(q_1, q_2, \ldots, q_{3N}),\\
y_{3N} &= y_{3N}(q_1, q_2, \ldots, q_{3N}),\\
z_{3N} &= z_{3N}(q_1, q_2, \ldots, q_{3N}).
\end{aligned}$$

The *generalized coordinates*, $q_1, q_2, \ldots, q_{3N}$, are the set of variables that can completely describe the position of the dynamical system. The *configuration space* is the space extended across the generalized coordinates. If the system of N particles has m constraint equations acting on it, the system can be represented uniquely by p *independent generalized coordinates* q_k, ($k = 1, 2, \ldots, p$), where $p = 3N - m = n - m$. The number p is called the number of degrees of freedom of the system.

The number of *degrees of freedom* is the minimum number of independent coordinates necessary to describe the dynamical system uniquely. The *generalized velocities*, denoted by $\dot{q}_k(t)$ $(k = 1, 2, \ldots, n)$, represent the rate of change of the generalized coordinates with respect to time.

The *state space* is the $2n$-dimensional space spanned by the generalized coordinates and generalized velocities.

The constraints are generally dominant as a result of contact between bodies, and they limit the motion of the bodies upon which they act. A *constraint equation* and a *constraint force* are related with a constraint. The constraint force is the joint reaction force and the constraint equation represents the kinematics of the contact.

Consider a smooth surface of equation

$$f(x,y,z,t) = 0, \tag{6.1}$$

where f has continuous second derivatives in all its variables. A particle P is subjected to a constraint of moving on the smooth surface described by Eq. 6.1. The constraint equation $f(x, y, z, t) = 0$ represents a *configuration constraint.*

The motion of the particle over the surface can be viewed as the motion of an otherwise free particle subjected to the constraint of moving on that particular surface. Hence, $f(x, y, z, t) = 0$ represents a constraint equation.

For a dynamical system with n generalized coordinates, a configuration constraint can be described as

$$f(q_1, q_2, \ldots, q_n, t) = 0. \tag{6.2}$$

The differential of the constraint f, given by Eq. 6.1, in terms of physical coordinates is

$$df = \frac{\partial f}{\partial x}dx + \frac{\partial f}{\partial y}dy + \frac{\partial f}{\partial z}dz + \frac{\partial f}{\partial t}dt = 0. \tag{6.3}$$

The differential of the constraint f, given by Eq. 6.2, in terms of the generalized coordinates is

$$df = \frac{\partial f}{\partial q_1}dq_1 + \frac{\partial f}{\partial q_2}dq_2 + \ldots + \frac{\partial f}{\partial q_n}dq_n + \frac{\partial f}{\partial t}dt = 0. \tag{6.4}$$

Equations 6.3 and 6.4 are called *constraint relations in Pfaffian form.* A constraint in Pfaffian form is a constraint that is represented in the form of differentials.

The *constraint equations in velocity form* (or *velocity constraints* or *motion constraints*) are obtained by dividing Eqs. 6.3 and 6.4 by dt

$$\frac{df}{dt} = \frac{\partial f}{\partial x}\dot{x} + \frac{\partial f}{\partial y}\dot{y} + \frac{\partial f}{\partial z}\dot{z} + \frac{\partial f}{\partial t} = 0, \tag{6.5}$$

$$\frac{df}{dt} = \frac{\partial f}{\partial q_1}\dot{q}_1 + \frac{\partial f}{\partial q_2}\dot{q}_2 + \ldots + \frac{\partial f}{\partial q_n}\dot{q}_n + \frac{\partial f}{\partial t} = 0. \tag{6.6}$$

The velocity constraint given by Eq. 6.5 can be represented as

$$a_x\dot{x} + a_y\dot{y} + a_z\dot{z} + a_0 = 0. \tag{6.7}$$

For a dynamical system with n generalized coordinates subjected to m constraints the velocity constraint given by Eq. 6.6 can be expressed as

$$\sum_{k=1}^{n} a_{jk}\dot{q}_k + a_{j0} = 0, \quad j = 1,2,...,m, \tag{6.8}$$

where a_{jk} and a_{j0} ($j = 1, 2, \ldots, m$; $k = 1, 2, \ldots, n$) are functions of the generalized coordinates and time.

A *holonomic* constraint is a constraint that can be represented as both a configuration constraint as well as velocity constraint. Constraints that do not have this property are called *non-holonomic* (non-holonomic constraints cannot be expressed as configuration constraints). When the constraint is non-holonomic, it can only be expressed in the form Eqs. 6.7 or 6.8, as an integrating factor does not exist to allow expression in the form of Eqs. 6.1 or 6.2.

A *scleronomic* system, $f(q_1, q_1, \ldots, q_n) = 0$, is an unconstrained dynamical system or a system subjected to a holonomic constraint that is not an explicit function of time. A *rhenomic* system is a system subjected to a holonomic constraint that is an explicit function of time.

6.2 Laws of Motion

Consider the motion of a system $\{S\}$ of ν particles $P_1, ..., P_\nu$ ($\{S\} = \{P_1, ..., P_\nu\}$) in an inertial reference frame (0). The equation of motion for the ith particle is

$$\mathbf{F}_i = m_i\,\mathbf{a}_i, \tag{6.9}$$

where $\mathbf{F}_i$ is the resultant of all contact and distance forces acting on P_i; m_i is the mass of P_i; and $\mathbf{a}_i$ is the acceleration of P_i in (0). Equation 6.9 is the expression of Newton's second law. The inertia force $\mathbf{F}_{\text{in}\,i}$ for P_i in (0) is defined as

$$\mathbf{F}_{\text{in}\,i} = -m_i\,\mathbf{a}_i, \tag{6.10}$$

then Eq. 6.9 is written as

$$\mathbf{F}_i + \mathbf{F}_{\text{in}\,i} = \mathbf{0}. \tag{6.11}$$

Equation 6.11 is the expression of D'Alembert's principle.

If $\{S\}$ is a holonomic system possessing n degrees of freedom, then the position vector $\mathbf{r}_i$ of P_i relative to a point O fixed in reference frame (0) is expressed as a vector function of n generalized coordinates $q_1, ..., q_n$ and time t

$$\mathbf{r}_i = \mathbf{r}_i(q_1, ..., q_n, t).$$

The velocity $\mathbf{v}_i$ of P_i in (0) has the form

$$\mathbf{v}_i = \sum_{r=1}^{n} \frac{\partial \mathbf{r}_i}{\partial q_r}\frac{\partial q_r}{\partial t} + \frac{\partial \mathbf{r}_i}{\partial t} = \sum_{r=1}^{n} \frac{\partial \mathbf{r}_i}{\partial q_r}\dot{q}_r + \frac{\partial \mathbf{r}_i}{\partial t}, \tag{6.12}$$

or as

$$\mathbf{v}_i = \sum_{r=1}^{n} (\mathbf{v}_i)_r \dot{q}_r + \frac{\partial \mathbf{r}_i}{\partial t},$$

where $(\mathbf{v}_i)_r$ is called the rth *partial velocity* of P_i in (0) and is defined as

$$(\mathbf{v}_i)_r = \frac{\partial \mathbf{r}_i}{\partial q_r} = \frac{\partial \mathbf{v}_i}{\partial \dot{q}_r}. \tag{6.13}$$

Next, replace Eq. 6.11 with

$$\sum_{i=1}^{\nu} (\mathbf{F}_i + \mathbf{F}_{\text{in}\,i}) \cdot (\mathbf{v}_i)_r = 0. \tag{6.14}$$

If a *generalized active force* Q_r and a *generalized inertia force* $K_{\text{in}\,r}$ are defined as

$$Q_r = \sum_{i=1}^{\nu} (\mathbf{v}_i)_r \cdot \mathbf{F}_i = \sum_{i=1}^{\nu} \frac{\partial \mathbf{r}_i}{\partial q_r} \cdot \mathbf{F}_i = \sum_{i=1}^{\nu} \frac{\partial \mathbf{v}_i}{\partial \dot{q}_r} \cdot \mathbf{F}_i, \tag{6.15}$$

and

$$K_{\text{in}\,r} = \sum_{i=1}^{\nu} (\mathbf{v}_i)_r \cdot \mathbf{F}_{\text{in}\,i} = \sum_{i=1}^{\nu} \frac{\partial \mathbf{r}_i}{\partial q_r} \cdot \mathbf{F}_{\text{in}\,i} = \sum_{i=1}^{\nu} \frac{\partial \mathbf{v}_i}{\partial \dot{q}_r} \cdot \mathbf{F}_{\text{in}\,i}, \tag{6.16}$$

then Eq. 6.14 can be written as

$$Q_r + K_{\text{in}\,r} = 0, \;\; r = 1, \ldots, n. \tag{6.17}$$

Equations 6.17 are *Kane's dynamical equations.*

Consider the generalized inertia force $K_{\text{in}\,r}$

$$\begin{aligned} K_{\text{in}\,r} &= \sum_{i=1}^{\nu} \mathbf{F}_{\text{in}\,i} \cdot (\mathbf{v}_i)_r = -\sum_{i=1}^{\nu} m_i \mathbf{a}_i \cdot (\mathbf{v}_i)_r = -\sum_{i=1}^{\nu} m_i \ddot{\mathbf{r}}_i \cdot \frac{\partial \mathbf{r}_i}{\partial q_r} \\ &= -\sum_{i=1}^{\nu} \left[\frac{d}{dt}\left(m_i \dot{\mathbf{r}}_i \cdot \frac{\partial \mathbf{r}_i}{\partial q_r} \right) - m_i \dot{\mathbf{r}}_i \cdot \frac{d}{dt}\left(\frac{\partial \mathbf{r}_i}{\partial q_r} \right) \right]. \end{aligned} \tag{6.18}$$

Now

$$\frac{d}{dt}\left(\frac{\partial \mathbf{r}_i}{\partial q_r} \right) = \sum_{k=1}^{n} \frac{\partial^2 \mathbf{r}_i}{\partial q_r \partial q_k}\dot{q}_k + \frac{\partial^2 \mathbf{r}_i}{\partial q_r \partial t} = \frac{\partial \mathbf{v}_i}{\partial q_r}, \tag{6.19}$$

and, furthermore, using Eq. 6.12

$$\frac{\partial \mathbf{v}_i}{\partial \dot{q}_r} = \frac{\partial \mathbf{r}_i}{\partial q_r}. \tag{6.20}$$

Substitution of Eq. 6.19 and Eq. 6.20 in Eq. 6.18 leads to

$$K_{\text{in}\,r} = -\sum_{i=1}^{\nu}\left[\frac{d}{dt}\left(m_i \mathbf{v}_i \cdot \frac{\partial \mathbf{v}_i}{\partial \dot{q}_r}\right) - m_i \mathbf{v}_i \cdot \frac{\partial \mathbf{v}_i}{\partial q_r}\right]$$

$$= -\left[\frac{d}{dt}\frac{\partial}{\partial \dot{q}_r}\left(\sum_{i=1}^{\nu}\frac{1}{2}m_i \mathbf{v}_i \cdot \mathbf{v}_i\right) - \frac{\partial}{\partial q_r}\left(\sum_{i=1}^{\nu}\frac{1}{2}m_i \mathbf{v}_i \cdot \mathbf{v}_i\right)\right].$$

The *kinetic energy* T of $\{S\}$ in reference frame (0) is defined as

$$T = \frac{1}{2}\sum_{i=1}^{\nu} m_i \mathbf{v}_i \cdot \mathbf{v}_i.$$

Therefore, the generalized inertia forces $K_{\text{in}\,r}$ are written as

$$K_{\text{in}\,r} = -\frac{d}{dt}\left(\frac{\partial T}{\partial \dot{q}_r}\right) + \frac{\partial T}{\partial q_r},$$

and Kane's dynamical equations can be written as

$$Q_r - \frac{d}{dt}\left(\frac{\partial T}{\partial \dot{q}_r}\right) + \frac{\partial T}{\partial q_r} = 0,$$

or

$$\frac{d}{dt}\left(\frac{\partial T}{\partial \dot{q}_r}\right) - \frac{\partial T}{\partial q_r} = Q_r.$$

The equations

$$\frac{d}{dt}\left(\frac{\partial T}{\partial \dot{q}_r}\right) - \frac{\partial T}{\partial q_r} = \sum_{i=1}^{\nu}\frac{\partial \mathbf{r}_i}{\partial q_r}\cdot \mathbf{F}_i, \quad r = 1,\dots,n, \tag{6.21}$$

or

$$\frac{d}{dt}\left(\frac{\partial T}{\partial \dot{q}_r}\right) - \frac{\partial T}{\partial q_r} = \sum_{i=1}^{\nu}\frac{\partial \mathbf{v}_i}{\partial \dot{q}_r}\cdot \mathbf{F}_i, \quad r = 1,\dots,n, \tag{6.22}$$

are known as *Lagrange's equations of motion* of the first kind.

6.3 Lagrange's Equations for Two-Link Robot Arm

Exercise

A two-link robot arm is considered in Fig. 5.7. The bars 1 and 2 are homogeneuos and have the lengths $L_1 = L_2 = L = 1$ m. The masses of the rigid links are $m_1 = m_2 = m = 1$ kg and the gravitational acceleration is $g = 9.81$ m/s^2. To characterize the instantaneous configuration of the system, two generalized coordinates $q_1(t)$ and $q_2(t)$ are employed. The generalized coordinates q_1 and q_2 denote the radian measure of the angles between the link 1 and 2 and the horizontal x-axis. The set of contact forces transmitted from 0 to 1 is replaced with a couple of torque $\mathbf{T}_{01} = T_{01z}\mathbf{k}$ applied to 1 at A, and the set of contact forces transmitted from 1 to 2 is replaced with a couple of torque $\mathbf{T}_{12} = T_{12z}\mathbf{k}$ applied to 2 at B.

The initial conditions, at $t = 0$ s, are $q_1(0) = \pi/18$ rad, $\dot{q}_1(0) = 0$ rad/s, $q_2(0) = \pi/6$ rad, and $\dot{q}_2(0) = 0$ rad/s. The robot arm can be brought from an initial state of rest to a final state of rest in such a way that q_1 and q_2 have the specified values $q_{1f} = \pi/6$ rad and $q_{2f} = \pi/3$ rad.

I. Direct Dynamics
The following feedback control laws are given

$$T_{01z} = -\beta_{01}\,\dot{q}_1 - \gamma_{01}\,(q_1 - q_{1f}) + 0.5\,g\,L_1\,m_1\,\cos q_1 + g\,L_1\,m_2\,\cos q_1,$$
$$T_{12z} = -\beta_{12}\,\dot{q}_2 - \gamma_{12}\,(q_2 - q_{2f}) + 0.5\,g\,L_2\,m_2\,\cos q_2.$$

The constant gains are: $\beta_{01} = 450$ N m s/rad, $\gamma_{01} = 300$ N m/rad, $\beta_{12} = 200$ N m s/rad, and $\gamma_{12} = 300$ N m/rad. Write a MATLAB® program for solving the equations of motion.

II. Inverse Dynamics
A desired motion of the robot arm is specified for a time interval $0 \le t \le T_p = 15$ s. The generalized coordinates can be established explicitly

$$q_r(t) = q_r(0) + \frac{q_r(T_p) - q_r(0)}{T_p}\left[t - \frac{T_p}{2\pi}\sin\left(\frac{2\pi t}{T_p}\right)\right], \quad r = 1,2,$$

with $q_r(T_p) = q_{rf}$. Find $T_{01z}(t)$ and $T_{12z}(t)$ for $0 \le t \le T_p = 15$ s.

Solution
The solution for the two-link robot arm will start with the dynamics when the the forces and moments are known and the equations are solved for the motion of the links.

I. Direct Dynamics
The position vector of the mass center of link 1 is

$$\mathbf{r}_{C_1} = 0.5\,L\cos q_1\,\imath + 0.5\,L\sin q_1\,\jmath,$$

and the position vector of the mass center of link 2 is

$$\mathbf{r}_{C_2} = (L\cos q_1 + 0.5\,L\cos q_2)\,\imath + (L\sin q_1 + 0.5\,L\sin q_2)\,\imath.$$

The velocity of C_1 is

$$\mathbf{v}_{C_1} = \frac{d\mathbf{r}_{C_1}}{dt} = \dot{\mathbf{r}}_{C_1} = -0.5\,L\dot{q}_1\,\sin q_1\,\imath + 0.5\,L\dot{q}_1\,\cos q_1\,\jmath,$$

and the velocity of C_2 is

$$\mathbf{v}_{C_2} = \frac{d\mathbf{r}_{C_2}}{dt} = \dot{\mathbf{r}}_{C_2}$$
$$= (-L\dot{q}_1 \sin q_1 - 0.5L\dot{q}_2 \sin q_2)\,\mathbf{ı} + (L\dot{q}_1 \cos q_1 + 0.5L\dot{q}_2 \cos q_2)\,\mathbf{ȷ}.$$

The angular velocity vectors of the links 1 and 2 are

$$\boldsymbol{\omega}_1 = \dot{q}_1\mathbf{k} \quad \text{and} \quad \boldsymbol{\omega}_2 = \dot{q}_2\mathbf{k}.$$

The MATLAB commands for the kinematics are:

```
syms t L1 L2 m1 m2 g
q1 = sym('q1(t)');
q2 = sym('q2(t)');
c1 = cos(q1); s1 = sin(q1);
c2 = cos(q2); s2 = sin(q2);
xB = L1*c1; yB = L1*s1; rB = [xB yB 0];
rC1 = rB/2; vC1 = diff(rC1,t);
xD = xB + L2*c2; yD = yB + L2*s2; rD = [xD yD 0];
rC2 = (rB + rD)/2; vC2 = diff(rC2,t);
omega1 = [0 0 diff(q1,t)];
omega2 = [0 0 diff(q2,t)];
```

Kinetic Energy
The kinetic energy of the link 1 that is in rotational motion is

$$T_1 = \frac{1}{2} I_A \boldsymbol{\omega}_1 \cdot \boldsymbol{\omega}_1 = \frac{1}{2} I_A \dot{q}_1^2 = \frac{1}{2}\frac{mL^2}{3}\dot{q}_1^2 = \frac{mL^2}{6}\dot{q}_1^2,$$

where I_A is the mass moment of inertia about the center of rotation A, $I_A = mL^2/3$. The kinetic energy of the bar 2 is due to the translation and rotation and can be expressed as

$$T_2 = \frac{1}{2} I_{C_2} \boldsymbol{\omega}_1 \cdot \boldsymbol{\omega}_1 + \frac{1}{2} m_2 \mathbf{v}_{C_2} \cdot \mathbf{v}_{C_2} = \frac{1}{2} I_{C_2} \dot{q}_2^2 + \frac{1}{2} m_2 \mathbf{v}_{C_2} \cdot \mathbf{v}_{C_2},$$

where I_{C_2} is the mass moment of inertia about the center of mass C_2, $I_{C_2} = mL^2/12$, and

$$\mathbf{v}_{C_2} \cdot \mathbf{v}_{C_2} = \mathbf{v}_{C_2}^2 = L^2\dot{q}_1^2 + \frac{1}{4}L^2\dot{q}_2^2 + L^2\dot{q}_1\dot{q}_2\cos(q_2 - q_1).$$

Equation 6.23 becomes

$$T_2 = \frac{1}{2}\frac{mL^2}{12}\dot{q}_2^2 + \frac{1}{2}mL^2\left[\dot{q}_1^2 + \frac{1}{4}\dot{q}_2^2 + \dot{q}_1\dot{q}_2\cos(q_2 - q_1)\right].$$

The total kinetic energy of the system is

$$T = T_1 + T_2 = \frac{mL^2}{6}\left[4\dot{q}_1^2 + 3\,\dot{q}_1\,\dot{q}_2\cos(q_2 - q_1) + \dot{q}_2^2\right].$$

The MATLAB commands for the kinetic energy are:

```
IA = Im1*L1^2/3; IC2 = m2*L2^2/12;
T1 = IA*omega1*omega1.'/2; % .' array transpose
T2 = m2*vC2*vC2.'/2 + IC2*omega2*omega2.'/2;
T2 = simple(T2); % simplest form of T2
T = expand(T1 + T2); % total kinetic energy
```

The MATLAB statements `A.'` is the array transpose of `A` and `simple(exp)` looks for simplest form of the symbolic expression `exp`. The MATLAB command `expand(exp)` expands trigonometric and algebraic functions.

The left-hand sides of Lagrange's equations $\partial T/\partial \dot{q}_i$, $i = 1,2$ are

$$\frac{\partial T}{\partial \dot{q}_1} = \frac{mL^2}{6}\left[8\dot{q}_1 + 3\dot{q}_2\cos(q_2 - q_1)\right] \text{ and } \frac{\partial T}{\partial \dot{q}_2} = \frac{mL^2}{6}\left[3\dot{q}_1\cos(q_2 - q_1) + 2\dot{q}_2\right].$$

To calculate the partial derivative of the kinetic energy `T` with respect to the variable `diff('q1(t)',t)` a MATLAB function, `deriv`, is created

```
function fout = deriv(f, g)
% deriv differentiates f with respect to g=g(t)
% the variable g=g(t) is a function of time
syms t x dx
lg = {diff(g, t), g};
lx = {dx, x};
f1 = subs(f, lg, lx);
f2 = diff(f1, x);
fout = subs(f2, lx, lg);
```

The function `deriv(f, g)` differentiates a symbolic expression `f` with respect to the variable `g`, where the variable `g` is a function of time `g = g(t)`. The statement `diff(f,'x')` differentiates `f` with respect to the free variable `x`. In MATLAB the free variable `x` cannot be a function of time and that is why the function `deriv` was introduced.

The partial derivatives of the kinetic energy T with respect to $\dot{q}_i$ or in MATLAB the partial derivatives of the kinetic energy `T` with respect to `diff('q1(t)',t)` and `diff('q2(t)',t)` are calculated with:

```
Tdq1 = deriv(T, diff(q1,t));
Tdq2 = deriv(T, diff(q1,t));
```

Next, the derivative of $\partial T/\partial \dot{q}_i$ with respect to time is calculated

$$\frac{d}{dt}\left(\frac{\partial T}{\partial \dot{q}_1}\right) = \frac{mL^2}{6}\left[8\ddot{q}_1 + 3\ddot{q}_2\cos(q_2-q_1) - 3\dot{q}_2(\dot{q}_2-\dot{q}_1)\sin(q_2-q_1)\right],$$
$$\frac{d}{dt}\left(\frac{\partial T}{\partial \dot{q}_2}\right) = \frac{mL^2}{6}\left[3\ddot{q}_1\cos(q_2-q_1) - 3\dot{q}_1(\dot{q}_2-\dot{q}_1)\sin(q_2-q_1) + 2\ddot{q}_2\right],$$

and in MATLAB the terms $\frac{d}{dt}\left(\frac{\partial T}{\partial \dot{q}_i}\right)$ are:

```
Tt1 = diff(Tdq1, t);
Tt2 = diff(Tdq2, t);
```

The partial derivative of the kinetic energy with respect to q_i are

$$\frac{\partial T}{\partial q_1} = \frac{mL^2}{6}3\dot{q}_1\dot{q}_2\sin(q_2-q_1) = \frac{mL^2}{2}\dot{q}_1\dot{q}_2\sin(q_2-q_1);$$
$$\frac{\partial T}{\partial q_2} = -\frac{mL^2}{6}3\dot{q}_1\dot{q}_2\sin(q_2-q_1) = -\frac{mL^2}{2}\dot{q}_1\dot{q}_2\sin(q_2-q_1),$$

and with MATLAB:

```
Tq1 = deriv(T, q1);
Tq2 = deriv(T, q2);
```

The left-hand side of Lagrange's equations, $\frac{d}{dt}\left(\frac{\partial T}{\partial \dot{q}_i}\right) - \frac{\partial T}{\partial q_i}$, with MATLAB are:

```
LHS1 = Tt1 - Tq1;
LHS2 = Tt2 - Tq2;
```

Generalized Active Forces

The gravity forces on links 1 and 2 at the mass centers C_1 and C_2

$$\mathbf{G}_1 = -m_1 g\mathbf{J} = -mg\mathbf{J} \text{ and } \mathbf{G}_2 = -m_2 g\mathbf{J} = -mg\mathbf{J}.$$

The torque transmitted from 0 to 1 at A is $\mathbf{T}_{01} = T_{01z}\mathbf{k}$ and the torque transmitted from 1 to 2 at B is $\mathbf{T}_{12} = T_{12z}\mathbf{k}$. The MATLAB commands for the net forces and moments are:

```
G1 = [0 -m1*g 0];
G2 = [0 -m2*g 0];
syms T01z T12z
T01 = [0 0 T01z];
T12 = [0 0 T12z];
```

There are two generalized forces. The generalized force associated to q_1 is

$$\begin{aligned} Q_1 &= \mathbf{G}_1 \cdot \frac{\partial \mathbf{r}_{C_1}}{\partial q_1} + \mathbf{T}_{01} \cdot \frac{\partial \omega_1}{\partial \dot{q}_1} - \mathbf{T}_{12} \cdot \frac{\partial \omega_1}{\partial \dot{q}_1} + \mathbf{G}_2 \cdot \frac{\partial \mathbf{r}_{C_2}}{\partial q_1} + \mathbf{T}_{12} \cdot \frac{\partial \omega_2}{\partial \dot{q}_1} \\ &= -mg\mathbf{J} \cdot (-0.5L \sin q_1 \mathbf{ı} + 0.5L \cos q_1 \mathbf{J}) + T_{01z} - T_{12z} \\ &\quad -mg\mathbf{J} \cdot (-L \sin q_1 \mathbf{ı} + L \cos q_1 \mathbf{J}) = -1.5mgL \cos q_1 + T_{01z} - T_{12z}. \end{aligned}$$

The generalized force associated to q_2 is

$$\begin{aligned} Q_2 &= \mathbf{G}_1 \cdot \frac{\partial \mathbf{r}_{C_1}}{\partial q_2} + \mathbf{T}_{01} \cdot \frac{\partial \omega_1}{\partial \dot{q}_2} - \mathbf{T}_{12} \cdot \frac{\partial \omega_1}{\partial \dot{q}_2} + \mathbf{G}_2 \cdot \frac{\partial \mathbf{r}_{C_2}}{\partial q_2} + \mathbf{T}_{12} \cdot \frac{\partial \omega_2}{\partial \dot{q}_2} \\ &= -mg\mathbf{J} \cdot (-0.5L \sin q_2 \mathbf{ı} + 0.5L \cos q_2 \mathbf{J}) + T_{12z} \\ &= -0.5mgL \cos q_2 + T_{12z}. \end{aligned}$$

The MATLAB commands for the partial derivatives of the position vectors of the mass centers,

$$\frac{\partial \mathbf{r}_{C_1}}{\partial q_1}, \frac{\partial \mathbf{r}_{C_2}}{\partial q_1}, \frac{\partial \mathbf{r}_{C_1}}{\partial q_2}, \frac{\partial \mathbf{r}_{C_2}}{\partial q_2}$$

are:

```
rC1_1 = deriv(rC1, q1);
rC2_1 = deriv(rC2, q1);
rC1_2 = deriv(rC1, q2);
rC2_2 = deriv(rC2, q2);
```

The MATLAB commands for the partial angular velocities,

$$\frac{\partial \omega_1}{\partial \dot{q}_1}, \frac{\partial \omega_2}{\partial \dot{q}_1}, \frac{\partial \omega_1}{\partial \dot{q}_2}, \frac{\partial \omega_2}{\partial \dot{q}_2}$$

are:

```
w1_1 = deriv(omega1, diff(q1,t));
w2_1 = deriv(omega2, diff(q1,t));
w1_2 = deriv(omega1, diff(q2,t));
w2_2 = deriv(omega2, diff(q2,t));
```

The generalized active forces are calculated with the MATLAB commands:

```
Q1 = rC1_1*G1.'+w1_1*T01.'+w1_1*(-T12.')+...
     rC2_1*G2.'+w2_1*T12.';

Q2 = rC1_2*G1.'+w1_2*T01.'+w1_2*(-T12.')+...
     rC2_2*G2.'+w2_2*T12.';
```

The two Lagrange's equations are

$$\frac{d}{dt}\left(\frac{\partial T}{\partial \dot{q}_1}\right) - \frac{\partial T}{\partial q_1} = Q_1,$$
$$1.333\,mL^2\,\ddot{q}_1 + 0.5\,mL^2\,\ddot{q}_2\cos(q_2 - q_1) - 0.5mL^2\dot{q}_2^2\sin(q_2 - q_1)$$
$$+1.5\,mgL\cos q_1 - T_{01z} + T_{12z} = 0;$$

$$\frac{d}{dt}\left(\frac{\partial T}{\partial \dot{q}_2}\right) - \frac{\partial T}{\partial q_2} = Q_2,$$
$$0.5\,mL^2\,\ddot{q}_1\cos(q_2 - q_1) + 0.333\,mL^2\,\ddot{q}_2 + 0.5\,mL^2\,\dot{q}_1^2\sin(q_2 - q_1)$$
$$+0.5\,mgL\cos q_2 - T_{12z} = 0, \tag{6.23}$$

or in MATLAB:

```
Lagrange1 = LHS1-Q1;
Lagrange2 = LHS2-Q2;
```

The feedback control laws are

$$T_{01z} = -\beta_{01}\,\dot{q}_1 - \gamma_{01}\,(q_1 - q_{1f}) + 0.5\,g\,L_1\,m_1\cos q_1 + g\,L_1\,m_2\cos q_1,$$
$$T_{12z} = -\beta_{12}\,\dot{q}_2 - \gamma_{12}\,(q_2 - q_{2f}) + 0.5\,g\,L_2\,m_2\cos q_2,$$

with $\beta_{01} = 450$ N m s/rad, $\gamma_{01} = 300$ N m/rad, $\beta_{12} = 200$ N m s/rad, and $\gamma_{12} = 300$ N m/rad.

The feedback control torques using MATLAB commands are:

```
b01 = 450; g01 = 300;
b12 = 200; g12 = 300;
q1f = pi/6;
q2f = pi/3;

T01zc = -b01*diff(q1,t)-g01*(q1-q1f)+ ...
        0.5*g*L1*m1*c1+g*L1*m2*c1;
T12zc = -b12*diff(q2,t)-g12*(q2-q2f)+0.5*g*L2*m2*c2;
tor = {T01z, T12z};
torf = {T01zc, T12zc};
```

The feedback control torques are introduced into Lagrange's equations:

```
Lagrang1 = subs(Lagrange1, tor, torf);
Lagrang2 = subs(Lagrange2, tor, torf);
```

The numerical data for L_1, L_2, m_1, m_2, and g are introduced in MATLAB with the lists:

```
data = {L1, L2, m1, m2, g };
datn = {1 , 1 , 1 , 1 , 9.81};
```

and are substituted into Lagrange's equations:

```
Lagran1 = subs(Lagrang1, data, datn);
Lagran2 = subs(Lagrang2, data, datn);
```

The two second-order Lagrange's equations have to be rewritten as a first-order system and two MATLAB lists are created:

```
ql={diff(q1,t,2),diff(q2,t,2),...
    diff(q1,t),diff(q2,t),q1,q2};
qf={'ddq1','ddq2','x(2)','x(4)','x(1)','x(3)'};
% ql                        qf
% ------------------------------
% diff('q1(t)',t,2)  -> 'ddq1'
% diff('q2(t)',t,2)  -> 'ddq2'
%   diff('q1(t)',t)  -> 'x(2)'
%   diff('q2(t)',t)  -> 'x(4)'
%           'q1(t)'  -> 'x(1)'
%           'q2(t)'  -> 'x(3)'
```

In the expression of Lagrange's equations:

`diff('q1(t)',t,2)` is replaced by `'ddq1'`,
`diff('q2(t)',t,2)` is replaced by `'ddq2'`,
`diff('q1(t)',t)` is replaced by `'x(2)'`,
`diff('q2(t)',t)` is replaced by `'x(4)'`,
`'q1(t)'` is replaced by `'x(1)'`, and
`'q2(t)'` is replaced by `'x(3)'`

or:

```
Lagra1 = subs(Lagran1, ql, qf);
Lagra2 = subs(Lagran2, ql, qf);
```

Lagrange's equations are solved in terms of `'ddq1'` ($\ddot{q}_1$) and `'ddq2'` ($\ddot{q}_2$):

```
sol = solve(Lagra1,Lagra2,'ddq1, ddq2');
Lagr1 = sol.ddq1; Lagr2 = sol.ddq2;
```

The system of differential equations is solved numerically by m-file functions. The function file, RR_Lagr.m is created using the statements:

```
dx2dt = char(Lagr1);
dx4dt = char(Lagr2);

fid = fopen('RR_Lagr.m','w+');
fprintf(fid,'function dx = RR_Lagr(t,x)\n');
fprintf(fid,'dx = zeros(4,1);\n');
fprintf(fid,'dx(1) = x(2);\n');
fprintf(fid,'dx(2) = ');
fprintf(fid,dx2dt);
fprintf(fid,';\n');
fprintf(fid,'dx(3) = x(4);\n');
fprintf(fid,'dx(4) = ');
fprintf(fid,dx4dt);
fprintf(fid,';');
fclose(fid); cd(pwd);
```

The ode45 solver is used for the system of differential equations

```
t0 = 0; tf = 15; time = [0 tf];
x0 = [pi/18 0 pi/6 0];
[t,xs] = ode45(@RR_Lagr, time, x0);
x1 = xs(:,1);
x2 = xs(:,2);
x3 = xs(:,3);
x4 = xs(:,4);
subplot(2,1,1),plot(t,x1*180/pi,'r'),...
xlabel('t (s)'),ylabel('q1 (deg)'),grid,...
subplot(2,1,2),plot(t,x3*180/pi,'b'),...
xlabel('t (s)'),ylabel('q2 (deg)'),grid
[ts,xs] = ode45(@RR_Lagr,0:1:5,x0);
fprintf('Results \n\n')
fprintf('t(s) q1(rad) dq1(rad/s) q2(rad) dq2(rad/s)\n')
[ts,xs]
```

A MATLAB computer program for the direct dynamics is given in the Appendix E.1.

II. Inverse Dynamics

The generalized coordinates are given explicitly for $0 \le t \le T_p = 15$ s

$$q_r(t) = q_r(0) + \frac{q_{rf}(T_p) - q_r(0)}{T_p}\left[t - \frac{T_p}{2\pi}\sin\left(\frac{2\pi t}{T_p}\right)\right], \quad r = 1,2. \tag{6.24}$$

The initial conditions, at $t = 0$ s, are $q_1(0) = \pi/18$ rad and $q_2(0) = \pi/6$ rad. The robot arm is brought from an initial state of rest to a final state of rest in such a way that q_1 and q_2 have the specified values $q_{1f}(T_p) = \pi/6$ rad and $q_{2f}(T_p) = \pi/3$ rad. Figure 6.1 shows the plots of $q_1(t)$ and $q_2(t)$ rad.

The MATLAB commands for finding Lagrage's equations are identical with the commands presented in *Direct Dynamics*:

```
........................

syms T01z T12z
T01 = [0 0 T01z];
T12 = [0 0 T12z];
Q1 = rC1_1*G1.'+w1_1*T01.'+w1_1*(-T12.')+...
     rC2_1*G2.'+w2_1*T12.';
```

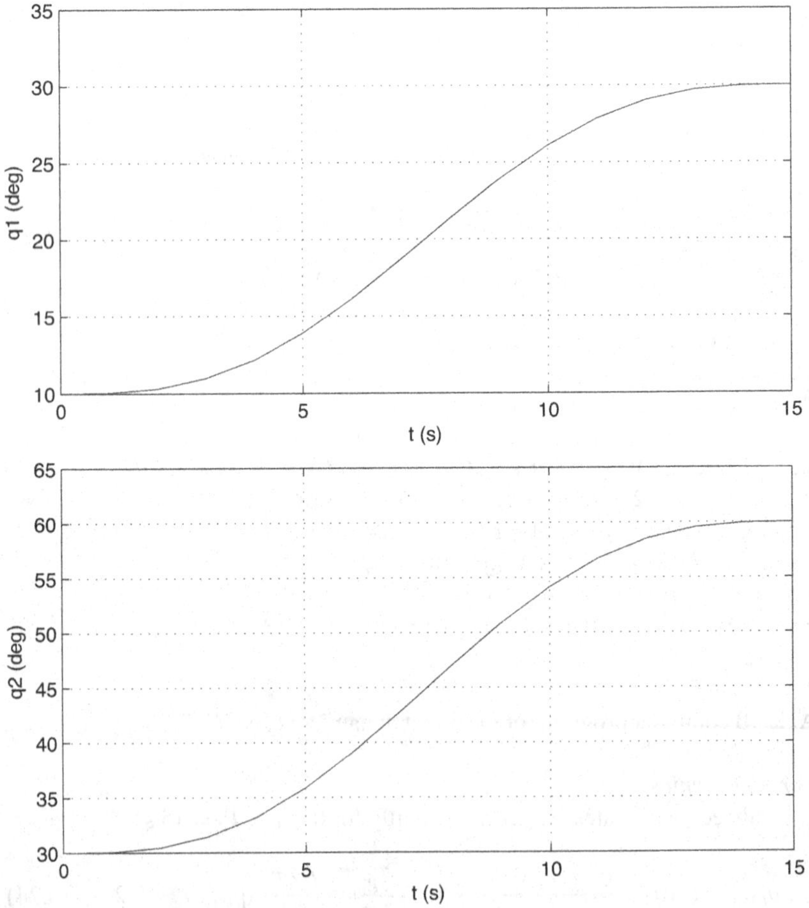

Fig. 6.1 Generalized coordinates $q_1(t)$ and $q_2(t)$

```
Q2 = rC1_2*G1.'+w1_2*T01.'+w1_2*(-T12.')+...
     rC2_2*G2.'+w2_2*T12.';
Lagrange1 = LHS1-Q1; Lagrange2 = LHS2-Q2;
data = {L1, L2, m1, m2, g};
datn = {1 , 1 , 1 , 1 , 9.81};
Lagr1 = subs(Lagrange1, data, datn);
Lagr2 = subs(Lagrange2, data, datn);
```

From Lagrange's equations of motions the torques T_{01z} and T_{12z} are calculated:

```
sol = solve(Lagr1,Lagr2,'T01z, T12z');
T01zc = sol.T01z;
T12zc = sol.T12z;
```

The generalized coordinates, q_1 and q_2, given by Eq. 6.24 and their derivatives, $\dot{q}_1, \dot{q}_2, \ddot{q}_1, \ddot{q}_2$, are substituted in the expressions of T_{01z} and T_{12z}:

```
q1f = pi/6 ; q2f = pi/3;
q1s = pi/18; q2s = pi/6;
Tp=15.;
q1n = q1s+(q1f-q1s)/Tp*(t-Tp/(2*pi)*sin(2*pi/Tp*t));
q2n = q2s+(q2f-q2s)/Tp*(t-Tp/(2*pi)*sin(2*pi/Tp*t));
dq1n = diff(q1n,t);
dq2n = diff(q2n,t);
ddq1n = diff(dq1n,t);
ddq2n = diff(dq2n,t);

ql={diff(q1,t,2),diff(q2,t,2),...
    diff(q1,t),diff(q2,t),q1,q2};
qn={ddq1n,ddq2n,dq1n,dq2n,q1n,q2n};
% ql                      qn
% ----------------------------
% diff('q1(t)',t,2)  -> ddq1n
% diff('q2(t)',t,2)  -> ddq2n
%   diff('q1(t)',t)  -> dq1n
%   diff('q2(t)',t)  -> dq2n
%           'q1(t)'  -> q1n
%           'q2(t)'  -> q1n
T01zt = subs(T01zc, ql, qn);
T12zt = subs(T12zc, ql, qn);
```

The MATLAB statement `ezplot(f, [min,max])` plots `f(t)` over the domain: `min < t < max`. The plots of $T_{01z}(t)$ and $T_{12z}(t)$ are obtained with the help of MATLAB function `ezplot`:

```
subplot(2,1,1), ezplot(T01zt,[0,Tp]),...
title(''), xlabel('t (s)'), ylabel('T01z (N m)'),grid

subplot(2,1,2),ezplot(T12zt,[0,Tp]),...
title(''), xlabel('t (s)'), ylabel('T12z (N m)'),grid
```

Another way of plotting $T_{01z}(t)$ and $T_{12z}(t)$ is:

```
time = 0:1:Tp;
T01t = subs(T01zt,'t',time);
T12t = subs(T12zt,'t',time);
subplot(2,1,1),plot(time,T01t),...
xlabel('t (s)'),ylabel('T01z (N m)'),grid
subplot(2,1,2),plot(time,T12t),...
xlabel('t (s)'),ylabel('T12z (N m)'),grid
```

Figure 6.2 shows the control torques and the MATLAB program is given in Appendix E.2.

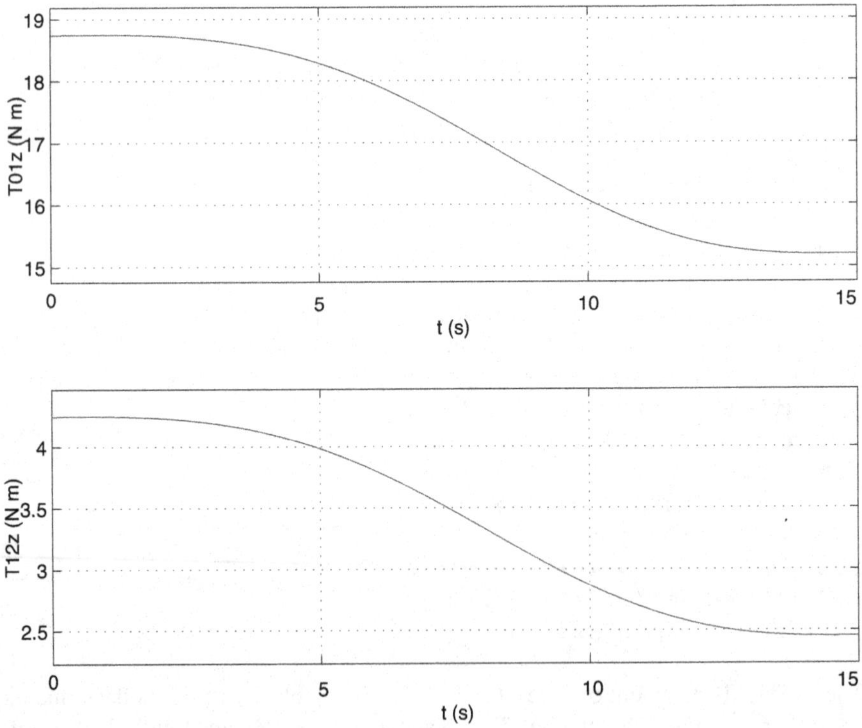

Fig. 6.2 Control torques

6.4 Rotation Transformation

Two orthogonal reference frames, $Oxyz$ and $O'x'y'z'$, are considered. The unit vectors of the reference frame $Oxyz$ are $\mathbf{ı}$, $\mathbf{ȷ}$, $\mathbf{k}$ and the unit vectors of the reference frame $O'x'y'z'$ are $\mathbf{ı}'$, $\mathbf{ȷ}'$, $\mathbf{k}'$. The origins of the reference frames may coincide because only the orientation of the axes is of interest $O = O'$.

The angles between the x'-axis and each of the x, y, z axes are the direction angles α, β, and γ $(O < \alpha, \beta, \gamma < \pi)$ as shown in Fig. 6.3. The unit vector $\mathbf{ı}'$ can be expressed in terms of $\mathbf{ı}$, $\mathbf{ȷ}$, $\mathbf{k}$ and the direction angles

$$\mathbf{ı}' = (\mathbf{ı}' \cdot \mathbf{ı})\mathbf{ı} + (\mathbf{ı}' \cdot \mathbf{ȷ})\mathbf{ȷ} + (\mathbf{ı}' \cdot \mathbf{k})\mathbf{k} = \cos\alpha\, \mathbf{ı} + \cos\beta\, \mathbf{ȷ} + \cos\gamma\, \mathbf{k}.$$

The cosines of the direction angles are the direction cosines and $\cos^2\alpha + \cos^2\beta + \cos^2\gamma = 1$.

With the notations $\cos\alpha = a_{x'x}$, $\cos\beta = a_{x'y}$, and $\cos\gamma = a_{x'z}$ the unit vector $\mathbf{ı}'$ is

$$\mathbf{ı}' = a_{x'x}\, \mathbf{ı} + a_{x'y}\, \mathbf{ȷ} + a_{x'z}\, \mathbf{k}.$$

In a similar way, the unit vectors $\mathbf{ȷ}'$ and $\mathbf{k}'$ are

$$\mathbf{ȷ}' = a_{y'x}\, \mathbf{ı} + a_{y'y}\, \mathbf{ȷ} + a_{y'z}\, \mathbf{k},$$
$$\mathbf{k}' = a_{z'x}\, \mathbf{ı} + a_{z'y}\, \mathbf{ȷ} + a_{z'z}\, \mathbf{k},$$

where $a_{r's} = a_{rs'}$ are the cosine of the angle between axis r' and axis s, with r and r representing x, y, or z. In matrix form

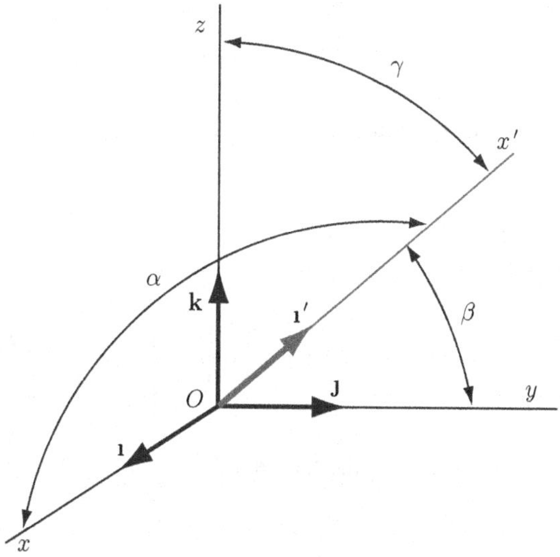

Fig. 6.3 Direction angles α, β, and γ

$$\begin{bmatrix} \mathbf{\imath}' \\ \mathbf{\jmath}' \\ \mathbf{k}' \end{bmatrix} = \mathbf{R} \begin{bmatrix} \mathbf{\imath} \\ \mathbf{\jmath} \\ \mathbf{k} \end{bmatrix},$$

where

$$\mathbf{R} = \begin{bmatrix} a_{x'x} & a_{x'y} & a_{x'z} \\ a_{y'x} & a_{y'y} & a_{y'z} \\ a_{z'x} & a_{z'y} & a_{z'z} \end{bmatrix}.$$

The matrix $\mathbf{R}$ is the *rotation transformation matrix* from xyz to $x'y'z'$. The unit vectors $\mathbf{\imath}, \mathbf{\jmath}, \mathbf{k}$ are an orthogonal set of unit vectors and the unit vectors $\mathbf{\imath}', \mathbf{\jmath}', \mathbf{k}'$ are an orthogonal set too. Using these properties it results that

$$\mathbf{R} \cdot \mathbf{R}^T = \mathbf{I},$$

where $\mathbf{I}$ is the identity matrix. Multiplication of Eq. 6.25 by $\mathbf{R}^{-1}$ gives

$$\mathbf{R}^{-1} = \mathbf{R}^T.$$

The matrix R is an orthonormal matrix because $\mathbf{R}^{-1} = \mathbf{R}^T$.

Let $\mathbf{R}'$ be the transformation matrix from $\mathbf{\imath}, \mathbf{\jmath}, \mathbf{k}$ to $\mathbf{\imath}', \mathbf{\jmath}', \mathbf{k}'$

$$\begin{bmatrix} \mathbf{\imath} \\ \mathbf{\jmath} \\ \mathbf{k} \end{bmatrix} = \mathbf{R}' \begin{bmatrix} \mathbf{\imath}' \\ \mathbf{\jmath}' \\ \mathbf{k}' \end{bmatrix}. \tag{6.25}$$

The matrix $\mathbf{R}'$ is the inverse of the original transformation matrix $\mathbf{R}$, which is identical to the transpose of $\mathbf{R}$.

$$\mathbf{R}' = \mathbf{R}^{-1} = \mathbf{R}^T.$$

Any vector $\mathbf{p}$ is independent of the reference frame used to describe its components, so

$$\mathbf{p} = p_x\,\mathbf{\imath} + p_y\,\mathbf{\jmath} + p_z\,\mathbf{k} = p_{x'}\,\mathbf{\imath}' + p_{y'}\,\mathbf{\jmath}' + p_{z'}\,\mathbf{k}',$$

or in matrix form as

$$[\mathbf{\imath}'\ \mathbf{\jmath}'\ \mathbf{k}'] \begin{bmatrix} p_{x'} \\ p_{y'} \\ p_{z'} \end{bmatrix} = [\mathbf{\imath}\ \mathbf{\jmath}\ \mathbf{k}] \begin{bmatrix} p_x \\ p_y \\ p_z \end{bmatrix}.$$

Using Eq. 6.25 and the fact that the transpose of a product is the product of the transposes the following relation is obtained

$$[\mathbf{\imath}' \ \mathbf{\jmath}' \ \mathbf{k}'] \begin{bmatrix} p_{x'} \\ p_{y'} \\ p_{z'} \end{bmatrix} = [\mathbf{\imath}' \ \mathbf{\jmath}' \ \mathbf{k}'][\mathbf{R}']^T \begin{bmatrix} p_x \\ p_y \\ p_z \end{bmatrix}.$$

With $[\mathbf{R}']^T = \mathbf{R}$ the above equation leads to

$$\begin{bmatrix} p_{x'} \\ p_{y'} \\ p_{z'} \end{bmatrix} = \mathbf{R} \begin{bmatrix} p_x \\ p_y \\ p_z \end{bmatrix}.$$

When the reference frame $x'y'z'$ is the result of a simple rotation about one of the axes of the reference frame xyz the following transformation matrices are obtained (Fig. 6.4):

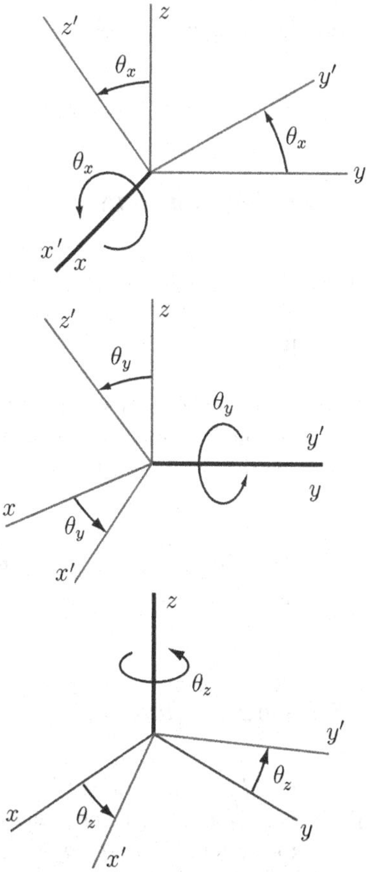

Fig. 6.4 $x'y'z'$ as a result of a simple rotation about one of the axes of xyz

- the reference frame xyz is rotated by an angle θ_x about the x-axis

$$\mathbf{R}(x,\theta_x) = \mathbf{R}(\theta_x) = \begin{bmatrix} 1 & 0 & 0 \\ 0 & \cos\theta_x & \sin\theta_x \\ 0 & -\sin\theta_x & \cos\theta_x \end{bmatrix},$$

- the reference frame xyz is rotated by an angle θ_y about the y-axis

$$\mathbf{R}(y,\theta_y) = \mathbf{R}(\theta_y) = \begin{bmatrix} \cos\theta_y & 0 & -\sin\theta_y \\ 0 & 1 & 0 \\ \sin\theta_y & 0 & \cos\theta_y \end{bmatrix},$$

- the reference frame xyz is rotated by an angle θ_y about the z-axis

$$\mathbf{R}(z,\theta_z) = \mathbf{R}(\theta_z) = \begin{bmatrix} \cos\theta_z & \sin\theta_z & 0 \\ -\sin\theta_z & \cos\theta_z & 0 \\ 0 & 0 & 1 \end{bmatrix}.$$

The following property holds

$$\mathbf{R}(s,-\theta_s) = \mathbf{R}^T(s,\theta_s), \quad s = x, y, z.$$

6.5 RRT Robot Arm

Figure 6.5 is a schematic representation of a RRT robot arm consisting of three links 1, 2, and 3. Let m_1, m_2, m_3 be the masses of 1, 2, 3, respectively. Link 1 can be rotated at A in a "fixed" reference frame (0) of unit vectors $[\mathbf{\imath}_0, \mathbf{\jmath}_0, \mathbf{k}_0]$ about a vertical axis $\mathbf{\imath}_0$. The unit vector $\mathbf{\imath}_0$ is fixed in 1. The link 1 is connected to link 2 at the pin joint B. The element 2 rotates relative to 1 about an axis fixed in both 1 and 2, passing through B, and perpendicular to the axis of 1. The last link 3 is connected to 2 by means of a slider joint. The mass centers of links 1, 2, and 3 are C_1, C_2, and C_3, respectively. The distances $L_1 = AC_1$, $L_B = AB = 2L_1$, and $L_2 = BC_2$ are indicated in Fig. 6.5a The length of link 1 is $2L_1$ and the length of link 2 is $2L_2$. The reference frame (1) of the unit vectors $[\mathbf{\imath}_1, \mathbf{\jmath}_1, \mathbf{k}_1]$ is attached to link 1, and the reference frame (2) of the unit vectors $[\mathbf{\imath}_2, \mathbf{\jmath}_2, \mathbf{k}_2]$ is attached to link 2, as shown in Fig. 6.5b.

6.5.1 Direct Dynamics

Generalized Coordinates and Transformation Matrices
The generalized coordinates (quantities associated with the the instantaneous position of the system) are $q_1(t)$, $q_2(t)$, $q_3(t)$. The first generalized coordinate q_1 denotes the radian measure of the angle between the axes of (1) and (0).

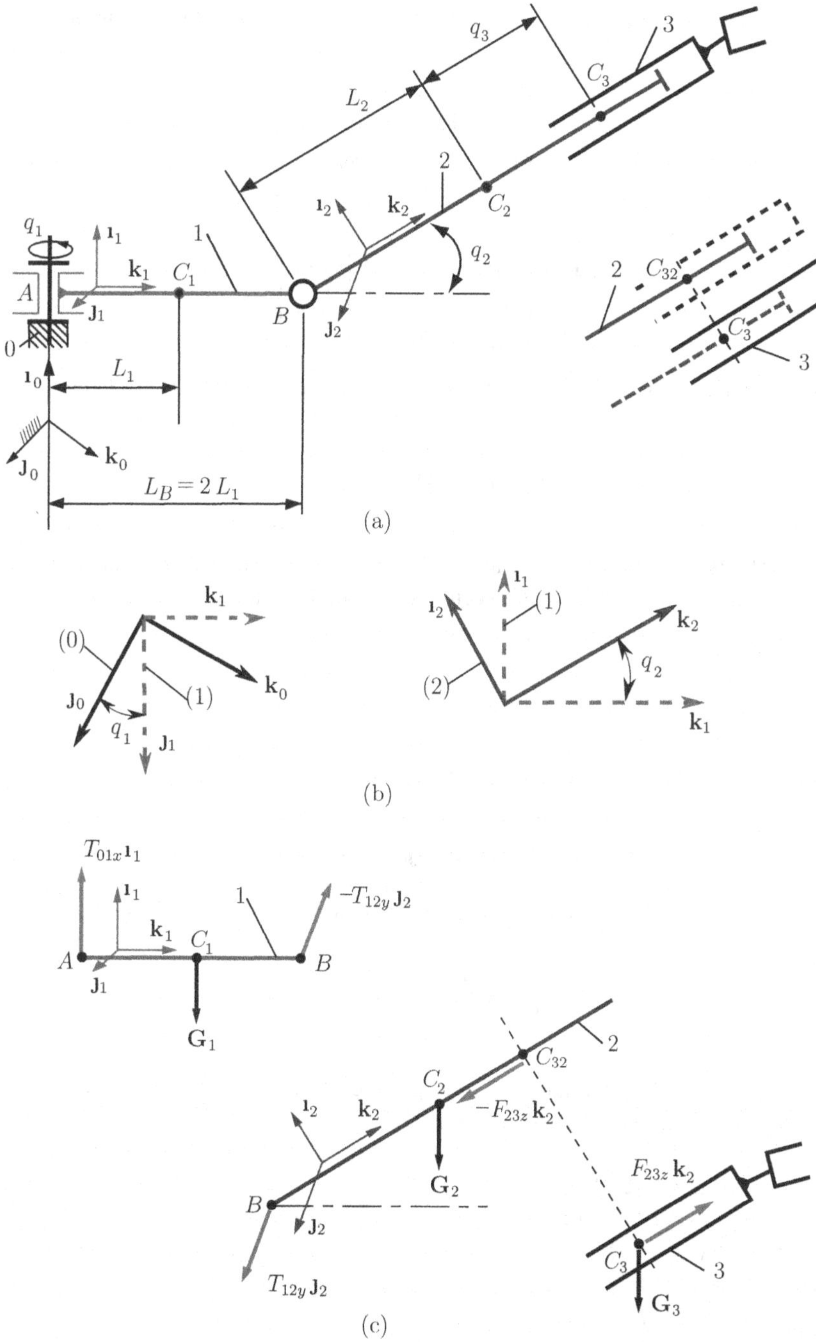

Fig. 6.5 RRT robot arm

The unit vectors $\mathbf{ı}_1$, $\mathbf{ȷ}_1$, and $\mathbf{k}_1$ can be expressed as functions of $\mathbf{ı}_0$, $\mathbf{ȷ}_0$, and $\mathbf{k}_0$

$$\begin{aligned}\mathbf{ı}_1 &= \mathbf{ı}_0,\\ \mathbf{ȷ}_1 &= c_1\,\mathbf{ȷ}_0 + s_1\,\mathbf{k}_0,\\ \mathbf{k}_1 &= -s_1\,\mathbf{ȷ}_0 + c_1\,\mathbf{k}_0,\end{aligned} \tag{6.26}$$

or

$$\begin{bmatrix}\mathbf{ı}_1\\ \mathbf{ȷ}_1\\ \mathbf{k}_1\end{bmatrix} = \begin{bmatrix}1 & 0 & 0\\ 0 & c_1 & s_1\\ 0 & -s_1 & c_1\end{bmatrix}\begin{bmatrix}\mathbf{ı}_0\\ \mathbf{ȷ}_0\\ \mathbf{k}_0\end{bmatrix},$$

where $s_1 = \sin q_1$ and $c_1 = \cos q_1$. The transformation matrix from (1) to (0) is

$$R_{10} = \begin{bmatrix}1 & 0 & 0\\ 0 & c_1 & s_1\\ 0 & -s_1 & c_1\end{bmatrix}. \tag{6.27}$$

The second generalized coordinate also designates a radian measure of the rotation angle between (1) and (2). The unit vectors $\mathbf{ı}_2$, $\mathbf{ȷ}_2$ and $\mathbf{k}_2$ can be expressed as

$$\begin{aligned}\mathbf{ı}_2 &= c_2\mathbf{ı}_1 - s_2\mathbf{k}_1\\ &= c_2\mathbf{ı}_0 + s_1s_2\,\mathbf{ȷ}_0 - c_1s_2\,\mathbf{k}_0,\\ \mathbf{ȷ}_2 &= \mathbf{ȷ}_1,\\ &= c_1\,\mathbf{ȷ}_0 + s_1\,\mathbf{k}_0,\\ \mathbf{k}_2 &= s_2\mathbf{ı}_1 + c_2\mathbf{k}_1\\ &= s_2\mathbf{ı}_0 - c_2s_1\,\mathbf{ȷ}_0 + c_1c_2\,\mathbf{k}_0,\end{aligned} \tag{6.28}$$

where $s_2 = \sin q_2$ and $c_2 = \cos q_2$. The transformation matrix from (2) to (1) is

$$R_{21} = \begin{bmatrix}c_2 & 0 & -s_2\\ 0 & 1 & 0\\ s_2 & 0 & c_2\end{bmatrix}. \tag{6.29}$$

The last generalized coordinate q_3 is the distance from C_2 to C_3. The MATLAB commands for the transformation matrices are:

```
q1 = sym('q1(t)');
q2 = sym('q2(t)');
q3 = sym('q3(t)');
c1 = cos(q1);
s1 = sin(q1);
c2 = cos(q2);
s2 = sin(q2);
```

```
% transformation matrix from RF1 to RF0
R10 = [[1 0 0]; [0 c1 s1]; [0 -s1 c1]];
% transformation matrix from RF2 to RF1
R21 = [[c2 0 -s2]; [0 1 0]; [s2 0 c2]];
```

Angular Velocities

Next, the angular velocity of the links 1, 2, and 3 will be expressed in the fixed reference frame (0). The angular velocity of 1 in (0) is

$$\boldsymbol{\omega}_{10} = \dot{q}_1 \mathbf{\imath}_1. \tag{6.30}$$

The angular velocity of the link 2 with respect to (1) is

$$\boldsymbol{\omega}_{21} = \dot{q}_2 \mathbf{\jmath}_2.$$

The angular velocity of the link 2 with respect to the fixed reference frame (0) is

$$\boldsymbol{\omega}_{20} = \boldsymbol{\omega}_{10} + \boldsymbol{\omega}_{21} = \dot{q}_1 \mathbf{\imath}_1 + \dot{q}_2 \mathbf{\jmath}_2.$$

With $\mathbf{\imath}_0 = \mathbf{\imath}_1 = c_2 \mathbf{\imath}_2 + s_2 \mathbf{k}_2$ the angular velocity of the link 2 in the reference frame (0) written in terms of the reference frame (2) is

$$\boldsymbol{\omega}_{20} = \dot{q}_1 (c_2 \mathbf{\imath}_2 + s_2 \mathbf{k}_2) + \dot{q}_2 \mathbf{\jmath}_2 = \dot{q}_1 c_2 \mathbf{\imath}_2 + \dot{q}_2 \mathbf{\jmath}_2 + \dot{q}_1 s_2 \mathbf{k}_2. \tag{6.31}$$

The link 3 has the same rotational motion as link 2, i.e., $\boldsymbol{\omega}_{30} = \boldsymbol{\omega}_{20}$.

Angular Accelerations

The angular acceleration of the link 1 in the reference frame (0) is

$$\boldsymbol{\alpha}_{10} = \ddot{q}_1 \mathbf{\imath}_1. \tag{6.32}$$

The angular acceleration of the link 2 with respect to the reference frame (0) is

$$\boldsymbol{\alpha}_{20} = \frac{d}{dt}\boldsymbol{\omega}_{20} = \frac{{}^{(2)}d}{dt}\boldsymbol{\omega}_{20} + \boldsymbol{\omega}_{20} \times \boldsymbol{\omega}_{20} = \frac{{}^{(2)}d}{dt}\boldsymbol{\omega}_{20},$$

where $\frac{{}^{(2)}d}{dt}$ represents the derivative with respect to time in reference frame (2), $[\mathbf{\imath}_2, \mathbf{\jmath}_2, \mathbf{k}_2]$. The angular acceleration of the link 2 is

$$\begin{aligned}\boldsymbol{\alpha}_{20} &= \frac{{}^{(2)}d}{dt}(\dot{q}_1 c_2 \mathbf{\imath}_2 + \dot{q}_2 \mathbf{\jmath}_2 + \dot{q}_1 s_2 \mathbf{k}_2) \\ &= (\ddot{q}_1 c_2 - \dot{q}_1 \dot{q}_2 s_2)\mathbf{\imath}_2 + \ddot{q}_2 \mathbf{\jmath}_2 + (\ddot{q}_1 s_2 + \dot{q}_1 \dot{q}_2 c_2)\mathbf{k}_2. \end{aligned} \tag{6.33}$$

The link 3 has the same angular acceleration as link 2, i.e., $\boldsymbol{\alpha}_{30} = \boldsymbol{\alpha}_{20}$.

The MATLAB commands for the angular velocities and accelerations are:

```
% angular velocity of link 1 in RF0
% expressed in terms of RF1 {i1,j1,k1}
w10 = [diff(q1,t) 0 0 ];

% angular velocity of link 2 in RF0
% expressed in terms of RF1 {i1,j1,k1}
w201 = [diff(q1,t) diff(q2,t) 0];

% angular velocity of link 2 in RF0
% expressed in terms of RF2 {i2,j2,k2}
w20 = w201 * transpose(R21);

% angular acceleration of link 1
% in RF0 expressed in terms of RF1 {i1,j1,k1}
alpha10 = diff(w10,t);

% angular acceleration of link 2 in RF0
% expressed in terms of RF2 {i2,j2,k2}
alpha20 = diff(w20,t);
```

Linear Velocities
The position vector of C_1, the mass center of link 1, is

$$\mathbf{r}_{C_1} = L_1\mathbf{k}_1,$$

and the velocity of C_1 in (0) is

$$\begin{aligned}\mathbf{v}_{C_1} &= \frac{d}{dt}\mathbf{r}_{C_1} = \frac{^{(1)}d}{dt}\mathbf{r}_{C_1} + \boldsymbol{\omega}_{10}\times\mathbf{r}_{C_1}\\ &= \mathbf{0} + \begin{vmatrix}\mathbf{1}_1 & \mathbf{J}_1 & \mathbf{k}_1\\ \dot{q}_1 & 0 & 0\\ 0 & 0 & L_1\end{vmatrix} = -\dot{q}_1 L_1\mathbf{J}_1.\end{aligned} \tag{6.34}$$

With MATLAB the position and velocity vectors of C_1 are:

```
% position vector of mass center C1 of link 1
% in RF0 expressed in terms of RF1 {i1,j1,k1}
rC1 = [0 0 L1];

% linear velocity of mass center C1 of link 1
% in RF0 expressed in terms of RF1 {i1,j1,k1}
vC1 = diff(rC1,t) + cross(w10, rC1);
```

The position vector of C_2, the mass center of link 2, is

$$\begin{aligned}\mathbf{r}_{C_2} &= L_B\mathbf{k}_1 + L_2\mathbf{k}_2 = L_B(-s_2\mathbf{\imath}_2 + c_2\mathbf{k}_2) + L_2\mathbf{k}_2\\ &= -L_B s_2\mathbf{\imath}_2 + (L_B c_2 + L_2)\mathbf{k}_2,\end{aligned}$$

where $L_B = 2L_1$. The velocity of C_2 in (0) is

$$\begin{aligned}\mathbf{v}_{C_2} &= \frac{d}{dt}\mathbf{r}_{C_2} = \frac{^{(2)}d}{dt}\mathbf{r}_{C_2} + \boldsymbol{\omega}_{20}\times\mathbf{r}_{C_2}\\ &= -L_B c_1\dot{q}_2\mathbf{\imath}_2 - L_B c_2\dot{q}_2\mathbf{k}_2 + \begin{vmatrix}\mathbf{\imath}_2 & \mathbf{\jmath}_2 & \mathbf{k}_2\\ \dot{q}_1 c_2 & \dot{q}_2 & \dot{q}_1 s_2\\ -L_B s_2 & 0 & L_B c_2 + L_2\end{vmatrix}\\ &= L_2\dot{q}_2\mathbf{\imath}_2 - (L_B + L_2 c_2)\dot{q}_1\mathbf{\jmath}_2.\end{aligned} \tag{6.35}$$

The position and velocity vectors of C_2, with MATLAB, are:

```
% position vector of mass center C2 of link 2
% in RF0 expressed in terms of RF2 {i2,j2,k2}
rC2 = [0 0 2*L1]*transpose(R21) + [0 0 L2];

% linear velocity of mass center C2 of link 2
% in RF0 expressed in terms of RF2 {i2,j2,k2}
vC2 = simple(diff(rC2,t) + cross(w20,rC2));
```

The position vector of C_3 with respect to reference frame (0) is

$$\begin{aligned}\mathbf{r}_{C_3} &= \mathbf{r}_{C_2} + q_3\mathbf{k}_2\\ &= -L_B s_2\mathbf{\imath}_2 + (L_B c_2 + L_2 + q_3)\mathbf{k}_2,\end{aligned}$$

and the velocity of this mass center in (0) is

$$\begin{aligned}\mathbf{v}_{C_3} &= \frac{d}{dt}\mathbf{r}_{C_3} = \frac{^{(2)}d}{dt}\mathbf{r}_{C_3} + \boldsymbol{\omega}_{20}\times\mathbf{r}_{C_3}\\ &= -L_B c_2\dot{q}_2\mathbf{\imath}_2 - (L_B c_2\dot{q}_2 + \dot{q}_3)\mathbf{k}_2 + \begin{vmatrix}\mathbf{\imath}_2 & \mathbf{\jmath}_2 & \mathbf{k}_2\\ \dot{q}_1 c_2 & \dot{q}_2 & \dot{q}_1 s_2\\ -L_B s_2 & 0 & L_B c_2 + L_2 + q_3\end{vmatrix}\\ &= (L_2 + q_3)\dot{q}_2\mathbf{\imath}_2 - (L_B + L_2 c_2 + c_2 q_2)\dot{q}_1\mathbf{\jmath}_2 + \dot{q}_3\mathbf{k}_2.\end{aligned} \tag{6.36}$$

The position and velocity vectors of C_3, with MATLAB, are:

```
% position vector of mass center C3 of link 3 in RF0
% expressed in terms of RF2 {i2,j2,k2}
rC3 = rC2 + [0 0 q3];
```

```
% linear velocity of mass center C3 of link 3 in RF0
% expressed in terms of RF2 {i2,j2,k2}
vC3 = simple(diff(rC3,t) + cross(w20,rC3));
```

There is a point C_{32} on link 2 ($C_{32} \in$ link 2) that instantaneously coincides with C_3, ($C_3 \in$ link 3). The velocity of point C_{32} is

$$\begin{aligned}\mathbf{v}_{C_{32}} &= \mathbf{v}_{C_2} + \boldsymbol{\omega}_{20} \times \mathbf{r}_{C_2C_3} = \mathbf{v}_{C_2} + \boldsymbol{\omega}_{20} \times q_3\mathbf{k}_2 \\ &= (L_2 + q_3)\dot{q}_2\mathbf{\imath}_2 - (L_B + L_2c_2 + c_2q_2)\dot{q}_1\mathbf{J}_2. \qquad (6.37)\end{aligned}$$

The point C_{32} of link 2 is superposed with the point C_3 of link 3. The velocity of mass center C_3 of link 3 in (0) can be computed in terms of the velocity of C_{32} using the relation

$$\mathbf{v}_{C_3} = \mathbf{v}_{C_{32}} + \dot{q}_3\mathbf{k}_2.$$

The velocity vector of C_{32}, with MATLAB, is:

```
vC32 = simple(vC2 + cross(w20,[0 0 q3]));
```

Linear Accelerations

The acceleration of C_1 is

$$\begin{aligned}\mathbf{a}_{C_1} &= \frac{d}{dt}\mathbf{v}_{C_1} = \frac{^{(1)}d}{dt}\mathbf{v}_{C_1} + \boldsymbol{\omega}_{10} \times \mathbf{v}_{C_1} = -L_1\ddot{q}_1\mathbf{J} + \begin{vmatrix} \mathbf{\imath}_1 & \mathbf{J}_1 & \mathbf{k}_1 \\ \dot{q}_1 & 0 & 0 \\ 0 & -L_1\dot{q}_1 & 0 \end{vmatrix} \\ &= -L_1\ddot{q}_1\mathbf{J}_1 - L_1\dot{q}_1^2\mathbf{k}_1. \qquad (6.38)\end{aligned}$$

The linear acceleration of the mass center C_2 is

$$\mathbf{a}_{C_2} = \frac{d}{dt}\mathbf{v}_{C_2} = \frac{^{(2)}d}{dt}\mathbf{v}_{C_2} + \boldsymbol{\omega}_{20} \times \mathbf{v}_{C_2}. \qquad (6.39)$$

The linear acceleration of C_2 is symbolically calculated in the program given in Appendix E.3. The acceleration of C_3 is

$$\mathbf{a}_{C_3} = \frac{d}{dt}\mathbf{v}_{C_3} = \frac{^{(2)}d}{dt}\mathbf{v}_{C_3} + \boldsymbol{\omega}_{20} \times \mathbf{v}_{C_3}. \qquad (6.40)$$

The linear acceleration of C_1, C_2, and C_3 are symbolically calculated with MATLAB:

```
aC1 = simple(diff(vC1,t)+cross(w10,vC1));
aC2 = simple(diff(vC2,t)+cross(w20,vC2));
aC3 = simple(diff(vC3,t)+cross(w20,vC3));
```

Generalized Forces

Remark: If a set of contact and/or body forces acting on a rigid body is equivalent to a couple of torque **T** together with force **R** applied at a point P of the rigid body, then the contribution of this set of forces to the generalized force, Q_r, is given by

$$Q_r = \frac{\partial \boldsymbol{\omega}}{\partial \dot{q}_r} \cdot \mathbf{T} + \frac{\partial \mathbf{v}_P}{\partial \dot{q}_r} \cdot \mathbf{R}, \quad r = 1, 2, ...,$$

where $\boldsymbol{\omega}$ is the angular velocity of the rigid body in (0), $\mathbf{v}_P$ is the velocity of P in (0), and r represents the generalized coordinates.

In the case of the robotic arm, there are two kinds of forces that contribute to the generalized forces Q_1, Q_2, and Q_3 namely, contact forces applied in order to drive the links 1, 2, and 3, and gravitational forces exerted on 1, 2, and 3 by the Earth. The set of contact forces transmitted from 0 to 1 can be replaced with a couple of torque $\mathbf{T}_{01}$ applied to 1 at A, Fig. 6.5c. Similarly, the set of contact forces transmitted from 1 to 2 can be replaced with a couple of torque $\mathbf{T}_{12}$ applied to 2 at B, Fig. 6.5c. The law of action and reaction then guarantees that the set of contact forces transmitted from 1 to 2 is equivalent to a couple of torque $-\mathbf{T}_{12}$ to 1 at B. Next, the set of contact forces exerted by link 2 on link 3 can be replaced with a force $\mathbf{F}_{23}$ applied to 3 at C_3, Fig. 6.5c. The law of action and reaction guarantees that the set of contact forces transmitted from 3 to 2 is equivalent to a force $-\mathbf{F}_{23}$ applied to 2 at C_{32}. The point C_{32} ($C_{32} \in$ link 2) instantaneously coincides with C_3, ($C_3 \in$ link 3). The expressions $\mathbf{T}_{01}$, $\mathbf{T}_{12}$, and $\mathbf{F}_{23}$ are

$$\mathbf{T}_{01} = T_{01x}\mathbf{1}_1 + T_{01y}\mathbf{J}_1 + T_{01z}\mathbf{k}_1, \quad \mathbf{T}_{12} = T_{12x}\mathbf{1}_2 + T_{12y}\mathbf{J}_2 + T_{12z}\mathbf{k}_2, \quad \text{and}$$
$$\mathbf{F}_{23} = F_{23x}\mathbf{1}_2 + F_{23y}\mathbf{J}_2 + F_{23z}\mathbf{k}_2.$$

The MATLAB statements for the contact torques and contact force are:

```
syms T01x T01y T01z T12x T12y T12z F23x F23y F23z
% contact torque of 0 that acts on link 1
% in RF0 expressed in terms of RF1 {i1,j1,k1}
T01 = [T01x T01y T01z];
% contact torque of link 1 that acts on link 2
% in RF0 expressed in terms of RF2 {i2,j2,k2}
T12 = [T12x T12y T12z];
% contact force of link 2 that acts on link 3 at C3
% in RF0 expressed in terms of RF2 {i2,j2,k2}
F23 = [F23x F23y F23z];
```

The external gravitational forces exerted on the links 1, 2, and 3 by the Earth, can be denoted by $\mathbf{G}_1$, $\mathbf{G}_2$, and $\mathbf{G}_3$ respectively, Fig. 6.5c, and can be expressed as

$$\mathbf{G}_1 = -m_1 g \mathbf{1}_1, \quad \mathbf{G}_2 = -m_2 g \mathbf{1}_1 = -m_2 g (c_2 \mathbf{1}_2 + s_2 \mathbf{k}_2), \quad \text{and}$$
$$\mathbf{G}_3 = -m_3 g \mathbf{1}_1 = -m_3 g (c_2 \mathbf{1}_2 + s_2 \mathbf{k}_2).$$

The reason for replacing $\mathbf{\imath}_1$ with $c_2\mathbf{\imath}_2 + s_2\mathbf{k}_2$ in connection with the forces $\mathbf{G}_2$ and $\mathbf{G}_3$ is that they are soon to be dot-multiplied with $\frac{\partial \mathbf{v}_{C_2}}{\partial \dot{q}_r}$ and $\frac{\partial \mathbf{v}_{C_3}}{\partial \dot{q}_r}$ That have been expressed in terms of $\mathbf{\imath}_2, \mathbf{\jmath}_2$, and $\mathbf{k}_2$.

The MATLAB statements for the gravitational forces are:

```
% gravitational force that acts on link 1 at C1
% RF0 expressed in terms of RF1 {i1,j1,k1}
G1 = [-m1*g 0 0]
% gravitational force that acts on link 2 at C2
% in RF0 expressed in terms of RF2 {i2,j2,k2}
G2 = [-m2*g 0 0]*transpose(R21)
% gravitational force that acts on link 3 at C3
% in RF0 expressed in terms of RF2 {i2,j2,k2}
G3 = [-m3*g 0 0]*transpose(R21)
```

One can express $(Q_r)_1$, the contribution to the generalized active force Q_r of all the forces and torques acting on the particles of the link 1, as

$$(Q_r)_1 = \frac{\partial \boldsymbol{\omega}_{10}}{\partial \dot{q}_r} \cdot (\mathbf{T}_{01} - \mathbf{T}_{12}) + \frac{\partial \mathbf{v}_{C_1}}{\partial \dot{q}_r} \cdot \mathbf{G}_1, \ r = 1,2,3.$$

The contribution $(Q_r)_2$ to the generalized active force of all the forces and torques acting on the link 2 is

$$(Q_r)_2 = \frac{\partial \boldsymbol{\omega}_{20}}{\partial \dot{q}_r} \cdot \mathbf{T}_{12} + \frac{\partial \mathbf{v}_{C_2}}{\partial \dot{q}_r} \cdot \mathbf{G}_2 + \frac{\partial \mathbf{v}_{C_{32}}}{\partial \dot{q}_r} \cdot (-\mathbf{F}_{23}), \ r = 1,2,3,$$

where $\mathbf{v}_{C_{32}} = \mathbf{v}_{C_3} - \dot{q}_3\mathbf{k}_2$.

The contribution $(Q_r)_3$, to the generalized active force of all the forces and torques acting on the link 3 is

$$(Q_r)_3 = \frac{\partial \mathbf{v}_{C_3}}{\partial \dot{q}_r} \cdot \mathbf{G}_3 + \frac{\partial \mathbf{v}_{C_3}}{\partial \dot{q}_r} \cdot \mathbf{F}_{23}, \ r = 1,2,3.$$

The generalized active force Q_r of all the forces and torques acting on the links 1, 2, and 3 are

$$\begin{aligned} Q_r &= (Q_r)_1 + (Q_r)_2 + (Q_r)_3 \\ &= \frac{\partial \boldsymbol{\omega}_{10}}{\partial \dot{q}_r} \cdot (\mathbf{T}_{01} - \mathbf{T}_{12}) + \frac{\partial \mathbf{v}_{C_1}}{\partial \dot{q}_r} \cdot \mathbf{G}_1 + \frac{\partial \boldsymbol{\omega}_{20}}{\partial \dot{q}_r} \cdot \mathbf{T}_{12} + \frac{\partial \mathbf{v}_{C_2}}{\partial \dot{q}_r} \cdot \mathbf{G}_2 + \frac{\partial \mathbf{v}_{C_{32}}}{\partial \dot{q}_r} \cdot (-\mathbf{F}_{23}) \\ &+ \frac{\partial \mathbf{v}_{C_3}}{\partial \dot{q}_r} \cdot \mathbf{G}_3 + \frac{\partial \mathbf{v}_{C_3}}{\partial \dot{q}_r} \cdot \mathbf{F}_{23}, \qquad r = 1, 2, 3. \end{aligned}$$

The generalized forces $Q_r, r = 1,2,3$ are symbolically calculated in the program given in Appendix E.3 and have the values

$$Q_1 = T_{01x},$$
$$Q_2 = T_{12y} - g\,m_2\,L_2\,c_2 - g\,m_3\,c_2\,(L_2 + q_3),$$
$$Q_3 = F_{23z} - g\,m_3\,s_2. \tag{6.41}$$

The MATLAB statements for the partial velocity $\dfrac{\partial \omega_{10}}{\partial \dot{q}_r}$, $r = 1, 2, 3$ are:

```
w1_1 = deriv(w10, diff(q1,t));
w1_2 = deriv(w10, diff(q2,t));
w1_3 = deriv(w10, diff(q3,t));
```

The MATLAB statements for the partial velocity $\dfrac{\partial \omega_{20}}{\partial \dot{q}_r}$, $r = 1, 2, 3$ are:

```
w2_1 = deriv(w20, diff(q1,t));
w2_2 = deriv(w20, diff(q2,t));
w2_3 = deriv(w20, diff(q3,t));
```

The MATLAB statements for the partial velocity $\dfrac{\partial \mathbf{v}_{C_2}}{\partial \dot{q}_r}$, $r = 1, 2, 3$ are:

```
vC1_1 = deriv(vC1, diff(q1,t));
vC1_2 = deriv(vC1, diff(q2,t));
vC1_3 = deriv(vC1, diff(q3,t));
```

The MATLAB statements for the partial velocity $\dfrac{\partial \mathbf{v}_{C_2}}{\partial \dot{q}_r}$, $r = 1, 2, 3$ are:

```
vC2_1 = deriv(vC2, diff(q1,t));
vC2_2 = deriv(vC2, diff(q2,t));
vC2_3 = deriv(vC2, diff(q3,t));
```

The MATLAB statements for the partial velocity $\dfrac{\partial \mathbf{v}_{C_{32}}}{\partial \dot{q}_r}$, $r = 1, 2, 3$ are:

```
vC32_1 = deriv(vC32, diff(q1,t));
vC32_2 = deriv(vC32, diff(q2,t));
vC32_3 = deriv(vC32, diff(q3,t));
```

The MATLAB statements for the partial velocity $\dfrac{\partial \mathbf{v}_{C_3}}{\partial \dot{q}_r}$, $r = 1, 2, 3$ are:

```
vC3_1 = deriv(vC3, diff(q1,t));
vC3_2 = deriv(vC3, diff(q2,t));
vC3_3 = deriv(vC3, diff(q3,t));
```

The generalized active force Q_1 is

$$Q_1 = \frac{\partial \omega_{10}}{\partial \dot{q}_1} \cdot (\mathbf{T}_{01} - \mathbf{T}_{12}) + \frac{\partial \mathbf{v}_{C_1}}{\partial \dot{q}_1} \cdot \mathbf{G}_1 + \frac{\partial \omega_{20}}{\partial \dot{q}_1} \cdot \mathbf{T}_{12} + \frac{\partial \mathbf{v}_{C_2}}{\partial \dot{q}_1} \cdot \mathbf{G}_2 + \frac{\partial \mathbf{v}_{C_{32}}}{\partial \dot{q}_1} \cdot (-\mathbf{F}_{23})$$
$$+ \frac{\partial \mathbf{v}_{C_3}}{\partial \dot{q}_1} \cdot \mathbf{G}_3 + \frac{\partial \mathbf{v}_{C_3}}{\partial \dot{q}_1} \cdot \mathbf{F}_{23},$$

and the MATLAB statement for the generalized active force Q_1 is:

```
% generalized active force Q1
Q1 = w1_1*T01.' + vC1_1*G1.' +...
     w1_1*transpose(R21)*(-T12.') +...
     w2_1*T12.' + vC2_1*G2.' + vC32_1*(-F23.') +...
     vC3_1*F23.' + vC3_1*G3.'
```

The generalized active force Q_2 is:

$$Q_2 = \frac{\partial \omega_{10}}{\partial \dot{q}_2} \cdot (\mathbf{T}_{01} - \mathbf{T}_{12}) + \frac{\partial \mathbf{v}_{C_1}}{\partial \dot{q}_2} \cdot \mathbf{G}_1 + \frac{\partial \omega_{20}}{\partial \dot{q}_2} \cdot \mathbf{T}_{12} + \frac{\partial \mathbf{v}_{C_2}}{\partial \dot{q}_2} \cdot \mathbf{G}_2 + \frac{\partial \mathbf{v}_{C_{32}}}{\partial \dot{q}_2} \cdot (-\mathbf{F}_{23})$$
$$+ \frac{\partial \mathbf{v}_{C_3}}{\partial \dot{q}_2} \cdot \mathbf{G}_3 + \frac{\partial \mathbf{v}_{C_3}}{\partial \dot{q}_2} \cdot \mathbf{F}_{23},$$

and the MATLAB statement for the generalized active force Q_2 is:

```
% generalized active force Q2
Q2 = w1_2*T01.' + vC1_2*G1.' +...
     w1_2*transpose(R21)*(-T12.') +...
     w2_2*T12.' + vC2_2*G2.' + vC32_2*(-F23.') +...
     vC3_2*F23.' + vC3_2*G3.'
```

The generalized active force Q_3 is

$$Q_3 = \frac{\partial \omega_{10}}{\partial \dot{q}_3} \cdot (\mathbf{T}_{01} - \mathbf{T}_{12}) + \frac{\partial \mathbf{v}_{C_1}}{\partial \dot{q}_3} \cdot \mathbf{G}_1 + \frac{\partial \omega_{20}}{\partial \dot{q}_3} \cdot \mathbf{T}_{12} + \frac{\partial \mathbf{v}_{C_2}}{\partial \dot{q}_3} \cdot \mathbf{G}_2 + \frac{\partial \mathbf{v}_{C_{32}}}{\partial \dot{q}_3} \cdot (-\mathbf{F}_{23})$$
$$+ \frac{\partial \mathbf{v}_{C_3}}{\partial \dot{q}_3} \cdot \mathbf{G}_3 + \frac{\partial \mathbf{v}_{C_3}}{\partial \dot{q}_3} \cdot \mathbf{F}_{23},$$

and the MATLAB statement for the generalized active force Q_3 is

```
% generalized active force Q3
Q3 = w1_3*T01.' + vC1_3*G1.' +...
     w1_3*transpose(R21)*(-T12.') +...
     w2_3*T12.' + vC2_3*G2.' + vC32_3*(-F23.') +...
     vC3_3*F23.' + vC3_3*G3.'
```

Kinetic Energy
The total kinetic energy of the robot arm in the reference frame (0) is

$$T = \sum_{i=1}^{3} T_i.$$

The kinetic energy of the link i, $i = 1,2,3$, is

$$T_i = \frac{1}{2} m_i \mathbf{v}_{C_i} \cdot \mathbf{v}_{C_i} + \frac{1}{2} \boldsymbol{\omega}_{i0} \cdot (\bar{I}_i \cdot \boldsymbol{\omega}_{i0}).$$

Remark: The kinetic energy for a rigid body is

$$T_{\text{rigid body}} = \frac{1}{2} m \mathbf{v}_C \cdot \mathbf{v}_C + \frac{1}{2} \boldsymbol{\omega} \cdot (\bar{I}_C \cdot \boldsymbol{\omega}),$$

where m is the mass of the rigid body, $\mathbf{v}_C$ is the velocity of the mass center of the rigid body in (0), $\boldsymbol{\omega} = \omega_x \mathbf{\imath} + \omega_y \mathbf{\jmath} + \omega_z \mathbf{k}$ is the angular velocity of the rigid body in (0), and $\bar{I} = (I_x \mathbf{\imath})\mathbf{\imath} + (I_y \mathbf{\jmath})\mathbf{\jmath} + (I_z \mathbf{k})\mathbf{k}$ is the central inertia dyadic of the rigid body. The central principal axes of the rigid body are parallel to $\mathbf{\imath}$, $\mathbf{\jmath}$, $\mathbf{k}$ and the associated moments of inertia have the values I_x, I_y, I_z, respectively. The inertia matrix associated with $\bar{I}$ is

$$\bar{I} \rightarrow \begin{bmatrix} I_x & 0 & 0 \\ 0 & I_y & 0 \\ 0 & 0 & I_z \end{bmatrix}.$$

The dot product of the vector $\boldsymbol{\omega}$ with the dyadic $\bar{I}$ is

$$\boldsymbol{\omega} \cdot \bar{I} = \bar{I} \cdot \boldsymbol{\omega} = \omega_x I_x \mathbf{\imath} + \omega_y I_y \mathbf{\jmath} + \omega_z I_z \mathbf{k}.$$

The central moments of inertia of links 1 and 2 are calculated using Fig. 6.6. The central principal axes of 1 are parallel to $\mathbf{\imath}_1$, $\mathbf{\jmath}_1$, $\mathbf{k}_1$ and the associated moments of inertia have the values I_{1x}, I_{1y}, I_{1z}, respectively. The inertia matrix associated with link 1 is

$$\bar{I}_1 \rightarrow \begin{bmatrix} I_{1x} & 0 & 0 \\ 0 & I_{1y} & 0 \\ 0 & 0 & I_{1z} \end{bmatrix} = \begin{bmatrix} \frac{m_1(2L_1)^2}{12} & 0 & 0 \\ 0 & \frac{m_1(2L_1)^2}{12} & 0 \\ 0 & 0 & 0 \end{bmatrix} = \begin{bmatrix} \frac{m_1 L_1^2}{3} & 0 & 0 \\ 0 & \frac{m_1 L_1^2}{3} & 0 \\ 0 & 0 & 0 \end{bmatrix}.$$

The central principal axes of 2 and 3 are parallel to $\mathbf{\imath}_2$, $\mathbf{\jmath}_2$, $\mathbf{k}_2$ and the associated moments of inertia have values I_{2x}, I_{2y}, I_{2z}, and I_{3x}, I_{3y}, I_{3z} respectively.

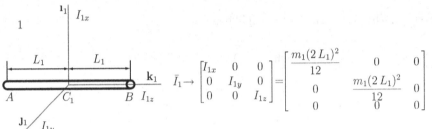

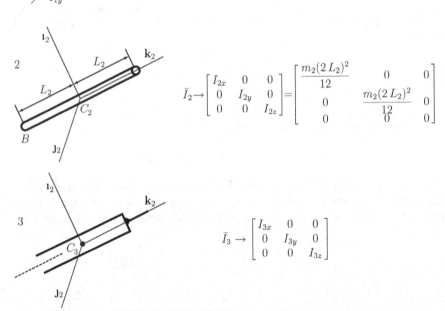

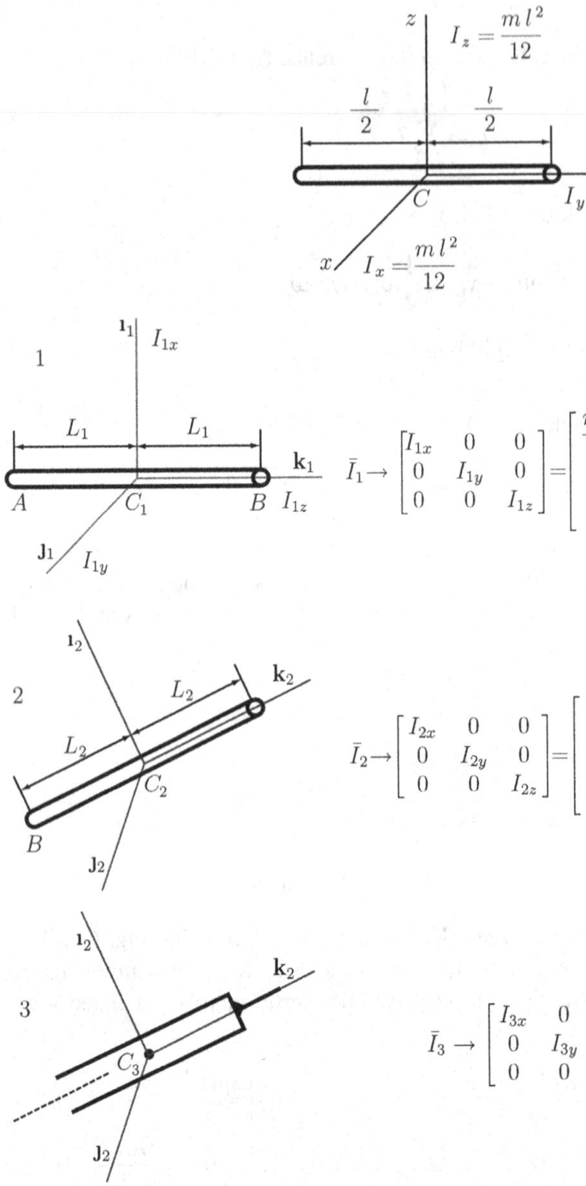

$$\bar{I}_1 \to \begin{bmatrix} I_{1x} & 0 & 0 \\ 0 & I_{1y} & 0 \\ 0 & 0 & I_{1z} \end{bmatrix} = \begin{bmatrix} \frac{m_1(2L_1)^2}{12} & 0 & 0 \\ 0 & \frac{m_1(2L_1)^2}{12} & 0 \\ 0 & 0 & 0 \end{bmatrix}$$

$$\bar{I}_2 \to \begin{bmatrix} I_{2x} & 0 & 0 \\ 0 & I_{2y} & 0 \\ 0 & 0 & I_{2z} \end{bmatrix} = \begin{bmatrix} \frac{m_2(2L_2)^2}{12} & 0 & 0 \\ 0 & \frac{m_2(2L_2)^2}{12} & 0 \\ 0 & 0 & 0 \end{bmatrix}$$

$$\bar{I}_3 \to \begin{bmatrix} I_{3x} & 0 & 0 \\ 0 & I_{3y} & 0 \\ 0 & 0 & I_{3z} \end{bmatrix}$$

Fig. 6.6 Central moments of inertia

The inertia matrix associated with link 2 is

$$\bar{I}_2 \rightarrow \begin{bmatrix} I_{2x} & 0 & 0 \\ 0 & I_{2y} & 0 \\ 0 & 0 & I_{2z} \end{bmatrix} = \begin{bmatrix} \frac{m_2(2L_2)^2}{12} & 0 & 0 \\ 0 & \frac{m_2(2L_2)^2}{12} & 0 \\ 0 & 0 & 0 \end{bmatrix} = \begin{bmatrix} \frac{m_2L_2^2}{3} & 0 & 0 \\ 0 & \frac{m_2L_2^2}{3} & 0 \\ 0 & 0 & 0 \end{bmatrix}.$$

The inertia matrix associated with the slider 3 is

$$\bar{I}_3 \rightarrow \begin{bmatrix} I_{3x} & 0 & 0 \\ 0 & I_{3y} & 0 \\ 0 & 0 & I_{3z} \end{bmatrix}.$$

The MATLAB commands for inertia matrices associated with the central inertia dyadics are:

```
% inertia mat. associated with central inertia dyadic
% for link 1 expressed in terms of RF1 {i1,j1,k1}
I1 = [m1*(2*L1)^2/12 0 0; 0 m1*(2*L1)^2/12 0; 0 0 0];
% inertia mat. associated with central inertia dyadic
% for link 2 expressed in terms of RF2 i2,j2,k2
I2 = [m2*(2*L2)^2/12 0 0; 0 m2*(2*L2)^2/12 0; 0 0 0];
% inertia mat. associated with central inertia dyadic
% for link 3 expressed in terms of RF2 {i2,j2,k2}
syms I3x I3y I3z real
I3 = [I3x 0 0; 0 I3y 0; 0 0 I3z];
```

The kinetic energy of link 1 is

$$T_1 = \frac{1}{2}m_1\mathbf{v}_{C_1}\cdot\mathbf{v}_{C_1} + \frac{1}{2}\boldsymbol{\omega}_{10}\cdot(\bar{I}_1\cdot\boldsymbol{\omega}_{10}) = \frac{1}{2}m_1L_1\dot{q}_1^2 + \frac{1}{6}m_1L_1\dot{q}_1^2 = \frac{2}{3}m_1L_1\dot{q}_1^2.$$

The kinetic energy of bar 2 is

$$\begin{aligned} T_2 &= \frac{1}{2}m_2\mathbf{v}_{C_2}\cdot\mathbf{v}_{C_2} + \frac{1}{2}\boldsymbol{\omega}_{20}\cdot(\bar{I}_2\cdot\boldsymbol{\omega}_{20}) \\ &= \frac{m_2}{3}\left[\left(6L_1^2 + L_2^2 + 6L_1L_2c_2 + L_2^2\cos 2q_2\right)\dot{q}_1^2 + 2L_2^2\dot{q}_2^2\right]. \end{aligned}$$

The kinetic energy of link 3 is

$$\begin{aligned} T_3 &= \frac{1}{2}m_3\mathbf{v}_{C_3}\cdot\mathbf{v}_{C_3} + \frac{1}{2}\boldsymbol{\omega}_{20}\cdot(\bar{I}_3\cdot\boldsymbol{\omega}_{20}) \\ &= \frac{1}{2}\{I_{3x}c_2^2\dot{q}_1^2 + I_{3z}s_2^2\dot{q}_1^2 + I_{3y}\dot{q}_2^2 \\ &+ m_3\left[(2L_1 + L_2c_2 + c_2q_3)^2\dot{q}_1^2 + (L_2+q_3)^2\dot{q}_2^2 + \dot{q}_3^2\right]\}. \end{aligned}$$

The total kinetic energy of the robot arm is

$$T = T_1 + T_2 + T_3,$$

and is symbolically calculated in the program given in Appendix E.3. The MATLAB commands for the total kinetic energy of the robot arm are:

```
T1 = (1/2)*m1*vC1*vC1.' + (1/2)*w10*I1*w10.'
T2 = (1/2)*m2*vC2*vC2.' + (1/2)*w20*I2*w20.'
T3 = (1/2)*m3*vC3*vC3.' + (1/2)*w20*I3*w20.'
T = expand(T1 + T2 + T3);    % total kinetic energy
```

Lagrange's Equations of Motion
The left-hand sides of Lagrange's equations are

$$\frac{d}{dt}\left(\frac{\partial T}{\partial \dot{q}_r}\right) - \frac{\partial T}{\partial q_r}, \quad r = 1, 2, 3.$$

To arrive at the dynamical equations governing the robot arm, all that remains to be done is to substitute into Lagrange's equations, namely,

$$\frac{d}{dt}\left(\frac{\partial T}{\partial \dot{q}_r}\right) - \frac{\partial T}{\partial q_r} = Q_r, \quad r = 1, 2, 3.$$

The left-hand side of Lagrange's equations are symbolically calculated in MATLAB with:

```
% deriv(f, g(t)) differentiates f
% with respect to g(t)
Tdq1 = deriv(T, diff(q1,t));
Tdq2 = deriv(T, diff(q2,t));
Tdq3 = deriv(T, diff(q3,t));

Tt1 = diff(Tdq1, t);
Tt2 = diff(Tdq2, t);
Tt3 = diff(Tdq3, t);

Tq1 = deriv(T, q1);
Tq2 = deriv(T, q2);
Tq3 = deriv(T, q3);

LHS1 = Tt1 - Tq1;
LHS2 = Tt2 - Tq2;
LHS3 = Tt3 - Tq3;
```

Lagrange's equations are symbolically calculated in MATLAB with:

```
Lagrange1 = LHS1-Q1;
Lagrange2 = LHS2-Q2;
Lagrange3 = LHS3-Q3;
```

The following feedback control laws are used

$$
\begin{aligned}
T_{01x} &= -\beta_{01}\dot{q}_1 - \gamma_{01}(q_1 - q_{1f}), \\
T_{12y} &= -\beta_{12}\dot{q}_2 - \gamma_{12}(q_2 - q_{2f}) + g\,m_2\,L_2\,c_2 + g\,m_3\,c_2\,(L_2 + q_3), \\
F_{23z} &= -\beta_{23}\dot{q}_3 - \gamma_{23}(q_3 - q_{3f}) + g\,m_3\,s_2.
\end{aligned} \tag{6.42}
$$

The constant gains are: $\beta_{01} = 450$ N m s/rad, $\gamma_{01} = 300$ N m/rad, $\beta_{12} = 200$ N m s/rad, $\gamma_{12} = 300$ N m/rad, $\beta_{23} = 150$ N s/m, and $\gamma_{23} = 50$ N/m.

The MATLAB commands for the control torques are:

```
q1f=pi/3; q2f=pi/3; q3f=0.3;
b01=450; g01=300;
b12=200; g12=300;
b23=150; g23=50;
T01xc = -b01*diff(q1,t)-g01*(q1-q1f);
T12yc = -b12*diff(q2,t)-g12*(q2-q2f)+...
        g*(m2*L2+m3*(L2+q3))*c2;
F23zc = -b23*diff(q3,t)-g23*(q3-q3f)+g*m3*s2;
tor = {T01x, T12y, F23z};
torf = {T01xc,T12yc,F23zc};
```

Lagrange's equations with the feedback control laws are:

```
Lagrang1 = subs(Lagrange1, tor, torf);
Lagrang2 = subs(Lagrange2, tor, torf);
Lagrang3 = subs(Lagrange3, tor, torf);
```

Lagrange's equations with the numerical values for input data are:

```
data = {L1, L2, I3x, I3y, I3z, m1, m2, m3, g};
datn = {0.4, 0.4, 5, 4, 1, 90, 60, 40, 9.81};

Lagran1 = subs(Lagrang1, data, datn);
Lagran2 = subs(Lagrang2, data, datn);
Lagran3 = subs(Lagrang3, data, datn);
```

The three second-order Lagrange's equations have to be rewritten as a first-order system:

```
ql = {diff(q1,t,2), diff(q2,t,2), diff(q3,t,2), ...
      diff(q1,t), diff(q2,t), diff(q3,t), q1, q2, q3};
qf = {'ddq1', 'ddq2', 'ddq3',...
      'x(2)', 'x(4)', 'x(6)', 'x(1)', 'x(3)', 'x(5)'};

% ql                          qf
%------------------------------
% diff('q1(t)',t,2)  -> 'ddq1'
% diff('q2(t)',t,2)  -> 'ddq2'
% diff('q3(t)',t,2)  -> 'ddq3'
%   diff('q1(t)',t)  -> 'x(2)'
%   diff('q2(t)',t)  -> 'x(4)'
%   diff('q3(t)',t)  -> 'x(6)'
%           'q1(t)'  -> 'x(1)'
%           'q2(t)'  -> 'x(3)'
%           'q3(t)'  -> 'x(5)'

Lagra1 = subs(Lagran1, ql, qf);
Lagra2 = subs(Lagran2, ql, qf);
Lagra3 = subs(Lagran3, ql, qf);

% solve e.o.m. for ddq1, ddq2, ddq3
sol = solve(Lagra1,Lagra2,Lagra3,'ddq1,ddq2,ddq3');
Lagr1 = sol.ddq1;
Lagr2 = sol.ddq2;
Lagr3 = sol.ddq3;

dx2dt = char(Lagr1);
dx4dt = char(Lagr2);
dx6dt = char(Lagr3);
```

The system of differential equations is solved numerically by m-file functions. The function file, `RRT_Lagr.m` is created using the statements:

```
fid = fopen('RRT_Lagr.m','w+');
fprintf(fid,'function dx = RRT_Lagr(t,x)\n');
fprintf(fid,'dx = zeros(6,1);\n');
fprintf(fid,'dx(1) = x(2);\n');
fprintf(fid,'dx(2) = ');
fprintf(fid,dx2dt);
fprintf(fid,';\n');
fprintf(fid,'dx(3) = x(4);\n');
fprintf(fid,'dx(4) = ');
fprintf(fid,dx4dt);
fprintf(fid,';\n');
```

```
fprintf(fid,'dx(5) = x(6);\n');
fprintf(fid,'dx(6) = ');
fprintf(fid,dx6dt);
fprintf(fid,';');
fclose(fid);
cd(pwd);
```

The `ode45` solver is used for the system of differential equations:

```
t0 = 0;
tf = 15;
time = [0 tf];
x0 = [pi/18 0 pi/6 0 0.25 0];

[t,xs] = ode45(@RRT_Lagr, time, x0);

x1 = xs(:,1);
x2 = xs(:,2);
x3 = xs(:,3);
x4 = xs(:,4);
x5 = xs(:,5);
x6 = xs(:,6);

subplot(3,1,1),...
plot(t,x1*180/pi,'r'),...
xlabel('t (s)'),ylabel('q1 (deg)'),grid,...
subplot(3,1,2),...
plot(t,x3*180/pi,'b'),...
xlabel('t (s)'),ylabel('q2 (deg)'),grid,...
subplot(3,1,3),...
plot(t,x5,'g'),...
xlabel('t (s)'),ylabel('q3 (m)'),grid

[ts,xs] = ode45(@RRT_Lagr,0:1:5,x0);

fprintf('Results \n\n')
fprintf...
('t(s) q1(rad) q2(rad) q3(m) \n')

[ts,xs(:,1),xs(:,3),xs(:,5)]
```

Figure 6.7 shows the plots of $q_1(t)$, $q_2(t)$, $q_3(t)$ and a MATLAB computer program for the direct dynamics is given in Appendix E.3.

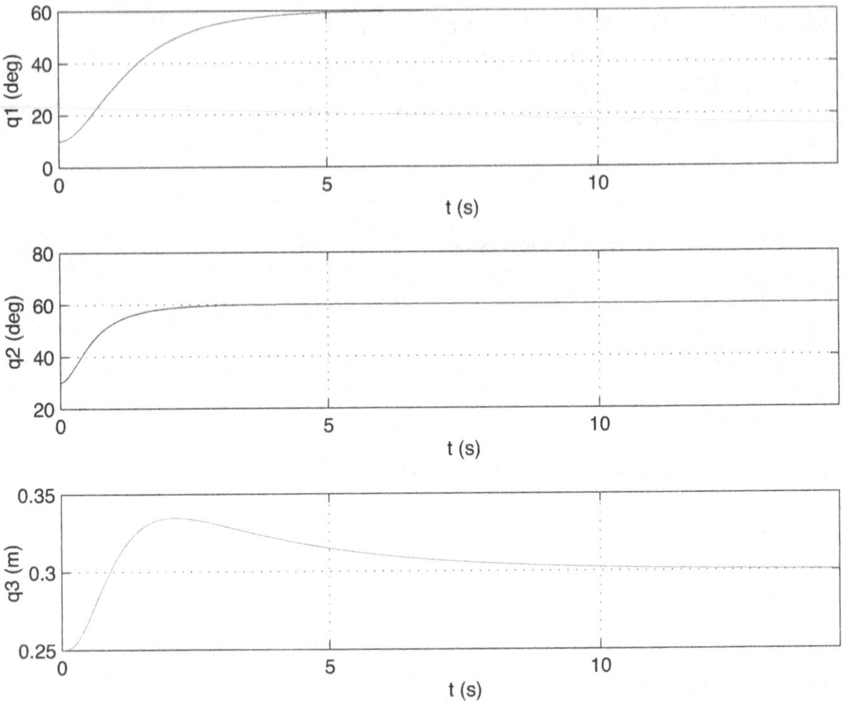

Fig. 6.7 Solution plots for the generalized coordinates $q_1(t)$, $q_2(t)$, and $q_3(t)$

6.5.2 Inverse Dynamics

A desired motion of the robot arm is specified for a time interval $0 \le t \le T_p = 15$ s. The generalized coordinates can be established explicitly

$$q_r(t) = q_r(0) + \frac{q_r(T_p) - q_r(0)}{T_p}\left[t - \frac{T_p}{2\pi}\sin\left(\frac{2\pi t}{T_p}\right)\right], \quad r = 1,2,3, \qquad (6.43)$$

with $q_r(T_p) = q_{rf}$.

The initial conditions, at $t = 0$ s, are $q_1(0) = \pi/18$ rad, $q_2(0) = \pi/6$ rad, $q_3(0) = 0.25$ m, and $\dot{q}_1(0) = \dot{q}_2(0) = \dot{q}_3(0) = 0$. The robot arm can be brought from an initial state of rest in reference frame (0) to a final state of rest in (0) in such a way that q_1, q_2, and q_3 have specified values $q_1(T_p) = q_{1f} = \pi/3$ rad, $q_2(T_p) = q_{2f} = \pi/3$ rad, and $q_3(T_p) = q_{3f} = 0.3$ m. Figure 6.8 shows the plots of $q_1(t)$, $q_2(t)$, and $q_3(t)$ rad.

The MATLAB commands for finding the Lagrage's equations are identical with the previous commands presented in Sect. 6.5.1:

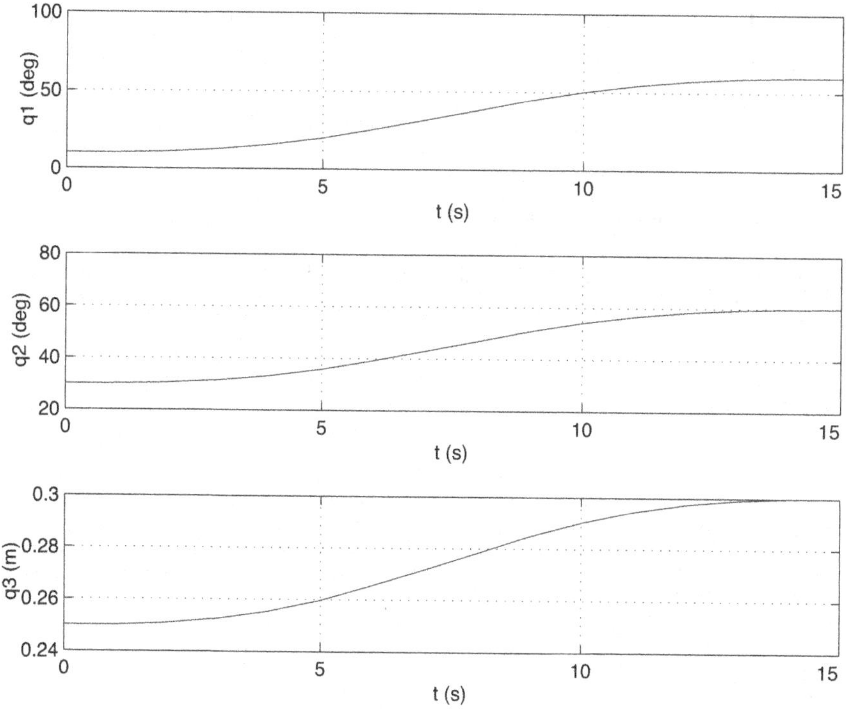

Fig. 6.8 Generalized coordinates $q_1(t)$, $q_2(t)$, and $q_3(t)$

....

```
syms T01x T01y T01z T12x T12y T12z F23x F23y F23z
T01 = [T01x T01y T01z];
T12 = [T12x T12y T12z];
F23 = [F23x F23y F23z];
```

....

```
Lagrange1=LHS1-Q1;
Lagrange2=LHS2-Q2;
Lagrange3=LHS3-Q3;
data = {L1, L2, I3x, I3y, I3z, m1, m2, m3, g};
datn = {0.4, 0.4, 5, 4, 1, 90, 60, 40, 9.81};

Lagra1 = subs(Lagrange1, data, datn);
Lagra2 = subs(Lagrange2, data, datn);
Lagra3 = subs(Lagrange3, data, datn);
```

From the Lagrange's equations of motions the torques T_{01x} and T_{12y} and the force F_{23z} are calculated:

```
sol = solve(Lagra1,Lagra2,Lagra3,'T01x, T12y, F23z');
T01xc = simple(sol.T01x);
T12yc = simple(sol.T12y);
F23zc = simple(sol.F23z);
```

The generalized coordinates, q_1, q_2, and q_3 given by Eq. 6.43 and their derivatives, $\dot{q}_1, \dot{q}_2, \dot{q}_3, \ddot{q}_1, \ddot{q}_2, \ddot{q}_3$ are substituted in the expressions of T_{01x}, T_{12y}, and F_{23z}:

```
q1s = pi/18; q2s = pi/6; q3s = 0.25;
q1f = pi/3 ; q2f = pi/3; q3f = 0.3;
Tp=15.;
q1t = q1s+(q1f-q1s)/Tp*(t-Tp/(2*pi)*sin(2*pi/Tp*t));
q2t = q2s+(q2f-q2s)/Tp*(t-Tp/(2*pi)*sin(2*pi/Tp*t));
q3t = q3s+(q3f-q3s)/Tp*(t-Tp/(2*pi)*sin(2*pi/Tp*t));
dq1t = diff(q1t,t);
dq2t = diff(q2t,t);
dq3t = diff(q3t,t);
ddq1t = diff(dq1t,t);
ddq2t = diff(dq2t,t);
ddq3t = diff(dq3t,t);

ql = {diff(q1,t,2), diff(q2,t,2), diff(q3,t,2), ...
    diff(q1,t), diff(q2,t), diff(q3,t), q1, q2, q3};
qn = {ddq1t,ddq2t,ddq3t,dq1t,dq2t,dq3t,q1t,q2t,q3t};

T01xt = subs(T01xc, ql, qn);
T12yt = subs(T12yc, ql, qn);
F23zt = subs(F23zc, ql, qn);
```

The plots and the values of $T_{01x}(t)$, $T_{12y}(t)$, and $F_{23z}(t)$ are obtained with the MATLAB commands:

```
time = 0:1:Tp;
T01t = subs(T01xt,'t',time);
T12t = subs(T12yt,'t',time);
F23t = subs(F23zt,'t',time);
subplot(3,1,1), plot(time,T01t),...
xlabel('t (s)'), ylabel('T01x (N m)'), grid,...
subplot(3,1,2), plot(time,T12t),...
xlabel('t (s)'), ylabel('T12y (N m)'), grid,...
subplot(3,1,3), plot(time,F23t),...
xlabel('t (s)'), ylabel('F23z (N)'), grid
```

```
fprintf('t(s) T01x(Nm) T12y(Nm) F23z(N) \n')
[time' T01t' T12t' F23t']
```

Another way of plotting $T_{01x}(t)$, $T_{12y}(t)$, and $F_{23z}(t)$ is:

```
subplot(3,1,1), ezplot(T01xt,[0,Tp]),...
title(''), xlabel('t (s)'), ylabel('T01x (N m)'),...
grid,...
subplot(3,1,2), ezplot(T12yt,[0,Tp]),...
title(''), xlabel('t (s)'), ylabel('T12y (N m)'),...
grid,...
subplot(3,1,3), ezplot(F23zt,[0,Tp]),...
title(''), xlabel('t (s)'), ylabel('F23z (N)'), grid
```

Figure 6.9 shows the control torques and force and the MATLAB program is given in Appendix E.4.

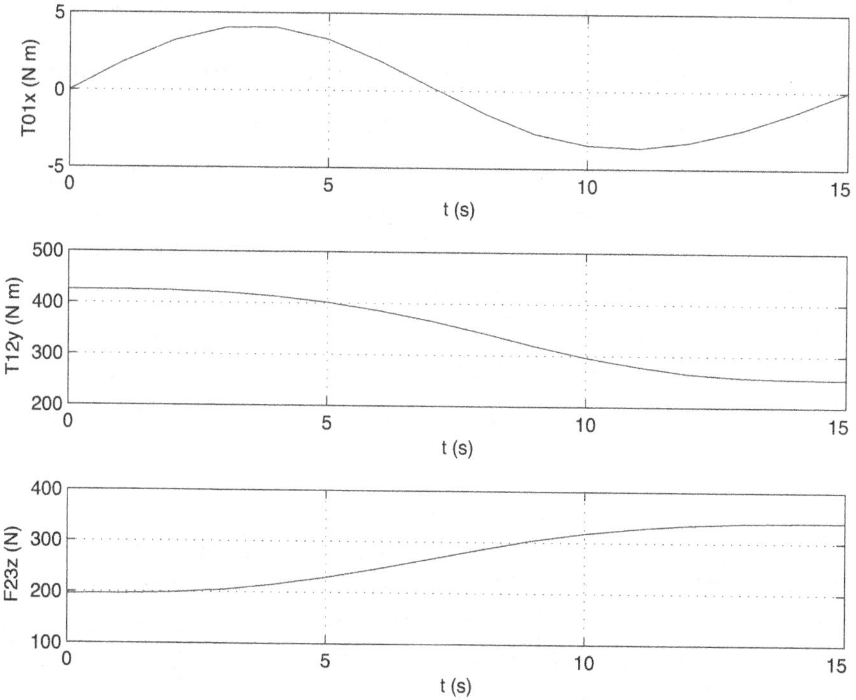

Fig. 6.9 Solution plots for the control torques and force

6.5.3 Kane's Dynamical Equations

The generalized coordinates q_i and the generalized speeds u_i are introduced:

```
% generalized coordinates q1, q2, q3
q1=sym('q1(t)'); q2=sym('q2(t)'); q3=sym('q3(t)');
% generalized speeds u1, u2, u3
u1=sym('u1(t)'); u2=sym('u2(t)'); u3=sym('u3(t)');
```

The generalized speeds, u_1, u_2, u_3, are associated with the motion of a system, and can be introduced as $\dot{q}_i = u_i$, or:

```
dq1 = u1;
dq2 = u2;
dq3 = u3;

qt = diff(q1,t), diff(q2,t), diff(q3,t);
qu = dq1, dq2, dq3;
```

The velocities and the accelerations of the robot need to be expressed in terms of q_i, u_i and $\dot{u}_i$:

```
c1=cos(q1); s1=sin(q1); c2=cos(q2); s2=sin(q2);
R10 = [[1 0 0]; [0 c1 s1]; [0 -s1 c1]];
R21 = [[c2 0 -s2]; [0 1 0]; [s2 0 c2]];

w10 = [dq1, 0, 0 ]
w201 = [dq1, dq2, 0];
w20 = w201 * transpose(R21)
alpha10 = diff(w10,t);
alpha20 = subs(diff(w20,t), qt, qu);

rC1 = [0 0 L1];
vC1 = diff(rC1,t) + cross(w10, rC1);
rC2 = [0 0 2*L1]*transpose(R21) + [0 0 L2];
vC2 = subs(diff(rC2, t), qt, qu) + cross(w20,rC2);

rC3 = rC2 + [0 0 q3];
vC3 = subs(diff(rC3, t), qt, qu) + cross(w20,rC3);
vC32 = vC2 + cross(w20,[0 0 q3]);

aC1 = diff(vC1,t) + cross(w10,vC1);
aC2 = diff(vC2,t) + cross(w20,vC2);
aC3 = subs(diff(vC3,t), qt, qu) + cross(w20,vC3);
```

The gravitational forces and the external moments and force are:

```
G1 = [-m1*g 0 0];
G2 = [-m2*g 0 0]*transpose(R21);
G3 = [-m3*g 0 0]*transpose(R21);
T01 = [T01x T01y T01z];
T12 = [T12x T12y T12z];
F23 = [F23x F23y F23z];
```

The partial velocities with respect to `u1, u2, u3` are calculated using the function `deriv`:

```
w1_1 = deriv(w10, u1); w2_1 = deriv(w20, u1);
w1_2 = deriv(w10, u2); w2_2 = deriv(w20, u2);
w1_3 = deriv(w10, u3); w2_3 = deriv(w20, u3);

vC1_1 = deriv(vC1, u1); vC2_1 = deriv(vC2, u1);
vC1_2 = deriv(vC1, u2); vC2_2 = deriv(vC2, u2);
vC1_3 = deriv(vC1, u3); vC2_3 = deriv(vC2, u3);

vC32_1 = deriv(vC32, u1); vC3_1 = deriv(vC3, u1);
vC32_2 = deriv(vC32, u2); vC3_2 = deriv(vC3, u2);
vC32_3 = deriv(vC32, u3); vC3_3 = deriv(vC3, u3);
```

The generalized active forces are:

```
Q1 = w1_1*T01.' + vC1_1*G1.' +...
     w1_1*transpose(R21)*(-T12.') +...
     w2_1*T12.' + vC2_1*G2.' + vC32_1*(-F23.') +...
     vC3_1*F23.' + vC3_1*G3.';

Q2 = w1_2*T01.' + vC1_2*G1.' +...
     w1_2*transpose(R21)*(-T12.') +...
     w2_2*T12.' + vC2_2*G2.' + vC32_2*(-F23.') +...
     vC3_2*F23.' + vC3_2*G3.';

Q3 = w1_3*T01.' + vC1_3*G1.' +...
     w1_3*transpose(R21)*(-T12.') +...
     w2_3*T12.' + vC2_3*G2.' + vC32_3*(-F23.') +...
     vC3_3*F23.' + vC3_3*G3.';
```

Generalized inertia forces

To explain what the *generalized inertia forces* are, a system $\{S\}$ formed by ν particles $P_1, ..., P_\nu$ and having masses $m_1, ..., m_\nu$ is considered. Suppose that n generalized speeds u_r, $r = 1, ..., n$ have been introduced. (For the robotic arm $u_r = \dot{q}_r$, $r =$

$1,\dots,n$.) Let $\mathbf{v}_{P_j}$ and $\mathbf{a}_{P_j}$ denote, respectively, the velocity of P_j and the acceleration of P_j in a reference frame (0).

Define $\mathbf{F}_{\text{in}\,j}$, called the inertia force for P_j, as

$$\mathbf{F}_{\text{in}\,j} = -m_j\,\mathbf{a}_{P_j}.$$

The quantities $K_{\text{in}\,1},\dots,K_{\text{in}\,n}$, defined as

$$K_{\text{in}\,r} = \sum_{j=1}^{\nu} \frac{\partial \mathbf{v}_{P_j}}{\partial u_r} \cdot \mathbf{F}_{\text{in}\,j},\ \ r = 1,\dots,n,$$

are called *generalized inertia forces* for $\{S\}$.

The contribution to $K_{\text{in}\,r}$, made by the particles of a rigid body RB belonging to $\{S\}$, are

$$(K_{\text{in}\,r})_R = \frac{\partial \mathbf{v}_C}{\partial u_r} \cdot \mathbf{F}_{\text{in}} + \frac{\partial \boldsymbol{\omega}}{\partial u_r} \cdot \mathbf{M}_{\text{in}},\ \ r = 1,\dots,n,$$

where $\mathbf{v}_C$ is the velocity of the center of gravity of RB in (0), and $\boldsymbol{\omega} = \omega_x\mathbf{\imath} + \omega_y\mathbf{\jmath} + \omega_z\mathbf{k}$ is the angular velocity of RB in (0).

The inertia force for the rigid body RB is

$$\mathbf{F}_{\text{in}} = -m\,\mathbf{a}_C,$$

where m is the mass of RB, and $\mathbf{a}_C$ is the acceleration of the mass center of RB in the fixed reference frame. The inertia moment $\mathbf{M}_{\text{in}}$ for RB is

$$\mathbf{M}_{\text{in}} = -\boldsymbol{\alpha} \cdot \bar{I} - \boldsymbol{\omega} \times (\bar{I} \cdot \boldsymbol{\omega}),$$

where $\boldsymbol{\alpha} = \dot{\boldsymbol{\omega}} = \alpha_x\mathbf{\imath} + \alpha_y\mathbf{\jmath} + \alpha_z\mathbf{k}$ is the angular acceleration of RB in (0), and $\bar{I} = (I_x\mathbf{\imath})\mathbf{\imath} + (I_y\mathbf{\jmath})\mathbf{\jmath} + (I_z\mathbf{k})\mathbf{k}$ is the central inertia dyadic of RB. The central principal axes of RB are parallel to $\mathbf{\imath}, \mathbf{\jmath}, \mathbf{k}$ and the associated moments of inertia have the values I_x, I_y, I_z, respectively. The inertia matrix associated with $\bar{I}$ is

$$\bar{I} \rightarrow \begin{bmatrix} I_x & 0 & 0 \\ 0 & I_y & 0 \\ 0 & 0 & I_z \end{bmatrix}.$$

The dot product of the vector $\boldsymbol{\alpha}$ with the dyadic $\bar{I}$ is

$$\boldsymbol{\alpha} \cdot \bar{I} = \bar{I} \cdot \boldsymbol{\alpha} = \alpha_x I_x \mathbf{\imath} + \alpha_y I_y \mathbf{\jmath} + \alpha_z I_z \mathbf{k},$$

and the cross product between a vector and a dyadic is

$$\boldsymbol{\omega} \times (\bar{I} \cdot \boldsymbol{\omega}) = \begin{vmatrix} \mathbf{\imath} & \mathbf{\jmath} & \mathbf{k} \\ \omega_x & \omega_y & \omega_z \\ \omega_x I_x & \omega_y I_y & \omega_z I_z \end{vmatrix}$$
$$= -\omega_y\omega_z(I_y - I_z)\mathbf{\imath} - \omega_z\omega_x(I_z - I_x)\mathbf{\jmath} - \omega_x\omega_y(I_x - I_y)\mathbf{k}.$$

The inertia moment of 1 in (0) can be written as

$$\mathbf{M}_{\text{in}\,1} = -\boldsymbol{\alpha}_{10} \cdot \bar{I}_1 - \boldsymbol{\omega}_{10} \times (\bar{I}_1 \cdot \boldsymbol{\omega}_{10}) = -I_{1x}\ddot{q}_1 \mathbf{1}_1.$$

The inertia moment of 2 in (0) is

$$\mathbf{M}_{\text{in}\,2} = -\boldsymbol{\alpha}_{20} \cdot \bar{I}_2 - \boldsymbol{\omega}_{20} \times (\bar{I}_2 \cdot \boldsymbol{\omega}_{20}).$$

Similarly the inertia moment of 3 in (0) is

$$\mathbf{M}_{\text{in}\,3} = \boldsymbol{\alpha}_{20} \cdot \bar{I}_3 - \boldsymbol{\omega}_{20} \times (\bar{I}_3 \cdot \boldsymbol{\omega}_{20}).$$

The inertia force for link $j = 1, 2, 3$ is

$$\mathbf{F}_{\text{in}\,j} = -m_j\,\mathbf{a}_{C_j}.$$

The contribution of link $j = 1, 2, 3$ to the generalized inertia force $K_{\text{in}\,r}$ is

$$(K_{\text{in}\,r})_j = \frac{\partial \mathbf{v}_{C_j}}{\partial u_r} \cdot \mathbf{F}_{\text{in}\,j} + \frac{\partial \boldsymbol{\omega}_{j0}}{\partial u_r} \cdot \mathbf{M}_{\text{in}\,j},\ r = 1,2,3.$$

The three generalized inertia forces are computed with

$$\begin{aligned} K_{\text{in}\,r} &= \sum_{j=1}^{3} (K_{\text{in}\,r})_j \\ &= \sum_{j=1}^{3} \left(\frac{\partial \mathbf{v}_{C_j}}{\partial u_r} \cdot \mathbf{F}_{\text{in}\,j} + \frac{\partial \boldsymbol{\omega}_{j0}}{\partial u_r} \cdot \mathbf{M}_{\text{in}\,j} \right),\ r = 1, 2, 3, \end{aligned}$$

or

$$\begin{aligned} K_{\text{in}\,r} = & \frac{\partial \mathbf{v}_{C_1}}{\partial u_r} \cdot (-m_1\,\mathbf{a}_{C_1}) + \frac{\partial \boldsymbol{\omega}_{10}}{\partial u_r} \cdot \mathbf{M}_{\text{in}\,1} + \frac{\partial \mathbf{v}_{C_2}}{\partial u_r} \cdot (-m_2\,\mathbf{a}_{C_2}) + \frac{\partial \boldsymbol{\omega}_{20}}{\partial u_r} \cdot \mathbf{M}_{\text{in}\,2} \\ & + \frac{\partial \mathbf{v}_{C_3}}{\partial u_r} \cdot (-m_3\,\mathbf{a}_{C_3}) + \frac{\partial \boldsymbol{\omega}_{30}}{\partial u_r} \cdot \mathbf{M}_{\text{in}\,3},\ r = 1, 2, 3. \end{aligned}$$

The generalized inertia forces for the RRT robot arm are calculated with the following MATLAB commands:

```
I1 = [m1*(2*L1)^2/12 0 0; 0 m1*(2*L1)^2/12 0; 0 0 0];
I2 = [m2*(2*L2)^2/12 0 0; 0 m2*(2*L2)^2/12 0; 0 0 0];
I3 = [I3x 0 0; 0 I3y 0; 0 0 I3z];

% inertia force for link 1
% expressed in terms of RF1 {i1,j1,k1}
Fin1= -m1*aC1;
```

```
% inertia force for link 2
% expressed in terms of RF2 {i2,j2,k2}
Fin2= -m2*aC2;
% inertia force for link 3
% expressed in terms of RF2 {i2,j2,k2}
Fin3= -m3*aC3;
% inertia moment for link 1
% expressed in terms of RF1 {i1,j1,k1}
Min1 = -alpha10*I1-cross(w10,w10*I1);
% inertia moment for link 2
% expressed in terms of RF2 {i2,j2,k2}
Min2 = -alpha20*I2-cross(w20,w20*I2);
% inertia moment for link 3
% expressed in terms of RF2 {i2,j2,k2}
Min3 = -alpha20*I3-cross(w20,w20*I3);

% generalized inertia forces corresponding to q1
Kin1 = w1_1*Min1.' + vC1_1*Fin1.' + ...
       w2_1*Min2.' + vC2_1*Fin2.' + ...
       w2_1*Min3.' + vC3_1*Fin3.';

% generalized inertia forces corresponding to q2
Kin2 = w1_2*Min1.' + vC1_2*Fin1.' + ...
       w2_2*Min2.' + vC2_2*Fin2.' + ...
       w2_2*Min3.' + vC3_2*Fin3.';

% generalized inertia forces corresponding to q3
Kin3 = w1_3*Min1.' + vC1_3*Fin1.' + ...
       w2_3*Min2.' + vC2_3*Fin2.' + ...
       w2_3*Min3.' + vC3_3*Fin3.';
```

To arrive at the dynamical equations governing the robot arm, all that remains to be done is to substitute into Kane's dynamical equations, namely,

$$K_{\text{in}r} + Q_r = 0, \ \ r = 1, 2, 3. \tag{6.44}$$

Kane's dynamical equations in MATLAB are:

```
Kane1 = Kin1 + Q1;
Kane2 = Kin2 + Q2;
Kane3 = Kin3 + Q3;
```

Using the same feedback control laws (the same as these used for Lagrange's equations) Kane's equations have to be rewritten:

```
q1f=pi/3; q2f=pi/3; q3f=0.3;
b01=450; g01=300; b12=200; g12=300; b23=150; g23=50;

T01xc = -b01*dq1-g01*(q1-q1f);
T12yc = -b12*dq2-g12*(q2-q2f)+g*(m2*L2+m3*(L2+q3))*c2;
F23zc = -b23*dq3-g23*(q3-q3f)+g*m3*s2;

tor = {T01x, T12y, F23z};
torf = {T01xc,T12yc,F23zc};

Kan1 = subs(Kane1, tor, torf);
Kan2 = subs(Kane2, tor, torf);
Kan3 = subs(Kane3, tor, torf);
```

The Kane's dynamical equations can be expressed in terms of $\dot{u}_1$, $\dot{u}_2$, and $\dot{u}_3$:

```
data = {L1, L2, I3x, I3y, I3z, m1, m2, m3, g};
datn = {0.4, 0.4, 5, 4, 1, 90, 60, 40, 9.81};

Ka1 = subs(Kan1, data, datn);
Ka2 = subs(Kan2, data, datn);
Ka3 = subs(Kan3, data, datn);

ql = {diff(u1,t), diff(u2,t), diff(u3,t) ...
      u1, u2, u3, q1, q2, q3};
qx = {'du1', 'du2', 'du3',...
      'x(4)', 'x(5)', 'x(6)', 'x(1)', 'x(2)', 'x(3)'};

Du1 = subs(Ka1, ql, qx);
Du2 = subs(Ka2, ql, qx);
Du3 = subs(Ka3, ql, qx);

% solve for du1, du2, du3
sol = solve(Du1, Du2, Du3,'du1, du2, du3');
sdu1 = sol.du1;
sdu2 = sol.du2;
sdu3 = sol.du3;
```

The system of differential equations is solved numerically by m-file functions. The function file, `RRT_Kane.m` is created using the statements:

```
% system of ODE
dx1 = char('x(4)');
dx2 = char('x(5)');
dx3 = char('x(6)');
```

```
dx4 = char(sdu1);
dx5 = char(sdu2);
dx6 = char(sdu3);

fid = fopen('RRT_Kane.m','w+');
fprintf(fid,'function dx = RRT_Kane(t,x)\n');
fprintf(fid,'dx = zeros(6,1);\n');
fprintf(fid,'dx(1) = '); fprintf(fid,dx1);
fprintf(fid,';\n');
fprintf(fid,'dx(2) = '); fprintf(fid,dx2);
fprintf(fid,';\n');
fprintf(fid,'dx(3) = '); fprintf(fid,dx3);
fprintf(fid,';\n');
fprintf(fid,'dx(4) = '); fprintf(fid,dx4);
fprintf(fid,';\n');
fprintf(fid,'dx(5) = '); fprintf(fid,dx5);
fprintf(fid,';\n');
fprintf(fid,'dx(6) = '); fprintf(fid,dx6);
fprintf(fid,'; ');
fclose(fid); cd(pwd);
```

The `ode45` solver is used to solve the system of first-order differential equations:

```
t0 = 0; tf = 15; time = [0 tf];
x0 = [pi/18 pi/6 0.25 0 0 0];
[t,xs] = ode45(@RRT_Kane, time, x0);
x1 = xs(:,1);
x2 = xs(:,2);
x3 = xs(:,3);
x4 = xs(:,4);
x5 = xs(:,5);
x6 = xs(:,6);
subplot(3,1,1), plot(t,x1*180/pi,'r'),...
xlabel('t (s)'), ylabel('q1 (deg)'), grid,...
subplot(3,1,2), plot(t,x2*180/pi,'b'),...
xlabel('t (s)'), ylabel('q2 (deg)'), grid,...
subplot(3,1,3), plot(t,x3,'g'),...
xlabel('t (s)'), ylabel('q3 (m)'), grid
```

The MATLAB computer program for the direct dynamics using Kane's dynamical equations is given in Appendix E.5.

6.6 RRTR Robot Arm

Figure 6.10 is a schematic representation of a robot (Kane and Levinson, 1983), with four links 1, 2, 3, and 4. The mass center of the link i is designated C_i, $i = 1,2,3,4$. The dimensions $AC_2 = L_1, AC_1 = L_2$ and $C_3C_4 = L_3$ are shown in the figure. Link 1 rotates in a "fixed" Newtonian reference frame (0) about a vertical axis fixed in both (0) and 1. The reference frame (0), RF0, has the unit vectors $[\mathbf{\imath}_0, \mathbf{\jmath}_0, \mathbf{k}_0]$ as shown in Fig. 6.10. The reference frame (1), RF1, of unit vectors $[\mathbf{\imath}_1, \mathbf{\jmath}_1, \mathbf{k}_1]$ is attached to link 1. The vertical unit vectors $\mathbf{\jmath}_0$ and $\mathbf{\jmath}_1$ are fixed in both (0) and 1.

The first generalized coordinate q_1 denotes the radian measure of the angle between the axes of (0) and (1). Link 1 supports link 2, and link 2 rotates relative to 1 about a horizontal axis fixed in both 1 and 2. The reference frame (2), RF2, of unit vectors $[\mathbf{\imath}_2, \mathbf{\jmath}_2, \mathbf{k}_2]$ is fixed in 2. The horizontal unit vectors $\mathbf{\imath}_1$ and $\mathbf{\imath}_2$ are fixed in both 1 and 2. The mass center C_2 is a point fixed in both 1 and 2.

The second generalized coordinate q_2 denotes the radian measure of the angle between the axes of (1) and (2). The link 2 supports link 3, and link 3 has a translational motion relative to 2.

The generalized coordinate q_4 is the distance between the mass centers, C_2 and C_3, of 2 and 3, respectively. The link 3 supports link 4, and link 4 rotates relative to 3 about an axis fixed in both 3 and 4. The reference frame (4), RF4, of unit vectors $[\mathbf{\imath}_4, \mathbf{\jmath}_4, \mathbf{k}_4]$ is fixed in 4. The unit vectors $\mathbf{\jmath}_2$ and $\mathbf{\jmath}_4$ are fixed in both 3 and 4. The mass center C_4 is a point fixed in both 3 and 4.

The generalized coordinate q_3 is the radian measure of the rotation angle between 3 and 4. The reference frame (4), $[\mathbf{\imath}_4, \mathbf{\jmath}_4, \mathbf{k}_4]$ is fixed in 4 in such a way that $\mathbf{\imath}_0 = \mathbf{\imath}_4, \mathbf{\jmath}_0 = \mathbf{\jmath}_4, \mathbf{k}_4 = \mathbf{k}_0$ when $q_1 = q_2 = q_3 = 0$.

The generalized speeds, u_1, u_2, u_3, u_4, are associated with the motion of a system, and can be introduced as

$$
\begin{aligned}
u_1 &= \boldsymbol{\omega}_{40} \cdot \mathbf{\imath}_4, \\
u_2 &= \boldsymbol{\omega}_{40} \cdot \mathbf{\jmath}_4, \\
u_3 &= \boldsymbol{\omega}_{40} \cdot \mathbf{k}_4, \\
u_4 &= \dot{q}_4,
\end{aligned} \tag{6.45}
$$

where $\boldsymbol{\omega}_{40}$ denotes the angular velocity of 4 in (0). One may verify that

$$
\boldsymbol{\omega}_{40} = (\dot{q}_1 s_2 s_3 + \dot{q}_2 c_3)\mathbf{\imath}_4 + (\dot{q}_1 c_2 + \dot{q}_3)\mathbf{\jmath}_4 + (-\dot{q}_1 s_2 c_3 + \dot{q}_2 s_3)\mathbf{k}_4, \tag{6.46}
$$

where s_i and c_i denote $\sin q_i$ and $\cos q_i$, $i = 1, 2, 3$, respectively. Substitution into Eq. 6.45 then yields

$$
\begin{aligned}
u_1 &= \dot{q}_1 s_2 s_3 + \dot{q}_2 c_3, \\
u_2 &= \dot{q}_1 c_2 + \dot{q}_3, \\
u_3 &= -\dot{q}_1 s_2 c_3 + \dot{q}_2 s_3, \\
u_4 &= \dot{q}_4.
\end{aligned} \tag{6.47}
$$

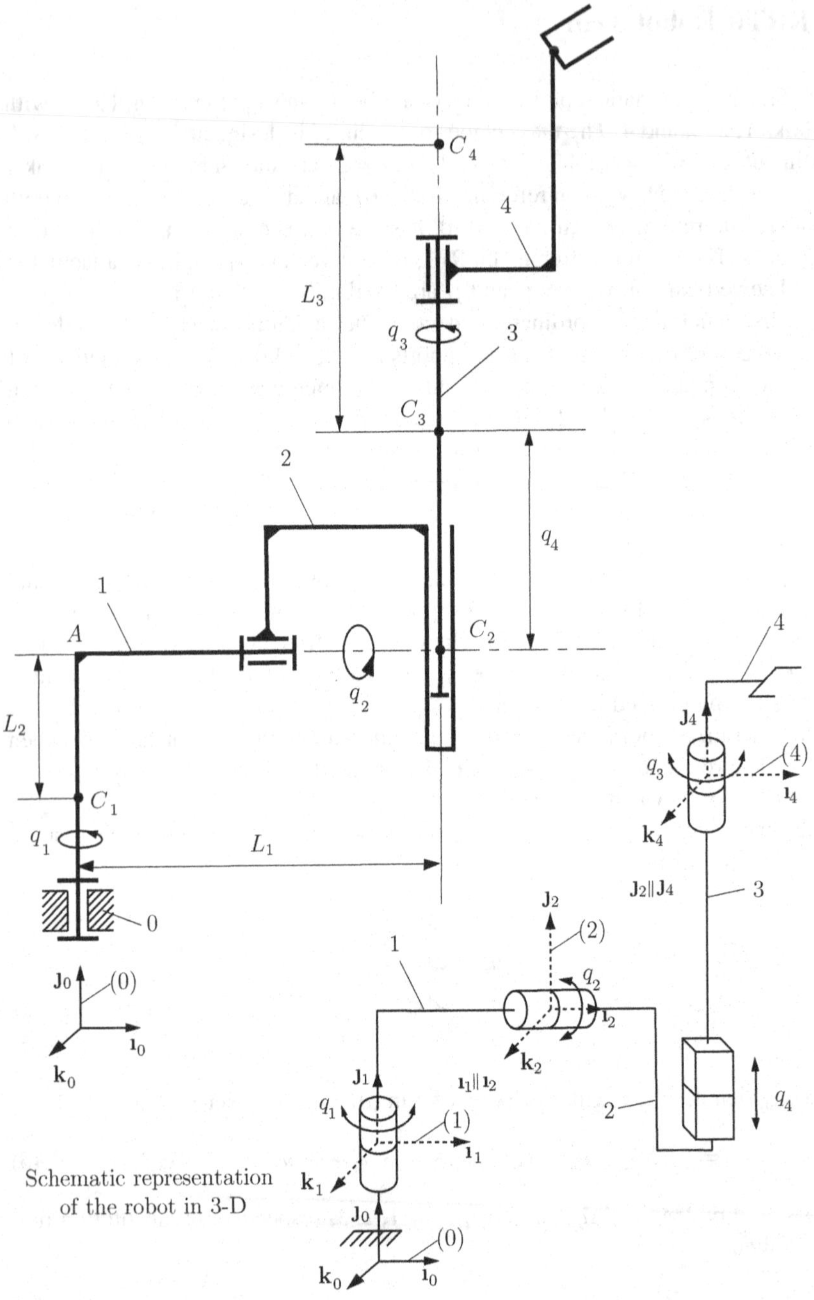

Fig. 6.10 RRTR robot arm

Equation 6.47 can be solved uniquely for $\dot{q}_1, \dot{q}_2, \dot{q}_3, \dot{q}_4$. Specifically,

$$
\begin{aligned}
\dot{q}_1 &= (u_1 s_3 - u_3 c_3)/s_2, \\
\dot{q}_2 &= (u_1 c_3 + u_3 s_3), \\
\dot{q}_3 &= u_2 + (u_3 c_3 - u_1 s_3) c_2/s_2, \\
\dot{q}_4 &= u_4,
\end{aligned} \tag{6.48}
$$

with singularities at $q_2 = 0°$ and $q_2 = 180°$ degrees posing no problem. Thus, u_r as defined in Eq. 6.45 form a set of generalized speeds for the robot. The mass of link r is m_r, $r = 1, 2, 3, 4$. The central inertia dyadic of link r, can be expressed as

$$\bar{I}_r = (I_{rx}\mathbf{\imath}_r)\,\mathbf{\imath}_r + (I_{ry}\mathbf{J}_r)\,\mathbf{J}_r + (I_{rz}\mathbf{k}_r)\,\mathbf{k}_r, \tag{6.49}$$

where I_{rx}, I_{ry}, I_{rz} are the central principal moments of inertia of $r = 1, 2, 3, 4$.

In the case of the robot, there are two kinds of forces that contribute to the generalized active forces Q_r, $r = 1, 2, 3, 4$ namely, contact forces applied in order to drive links 1, 2, 3, 4, and gravitational forces exerted on the links.

Considering, the contact forces, the set of such forces transmitted from the Newtonian frame (0) to link 1 (through bearings and by means of a motor) is replaced with a couple of torque $\mathbf{T}_{01}$ together with a force $\mathbf{F}_{01}$ applied to 1 at C_1.

Similarly, the set of contact forces transmitted from 2 to 1 is replaced with a couple of torque $\mathbf{T}_{21}$ together with a force $\mathbf{F}_{21}$ applied to 1 at C_2 (which is a point fixed in 1). The law of action and reaction then guarantees that the set of contact forces transmitted from 1 to 2 is equivalent to a couple of torque $-\mathbf{T}_{21}$ together with the force $-\mathbf{F}_{21}$ applied to 2 at C_2.

Next, the set of contact forces exerted on 2 by 3 is replaced with a couple of torque $\mathbf{T}_{32}$ together with a force $\mathbf{F}_{32}$ applied to 2 at C_{32}, the point of 2 instantaneously coinciding with C_3. The set of forces exerted by 2 on 3 is, therefore equivalent to a couple torque $-\mathbf{T}_{32}$ together with the force $-\mathbf{F}_{32}$ applied to 3 at C_3.

Similarly, torque $\mathbf{T}_{43}$ and forces $\mathbf{F}_{43}$ come into play in connection with the interactions of 3 and 4 at C_4. As for gravitational forces exerted on the links of the robot by the Earth, these are denoted by $\mathbf{G}_r$, $r = 1, 2, 3, 4$, respectively.

The following notations are introduced

$$
\begin{aligned}
T_{01y} &= \mathbf{T}_{01} \cdot \mathbf{J}_1 = k_1(q_{1f} - q_1) - k_2 \dot{q}_1, \\
T_{21x} &= \mathbf{T}_{21} \cdot \mathbf{\imath}_2 = k_3(q_2 - q_{2f}) + k_4 \dot{q}_2 + g[(m_3 + m_4) q_4 + m_4 L_3] s_2, \\
T_{43y} &= \mathbf{T}_{43} \cdot \mathbf{J}_3 = \mathbf{T}_{43} \cdot \mathbf{J}_2 = k_5(q_3 - q_{3f}) + k_6 \dot{q}_3, \\
F_{32y} &= \mathbf{F}_{32} \cdot \mathbf{J}_3 = \mathbf{F}_{32} \cdot \mathbf{J}_2 = k_7(q_4 - q_{4f}) + k_8 \dot{q}_4 - g(m_3 + m_4) c_2,
\end{aligned} \tag{6.50}
$$

where $k_1, \ldots, k_8$ are constant gains k_1 = 3.0 N m, k_2 = 5.0 N m s, k_3 = 1.0 N m, k_4 = 3.0 N m s, k_5 = 0.3 N m, k_6 = 0.6 N m s, k_7 = 30 N m, k_8 = 41 N s, and $q_{rf} = \pi/3$ rad $r = 1, 2, 3$, while q_{4f} = 0.1 m.

The initial numerical data are L_1 = 0.1 m, L_2 = 0.1 m, L_3 = 0.7 m, m_1 = 9 kg, m_2 = 6 kg, m_3 = 4 kg, m_4 = 1 kg, I_{1x} = 0.01 kg m^2, I_{1y} = 0.02 kg m^2, I_{1z} = 0.01 kg m^2, I_{2x} = 0.06 kg m^2, I_{2y} = 0.01 kg m^2, I_{2z} = 0.05 kg m^2, I_{3x} = 0.4 kg m^2, I_{3y} = 0.01 kg m^2, I_{3z} = 0.4 kg m^2, I_{4x} = 0.0005 kg m^2, I_{4y} = 0.001 kg m^2, and I_{4z} = 0.001 kg m^2.

Kane's Dynamical Equations
The numerical data are introduced with the MATLAB statements:

```
L1=0.1; L2=0.1; L3=0.7;
m1=9.; m2=6.; m3=4.; m4=1.;
I1x=0.01; I1y=0.02; I1z=0.01;
I2x=0.06; I2y=0.01; I2z=0.05;
I3x=0.4; I3y=0.01; I3z=0.4;
I4x=0.0005; I4y=0.001; I4z=0.001;
k1=3.; k2=5.; k3=1.; k4=3.; k5=0.3;
k6=0.6; k7=30.; k8=41.; g=9.8;
q1f=pi/3; q2f=pi/3; q3f=pi/3; q4f=0.1;
```

The MATLAB commands for the generalized coordinates are:

```
syms t real
q1 = sym('q1(t)');
q2 = sym('q2(t)');
q3 = sym('q3(t)');
q4 = sym('q4(t)');
```

The transformation matrix from RF2 to RF1 is

$$R_{21} = \begin{bmatrix} 1 & 0 & 0 \\ 0 & c_2 & s_2 \\ 0 & -s_2 & c_2 \end{bmatrix},$$

and the transformation matrix from RF4 to RF2 is

$$R_{42} = \begin{bmatrix} c_3 & 0 & -s_3 \\ 0 & 1 & 0 \\ s_3 & 0 & c_3 \end{bmatrix}.$$

The MATLAB commands for the transformation matrices are:

```
c2 = cos(q2); s2 = sin(q2);
c3 = cos(q3); s3 = sin(q3);
% transformation matrix from RF2 to RF1
R21 = [[1 0 0]; [0 c2 s2]; [0 -s2 c2]];
% transformation matrix from RF4 to RF2
R42 = [[c3 0 -s3]; [0 1 0]; [s3 0 c3]];
```

The generalized speeds, u_1, u_2, u_3, u_4, are associated with the motion of a system, and can be introduced with:

```
u1 = sym('u1(t)');
u2 = sym('u2(t)');
u3 = sym('u3(t)');
u4 = sym('u4(t)');
```

Equation 6.48 represents the derivatives, $\dot{q}_1, \dot{q}_2, \dot{q}_3, \dot{q}_4$, of the generalized coordinates, function of the generalized speeds:

```
dq1 = (u1*s3-u3*c3)/s2;
dq2 = u1*c3+u3*s3;
dq3 = u2+(u3*c3-u1*s3)*c2/s2;
dq4 = u4;
```

The Kane's dynamical equations are expressed in terms of the generalized coordinates `q1, q2, q3, q4`, the generalized speeds `u1, u2, u3, u4`, and the time derivatives of the generalized speeds. Two lists are introduced for the derivatives of the generalized coordinates and the derivatives of the generalized speeds:

```
qt={diff(q1,t),diff(q2,t),diff(q3,t),diff(q4,t),...
    diff(u1,t),diff(u2,t),diff(u3,t),diff(u4,t)};
ut={dq1,dq2,dq3,dq4, 'du1', 'du2', 'du3', 'du4'};

% qt                      ut
%-------------------------
% diff('q1(t)',t)  ->  dq1
% diff('q2(t)',t)  ->  dq2
% diff('q3(t)',t)  ->  dq3
% diff('q4(t)',t)  ->  dq4
% diff('u1(t)',t)  -> 'du1'
% diff('u2(t)',t)  -> 'du2'
% diff('u3(t)',t)  -> 'du3'
% diff('u4(t)',t)  -> 'du4'
```

Angular Velocities and Accelerations
The angular velocities of each link 1, 2, 3, 4, in RF0, involving the generalized speeds, are expressed using a vector basis fixed in the body under consideration. The angular velocity of link 1 with respect to RF0 expressed in terms of RF1 $[\mathbf{\imath}_1, \mathbf{J}_1, \mathbf{k}_1]$ is $\boldsymbol{\omega}_{10} = \dot{q}_1 \mathbf{J}_1$, or:

```
w101 = [0, dq1, 0];
```

The angular velocity of link 1 relative to RF0 expressed in terms of RF2 $[\mathbf{\imath}_2, \mathbf{\jmath}_2, \mathbf{k}_2]$ is:

```
w102 = w101*transpose(R21);
```

The angular velocity of link 2 relative to RF1 expressed in terms of RF1 $[\mathbf{\imath}_1, \mathbf{\jmath}_1, \mathbf{k}_1]$ is $\boldsymbol{\omega}_{21} = \dot{q}_2 \mathbf{\imath}_1$, or:

```
w211 = [dq2, 0, 0];
```

The angular velocity of link 2 relative to RF1 expressed in terms of RF2 $[\mathbf{\imath}_2, \mathbf{\jmath}_2, \mathbf{k}_2]$ is $\boldsymbol{\omega}_{21} = \dot{q}_2 \mathbf{\imath}_2$, or:

```
w212 = [dq2, 0, 0];
```

and the angular velocity of link 2 with respect to RF0 expressed in terms of RF2 $[\mathbf{\imath}_2, \mathbf{\jmath}_2, \mathbf{k}_2]$ is $\boldsymbol{\omega}_{20} = \boldsymbol{\omega}_{10} + \boldsymbol{\omega}_{21}$, or:

```
w202 = w102 + w212;
```

The angular velocity of link 2 relative to RF0 expressed in terms of RF4 $[\mathbf{\imath}_4, \mathbf{\jmath}_4, \mathbf{k}_4]$ is:

```
w204 = w202*transpose(R42);
```

The angular velocity of link 3 with respect to RF0 expressed in terms of RF2 $[\mathbf{\imath}_2, \mathbf{\jmath}_2, \mathbf{k}_2]$ is $\boldsymbol{\omega}_{30} = \boldsymbol{\omega}_{20}$, or:

```
w302 = w202;
```

The angular velocity of link 4 relative to RF2 expressed in terms of RF2 $[\mathbf{\imath}_2, \mathbf{\jmath}_2, \mathbf{k}_2]$ is $\boldsymbol{\omega}_{42} = \dot{q}_3 \mathbf{\jmath}_2$, or:

```
w422 = [0, dq3, 0];
```

The angular velocity of link 4 with respect to RF2 expressed in terms of RF4 $[\mathbf{\imath}_4, \mathbf{\jmath}_4, \mathbf{k}_4]$ is:

```
w424 = w422*transpose(R42);
```

The anglar velocity of link 4 with respect to RF0 expressed in terms of RF4 $[\mathbf{\imath}_4, \mathbf{\jmath}_4, \mathbf{k}_4]$ is $\boldsymbol{\omega}_{40} = \boldsymbol{\omega}_{20} + \boldsymbol{\omega}_{42}$, or:

```
w404 = w204 + w424;
```

The angular acceleration of link 1 with respect to RF0 expressed in terms of RF1 $[\mathbf{\imath}_1, \mathbf{\jmath}_1, \mathbf{k}_1]$ is $\boldsymbol{\alpha}_{10} = \dot{\boldsymbol{\omega}}_{10}$, or:

```
alpha101 = subs(diff(w101, t), qt, ut);
```

The angular acceleration of link 2 with respect to RF0 expressed in terms of RF2 $[\mathbf{\imath}_2, \mathbf{\jmath}_2, \mathbf{k}_2]$ is $\boldsymbol{\alpha}_{20} = \dot{\boldsymbol{\omega}}_{20}$, or:

```
alpha202 = subs(diff(w202, t), qt, ut);
```

The angular acceleration of link 3 with respect to RF0 expressed in terms of RF2 $[\mathbf{\imath}_2, \mathbf{\jmath}_2, \mathbf{k}_2]$ is:

```
alpha302 = alpha202;
```

The anglar acceleration of link 4 with respect to RF0 expressed in terms of RF4 $[\mathbf{\imath}_4, \mathbf{\jmath}_4, \mathbf{k}_4]$ is $\boldsymbol{\alpha}_{40} = \dot{\boldsymbol{\omega}}_{40}$, or:

```
alpha404 = subs(diff(w404, t), qt, ut);
```

Remark: The angular velocity $\boldsymbol{\omega}_{10}$ (`w101`) was expressed in terms of RF1, $\boldsymbol{\omega}_{20}$ (`w202`) was expressed in terms of RF2, $\boldsymbol{\omega}_{30}$ (`w302`) was expressed in terms of RF3=RF2, and $\boldsymbol{\omega}_{40}$ (`w404`) was expressed in terms of RF4.

The partial velocity $\dfrac{\partial \boldsymbol{\omega}_{10}}{\partial u_r}$ will be in terms of RF1, $\dfrac{\partial \boldsymbol{\omega}_{20}}{\partial u_r}$ will be in terms of RF2, $\dfrac{\partial \boldsymbol{\omega}_{30}}{\partial u_r}$ will be in terms of RF3, and $\dfrac{\partial \boldsymbol{\omega}_{40}}{\partial u_r}$ will be in terms of RF4.

The central principal axes of inertia of link 1 are parallel to $[\mathbf{\imath}_1, \mathbf{\jmath}_1, \mathbf{k}_1]$, link 2 are parallel to $[\mathbf{\imath}_2, \mathbf{\jmath}_2, \mathbf{k}_2]$, link 3 are parallel to $[\mathbf{\imath}_2, \mathbf{\jmath}_2, \mathbf{k}_2]$, and link 4 are parallel to $[\mathbf{\imath}_4, \mathbf{\jmath}_4, \mathbf{k}_4]$.

The angular velocities of each link are expressed using a vector basis fixed in the body under consideration because the central principal axes of inertia of the body are parallel to the vector basis fixed in the body.

For the velocities of C_1, C_2, C_3, C_4, the mass centers of links 1, 2, 3, 4, it is not necessarily useful to work with RF1 in the case of velocity of C_1, with RF2 for the velocity of C_2, and so forth. As an alternative, it is convenient to use any vector basis that permits one to write the simplest expression.

Linear Velocities and Accelerations

The linear velocity of mass center C_1 of link 1 is zero since C_1 is fixed in RF0:

```
vC101 = [0, 0, 0];
```

The position vector from C_1, the mass center of link 1, to C_2, the mass center of link 2, expressed in terms of RF1 $[\mathbf{\imath}_1, \mathbf{\jmath}_1, \mathbf{k}_1]$ is $\mathbf{r}_{C_1C_2} = L_1\mathbf{\imath}_1 + L_2\mathbf{\jmath}_1$, or:

```
rC121 = [L1, L2, 0];
```

The linear velocity of mass center C_2 of link 2 with respect to RF0 expressed in terms of RF1 $[\mathbf{\imath}_1, \mathbf{\jmath}_1, \mathbf{k}_1]$ is

$$\mathbf{v}_{C_2} = \frac{d}{dt}\mathbf{r}_{C_1C_2} = \frac{^{(1)}d}{dt}\mathbf{r}_{C_1C_2} + \boldsymbol{\omega}_{10} \times \mathbf{r}_{C_1C_2},$$

or:

```
vC201 = diff(rC121, t) + cross(w101, rC121);
```

The velocity of C_2, which is "fixed" in RF1 can be computed also as:

```
vC201 = cross(w101, [L1, 0, 0]);
```

The linear velocity of the mass center C_2 of link 2 with respect to RF0 expressed in terms of RF2 $[\mathbf{\imath}_2, \mathbf{\jmath}_2, \mathbf{k}_2]$ is:

```
vC202 = vC201*transpose(R21);
```

The position vector from C_1, the mass center of link 1, to C_3, the mass center of link 3, expressed in terms of RF2 $[\mathbf{\imath}_2, \mathbf{\jmath}_2, \mathbf{k}_2]$ is $\mathbf{r}_{C_1C_3} = \mathbf{r}_{C_1C_2} + q_4\mathbf{\jmath}_2$, or:

```
rC132 = [L1, L2, 0]*transpose(R21) + [0, q4, 0];
```

The linear velocity of mass center C_3 of link 3 with respect to RF0 expressed in terms of RF2 $[\mathbf{\imath}_2, \mathbf{\jmath}_2, \mathbf{k}_2]$ is

$$\mathbf{v}_{C_3} = \frac{d}{dt}\mathbf{r}_{C_1C_3} = \frac{^{(2)}d}{dt}\mathbf{r}_{C_1C_3} + \boldsymbol{\omega}_{20} \times \mathbf{r}_{C_1C_3},$$

or:

```
vC302 = subs(diff(rC132,t),qt,ut)+cross(w202,rC132);
```

Another way of computing the previous velocity is:

```
vC302 = vC202+diff([0,q4,0],t]+cross(w202,[0,q4,0]);
```

The linear velocity of point C_{32} of link 2 with respect to RF0 expressed in terms of RF2 $[\mathbf{\imath}_2, \mathbf{\jmath}_2, \mathbf{k}_2]$ is

$$\mathbf{v}_{C_{32}} = \mathbf{v}_{C_2} + \boldsymbol{\omega}_{20} \times \mathbf{r}_{C_2C_3}.$$

The MATLAB command for $\mathbf{v}_{C_{32}}$ is:

```
vC3202 = vC202 + cross(w202, [0, q4, 0]);
```

The point C_{32}, of link 2, is superposed with the mass center C_3, of link 3. The linear velocity of mass center C_4 of link 4 with respect to RF0 expressed in terms of RF2 $[\mathbf{\imath}_2, \mathbf{\jmath}_2, \mathbf{k}_2]$ is

$$\mathbf{v}_{C_4} = \mathbf{v}_{C_3} + \boldsymbol{\omega}_{20} \times L_3 \mathbf{\jmath}_2,$$

or:

```
vC402 = vC302 + cross(w202, [0, L3, 0]);
```

The linear acceleration of mass center C_1 of link 1 with respect to RF0 expressed in terms of RF1 $[\mathbf{\imath}_1, \mathbf{\jmath}_1, \mathbf{k}_1]$ is:

```
aC101 = [0, 0, 0];
```

The linear acceleration of mass center C_2 of link 2 with respect to RF0 expressed in terms of RF1 $[\mathbf{\imath}_1, \mathbf{\jmath}_1, \mathbf{k}_1]$ is

$$\mathbf{a}_{C_2} = \frac{d}{dt}\mathbf{v}_{C_2} = \frac{^{(1)}d}{dt}\mathbf{v}_{C_2} + \boldsymbol{\omega}_{10} \times \mathbf{v}_{C_1},$$

or:

```
aC201 = subs(diff(vC201, t),qt,ut)+cross(w101,vC201);
```

The linear acceleration of mass center C_3 of link 3 with respect to RF0 expressed in terms of RF2 $[\mathbf{\imath}_2, \mathbf{\jmath}_2, \mathbf{k}_2]$ is

$$\mathbf{a}_{C_3} = \frac{d}{dt}\mathbf{v}_{C_3} = \frac{^{(2)}d}{dt}\mathbf{v}_{C_3} + \boldsymbol{\omega}_{20} \times \mathbf{v}_{C_3},$$

or:

```
aC302 = subs(diff(vC302,t),qt,ut)+cross(w202,vC302);
```

The linear acceleration of mass center C_4 of link 4 with respect to RF0 expressed in terms of RF2 $[\mathbf{\imath}_2, \mathbf{\jmath}_2, \mathbf{k}_2]$ is

$$\mathbf{a}_{C_4} = \frac{d}{dt}\mathbf{v}_{C_4} = \frac{^{(2)}d}{dt}\mathbf{v}_{C_4} + \boldsymbol{\omega}_{40} \times \mathbf{v}_{C_4},$$

The MATLAB command for $\mathbf{a}_{C_4}$ is:

```
aC402 = subs(diff(vC402,t),qt,ut)+cross(w402,vC402);
```

The partial velocities with respect to u_r (`u1, u2, u3, u4`) are

$$\frac{\partial \omega_{10}}{\partial u_r}, \frac{\partial \omega_{20}}{\partial u_r}, \frac{\partial \omega_{40}}{\partial u_r},$$
$$\frac{\partial \mathbf{v}_{C_1}}{\partial u_r}, \frac{\partial \mathbf{v}_{C_2}}{\partial u_r}, \frac{\partial \mathbf{v}_{C_3}}{\partial u_r}, \frac{\partial \mathbf{v}_{C_4}}{\partial u_r}, \frac{\partial \mathbf{v}_{C_{32}}}{\partial u_r},$$
$$r = 1, 2, 3, 4,$$

and in MATLAB are calculated using the function `deriv`:

```
w1_1 = deriv(w101, u1); w1_2 = deriv(w101, u2);
w1_3 = deriv(w101, u3); w1_4 = deriv(w101, u4);
w2_1 = deriv(w202, u1); w2_2 = deriv(w202, u2);
w2_3 = deriv(w202, u3); w2_4 = deriv(w202, u4);
w4_1 = deriv(w404, u1); w4_2 = deriv(w404, u2);
w4_3 = deriv(w404, u3); w4_4 = deriv(w404, u4);

vC1_1 = deriv(vC101, u1); vC1_2 = deriv(vC101, u2);
vC1_3 = deriv(vC101, u3); vC1_4 = deriv(vC101, u4);
vC2_1 = deriv(vC201, u1); vC2_2 = deriv(vC201, u2);
vC2_3 = deriv(vC201, u3); vC2_4 = deriv(vC201, u4);

vC3_1 = deriv(vC302, u1); vC3_2 = deriv(vC302, u2);
vC3_3 = deriv(vC302, u3); vC3_4 = deriv(vC302, u4);
vC4_1 = deriv(vC402, u1); vC4_2 = deriv(vC402, u2);
vC4_3 = deriv(vC402, u3); vC4_4 = deriv(vC402, u4);

vC32_1 = deriv(vC3202, u1);
vC32_2 = deriv(vC3202, u2);
vC32_3 = deriv(vC3202, u3);
vC32_4 = deriv(vC3202, u4);
```

Generalized Active Forces
Link 1
The following symbolical variables are introduced:

```
syms F01x F01y F01z T01x T01y T01z real
syms F21x F21y F21z T21x T21y T21z real
```

The force applied by base 0 to link 1 at C_1 expressed in terms of RF1 is $\mathbf{F}_{01} = F_{01x}\mathbf{\imath}_1 + F_{01y}\mathbf{J}_1 + F_{01z}\mathbf{k}_1$, or:

```
F01 = [F01x F01y F01z];
```

The torque applied by base 0 to link 1 expressed in terms of RF1 is $\mathbf{T}_{01} = T_{01x}\mathbf{\imath}_1 + T_{01y}\mathbf{J}_1 + T_{01z}\mathbf{k}_1$, or:

```
T01 = [T01x T01y T01z];
```

The force applied by link 2 to link 1 at C_2 expressed in terms of RF2 is $\mathbf{F}_{21} = F_{21x}\mathbf{\imath}_2 + F_{21y}\mathbf{J}_2 + F_{21z}\mathbf{k}_2$, or:

```
F21 = [F21x F21y F21z];
```

The torque applied by link 2 to link 1 expressed in terms of RF2 is $\mathbf{T}_{21} = T_{21x}\mathbf{\imath}_2 + T_{21y}\mathbf{J}_2 + T_{21z}\mathbf{k}_2$, or:

```
T21 = [T21x T21y T21z];
```

The gravitational force that acts on link 1 at C_1 is $\mathbf{G}_1 = -m_1 g\mathbf{J}_0 = -m_1 g\mathbf{J}_1$, or:

```
G1 = [0 -m1*g 0];
```

The generalized active forces for link 1 are

$$Q_{1r} = \frac{\partial \omega_{10}}{\partial u_r} \cdot (\mathbf{T}_{01} + \mathbf{T}_{21}) + \frac{\partial \mathbf{v}_{C_1}}{\partial u_r} \cdot (\mathbf{G}_1 + \mathbf{F}_{01}) + \frac{\partial \mathbf{v}_{C_2}}{\partial u_r} \cdot \mathbf{F}_{21},\ r = 1, 2, 3, 4.$$

The MATLAB statements for the generalized active forces for link 1 are:

```
Q1a1=w1_1*(T01.'+R21.'*T21.')+vC1_1*(G1.'+F01.')+...
     vC2_1*R21.'*F21.';
Q1a2=w1_2*(T01.'+R21.'*T21.')+vC1_2*(G1.'+F01.')+...
     vC2_2*R21.'*F21.';
Q1a3=w1_3*(T01.'+R21.'*T21.')+vC1_3*(G1.'+F01.')+...
     vC2_3*R21.'*F21.';
Q1a4=w1_4*(T01.'+R21.'*T21.')+vC1_4*(G1.'+F01.')+...
     vC2_4*R21.'*F21.';
```

Link 2
The following symbolical variables are introduced:

```
syms F32x F32y F32z T32x T32y T32z real
```

The force applied by link 1 to link 2 at C_2 expressed in terms of RF2 is $-\mathbf{F}_{21}$. The torque applied by link 1 to link 2 expressed in terms of RF2 is $-\mathbf{T}_{21}$. The force applied by link 3 to link 2 at C_{32} expressed in terms of RF2 is $\mathbf{F}_{32} = F_{32x}\,\mathbf{1}_2 + F_{32y}\,\mathbf{J}_2 + F_{32z}\,\mathbf{k}_2$, or:

```
F32 = [F32x F32y F32z];
```

The torque applied by link 3 to link 2 expressed in terms of RF2 is $\mathbf{T}_{32} = T_{32x}\,\mathbf{1}_2 + T_{32y}\,\mathbf{J}_2 + T_{32z}\,\mathbf{k}_2$, or:

```
T32 = [T32x T32y T32z];
```

The gravitational force that acts on link 2 at C_2 expressed in terms of RF1 is $\mathbf{G}_2 = -m_2\,g\,\mathbf{J}_1$, or:

```
G2 = [0 -m2*g 0];
```

The generalized active forces for link 2 are

$$Q_{2r} = \frac{\partial \omega_{20}}{\partial u_r} \cdot (\mathbf{T}_{32} - \mathbf{T}_{21}) + \frac{\partial \mathbf{v}_{C_2}}{\partial u_r} \cdot (\mathbf{G}_2 - \mathbf{F}_{21}) + \frac{\partial \mathbf{v}_{C_{32}}}{\partial u_r} \cdot \mathbf{F}_{32},\ r = 1, 2, 3, 4.$$

The MATLAB statements for the generalized active forces for link 2 are:

```
Q2a1=w2_1*(T32.'-T21.')+vC2_1*(G2.'-R21.'*F21.')+...
    vC32_1*F32.';
Q2a2=w2_2*(T32.'-T21.')+vC2_2*(G2.'-R21.'*F21.')+...
    vC32_2*F32.';
Q2a3=w2_3*(T32.'-T21.')+vC2_3*(G2.'-R21.'*F21.')+...
    vC32_3*F32.';
Q2a4=w2_4*(T32.'-T21.')+vC2_4*(G2.'-R21.'*F21.')+...
    vC32_2*F32.';
```

Link 3
The following symbolical variables are introduced:

```
syms F43x F43y F43z T43x T43y T43z real
```

The force applied by link 2 to link 3 at C_3 expressed in terms of RF2 is $-\mathbf{F}_{32}$. The torque applied by link 2 to link 3 expressed in terms of RF2 is $-\mathbf{T}_{32}$. The force applied by link 4 to link 3 at C_4 expressed in terms of RF2 is $\mathbf{F}_{43} = F_{43x}\,\mathbf{1}_2 + F_{43y}\,\mathbf{J}_2 + F_{43z}\,\mathbf{k}_2$, or:

```
F43 = [F43x F43y F43z];
```

The torque applied by link 4 to link 3 expressed in terms of RF2 is $\mathbf{T}_{43} = T_{43x}\,\mathbf{l}_2 + T_{43y}\,\mathbf{J}_2 + T_{43z}\,\mathbf{k}_2$, or:

```
T43 = [T43x T43y T43z];
```

The gravitational force that acts on link 3 at C_3 expressed in terms of RF2 is:

```
G3 = [0 -m3*g 0]*transpose(R21);
```

The generalized active forces for link 3 are

$$Q_{3r} = \frac{\partial \omega_{20}}{\partial u_r} \cdot (\mathbf{T}_{43} - \mathbf{T}_{32}) + \frac{\partial \mathbf{v}_{C_3}}{\partial u_r} \cdot (\mathbf{G}_3 - \mathbf{F}_{32}) + \frac{\partial \mathbf{v}_{C_4}}{\partial u_r} \cdot \mathbf{F}_{43},\ r = 1, 2, 3, 4.$$

The MATLAB statements for the generalized active forces for link 3 are:

```
Q3a1=w2_1*(T43.'-T32.')+vC3_1*(G3.'-F32.')+...
     vC4_1*F43.';
Q3a2=w2_2*(T43.'-T32.')+vC3_2*(G3.'-F32.')+...
     vC4_2*F43.';
Q3a3=w2_3*(T43.'-T32.')+vC3_3*(G3.'-F32.')+...
     vC4_3*F43.';
Q3a4=w2_4*(T43.'-T32.')+vC3_4*(G3.'-F32.')+...
     vC4_4*F43.';
```

Link 4

The force applied by link 3 to link 4 at C_4 expressed in terms of RF2 is $-\mathbf{F}_{43}$. The torque applied by link 3 to link 4 expressed in terms of RF2 is $-\mathbf{T}_{43}$. The gravitational force that acts on link 4 at C_4 expressed in terms of RF2 is:

```
G4 = [0 -m4*g 0]*transpose(R21);
```

The generalized active forces for link 3 are

$$Q_{4r} = \frac{\partial \omega_{40}}{\partial u_r} \cdot (-\mathbf{T}_{43}) + \frac{\partial \mathbf{v}_{C_4}}{\partial u_r} \cdot (\mathbf{G}_4 - \mathbf{F}_{43}),\ r = 1, 2, 3, 4.$$

The MATLAB statements for the generalized active forces for link 4 are:

```
Q4a1 = w4_1*R42*(-T43).' + vC4_1*(G4.'-F43.');
Q4a2 = w4_2*R42*(-T43).' + vC4_2*(G4.'-F43.');
Q4a3 = w4_3*R42*(-T43).' + vC4_3*(G4.'-F43.');
Q4a4 = w4_4*R42*(-T43).' + vC4_4*(G4.'-F43.');
```

The total generalized active forces are

$$Q_r = Q_{1r} + Q_{2r} + Q_{3r} + Q_{4r}, \quad r = 1, 2, 3, 4,$$

or:

```
Q1 = simple(Q1a1+Q2a1+Q3a1+Q4a1);
Q2 = simple(Q1a2+Q2a2+Q3a2+Q4a2);
Q3 = simple(Q1a3+Q2a3+Q3a3+Q4a3);
Q4 = simple(Q1a4+Q2a4+Q3a4+Q4a4);
```

Generalized Inertia Forces
The central inertia dyadic of link p, $p = 1, 2, 3, 4$ is $\bar{I}_p = (I_{px}\mathbf{\imath}_p)\mathbf{\imath}_p + (I_{py}\mathbf{J}_p)\mathbf{J}_p + (I_{pz}\mathbf{k}_p)\mathbf{k}_p$. The central principal axes of link p are parallel to $\mathbf{\imath}_p, \mathbf{J}_p, \mathbf{k}_p$ and the associated moments of inertia have the values I_{px}, I_{py}, I_{pz}, respectively. The inertia matrix associated with $\bar{I}_p$ is

$$\bar{I}_p \rightarrow \begin{bmatrix} I_{px} & 0 & 0 \\ 0 & I_{py} & 0 \\ 0 & 0 & I_{pz} \end{bmatrix},$$

or in MATLAB:

```
% inertia matrix associated with
% central inertia dyadic of link 1
I1 = [I1x 0 0; 0 I1y 0; 0 0 I1z];
% inertia matrix associated with
% central inertia dyadic of link 2
I2 = [I2x 0 0; 0 I2y 0; 0 0 I2z];
% inertia matrix associated with
% central inertia dyadic of link 3
I3 = [I3x 0 0; 0 I3y 0; 0 0 I3z];
% inertia matrix associated with
% central inertia dyadic of link 4
I4 = [I4x 0 0; 0 I4y 0; 0 0 I4z];
```

Define $\mathbf{F}_{\text{in}\,p}$, the inertia force for link p, as

$$\mathbf{F}_{\text{in}\,p} = -m_r \mathbf{a}_{C_p}, \quad p = 1, 2, 3, 4,$$

or in MATLAB:

```
% inertia force Fin1 of link1 in RF0
% expressed in terms of RF1
Fin1 = -m1*aC101;
```

```
% inertia force Fin2 of link2 in RF0
% expressed in terms of RF1
Fin2 = -m2*aC201;
% inertia force Fin3 of link3 in RF0
% expressed in terms of RF2
Fin3 = -m3*aC302;
% inertia force Fin4 of link4 in RF0
% expressed in terms of RF2
Fin4 = -m4*aC402;
```

The inertia moment $\mathbf{M}_{\text{in}\,p}$ for link p is

$$\mathbf{M}_{\text{in}\,p} = -\boldsymbol{\alpha}_{p0} \cdot \bar{I}_p - \boldsymbol{\omega}_{p0} \times (\bar{I}_p \cdot \boldsymbol{\omega}_{p0}).$$

The MATLAB statements for the inertia moments are:

```
% inertia moment Min1 of link 1 in RF0
% expressed in terms of RF1
Min1 = -alpha101*I1-cross(w101,w101*I1);
% inertia moment Min2 of link 2 in RF0
% expressed in terms of RF2
Min2 = -alpha202*I2-cross(w202,w202*I2);
% inertia moment Min3 of link 3 in RF0
% expressed in terms of RF2
Min3 = -alpha302*I3-cross(w202,w202*I3);
% inertia moment Min4 of link 4 in RF0
% expressed in terms of RF4
Min4 = -alpha404*I4-cross(w404,w404*I4);
```

The generalized inertia force $K_{\text{in}\,r}$ is

$$K_{\text{in}\,r} = \sum_{p=1}^{4} \frac{\partial \boldsymbol{\omega}_{p0}}{\partial u_r} \cdot \mathbf{M}_{\text{in}\,p} + \sum_{p=1}^{4} \frac{\partial \mathbf{v}_{C_p}}{\partial u_r} \cdot \mathbf{F}_{\text{in}\,p}, \quad r = 1, 2, 3, 4.$$

The MATLAB commands for the generalized inertia forces are:

```
Kin1 = w1_1*Min1.' + vC1_1*Fin1.' + ...
       w2_1*Min2.' + vC2_1*Fin2.' + ...
       w2_1*Min3.' + vC3_1*Fin3.' + ...
       w4_1*Min4.' + vC4_1*Fin4.' ;
Kin2 = w1_2*Min1.' + vC1_2*Fin1.' + ...
       w2_2*Min2.' + vC2_2*Fin2.' + ...
       w2_2*Min3.' + vC3_2*Fin3.' + ...
       w4_2*Min4.' + vC4_2*Fin4.' ;
```

```
Kin3 = w1_3*Min1.' + vC1_3*Fin1.' + ...
       w2_3*Min2.' + vC2_3*Fin2.' + ...
       w2_3*Min3.' + vC3_3*Fin3.' + ...
       w4_3*Min4.' + vC4_3*Fin4.' ;
Kin4 = w1_4*Min1.' + vC1_4*Fin1.' + ...
       w2_4*Min2.' + vC2_4*Fin2.' + ...
       w2_4*Min3.' + vC3_4*Fin3.' + ...
       w4_4*Min4.' + vC4_4*Fin4.' ;
```

The dynamical equations governing the robot arm are $Q_r + K_{\text{in}r} = 0,\ r = 1,2,3,4$. Kane's dynamical equations in MATLAB are:

```
Kane1 = Q1 + Kin1;
Kane2 = Q2 + Kin2;
Kane3 = Q3 + Kin3;
Kane4 = Q4 + Kin4;
```

Using the feedback control laws Kane's equations have to be rewritten and in MATLAB:

```
T01yc = k1*(q1f-q1)-k2*dq1;
T21xc = k3*(q2-q2f)+k4*dq2+g*((m3+m4)*q4+m4*L3)*s2;
T43yc = k5*(q3-q3f)+k6*dq3;
F32yc = k7*(q4-q4f)+k8*dq4-g*(m3+m4)*c2;

tor = {T01y , T21x , T43y , F32y };
torf = {T01yc, T21xc, T43yc, F32yc };

Kan1 = subs(Kane1, tor, torf);
Kan2 = subs(Kane2, tor, torf);
Kan3 = subs(Kane3, tor, torf);
Kan4 = subs(Kane4, tor, torf);
```

Kane's dynamical equations are transformed into a first-order system of differential equations:

```
ql = {q1, q2, q3, q4, u1, u2, u3, u4,};
qe = {'x(1)','x(2)','x(3)','x(4)',...
      'x(5)','x(6)','x(7)','x(8)'};

% ql            qe
%-----------------
% 'q1(t)' -> 'x(1)'
% 'q2(t)' -> 'x(2)'
% 'q3(t)' -> 'x(3)'
```

```
% 'q4(t)'  -> 'x(4)'
% 'u1(t)'  -> 'x(5)'
% 'u2(t)'  -> 'x(6)'
% 'u3(t)'  -> 'x(7)'
% 'u4(t)'  -> 'x(8)'

e1 = subs(Kan1, ql, qe);
e2 = subs(Kan2, ql, qe);
e3 = subs(Kan3, ql, qe);
e4 = subs(Kan4, ql, qe);

% system of ODE
dx1 = char(subs(dq1, ql, qe));
dx2 = char(subs(dq2, ql, qe));
dx3 = char(subs(dq3, ql, qe));
dx4 = char(subs(dq4, ql, qe));
dx5 = char(du1c);
dx6 = char(du2c);
dx7 = char(du3c);
dx8 = char(du4c);
```

The system of differential equations is solved numerically by m-file functions. The function file, `RRTR.m` is created using the statements:

```
fid = fopen('RRTR.m','w+');
fprintf(fid,'function dx = RRTR(t,x)\n');
fprintf(fid,'dx = zeros(8,1);\n');
fprintf(fid,'dx(1) = '); fprintf(fid,dx1);
fprintf(fid,';\n');
fprintf(fid,'dx(2) = '); fprintf(fid,dx2);
fprintf(fid,';\n');
fprintf(fid,'dx(3) = '); fprintf(fid,dx3);
fprintf(fid,';\n');
fprintf(fid,'dx(4) = '); fprintf(fid,dx4);
fprintf(fid,';\n');
fprintf(fid,'dx(5) = '); fprintf(fid,dx5);
fprintf(fid,';\n');
fprintf(fid,'dx(6) = '); fprintf(fid,dx6);
fprintf(fid,';\n');
fprintf(fid,'dx(7) = '); fprintf(fid,dx7);
fprintf(fid,';\n');
fprintf(fid,'dx(8) = '); fprintf(fid,dx8);
fprintf(fid,';');
fclose(fid); cd(pwd);
```

The `ode45` solver is used for the system of differential equations:

```
t0 = 0; tf = 15; time = [0 tf];
q10=pi/6; q20=pi/12; q30=pi/10; q40=0.01;
u10=0; u20=0; u30=0; u40=0;
x0=[q10 q20 q30 q40 u10 u20 u30 u40];
[t,xs]=ode45(@RRTR, time, x0);
x1=xs(:,1); x2=xs(:,2); x3=xs(:,3); x4=xs(:,4);
x5=xs(:,5); x6=xs(:,6); x7=xs(:,7); x8=xs(:,8);
subplot(4,1,1),plot(t,x1*180/pi,'r'),...
xlabel('t (s)'),ylabel('q1 (deg)'),grid,...
subplot(4,1,2),plot(t,x2*180/pi,'b'),...
xlabel('t (s)'),ylabel('q2 (deg)'),grid,...
subplot(4,1,3),plot(t,x3*180/pi,'g'),...
xlabel('t (s)'),ylabel('q3 (deg)'),grid,...
subplot(4,1,4),plot(t,x4,'black'),...
xlabel('t (s)'),ylabel('q4 (m)'),grid
[ts,xs] = ode45(@RRTR,0:1:5,x0)
```

Figure 6.11 shows the plots of the generalized coordinates $q_1(t)$, $q_2(t)$, $q_3(t)$, and $q_4(t)$. The MATLAB computer program for the robot arm using Kane's dynamical equations is given in Appendix E.6.

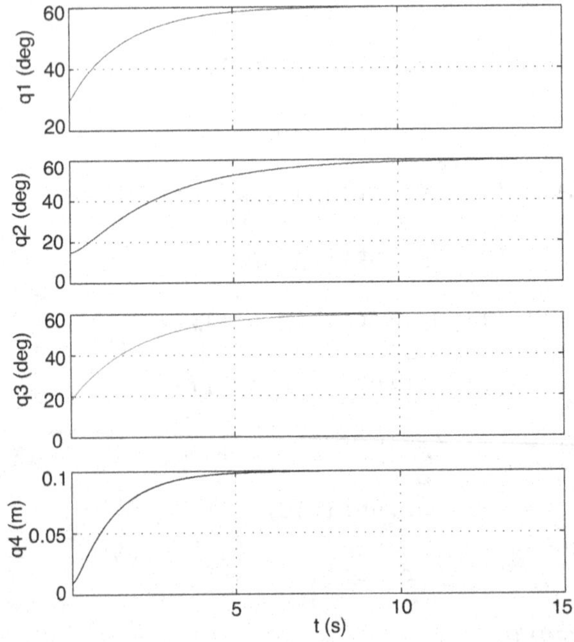

Fig. 6.11 Solution plots of the generalized coordinates $q_1(t)$, $q_2(t)$, $q_3(t)$, and $q_4(t)$

Chapter 7
Problems

7.1 Problem Set: Mechanisms

The dimensions of the planar mechanisms, shown in Figs. 7.1–7.15, are given in Tables 7.1–7.15, respectively. The angle of the driver link 1 with the horizontal axis is ϕ. The constant angular speed of the driver link 1 is n and is given in the tables.

1. Determine the type of motion (rotation, translation, and complex motion) for each link, the connectivity table, the structural diagram, the contour diagram, the independent contours, the number of degrees of freedom, the dyads, and the type of the dyads.
2. Find the positions of the joints and the angles of the links with the horizontal axis when the angle of the driver link 1 with the horizontal axis is ϕ. Write a MATLAB® program for the positions of the mechanism for the given angle ϕ.
3. Write a MATLAB program for the positions of the mechanism for a complete rotation of the driver link 1, $\phi \in [0°, ..., 360°]$, using different methods.
4. Write a MATLAB program for the path of a point on a link with general plane motion.
5. Write a MATLAB program for the animation (movie) of the mechanism for a complete rotation of the driver link 1, $\phi \in [0°, ..., 360°]$.
6. Find the velocities and the accelerations of the mechanism, using different methods, for the given position when the driver link 1 makes an angle ϕ with the horizontal axis. The constant angular speed of the driver link 1 is n. Write a MATLAB program for this.
7. The link bars of the mechanism are homogeneous rectangular prisms with the width $h = 0.01$ m and the depth $d = 0.001$ m. The sliders have the width $w_{Slider} = 0.050$ m, the height $h_{Slider} = 0.020$ m, and the same depth $d = 0.001$ m. The links of the mechanism are homogeneous and are made of steel having a mass density $\rho_{Steel} = 8000$ kg/m^3. The gravitational acceleration is $g = 9.807$ m/s^2. The driver link 1 has a constant angular speed n. The external force or moment applied on the last link 5 is opposed to the motion of the link and has the value: $|\mathbf{F}_{5ext}| =$

2000 N if the last link 5 has a translational motion or, $|\mathbf{M}_{5ext}| = 3000$ N m if the last link 5 has a rotational motion.
Determine the joint forces and the motor moment $\mathbf{M}_m$ required for the dynamic equilibrium of the considered mechanism when the the driver link 1 makes an angle ϕ with the horizontal axis. Write MATLAB programs for different methods of calculating the dynamical forces.

Table 7.1 Mechanism 1

No	AB [m]	AD [m]	BC [m]	CD [m]	ϕ [°]	n [rpm]
1	0.08	0.19	0.21	0.12	60	500
2	0.10	0.24	0.26	0.14	120	600
3	0.18	0.43	0.47	0.27	210	700
4	0.15	0.36	0.40	0.21	330	800

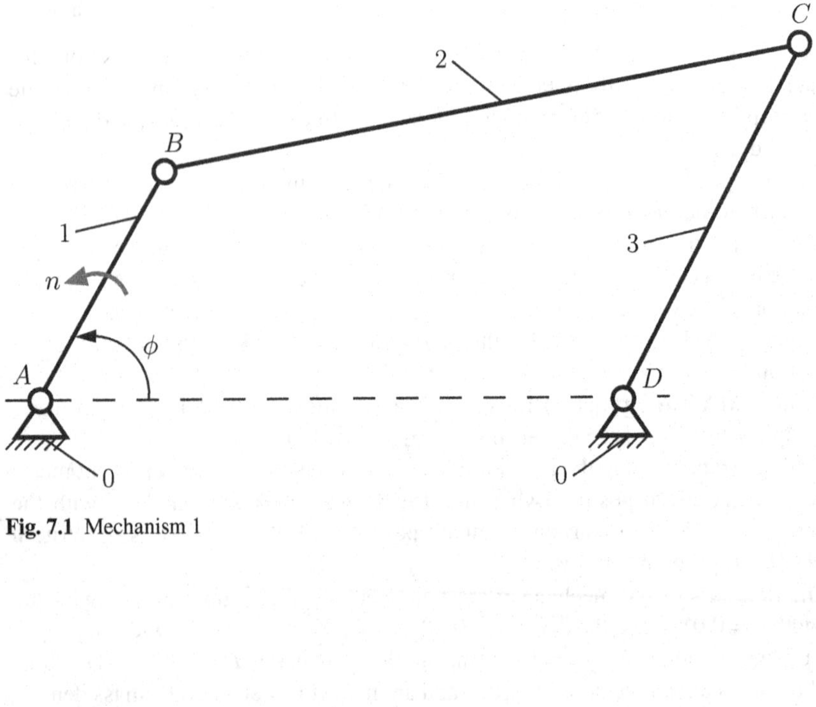

Fig. 7.1 Mechanism 1

Table 7.2 Mechanism 2

No	*AB* [m]	*BC* [m]	*CD* [m]	*CE* [m]	*EF* [m]	*a* [m]	*b* [m]	*c* [m]	ϕ [°]	*n* [rpm]
1	0.15	0.40	0.37	0.23	0.23	0.30	0.45	0.37	330	500
2	0.10	0.27	0.25	0.15	0.15	0.20	0.30	0.25	120	800
3	0.12	0.32	0.30	0.18	0.18	0.24	0.36	0.30	240	900
4	0.20	0.55	0.50	0.30	0.30	0.40	0.60	0.50	300	1000

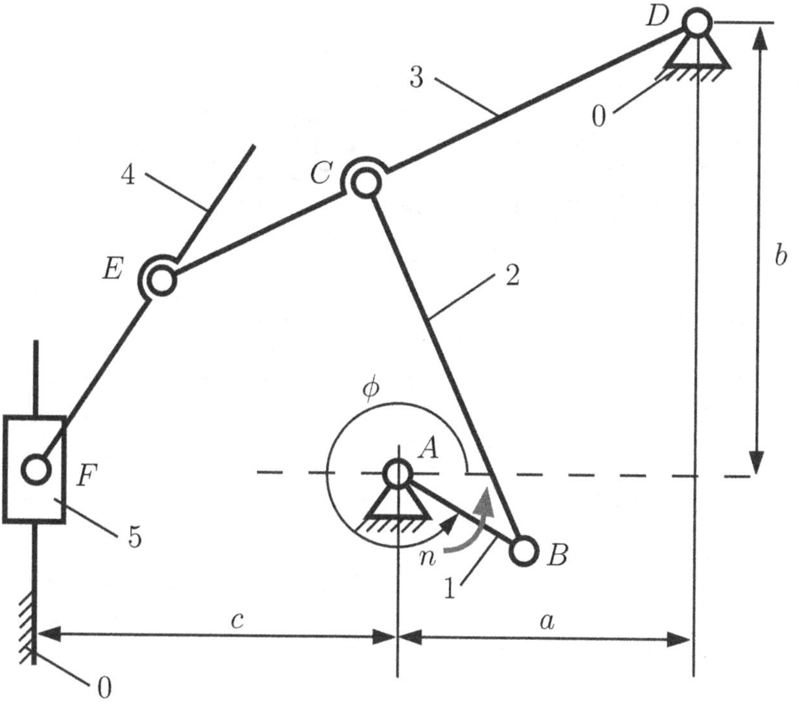

Fig. 7.2 Mechanism 2

Table 7.3 Mechanism 3

No	*AB* [m]	*BC* [m]	*CD* [m]	*DE* [m]	*EF* [m]	*a* [m]	*b* [m]	ϕ [°]	*n* [rpm]
1	0.02	0.03	0.03	0.03	0.06	0.03	0.01	30	600
2	0.05	0.20	0.20	0.31	0.30	0.10	0.06	150	700
3	0.09	0.11	0.07	0.11	0.12	0.035	0.025	240	900
4	0.22	0.27	0.17	0.25	0.28	0.09	0.055	330	1100

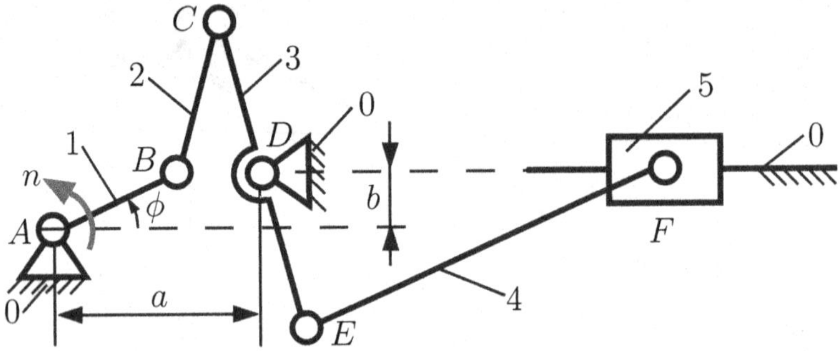

Fig. 7.3 Mechanism 3

Table 7.4 Mechanism 4

No	*AB* [m]	*BC* [m]	*CE* [m]	*CD* [m]	*EF* [m]	*a* [m]	*b* [m]	*c* [m]	ϕ [°]	*n* [rpm]
1	0.09	0.40	0.25	0.12	0.21	0.22	0.35	0.40	60	400
2	0.15	0.67	0.45	0.22	0.32	0.33	0.60	0.65	135	600
3	0.22	1.00	0.65	0.35	0.60	0.55	0.90	1.20	240	800
4	0.16	0.70	0.50	0.25	0.48	0.40	0.60	0.70	300	1000

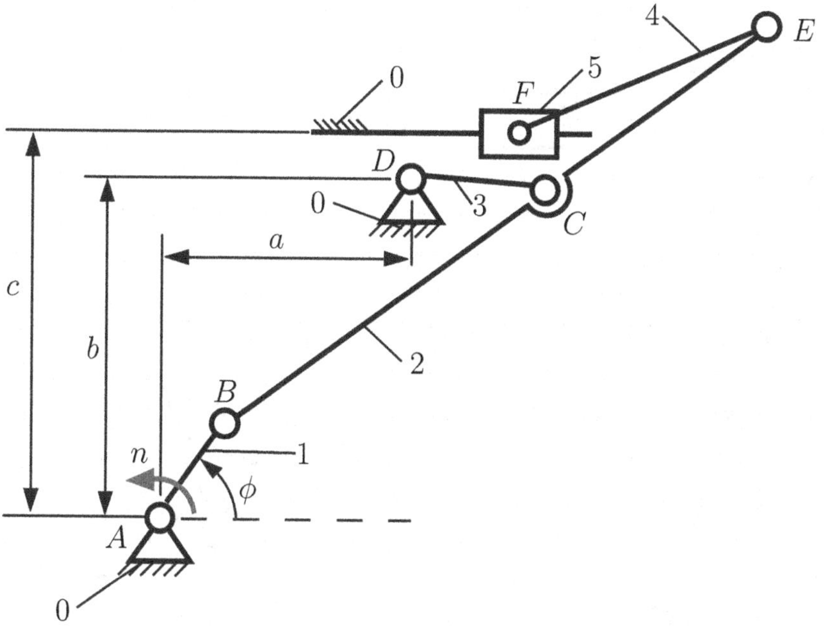

Fig. 7.4 Mechanism 4

Table 7.5 Mechanism 5

No	*AB* [m]	*BC* [m]	*CD* [m]	*CE* [m]	*EF* [m]	*a* [m]	*b* [m]	*c* [m]	ϕ [°]	*n* [rpm]
1	0.10	0.37	0.15	0.15	0.25	0.12	0.36	0.09	45	1000
2	0.08	0.30	0.12	0.12	0.20	0.10	0.30	0.08	135	1100
3	0.12	0.45	0.18	0.18	0.30	0.15	0.45	0.14	225	1200
4	0.06	0.24	0.09	0.09	0.18	0.075	0.225	0.070	315	1300

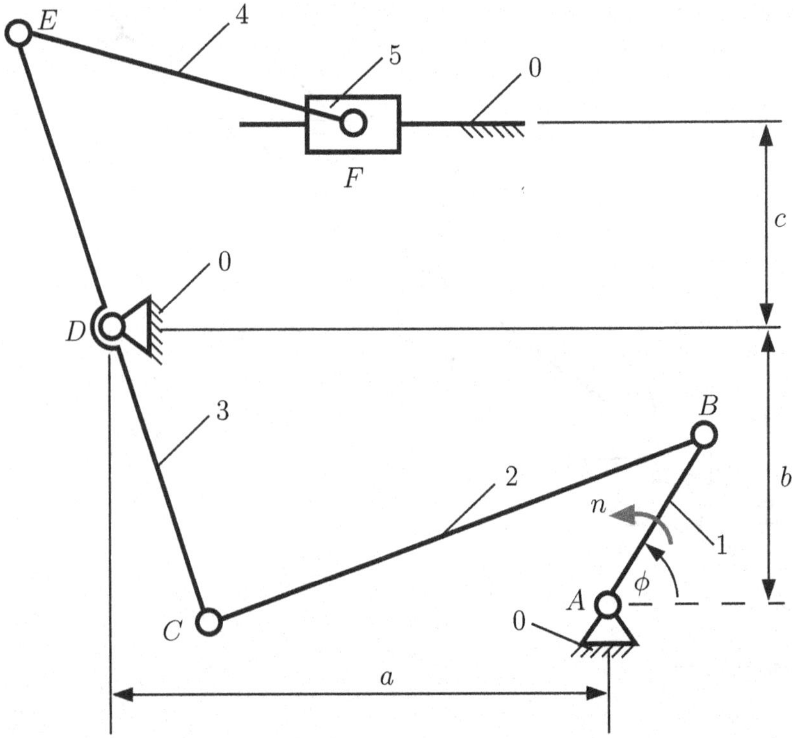

Fig. 7.5 Mechanism 5

Table 7.6 Mechanism 6

No	*AB* [m]	*AC* [m]	*BD* [m]	*DE* [m]	*EF* [m]	*a* [m]	*b* [m]	ϕ [°]	*n* [rpm]
1	0.06	0.18	0.27	0.12	0.08	0.02	0.01	225	600
2	0.08	0.20	0.30	0.16	0.11	0.025	0.13	45	750
3	0.05	0.12	0.18	0.10	0.065	0.017	0.085	135	950
4	0.04	0.10	0.16	0.09	0.063	0.013	0.065	315	1150

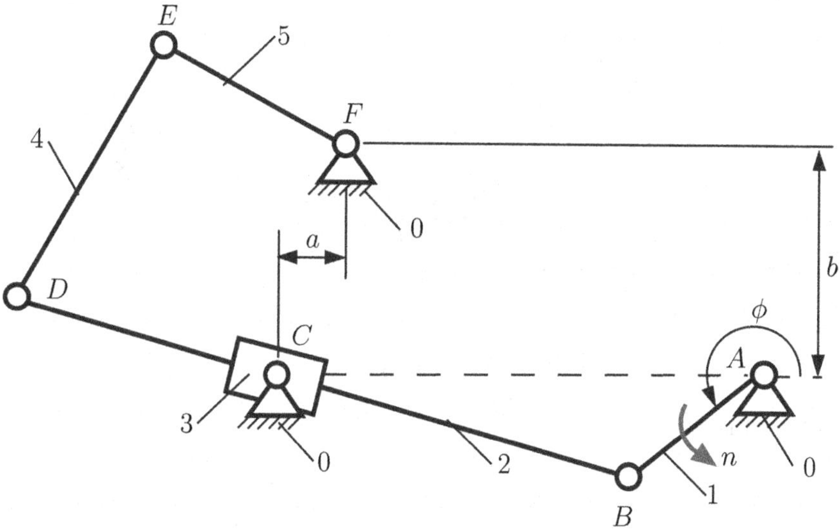

Fig. 7.6 Mechanism 6

Table 7.7 Mechanism 7

No	*AB* [m]	*AD* [m]	*BC* [m]	*a* [m]	ϕ [°]	*n* [rpm]
1	0.08	0.20	0.10	0.15	300	1000
2	0.06	0.15	0.08	0.50	210	1200
3	0.10	0.25	0.12	0.50	150	1300
4	0.12	0.30	0.15	0.50	30	1500

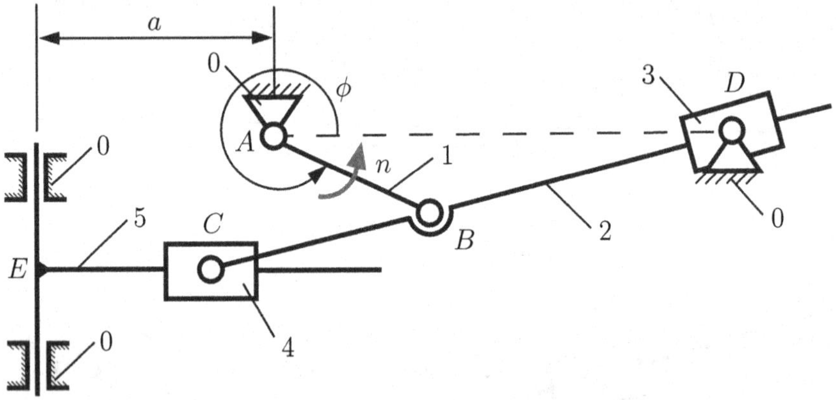

Fig. 7.7 Mechanism 7

Table 7.8 Mechanism 8

No	AB [m]	AD [m]	BC [m]	CE [m]	EF [m]	a [m]	ϕ [°]	n [rpm]
1	0.22	1.00	0.65	0.18	0.70	0.60	300	1000
2	0.25	1.10	0.70	0.20	0.80	0.65	120	1200
3	0.20	1.00	0.55	0.15	0.60	0.52	60	1500
4	0.18	0.80	0.40	0.10	0.50	0.43	225	1800

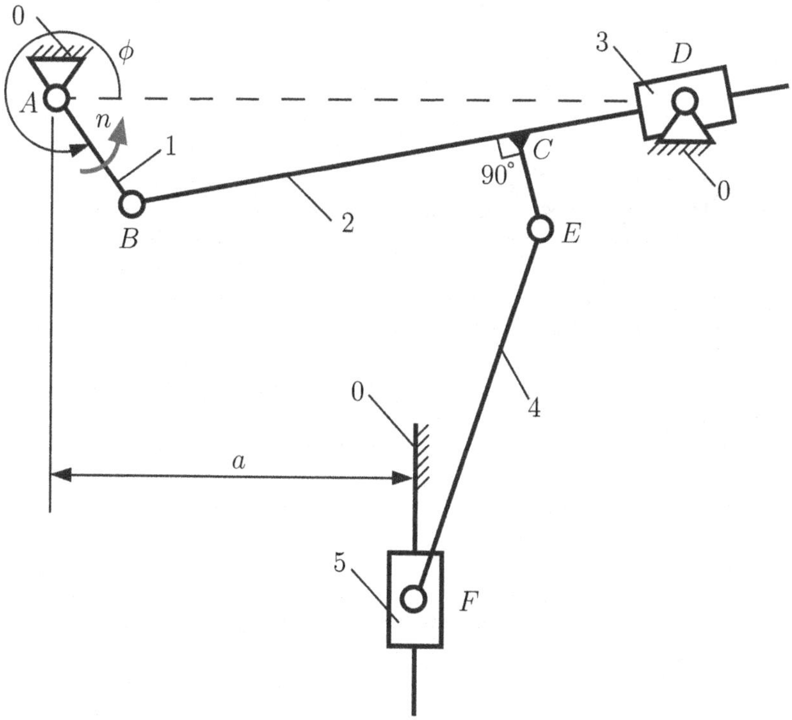

Fig. 7.8 Mechanism 8

Table 7.9 Mechanism 9

No	*AB* [m]	*AC* [m]	*CD* [m]	*DE* [m]	ϕ [°]	*n* [rpm]
1	0.22	0.08	0.20	0.60	240	500
2	0.18	0.08	0.20	0.50	30	600
3	0.15	0.05	0.18	0.45	120	700
4	0.08	0.03	0.07	0.20	330	800

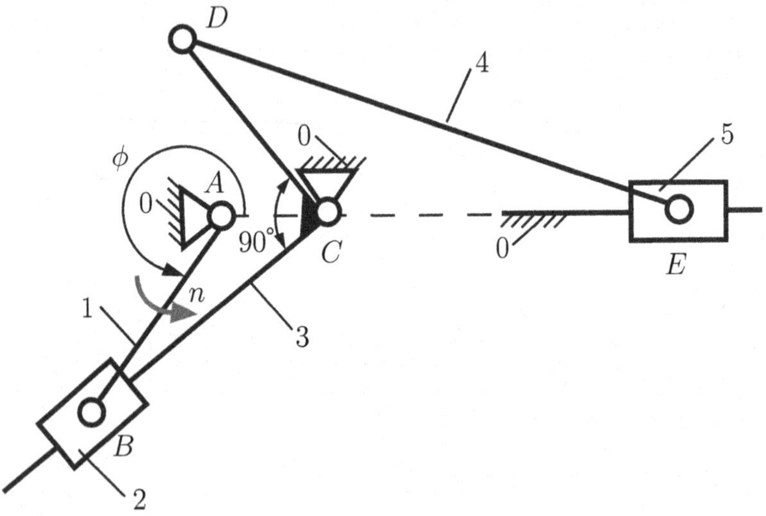

Fig. 7.9 Mechanism 9

Table 7.10 Mechanism 10

No	AB [m]	AE [m]	BC [m]	CD [m]	a [m]	ϕ [°]	n [rpm]
1	0.03	0.07	0.05	0.08	0.05	30	400
2	0.10	0.25	0.17	0.30	0.09	120	600
3	0.12	0.30	0.20	0.35	0.10	200	800
4	0.09	0.35	0.25	0.40	0.12	300	1000

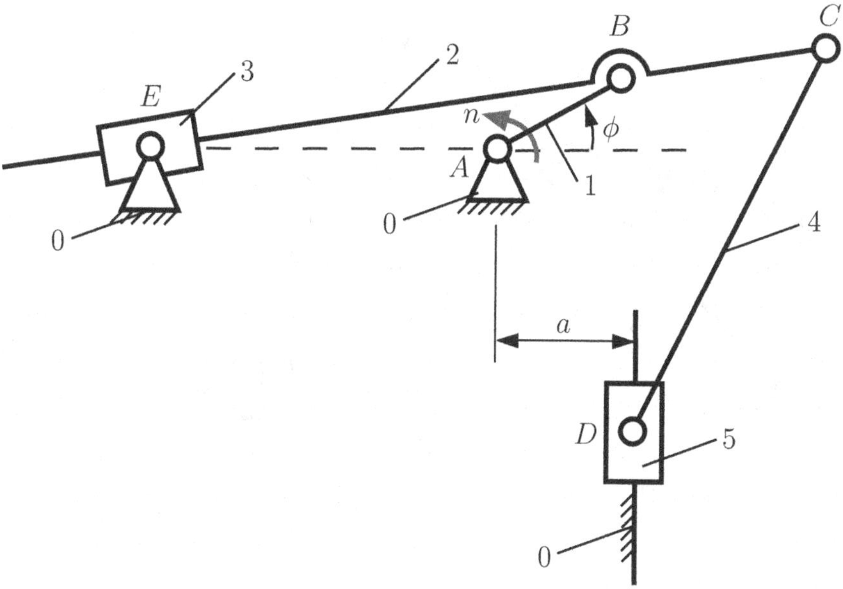

Fig. 7.10 Mechanism 10

Table 7.11 Mechanism 11

No	*AB* [m]	*CD* [m]	*DE* [m]	*AC* [m]	ϕ [°]	*n* [rpm]
1	0.08	0.14	0.40	0.04	45	300
2	0.10	0.16	0.45	0.03	135	400
3	0.25	0.40	1.00	0.10	225	500
4	0.22	0.30	0.90	0.06	315	600

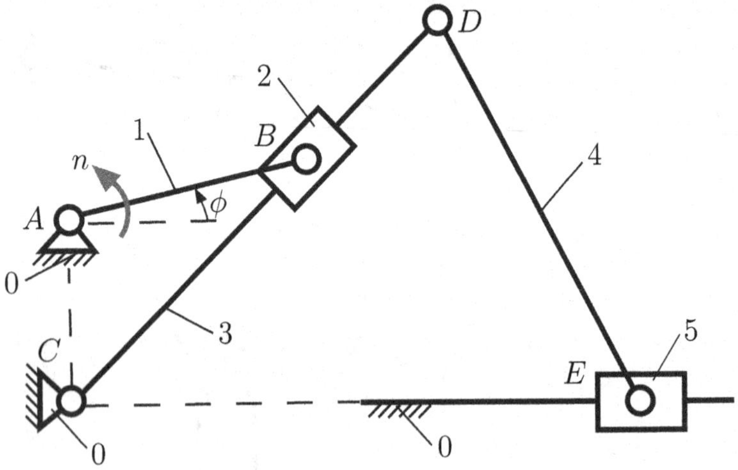

Fig. 7.11 Mechanism 11

Table 7.12 Mechanism 12

No	*AB* [m]	*AC* [m]	*BD* [m]	*a* [m]	*b* [m]	ϕ [°]	*n* [rpm]
1	0.12	0.30	0.50	0.08	0.15	60	500
2	0.10	0.30	0.50	0.07	0.12	150	700
3	0.09	0.30	0.45	0.06	0.10	240	900
4	0.08	0.25	0.40	0.05	0.09	330	1100

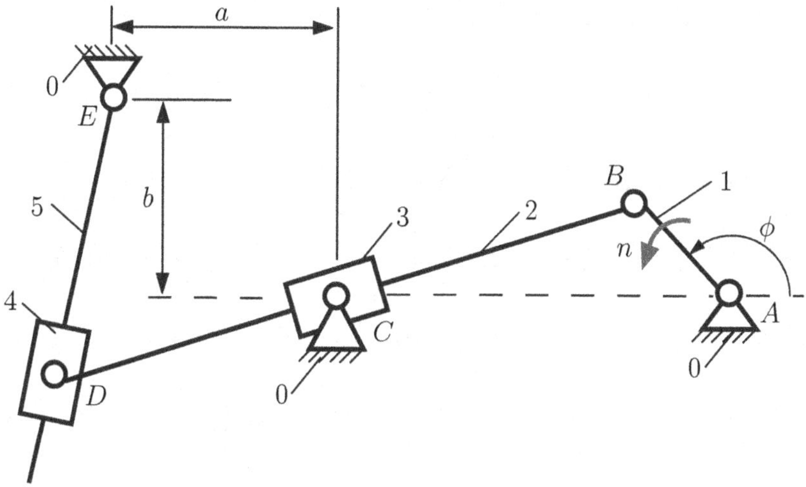

Fig. 7.12 Mechanism 12

Table 7.13 Mechanism 13

No	*AB* [m]	*BC* [m]	*CD* [m]	*a* [m]	*b* [m]	ϕ [°]	*n* [rpm]
1	0.20	0.21	0.39	0.30	0.25	45	500
2	0.18	0.17	0.35	0.27	0.26	135	1000
3	0.10	0.25	0.15	0.225	0.30	240	1500
4	0.22	0.23	0.45	0.30	0.32	315	2000

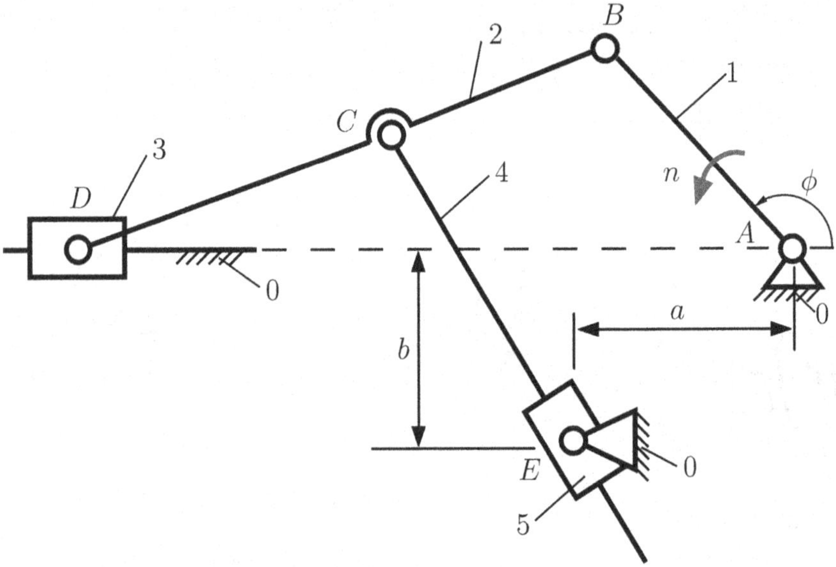

Fig. 7.13 Mechanism 13

Table 7.14 Mechanism 14

No	*AB* [m]	*AC* [m]	*CD* [m]	*DE* [m]	*a* [m]	ϕ [°]	*n* [rpm]
1	0.10	0.20	0.12	0.08	0.13	20	500
2	0.08	0.15	0.10	0.10	0.12	110	600
3	0.15	0.30	0.20	0.14	0.25	200	700
4	0.07	0.15	0.10	0.06	0.11	290	800

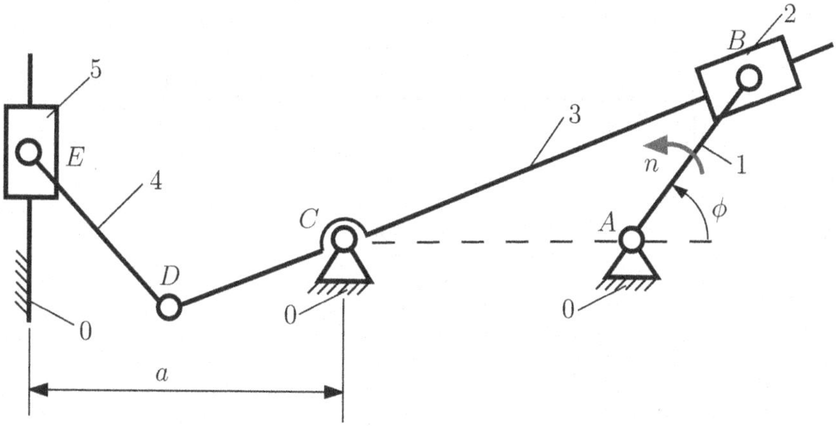

Fig. 7.14 Mechanism 14

Table 7.15 Mechanism 15

No	*AB* [m]	*AC* [m]	*a* [m]	ϕ [°]	*n* [rpm]
1	0.04	0.10	0.08	60	500
2	0.05	0.10	0.07	120	600
3	0.08	0.10	0.09	210	700
4	0.08	0.20	0.10	330	800

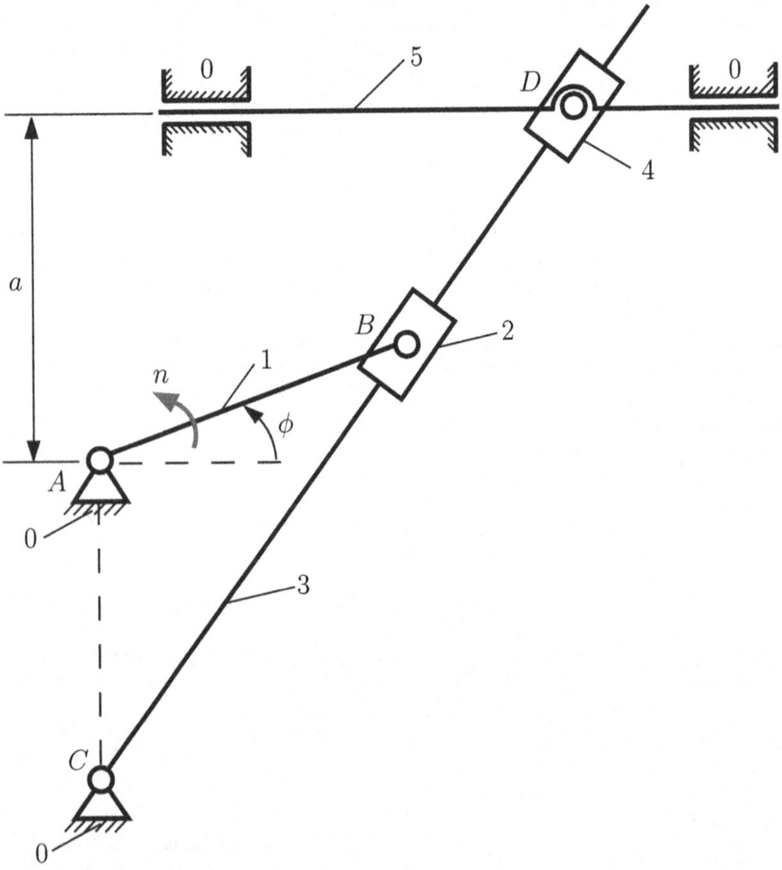

Fig. 7.15 Mechanism 15

7.2 Problem Set: Robots

Schematic representations of a robot arm consisting of three links 1, 2, and 3 are shown in Figs. 7.16–7.25. The mass centers of links 1, 2, and 3 are C_1, C_2, and C_3, respectively. The generalized coordinates (quantities associated with the instantaneous position of the system) are $q_1(t)$, $q_2(t)$, and $q_3(t)$.

The central principal axes of link p, $p = 1,2,3$ are parallel to $\mathbf{\imath}_p, \mathbf{\jmath}_p, \mathbf{k}_p$ and the associated moments of inertia have the values I_{px}, I_{py}, I_{pz}, respectively. The central inertia dyadic of link p is

$$\bar{I}_p = (I_{px}\mathbf{\imath}_p)\mathbf{\imath}_p + (I_{py}\mathbf{\jmath}_p)\mathbf{\jmath}_p + (I_{pz}\mathbf{k}_p)\mathbf{k}_p.$$

If the joint between link p and link $p+1$ is a

rotational joint consider a control vector moment

$$\mathbf{T}_{p,p+1} = T_{(p,p+1)x}\mathbf{\imath}_{p+1} + T_{(p,p+1)y}\mathbf{\jmath}_{p+1} + T_{(p,p+1)z}\mathbf{k}_{p+1},$$

translational joint consider a control vector force

$$\mathbf{F}_{p,p+1} = F_{(p,p+1)x}\mathbf{\imath}_{p+1} + F_{(p,p+1)y}\mathbf{\jmath}_{p+1} + F_{(p,p+1)z}\mathbf{k}_{p+1}.$$

Select suitable numerical values for the input numerical data.

1. Find the transformation matrices R_{ij}.
2. Calculate the angular velocities and accelerations of the links, $\boldsymbol{\omega}_{ij}$ and $\boldsymbol{\alpha}_{ij}$.
3. Determine the position vectors, $\mathbf{r}_{C_i}$, the velocities, $\mathbf{v}_{C_i}$, and the accelerations, $\mathbf{a}_{C_i}$ of the mass centers C_i.
4. Find the generalized (active) forces Q_i.
5. Write a MATLAB program for the symbolical calculation of Lagrange's equations of motion or/and Kane's dynamical equations.
6. Find the numerical solutions for inverse dynamics and direct dynamics.

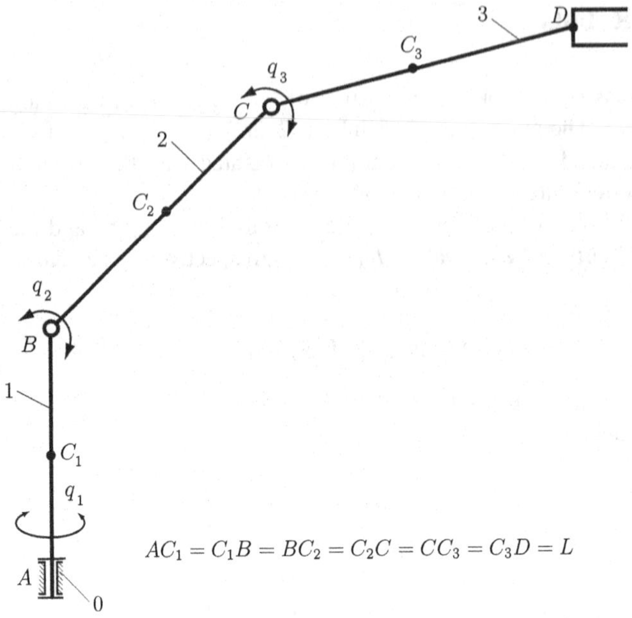

Fig. 7.16 Robot 1

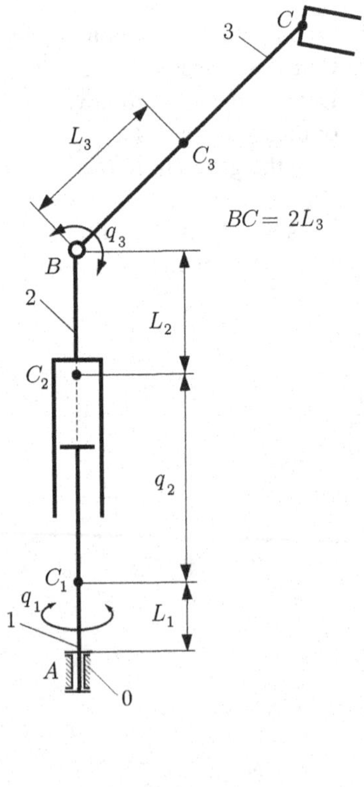

Fig. 7.17 Robot 2

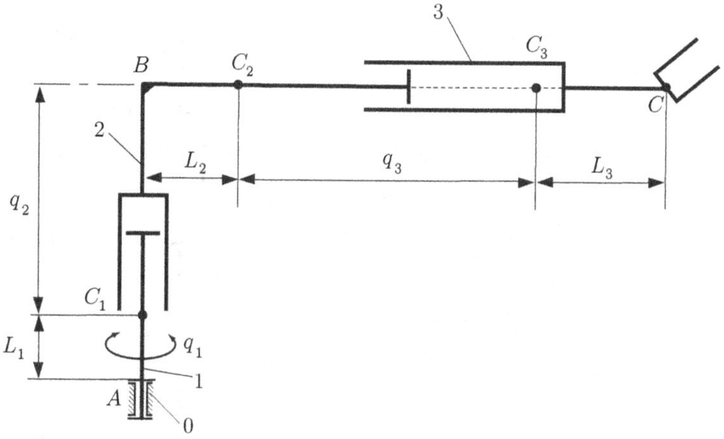

Fig. 7.18 Robot 3

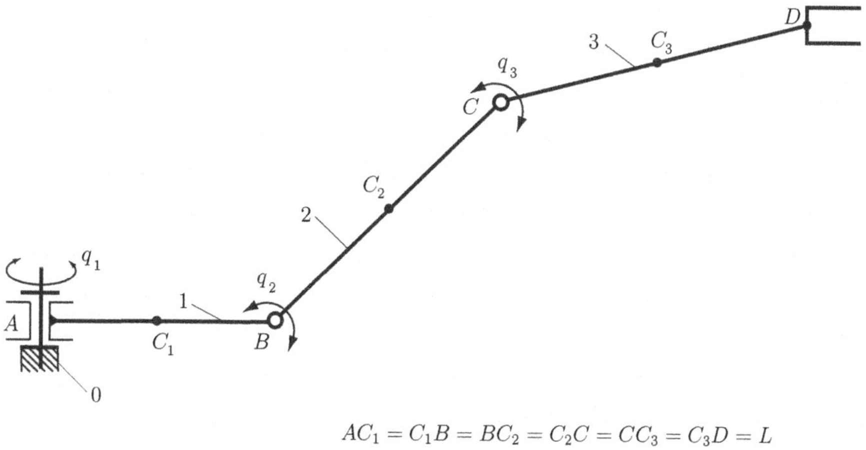

Fig. 7.19 Robot 4

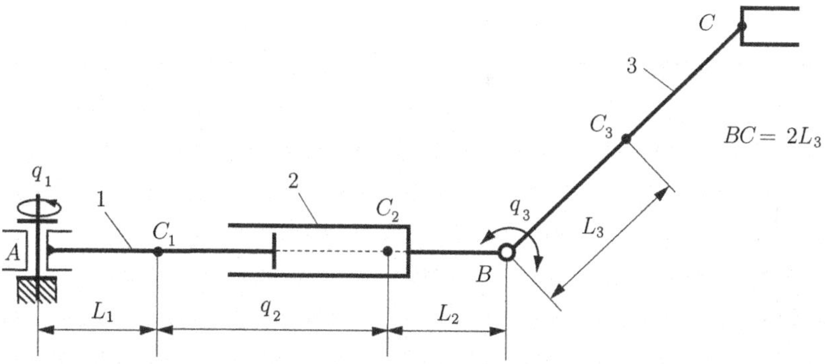

Fig. 7.20 Robot 5

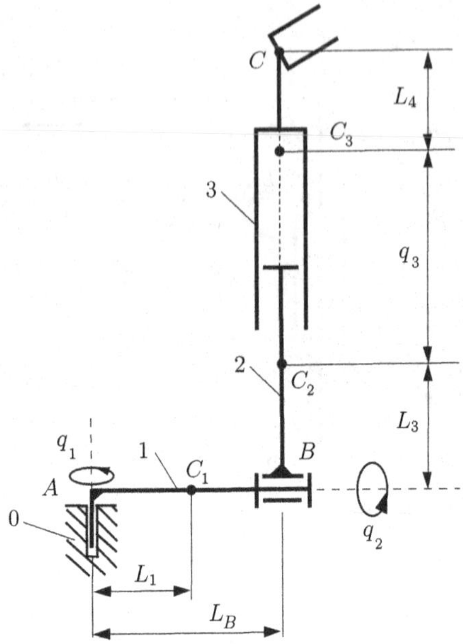

Fig. 7.21 Robot 6

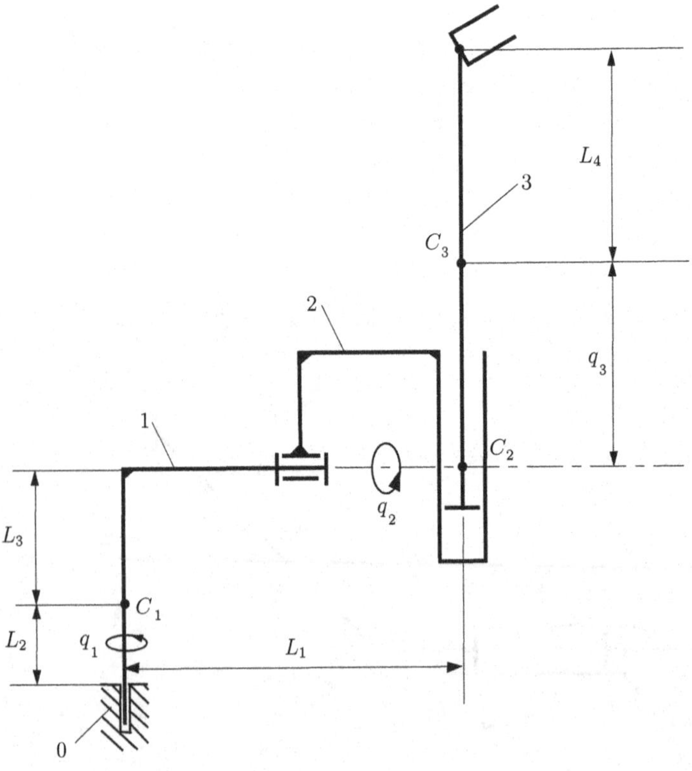

Fig. 7.22 Robot 7

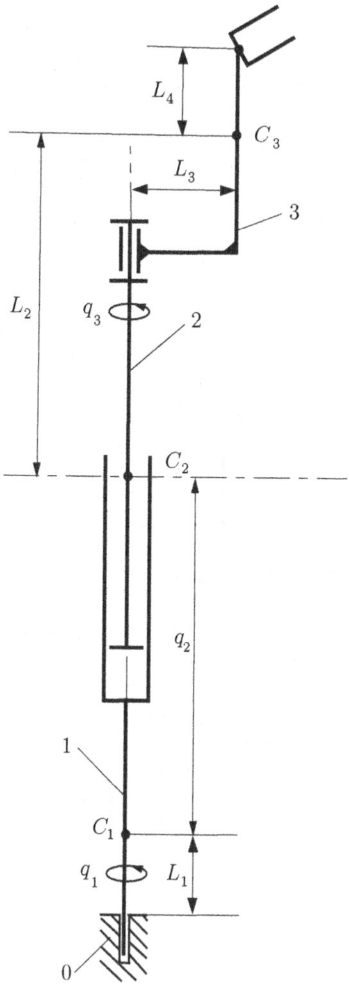

Fig. 7.23 Robot 8

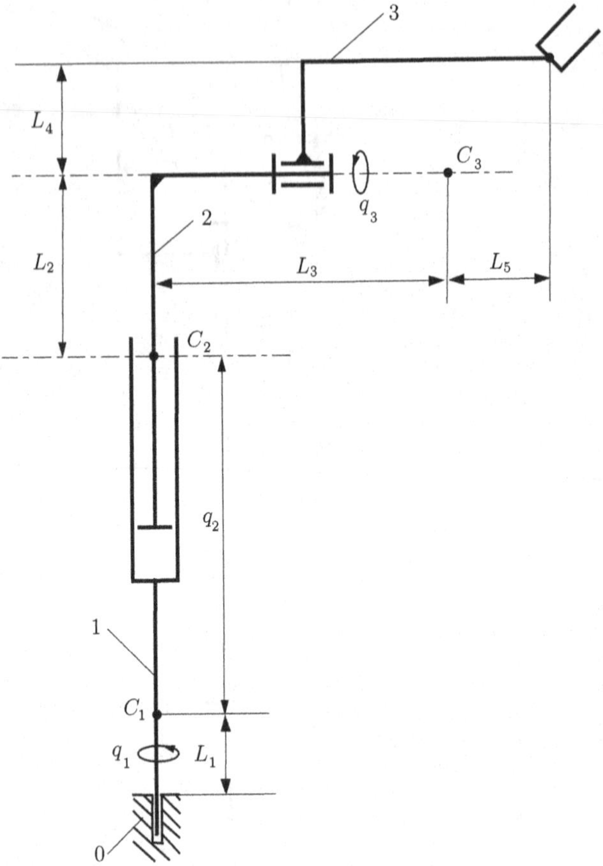

Fig. 7.24 Robot 9

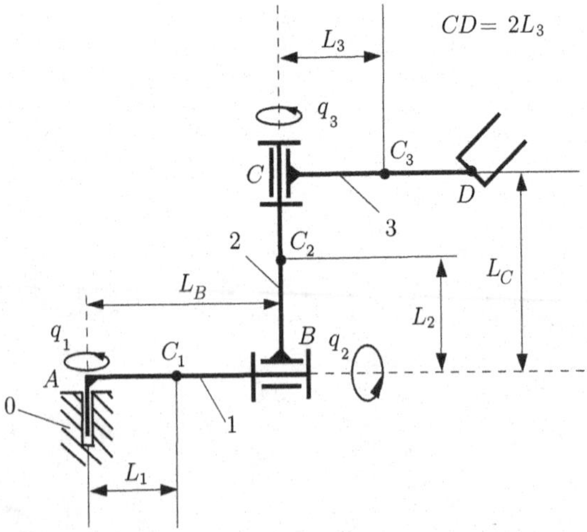

Fig. 7.25 Robot 10

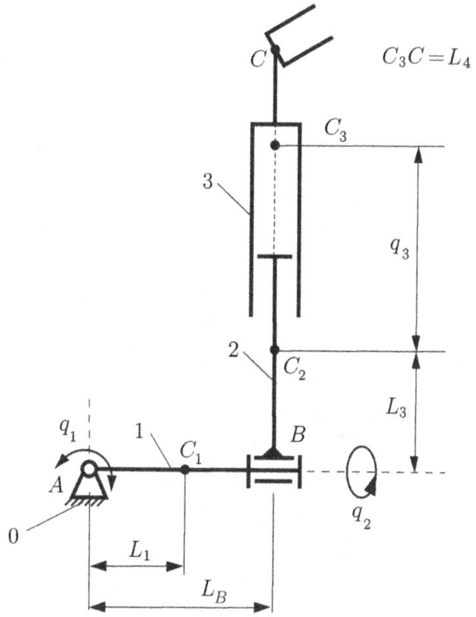

Fig. 7.26 Robot 11

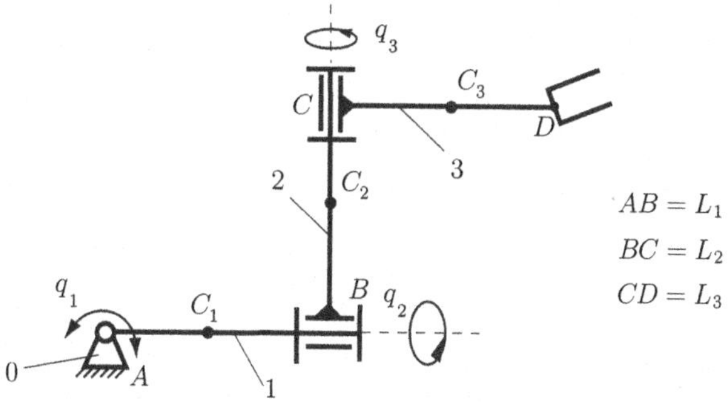

Fig. 7.27 Robot 12

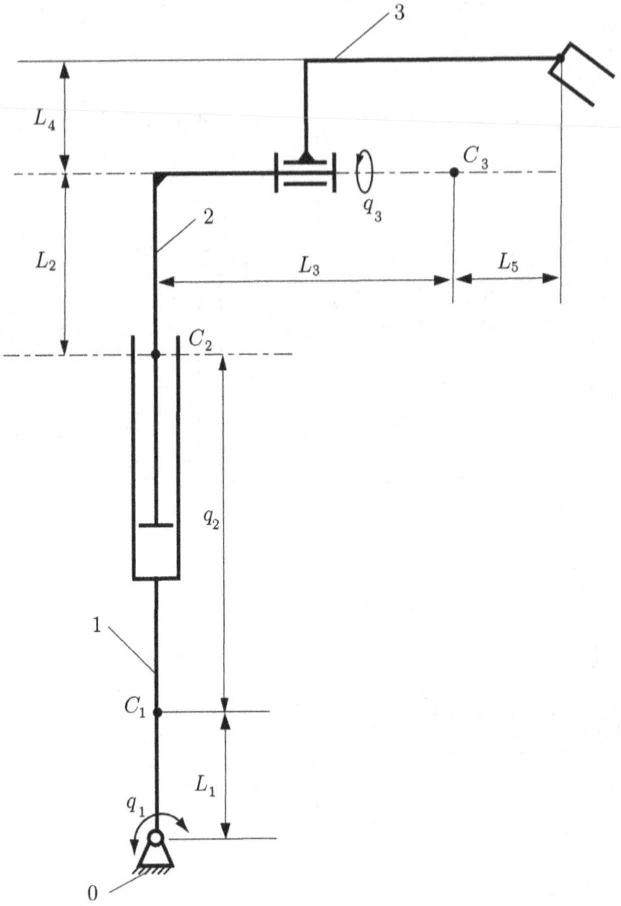

Fig. 7.28 Robot 13

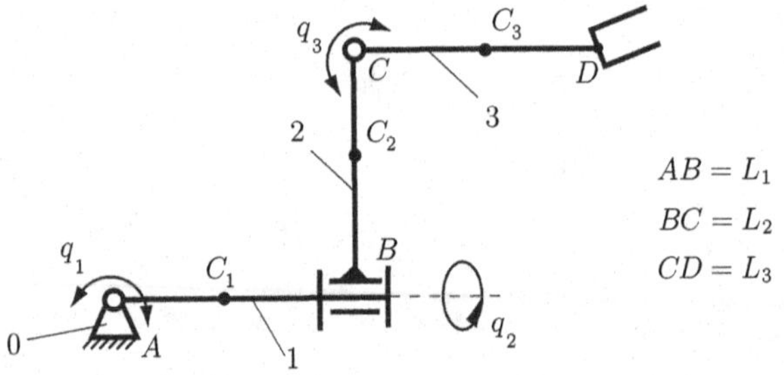

Fig. 7.29 Robot 14

Fig. 7.30 Robot 15

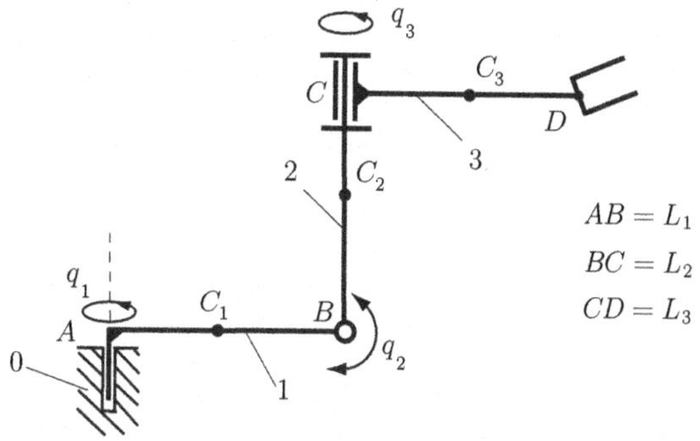

Appendix A
Programs of Chapter 2: Position Analysis

A.1 Slider-Crank (R-RRT) Mechanism

```
% A1
% Position analysis
% R-RRT

clear % clears all variables from the workspace
clc % clears the command window and homes the cursor
close all % closes all the open figure windows
% Input data
AB=0.5;
BC=1;
phi = pi/4;

% Position of joint A (origin)
xA = 0; yA = 0;

% Position of joint B - position of the driver link
xB = AB*cos(phi);
yB = AB*sin(phi);

% Position of joint C
yC = 0;
% Distance formula: BC=constant
eqnC = '( xB - xCsol )^2 + ( yB - yC )^2 = BC^2';
% Solve the above equation
solC = solve(eqnC, 'xCsol');

% solve symbolic solution of algebraic equations
% Two solutions for xC - vector form
```

```
% first  component of the vector solC
xC1=eval(solC(1));
% second component of the vector solC
xC2=eval(solC(2));
% eval executes string as an expression or statement
% Select the correct position for C
%  for the given input angle
if xC1 > xB xC = xC1; else xC = xC2; end
% if conditionally executes statements

% Angle of the link 2 with the horizontal
phi2 = atan((yB-yC)/(xB-xC));

fprintf('Results \n\n')
% Print the coordinates of B
fprintf('xB = %g (m)\n', xB)
fprintf('yB = %g (m)\n', yB)
% Print the coordinates of C
fprintf('xC = %g (m)\n', xC)
fprintf('yC = %g (m)\n', yC)
% Print the angle phi2
fprintf('phi2 = %g (degrees) \n', phi2*180/pi)

% Graphic of the mechanism
plot([xA,xB],[yA,yB],'r-o',...
     [xB,xC],[yB,yC],'b-o'),...
xlabel('x (m)'),...
ylabel('y (m)'),...
title('positions for \phi = 45 (deg)'),...
text(xA,yA,'  A'),...
text(xB,yB,'  B'),...
text(xC,yC,'  C'),...
axis([-0.2 1.4 -0.2 1.4]),...
grid
% the commas and ellipses (...) after the commands
% were used to execute the commands together
% end of program
```

Results:

```
xB = 0.353553 (m)
yB = 0.353553 (m)
xC = 1.28897 (m)
yC = 0 (m)
phi2 = -20.7048 (degrees)
```

A.2 Four-Bar (R-RRR) Mechanism

```
% A2
% Position analysis
% R-RRR

clear all; clc; close all

% Input data
AB=0.15; %(m)
BC=0.35; %(m)
CD=0.30; %(m)
CE=0.15; %(m)
xD=0.30; %(m)
yD=0.30; %(m)
phi = pi/4 ; %(rad)

xA = 0; yA = 0;
rA = [xA yA 0];
rD = [xD yD 0];

xB = AB*cos(phi); yB = AB*sin(phi); rB = [xB yB 0];

% Position of joint C
% Distance formula: BC=constant
eqnC1 = '(xCsol - xB)^2 + (yCsol - yB)^2 = BC^2 ';
% Distance formula: CD=constant
eqnC2 = '(xCsol - xD)^2 + (yCsol - yD)^2 = CD^2 ';

% Simultaneously solve above equations
solC = solve(eqnC1, eqnC2, 'xCsol, yCsol');

% Two solutions for xC - vector form
xCpositions = eval(solC.xCsol);
% Two solutions for yC - vector form
yCpositions = eval(solC.yCsol);

% Separate the solutions in scalar form

% first  component of the vector xCpositions
xC1 = xCpositions(1);
% second component of the vector xCpositions
xC2 = xCpositions(2);
```

```
% first  component of the vector yCpositions
yC1 = yCpositions(1);
% second component of the vector yCpositions
yC2 = yCpositions(2);

% Select the correct position for C
% for the given input angle

if xC1 < xD
    xC = xC1; yC=yC1;
else
    xC = xC2; yC=yC2;
end
rC = [xC yC 0]; % Position vector of C

% Position of joint E
% Distance formula: CE=constant
eqnE1='(xEsol-xC)^2+(yEsol-yC)^2=CE^2';
% Slope formula:
% E, C, and D are on the  same straight line
eqnE2='(yD-yC)/(xD-xC)=(yEsol-yC)/(xEsol-xC)';

solE = solve(eqnE1, eqnE2, 'xEsol, yEsol');

xEpositions = eval(solE.xEsol);
yEpositions = eval(solE.yEsol);

xE1 = xEpositions(1); xE2 = xEpositions(2);
yE1 = yEpositions(1); yE2 = yEpositions(2);

if xE1 < xC
    xE = xE1; yE=yE1;
else
    xE = xE2; yE=yE2;
end
rE = [xE yE 0]; % Position vector of E

% Angles of the links with the horizontal
phi2 = atan((yB-yC)/(xB-xC));
phi3 = atan((yD-yC)/(xD-xC));

fprintf('Results \n\n')

fprintf('rA = [ %g, %g, %g ] (m)\n', rA)
fprintf('rD = [ %g, %g, %g ] (m)\n', rD)
```

```
fprintf('rB = [ %g, %g, %g ] (m)\n', rB)
fprintf('rC = [ %g, %g, %g ] (m)\n', rC)
fprintf('rE = [ %g, %g, %g ] (m)\n', rE)

fprintf('phi2 = %g (degrees) \n', phi2*180/pi)
fprintf('phi3 = %g (degrees) \n', phi3*180/pi)

% Graphic of the mechanism

plot([xA,xB],[yA,yB],'k-o','LineWidth',1.5)
hold on % holds the current plot
plot([xB,xC],[yB,yC],'b-o','LineWidth',1.5)
hold on
plot([xD,xE],[yD,yE],'r-o','LineWidth',1.5)
% adds major grid lines to the current axes
grid on,...
xlabel('x (m)'), ylabel('y (m)'),...
title('positions for \phi = 45 (deg)'),...
text(xA,yA,'\leftarrow A = ground',...
'HorizontalAlignment','left'),...
text(xB,yB,'  B'),...
text(xC,yC,'\leftarrow C = ground',...
'HorizontalAlignment','left'),...
text(xD,yD,'\leftarrow D = ground',...
'HorizontalAlignment','left'),...
text(xE,yE,'  E'), axis([-0.2 0.45 -0.1 0.6])
% end of program
```

Results:

```
rA = [ 0, 0, 0 ] (m)
rD = [ 0.3, 0.3, 0 ] (m)
rB = [ 0.106066, 0.106066, 0 ] (m)
rC = [ 0.0400698, 0.449788, 0 ] (m)
rE = [ -0.0898952, 0.524681, 0 ] (m)
phi2 = -79.1312 (degrees)
phi3 = -29.9532 (degrees)
```

A.3 R-RTR-RTR Mechanism

```
% A3
% Position analysis - input angle phi
% R-RTR-RTR

clear all; clc; close all

% Input data
AB = 0.15; AC = 0.10; CD = 0.15; %(m)
phi = pi/6;  % (rad)
% Select the dimensions
DF=0.40; AG=0.30; % (m)

xA = 0; yA = 0; rA = [xA yA 0]; % Position of A
xC = 0; yC = AC; rC = [xC yC 0]; % Position of C
% Position of B
xB = AB*cos(phi); yB = AB*sin(phi);
rB = [xB yB 0];

% Position of joint D

% Distance formula: CD=constant
eqnD1='( xDsol - xC )^2 + ( yDsol - yC )^2 = CD^2';

% Slope formula:
% B, C, and D are on the  same straight line
eqnD2='(yB-yC)/(xB-xC)=(yDsol-yC)/(xDsol-xC)';

% Simultaneously solve above equations
solD = solve(eqnD1, eqnD2, 'xDsol, yDsol');
% solve symbolic solution of algebraic equations

% Two solutions for xD - vector form
xDpositions = eval(solD.xDsol);
% eval execute string as an expression or statement
% Two solutions for yD - vector form
yDpositions = eval(solD.yDsol);

% Separate the solutions in scalar form

% first  component of the vector xDpositions
xD1 = xDpositions(1);
% second component of the vector xDpositions
xD2 = xDpositions(2);
```

```
% first  component of the vector yDpositions
yD1 = yDpositions(1);
% second component of the vector yDpositions
yD2 = yDpositions(2);

% Select the correct position for D
% for the given input angle

if xD1 <= xC
    xD = xD1; yD=yD1;
else
    xD = xD2; yD=yD2;
end
rD = [xD yD 0]; % Position of D

% Angles of the links with the horizontal
phi2 = atan((yB-yC)/(xB-xC));
phi3 = phi2;
phi4 = atan(yD/xD)+pi;
phi5 = phi4;

% Positions of the points F and G

xF = xD + DF*cos(phi3);
yF = yD + DF*sin(phi3);
rF = [xF yF 0]; % Position vector of F

xG = AG*cos(phi5);
yG = AG*sin(phi5);
rG = [xG yG 0]; % Position vector of G

fprintf('Results \n\n')

fprintf('rA = [ %g, %g, %g ] (m)\n', rA)
fprintf('rC = [ %g, %g, %g ] (m)\n', rC)
fprintf('rB = [ %g, %g, %g ] (m)\n', rB)
fprintf('rD = [ %g, %g, %g ] (m)\n', rD)

fprintf('phi2 = phi3 = %g (degrees) \n',phi2*180/pi)
fprintf('phi4 = phi5 = %g (degrees) \n',phi4*180/pi)

fprintf('rF = [ %g, %g, %g] (m)\n', rF)
fprintf('rG = [ %g, %g, %g] (m)\n', rG)
```

```
% Graphic of the mechanism

plot([xA,xB],[yA,yB],'k-o','LineWidth',1.5)
hold on % holds the current plot
plot([xD,xC],[yD,yC],'b-o','LineWidth',1.5)
hold on
plot([xC,xB],[yC,yB],'b-o','LineWidth',1.5)
hold on
plot([xB,xF],[yB,yF],'b-o','LineWidth',1.5)
hold on
plot([xA,xD],[yA,yD],'r-o','LineWidth',1.5)
hold on
plot([xD,xG],[yD,yG],'r-o','LineWidth',1.5)

grid on,...

xlabel('x (m)'), ylabel('y (m)'),...

title('positions for \phi = 30 (deg)'),...

text(xA,yA,'\leftarrow A = ground',...
'HorizontalAlignment','left'),...
text(xB,yB,'  B'),...
text(xC,yC,'\leftarrow C = ground',...
'HorizontalAlignment','left'),...
text(xD,yD,'  D'),...
text(xF,yF,'  F'), text(xG,yG,'  G'),...

axis([-0.3 0.3 -0.1 0.3])

% end of program
```

Results:

```
rA = [ 0, 0, 0 ] (m)
rC = [ 0, 0.1, 0 ] (m)
rB = [ 0.129904, 0.075, 0 ] (m)
rD = [ -0.147297, 0.128347, 0 ] (m)
phi2 = phi3 = -10.8934 (degrees)
phi4 = phi5 = 138.933 (degrees)
rF = [ 0.245495, 0.0527544, 0] (m)
rG = [ -0.226182, 0.197083, 0] (m)
```

A.4 R-RTR-RTR Mechanism: Complete Rotation

```
% A4
% Position analysis - complete rotation
% R-RTR-RTR

clear all; clc; close all

% Input data
AB=0.15; AC=0.10; CD=0.15; %(m)
xA = 0; yA = 0; rA = [xA yA 0];
xC = 0; yC = AC; rC = [xC yC 0];

fprintf('Results \n\n')
fprintf('rA = [ %g, %g, %g ] (m)\n', rA)
fprintf('rC = [ %g, %g, %g ] (m)\n', rC)
fprintf('\n')

% complete rotation phi=0 to 2*pi step pi/3
for phi=0:pi/3:2*pi,
% for  repeat statements a specific number of times

fprintf('phi =  %g deegres \n', phi*180/pi)

% Position of joint B - position of the driver link
xB = AB*cos(phi); yB = AB*sin(phi); rB = [xB yB 0];
fprintf('rB = [ %g, %g, %g ] (m)\n', rB)

% Position of joint D
eqnD1='(xDsol - xC)^2 + (yDsol - yC)^2 = CD^2 ';
eqnD2='(yB-yC)/(xB-xC)=(yDsol-yC)/(xDsol-xC)';

% Simultaneously solve above equations
solD = solve(eqnD1, eqnD2, 'xDsol, yDsol');

xDpositions = eval(solD.xDsol);
yDpositions = eval(solD.yDsol);

% Separate the solutions in scalar form
xD1 = xDpositions(1); xD2 = xDpositions(2);
yD1 = yDpositions(1); yD2 = yDpositions(2);

% Select the correct position for D for the angle phi
% see the drawings for each quadrant
```

```
if(phi>=0 && phi<=pi/2)||(phi >= 3*pi/2 && phi<=2*pi)
 if xD1 <= xC xD=xD1; yD=yD1; else xD=xD2; yD=yD2;
   end
   else
 if xD1 >= xC xD=xD1; yD=yD1; else xD=xD2; yD=yD2;
   end
end
% &&  short-circuit logical AND
% ||  short-circuit logical OR

rD = [xD yD 0];
fprintf('rD = [ %g, %g, %g] (m)\n', rD)

% Angles of the links with the horizontal
phi2 = atan((yB-yC)/(xB-xC));
phi3 = phi2;
fprintf('phi2 = phi3 = %g (degrees) \n', phi2*180/pi)
phi4 = atan(yD/xD);
phi5 = phi4;
fprintf('phi4 = phi5 = %g (degrees) \n', phi4*180/pi)
fprintf('\n')

% Graphic of the mechanism

plot([xA,xB],[yA,yB],'k-o',[xB,xC],[yB,yC],'b-o',...
    [xC,xD],[yC,yD],'b-o')
hold on
plot([xD,xA],[yD,yA],'r-o')
xlabel('x (m)'),...
ylabel('y (m)'),...
title('positions for \phi=0 to 360 step 60(deg)'),...
text(xA,yA,'  A'),...
text(xB,yB,'  B'),...
text(xC,yC,'  C'),...
text(xD,yD,'  D'),...
axis([-0.3 0.3 -0.2 0.3])

end % end for
% end of program
```

Results:

```
rA = [ 0, 0, 0 ] (m)
rC = [ 0, 0.1, 0 ] (m)
```

```
phi = 0 deegres
rB = [ 0.15, 0, 0 ] (m)
rD = [ -0.124808, 0.183205, 0] (m)
phi2 = phi3 = -33.6901 (degrees)
phi4 = phi5 = -55.7355 (degrees)

phi = 60 deegres
rB = [ 0.075, 0.129904, 0 ] (m)
rD = [ -0.139333, 0.0444455, 0] (m)
phi2 = phi3 = 21.738 (degrees)
phi4 = phi5 = -17.692 (degrees)

phi = 120 deegres
rB = [ -0.075, 0.129904, 0 ] (m)
rD = [ 0.139333, 0.0444455, 0] (m)
phi2 = phi3 = -21.738 (degrees)
phi4 = phi5 = 17.692 (degrees)

phi = 180 deegres
rB = [ -0.15, 1.83697e-17, 0 ] (m)
rD = [ 0.124808, 0.183205, 0] (m)
phi2 = phi3 = 33.6901 (degrees)
phi4 = phi5 = 55.7355 (degrees)

phi = 240 deegres
rB = [ -0.075, -0.129904, 0 ] (m)
rD = [ 0.0465207, 0.242604, 0] (m)
phi2 = phi3 = 71.9325 (degrees)
phi4 = phi5 = 79.145 (degrees)

phi = 300 deegres
rB = [ 0.075, -0.129904, 0 ] (m)
rD = [ -0.0465207, 0.242604, 0] (m)
phi2 = phi3 = -71.9325 (degrees)
phi4 = phi5 = -79.145 (degrees)

phi = 360 deegres
rB = [ 0.15, -3.67394e-17, 0 ] (m)
rD = [ -0.124808, 0.183205, 0] (m)
phi2 = phi3 = -33.6901 (degrees)
phi4 = phi5 = -55.7355 (degrees)
```

A.5 R-RTR-RTR Mechanism: Complete Rotation Using Euclidian Distance Function

```
% A5
% Position analysis - complete rotation
% R-RTR-RTR
% Euclidian distance function
% the program uses the function: Dist(x1,y1,x2,y2)
% the function is defined in the program Dist.m
clear all; clc; close all

% Input data
AB=0.15; AC=0.10; CD=0.15; %(m)
xA = 0; yA = 0; rA = [xA yA 0];
xC = 0 ; yC = AC ; rC = [xC yC 0];

fprintf('Results \n\n')
fprintf('rA =  [ %g, %g, %g ] (m)\n', rA)
fprintf('rC =  [ %g, %g, %g ] (m)\n', rC)

% at the initial moment phi=0 => increment = 0
increment = 0 ;

% the step has to be small for this method
step=pi/6;
for phi=0:step:2*pi,
fprintf('phi =  %g deegres \n', phi*180/pi)

xB = AB*cos(phi); yB = AB*sin(phi); rB = [xB yB 0];
fprintf('rB =  [ %g, %g, %g ] (m)\n', rB)
eqnD1=' ( xDsol - xC )^2 + ( yDsol - yC )^2=CD^2';
eqnD2=' (yB-yC)/(xB-xC)=(yDsol-yC)/(xDsol-xC)';

solD = solve(eqnD1, eqnD2, 'xDsol, yDsol');
xDpositions = eval(solD.xDsol);
yDpositions = eval(solD.yDsol);
xD1 = xDpositions(1); xD2 = xDpositions(2);
yD1 = yDpositions(1); yD2 = yDpositions(2);

% select the correct position for D
%    only for increment == 0
% the selection process is automatic
%    for all the other steps
```

```
if increment == 0
   if xD1 <= xC xD=xD1; yD=yD1; else xD=xD2; yD=yD2;
   end
else
  dist1 = Dist(xD1,yD1,xDold,yDold);
  dist2 = Dist(xD2,yD2,xDold,yDold);
  if dist1<dist2 xD=xD1; yD=yD1; else xD=xD2; yD=yD2;
   end
end

xDold=xD;
yDold=yD;

increment=increment+1;

rD = [xD yD 0];
fprintf('rD =  [ %g, %g, %g ] (m)\n', rD)

phi2 = atan((yB-yC)/(xB-xC)); phi3 = phi2;
fprintf('phi2 = phi3 = %g (degrees) \n', phi2*180/pi)

phi4 = atan(yD/xD); phi5 = phi4;
fprintf('phi4 = phi5 = %g (degrees) \n', phi4*180/pi)
fprintf('\n')

% Graphic of the mechanism

plot([xA,xB],[yA,yB],'k-o',[xB,xC],[yB,yC],'b-o',...
[xC,xD],[yC,yD],'b-o')
hold on
plot([xD,xA],[yD,yA],'r-o')
xlabel('x (m)'),...
ylabel('y (m)'),...

title('positions for \phi=0 to 360 step 30(deg)'),...
text(xA,yA,'  A'),...
text(xB,yB,'  B'),...
text(xC,yC,'  C'),...
text(xD,yD,'  D'),...

axis([-0.3 0.3 -0.2 0.3])

end

% end of program
```

A.6 Path of a Point on a Link with General Plane Motion: R-RRT Mechanism

```
% A6
% Position analysis
% R-RRT
% Path of C2 (mass center of link BC)

clear all; clc; close all
AB = .5; BC = 1; xA = 0; yA = 0; yC = 0;

incr = 0 ;

for phi=0:pi/10:2*pi,

xB = AB*cos(phi); yB = AB*sin(phi);

xC = xB + sqrt(BC^2-yB^2);

incr = incr + 1;
xC2(incr)=(xB+xC)/2; yC2(incr)=(yB+yC)/2;

% Graphic of the mechanism
subplot(2,1,1),...
plot( [xA,xB],[yA,yB],'r-',...
[xB,xC],[yB,yC],'b-' ),...
hold on

xlabel('x (m)'), ylabel('y (m)'),...

title('Graphic of the mechanism'),...
text(xC2,yC2,'C2'),...
axis([-0.6 1.6 -0.6 0.6]),...

end % end for

% Path of C2 (mass center of link 2)
subplot(2,1,2),...
plot(xC2, yC2, '-ko'),...
xlabel('x (m)'), ylabel('y (m)'),...
title('Path described by C2'), grid

% end of program
```

A.7 Path of a Point on a Link with General Plane Motion: R-RRR Mechanism

```
% A7
% R-RRR
% Position analysis
% Path of C2 (mass center of link BC)

clear all; clc; close all

AB=0.15; %(m)
BC=0.35; %(m)
CD=0.30; %(m)
CE=0.15; %(m)

xA=0; yA=0;
xD=0.30; yD=0.30;

incr = 0;

for phi=0:pi/10:2*pi,

xB = AB*cos(phi); yB = AB*sin(phi);

eqnC1='(xCsol - xB)^2 +(yCsol - yB)^2 =BC^2';
eqnC2='(xCsol - xD)^2 +(yCsol - yD)^2 =CD^2';

solC = solve(eqnC1, eqnC2, 'xCsol, yCsol');
xCpositions = eval(solC.xCsol);
yCpositions = eval(solC.yCsol);

xC1 = xCpositions(1); xC2 = xCpositions(2);
yC1 = yCpositions(1); yC2 = yCpositions(2);

if xC1 < xD xC = xC1; yC = yC1;
else xC = xC2; yC=yC2; end

eqnE1='(xEsol-xC)^2 + (yEsol-yC)^2=CE^2 ';
eqnE2='(yD-yC)/(xD-xC)=(yEsol-yC)/(xEsol-xC)';

solE = solve(eqnE1, eqnE2, 'xEsol, yEsol');
xEpositions = eval(solE.xEsol);
yEpositions = eval(solE.yEsol);
```

```
xE1 = xEpositions(1); xE2 = xEpositions(2);
yE1 = yEpositions(1); yE2 = yEpositions(2);

if xE1 < xC xE = xE1; yE=yE1;
else xE = xE2; yE=yE2; end

incr = incr + 1;

XC2(incr) = ( xB + xC )/2;
YC2(incr) = ( yB + yC )/2;

% Graphic of the mechanism
subplot(2,1,1),...
plot([xA,xB],[yA,yB],'k-',...
[xB,xC],[yB,yC],'b-',...
[xD,xE],[yD,yE],'r-'),...
hold on,...
xlabel('x (m)'), ylabel('y (m)'), grid,...
axis([-0.2 0.35 -0.25 0.7]),...
title('Positions of the mechanism'),...
text(XC2,YC2,'C2'),...

end

% Path of C2 (mass center of link 2)
subplot(2,1,2),...
plot(XC2, YC2, '-ko'),...
xlabel('x (m)'), ylabel('y (m)'),...
title('Path described by C2'), grid

% end of program
```

Appendix B
Programs of Chapter 3: Velocity and Acceleration Analysis

B.1 Slider-Crank (R-RRT) Mechanism

```
% B1
% Velocity and acceleration analysis
% R-RRT

clear all; clc; close all
AB=1; BC=1;
phi = pi/6; % input angle
xA = 0; yA = 0; rA = [xA yA 0];
xB = AB*cos(phi); yB = AB*sin(phi);
rB = [xB yB 0];
yC = 0; xC = xB+sqrt(BC^2-(yC-yB)^2);
rC = [xC yC 0];
phi2 = atan((yB-yC)/(xB-xC));

fprintf('Results \n\n')
fprintf('phi = phi1 = %g (degrees) \n', phi*180/pi)
fprintf('rA = [ %g, %g, %g ] (m)\n', rA)
fprintf('rB = [ %g, %g, %g ] (m)\n', rB)
fprintf('rC = [ %g, %g, %g ] (m)\n', rC)
fprintf('phi2 = %g (degrees) \n', phi2*180/pi)

% Graphic of the mechanism
plot([0,xB],[0,yB],'r-o',[xB,xC],[yB,yC],'b-o'),...
xlabel('x (m)'), ylabel('y (m)'),...
title('positions for \phi = \pi/6 (rad)'),...
text(xA,yA,'  A'),text(xB,yB,'  B'),...
text(xC,yC,'  C')
```

```
fprintf('\n')
fprintf('Velocity and acceleration analysis \n\n')

% angular velocity of the driver link 1
omega1 = [0 0  1 ]; % (rad/s)
% velocity of A (fixed)
vA = [0 0 0 ]; % (m/s)
% A and B=B1 are two points on the rigid link 1
vB1 = vA + cross(omega1,rB); % velocity of B1
% between 1 & 2 there is a rotational joint B_R
vB2 = vB1;
vB = norm(vB1);
% norm() is the vector norm
fprintf('omega1 = [ %g, %g, %g ] (rad/s)\n', omega1)
fprintf('vB=vB1=vB2 = [ %g, %g, %g ] (m/s)\n', vB1)
fprintf('|vB|= %g (m/s)\n', vB)

% velocity of C
% sym  constructs symbolic numbers and variables
% sym('x','real') also assumes that x is real
omega2z=sym('omega2z','real');
vCx=sym('vCx','real');

omega2 = [ 0 0 omega2z ];
vC = [ vCx 0 0 ];
% vC = vB + omega2 x rBC (B2 & C points on link 2)
eqvC = vC - (vB2 + cross(omega2,rC-rB));
% vectorial equation
eqvCx = eqvC(1); % equation component on x-axis
eqvCy = eqvC(2); % equation component on y-axis

solvC = solve(eqvCx,eqvCy);

omega2zs=eval(solvC.omega2z);
vCxs=eval(solvC.vCx);

Omega2 = [0 0 omega2zs];
VC = [vCxs 0 0];

vCB = cross(Omega2,rC-rB);

% print the equations for calculating
% omega2z and vCx
fprintf('vC = vB + omega2 x rBC => \n')
qvCx=vpa(eqvCx,6);
```

```
% vpa(S,D) uses variable-precision arithmetic (vpa)
% to compute each element of S to D decimal digits
% of accuracy
fprintf('x-axis: %s = 0 \n', char(qvCx))
% char()  creates character array (string)
qvCy=vpa(eqvCy,6);
fprintf('y-axis: %s = 0 \n', char(qvCy))

fprintf('=>\n')
fprintf('omega2z =  %g (rad/s)\n', omega2zs)
fprintf('vCx =  %g (m/s)\n', vCxs)
fprintf('\n')

fprintf('omega2 = [ %g, %g, %g ] (rad/s)\n', Omega2)
fprintf('vC =   [ %g, %g, %g ] (m/s)\n', VC)
fprintf('vCB = [ %g, %g, %d ] (m/s)\n', vCB)
fprintf('\n')

% angular acceleration of the driver link 1
alpha1 = [0 0 -1 ]; % (rad/s^2)
fprintf('alpha1 = [%g, %g, %g] (rad/s^2)\n',alpha1)
aA = [0 0 0 ]; % (m/s^2) acceleration of A

% acceleration of B
aB1 = aA + cross(alpha1,rB) - dot(omega1,omega1)*rB;
aB2 = aB1;
aBn = - dot(omega1,omega1)*rB;
aBt = cross(alpha1,rB);
fprintf('aB=aB1=aB2 = [%g, %g, %g] (m/s^2)\n', aB1)
fprintf('aBn = [ %g, %g, %d ] (m/s^2)\n', aBn)
fprintf('aBt = [ %g, %g, %g ] (m/s^2)\n', aBt)
fprintf('\n')

% acceleration of C
alpha2z=sym('alpha2z','real');
aCx=sym('aCx','real');
alpha2 = [ 0 0 alpha2z ]; % alpha3z unknown
aC = [aCx 0 0 ]; % aCx unknown

eqaC=aC-(aB1+cross(alpha2,rC-rB)-...
     dot(Omega2,Omega2)*(rC-rB));
eqaCx = eqaC(1); % equation component on x-axis
eqaCy = eqaC(2); % equation component on y-axis
solaC = solve(eqaCx,eqaCy);
```

```
alpha2zs=eval(solaC.alpha2z);
aCxs=eval(solaC.aCx);

Alpha2 = [0 0 alpha2zs];
aCs = [aCxs 0 0];

aCB=cross(Alpha2,rC-rB)-dot(Omega2,Omega2)*(rC-rB);
aCBn=-dot(Omega2,Omega2)*(rC-rB);
aCBt=cross(Alpha2,rC-rB);

% print the equations for calculating alpha2z and aCx
fprintf...
('aC = aB + omega2 x rBC - (omega2.omega2)rBC =>\n')
qaCx=vpa(eqaCx,6);
fprintf('x-axis: %s = 0 \n', char(qaCx))
qaCy=vpa(eqaCy,6);
fprintf('y-axis: %s = 0 \n', char(qaCy))

fprintf('=>\n')
fprintf('alpha2z =  %g (rad/s^2)\n', alpha2zs)
fprintf('aCx   =  %g (m/s^2)\n', aCxs)
fprintf('\n')

fprintf('alpha2 = [%g, %g, %g] (rad/s^2)\n', Alpha2)

fprintf('aC = [ %g, %g, %g ] (m/s^2)\n', aCs)

fprintf('aCB = [ %g, %g, %d ] (m/s^2)\n', aCB)

fprintf('aCBn = [ %g, %g, %d ] (m/s^2)\n', aCBn)

fprintf('|aCBn| = %g (m/s^2)\n', norm(aCBn))

fprintf('aCBt = [ %g, %g, %d ] (m/s^2)\n', aCBt)

fprintf('|aCBt| = %g (m/s^2)\n', norm(aCBt))

fprintf('\n')

% end of program
```

Results:

```
phi = phi1 = 30 (degrees)
rA = [ 0, 0, 0 ] (m)
rB = [ 0.866025, 0.5, 0 ] (m)
rC = [ 1.73205, 0, 0 ] (m)
phi2 = -30 (degrees)

Velocity and acceleration analysis

omega1 = [ 0, 0, 1 ] (rad/s)
vB=vB1=vB2 = [ -0.5, 0.866025, 0 ] (m/s)
|vB|= 1 (m/s)
vC = vB + omega2 x rBC =>
x-axis: vCx+.500000-.500000*omega2z = 0
y-axis: -.866025-.866025*omega2z = 0
=>
omega2z =  -1 (rad/s)
vCx =  -1 (m/s)

omega2 = [ 0, 0, -1 ] (rad/s)
vC =  [ -1, 0, 0 ] (m/s)
vCB = [ -0.5, -0.866025, 0 ] (m/s)

alpha1 = [ 0, 0, -1 ] (rad/s^2)
aB=aB1=aB2 = [ -0.366025, -1.36603, 0 ] (m/s^2)
aBn = [ -0.866025, -0.5, 0 ] (m/s^2)
aBt = [ 0.5, -0.866025, 0 ] (m/s^2)

aC = aB + omega2 x rBC - (omega2.omega2)rBC =>
x-axis: aCx+1.23205-.500000*alpha2z = 0
y-axis: .866025-.866025*alpha2z = 0
=>
alpha2z =  1 (rad/s^2)
aCx  =  -0.732051 (m/s^2)

alpha2 = [ 0, 0, 1 ] (rad/s^2)
aC = [ -0.732051, 0, 0 ] (m/s^2)
aCB = [ -0.366025, 1.36603, 0 ] (m/s^2)
aCBn = [ -0.866025, 0.5, 0 ] (m/s^2)
|aCBn| = 1 (m/s^2)
aCBt = [ 0.5, 0.866025, 0 ] (m/s^2)
|aCBt| = 1 (m/s^2)
```

B.2 Four-Bar (R-RRR) Mechanism

```
% B2
% Velocity and acceleration analysis
% R-RRR

clear all; clc; close all
AB=0.15; BC=0.35; CD=0.30; CE=0.15;
xA = 0; yA = 0; rA = [xA yA 0];
xD=0.30; yD=0.30; rD = [xD yD 0];
phi = pi/4 ; %(rad)
xB=AB*cos(phi); yB=AB*sin(phi); rB=[xB yB 0];
eqnC1='(xCsol - xB)^2 +(yCsol - yB)^2 =BC^2';
eqnC2='(xCsol - xD)^2 +(yCsol - yD)^2 =CD^2';
solC = solve(eqnC1, eqnC2, 'xCsol, yCsol');
xCpositions = eval(solC.xCsol);
yCpositions = eval(solC.yCsol);
xC1 = xCpositions(1); xC2 = xCpositions(2);
yC1 = yCpositions(1); yC2 = yCpositions(2);
if xC1 < xD xC = xC1; yC = yC1;
else xC = xC2; yC=yC2; end
rC = [xC yC 0];
phi3=atan((yD-yC)/(xD-xC))+pi;
xE=xC+CE*cos(phi3); yE=yC+CE*sin(phi3);
rE = [xE yE 0];

fprintf('Results \n\n')
fprintf('phi = %g (deg)\n', phi*180/pi)
fprintf('rA = [ %g, %g, %g ] (m)\n', rA)
fprintf('rD = [ %g, %g, %g ] (m)\n', rD)
fprintf('rB = [ %g, %g, %g ] (m)\n', rB)
fprintf('rC = [ %g, %g, %g ] (m)\n', rC)
fprintf('rE = [ %g, %g, %g ] (m)\n', rE)

% Graphic of the mechanism
plot([xA,xB],[yA,yB],'k-',...
[xB,xC],[yB,yC],'b-',...
[xD,xE],[yD,yE],'r-'),...
hold on,...
xlabel('x (m)'), ylabel('y (m)'), grid,...
axis([-0.2 0.45 -0.1 0.6]),...
title('Four-bar (R-RRR) mechanism'),...
text(xA,yA,'  A'), text(xB,yB,'  B'),...
text(xC,yC,'  C'), text(xD,yD,'  D'),...
text(xE,yE,'  E')
```

```
fprintf('\n')
fprintf('Velocity and acceleration analysis \n\n')

n = 60; % (rpm)  driver link
omega1 = [0 0 pi*n/30]; alpha1 = [0 0 0];
fprintf('omega1 = [%g, %g, %g](rad/s)\n', omega1)
fprintf('alpha1 = [%g, %g, %g](rad/s^2)\n', alpha1)
fprintf('\n')

vA = [0 0 0]; aA = [0 0 0];
vB1 = vA + cross(omega1,rB); vB2 = vB1;
aB1 = aA + cross(alpha1,rB) - dot(omega1,omega1)*rB;
aB2 = aB1;
fprintf('vB=vB1=vB2= [ %g, %g, %g ] (m/s)\n', vB1)
fprintf('aB=aB1=aB2= [ %g, %g, %g ] (m/s^2)\n', aB1)
fprintf('\n')

vD = [0 0 0]; aD = [0 0 0];

% velocity of C
omega2z = sym('omega2z','real');
omega3z = sym('omega3z','real');
omega2 = [ 0 0 omega2z ];
omega3 = [ 0 0 omega3z ];
% B2 & C points on link 2
% vC = vC2 = vB + omega2 x rBC
% B3 & D points on link 3
% vC = vC3 = vD + omega3 x rDC
% vC = vC2 = vC3
eqvC=vB2 + cross(omega2,rC-rB)-...
     (vD + cross(omega3,rC-rD));
eqvCx = eqvC(1); % equation component on x-axis
eqvCy = eqvC(2); % equation component on y-axis
solvC = solve(eqvCx,eqvCy);
omega2zs=eval(solvC.omega2z);
omega3zs=eval(solvC.omega3z);
Omega2 = [0 0 omega2zs];
Omega3 = [0 0 omega3zs];

% print the equations for calculating
% omega2z and omega3z
fprintf...
('vC=vB + omega2 x rBC = vD + omega3 x rDC => \n')
qvCx=vpa(eqvCx,6);
```

```
fprintf('x-axis:\n')
fprintf('%s=0\n',char(qvCx))
qvCy=vpa(eqvCy,6);
fprintf('y-axis:\n')
fprintf('%s=0\n',char(qvCy))
fprintf('=>\n')
fprintf('omega2z = %g (rad/s)\n', omega2zs)
fprintf('omega3z = %g (rad/s)\n', omega3zs)
fprintf('\n')
fprintf('omega2 = [ %g, %g, %g ] (rad/s)\n', Omega2)
fprintf('omega3 = [ %g, %g, %g ] (rad/s)\n', Omega3)
fprintf('\n')

vC = vB2 + cross(Omega2,rC-rB);
fprintf('vC = [ %g, %g, %g ] (m/s)\n', vC)
fprintf('\n')

vE = vD + cross(Omega3,rE-rD);
fprintf('vE = [ %g, %g, %g ] (m/s)\n', vE)
fprintf('\n')

% acceleration of C
alpha2z = sym('alpha2z','real');
alpha3z = sym('alpha3z','real');
alpha2 = [ 0 0 alpha2z ];
alpha3 = [ 0 0 alpha3z ];
% aC=aC2=aB + alpha2 x rBC - (omega2.omega2)rBC
% aC=aC3=aD + alpha3 x rDC - (omega3.omega3)rDC
eqaC2 = aB2 + cross(alpha2,rC-rB) - ...
        dot(Omega2,Omega2)*(rC-rB);
eqaC3 = aD + cross(alpha3,rC-rD) -...
        dot(Omega3,Omega3)*(rC-rD);
eqaC = eqaC2 - eqaC3;
eqaCx = eqaC(1); % equation component on x-axis
eqaCy = eqaC(2); % equation component on y-axis
solaC = solve(eqaCx,eqaCy);
alpha2zs=eval(solaC.alpha2z);
alpha3zs=eval(solaC.alpha3z);
Alpha2 = [0 0 alpha2zs];
Alpha3 = [0 0 alpha3zs];

% print the equations for calculating
% alpha2z and alpha3z
fprintf...
('aC2=aB + alpha2 x rBC - (omega2.omega2)rBC \n')
```

```
fprintf...
('aC3=aD + alpha3 x rDC - (omega3.omega3)rDC \n')
fprintf('aC = aC2 = aC3 \n')
qaCx=vpa(eqaCx,6);
fprintf('x-axis:\n')
fprintf('%s=0\n',char(qaCx))
qaCy=vpa(eqaCy,6);
fprintf('y-axis:\n')
fprintf('%s=0\n', char(qaCy))
fprintf('=>\n')
fprintf('alpha2z =  %g (rad/s^2)\n', alpha2zs)
fprintf('alpha3z =  %g (rad/s^2)\n', alpha3zs)
fprintf('\n')
fprintf('alpha2 = [%g, %g, %g](rad/s^2)\n', Alpha2)
fprintf('alpha3 = [%g, %g, %g](rad/s^2)\n', Alpha3)
fprintf('\n')

aC = aB2 + cross(Alpha2,rC-rB) -...
     dot(Omega2,Omega2)*(rC-rB);
fprintf('aC = [ %g, %g, %g] (m/s^2)\n', aC)
fprintf('\n')

aE = aD + cross(Alpha3,rE-rD) -...
     dot(Omega3,Omega3)*(rE-rD);
fprintf('aE = [ %g, %g, %g] (m/s^2)\n', aE)

% end of program
```

Results:

```
phi = 45 (deg)
rA = [ 0, 0, 0 ] (m)
rD = [ 0.3, 0.3, 0 ] (m)
rB = [ 0.106066, 0.106066, 0 ] (m)
rC = [ 0.0400698, 0.449788, 0 ] (m)
rE = [ -0.0898952, 0.524681, 0 ] (m)

Velocity and acceleration analysis

omega1 = [0, 0, 6.28319](rad/s)
alpha1 = [0, 0, 0](rad/s^2)

vB=vB1=vB2= [ -0.666432, 0.666432, 0 ] (m/s)
aB=aB1=aB2= [ -4.18732, -4.18732, 0 ] (m/s^2)
```

```
vC=vB + omega2 x rBC = vD + omega3 x rDC =>
x-axis:
-.666432-.343722*omega2z+.149788*omega3z=0
y-axis:
.666432-.659962e-1*omega2z+.259930*omega3z=0
=>
omega2z = -3.43639 (rad/s)
omega3z = -3.43639 (rad/s)

omega2 = [ 0, 0, -3.43639 ] (rad/s)
omega3 = [ 0, 0, -3.43639 ] (rad/s)

vC = [ 0.514728, 0.893221, 0 ] (m/s)

vE = [ 0.772092, 1.33983, 0 ] (m/s)

aC2=aB + alpha2 x rBC - (omega2.omega2)rBC
aC3=aD + alpha3 x rDC - (omega3.omega3)rDC
aC = aC2 = aC3
x-axis:
-6.47744-.343722*alpha2z+.149788*alpha3z=0
y-axis:
-6.47744-.659962e-1*alpha2z+.259930*alpha3z=0
=>
alpha2z =  -8.97883 (rad/s^2)
alpha3z =  22.6402 (rad/s^2)

alpha2 = [0, 0, -8.97883](rad/s^2)
alpha3 = [0, 0, 22.6402](rad/s^2)

aC = [ -0.321767, -7.65368, 0] (m/s^2)

aE = [ -0.48265, -11.4805, 0] (m/s^2)
```

B.3 Inverted Slider-Crank Mechanism

```
% B3
% Velocity and acceleration analysis
% Inverted slider-crank mechanism

clear all; clc; close all
AC = 0.15; BC = 0.2;
xA = 0; yA = 0;
```

```
xC = AC; yC = 0;
phi = pi/3; phi1=phi;
rA = [ xA, yA, 0 ]; rC = [ xC, yC, 0 ];
AD = AC+BC;
xD=AD*cos(phi); yD=AD*sin(phi); rD=[xD,yD,0];
% position of B
eqB1 = 'xBsol*sin(phi) = yBsol*cos(phi)';
eqB2 = 'yBsol^2+(xC-xBsol)^2-BC^2 = 0';
solB = solve(eqB1, eqB2, 'xBsol, yBsol');
xBpositions = eval(solB.xBsol);
yBpositions = eval(solB.yBsol);
xB1 = xBpositions(1); xB2 = xBpositions(2);
yB1 = yBpositions(1); yB2 = yBpositions(2);
if (phi>=0 && phi<= pi)
    if yB1 >= 0 xB=xB1; yB=yB1;
    else xB=xB2; yB=yB2; end
end
if (phi>pi && phi<=2*pi)
   if yB1 < 0 xB=xB1; yB=yB1;
   else xB=xB2; yB=yB2; end
end
rB = [ xB, yB, 0 ];
phi3 = atan((yB-yC)/(xB-xC))+pi;
fprintf('Results \n\n')
fprintf('phi = %g (degrees)\n', phi*180/pi)
fprintf('rB = [ %g, %g, %g ] (m)\n', rB)
fprintf('rC = [ %g, %g, %g ] (m)\n', rC)
fprintf('rD = [ %g, %g, %g ] (m)\n', rD)
fprintf('phi3 = %g (degrees) \n', phi3*180/pi)

% Graphic of the mechanism
plot([xA,xD],[yA,yD],'k-o',[xB,xC],[yB,yC],'b-o'),
text(xA,yA,'  A'),text(xB,yB,'  B'),...
text(xC,yC,'  C'),text(xD,yD,'  D'),...
axis([-0.05 0.35 -0.05 0.35]);

fprintf('\n')
fprintf('Velocity and acceleration analysis \n\n')

n = 30; % (rpm)  driver link
omega1 = [ 0 0 pi*n/30 ]; omega2 = omega1;
alpha1 = [0 0 0 ]; alpha2 = alpha1;

fprintf('omega1 = [%g, %g, %g](rad/s)\n', omega1)
fprintf('alpha1 = [%g, %g, %g](rad/s^2)\n', alpha1)
```

```
fprintf('\n')

vA = [0 0 0 ]; aA = [0 0 0 ];
vC = [0 0 0 ]; aC = [0 0 0 ];

vB1 = vA + cross(omega1,rB);
aB1 = aA + cross(alpha1,rB) - dot(omega1,omega1)*rB;

fprintf('vB = vB1 = [ %g, %g, %g ] (m/s)\n', vB1)
fprintf('aB = aB1 = [ %g, %g, %g ] (m/s^2)\n', aB1)
fprintf('\n')

% angular velocity of link 3
omega3z = sym('omega3z','real'); % omega3z unknown
omega3 = [ 0 0 omega3z ];
vB21 = sym('vB21','real'); % vB21 unknown

% vB2B1 parallel to the sliding direction
vB2B1 = [ vB21*cos(phi1) vB21*sin(phi1) 0 ];
% vB2 = vB3 = vC + omega3 x (rB-rC)
vB3 = vC + cross(omega3,rB-rC);
vB2 = vB3;

eqvB = vB2 - ( vB1 + vB2B1 ); % vB2 = vB1 + vB2B1
eqvBx = eqvB(1); % equation component on x-axis
eqvBy = eqvB(2); % equation component on y-axis

solvB = solve(eqvBx,eqvBy);
omega3zs = eval(solvB.omega3z);
vB21s = eval(solvB.vB21);

Omega3 = [0 0 omega3zs];

VB21 = vB21s*[cos(phi1) sin(phi1) 0];
VB3 = vC + cross(Omega3,rB-rC);

% print the equations for calculating
% omega3z and vB21
fprintf('vB2 = vB1 + vB2B1 => \n')
qvBx=vpa(eqvBx,6);
fprintf('x-axis:\n')
fprintf('%s=0\n',char(qvBx))
qvBy=vpa(eqvBy,6);
fprintf('y-axis:\n')
fprintf('%s=0\n',char(qvBy))
```

```
fprintf('=>\n')
fprintf('omega3z = %g (rad/s)\n', omega3zs)
fprintf('vB21 = %g (m/s)\n\n', vB21s)

fprintf('omega3 = [%g, %g, %g](rad/s)\n', Omega3)
fprintf('vB2B1 = [ %g, %g, %d ] (m/s)\n', VB21)
fprintf('\n')
fprintf('vB3 = [ %g, %g, %d ] (m/s)\n', VB3)
fprintf('\n')

% angular acceleration of link 3
alpha3z = sym('alpha3z','real'); % alpha3z unknown
alpha3 = [ 0 0 alpha3z ];
aB21 = sym('aB21','real'); % aB21 unknown

% aB2B1 parallel to the sliding direction
aB2B1 = [ aB21*cos(phi1) aB21*sin(phi1) 0 ];
% aB2=aB3=aC+alpha3 x rCB-(omega3)^2 rCB
aB3=aC+cross(alpha3,rB-rC)-...
    dot(Omega3,Omega3)*(rB-rC);
aB2 = aB3;

% aB2B1cor = 2 omega1 x vB2B1
aB2B1cor = 2*cross(omega1,VB21);

% aB2 = aB1 + aB2B1 + aB2B1cor
eqaB = aB2 - ( aB1 + aB2B1 + aB2B1cor );
eqaBx = eqaB(1); % equation component on x-axis
eqaBy = eqaB(2); % equation component on y-axis

solaB = solve(eqaBx,eqaBy);

alpha3zs = eval(solaB.alpha3z);
aB21s = eval(solaB.aB21);

Alpha3 = [0 0 alpha3zs];
AB21 = aB21s*[cos(phi1) sin(phi1) 0];

AB3=aC+cross(Alpha3,rB-rC)-...
    dot(Omega3,Omega3)*(rB-rC);

% print the equations for calculating
% alpha3z and aB21
fprintf...
('aB2B1cor = [%g, %g, %d] (m/s^2)\n', aB2B1cor)
```

```
fprintf('\n')
fprintf...
('aB2 = aB1 + aB2B1 + aB2B1cor => \n')
qaBx=vpa(eqaBx,6);
fprintf('x-axis:\n')
fprintf('%s=0\n',char(qaBx))
qaBy=vpa(eqaBy,6);
fprintf('y-axis:\n')
fprintf('%s=0\n',char(qaBy))
fprintf('=>\n')

fprintf('alpha3z =  %g (rad/s^2)\n', alpha3zs)
fprintf('aB21 =  %g (m/s^2)\n', aB21s)
fprintf('\n')

fprintf...
('alpha3 = [ %g, %g, %g ] (rad/s^2)\n', Alpha3)
fprintf('aB2B1 = [ %g, %g, %d ] (m/s^2)\n', AB21)
fprintf('aB3 = [ %g, %g, %d ] (m/s^2)\n', AB3)
fprintf('\n')

omega23 = omega2 - Omega3;
alpha23 = alpha2 - Alpha3;

fprintf...
('omega23 = [ %g, %g, %g ] (rad/s)\n', omega23)
fprintf...
('alpha23 = [ %g, %g, %g ] (rad/s^2)\n', alpha23)

% end of program
```

Results:

```
phi = 60 (degrees)
rB = [ 0.113535, 0.196648, 0 ] (m)
rC = [ 0.15, 0, 0 ] (m)
rD = [ 0.175, 0.303109, 0 ] (m)
phi3 = 100.505 (degrees)

Velocity and acceleration analysis

omega1 = [0, 0, 3.14159](rad/s)
alpha1 = [0, 0, 0](rad/s^2)

vB = vB1 = [ -0.617787, 0.356679, 0 ] (m/s)
aB = aB1 = [ -1.12054, -1.94083, 0 ] (m/s^2)
```

```
vB2 = vB1 + vB2B1 =>
x-axis:
-.196648*omega3z+.617787-.500000*vB21=0
y-axis:
-.364655e-1*omega3z-.356679-.866025*vB21=0
=>
omega3z = 4.69102 (rad/s)
vB21 = -0.609381 (m/s)

omega3 = [0, 0, 4.69102](rad/s)
vB2B1 = [ -0.30469, -0.527739, 0 ] (m/s)

vB3 = [ -0.922477, -0.17106, 0 ] (m/s)

aB2B1cor = [3.31588, -1.91443, 0] (m/s^2)

aB2 = aB1 + aB2B1 + aB2B1cor =>
x-axis:
-.196648*alpha3z-1.39290-.500000*aB21=0
y-axis:
-.364655e-1*alpha3z-.472095-.866025*aB21=0
=>
alpha3z =  -6.38024 (rad/s^2)
aB21 =  -0.276477 (m/s^2)

alpha3 = [ 0, 0, -6.38024 ] (rad/s^2)
aB2B1 = [ -0.138239, -0.239436, 0 ] (m/s^2)
aB3 = [ 2.0571, -4.0947, 0 ] (m/s^2)

omega23 = [ 0, 0, -1.54942 ] (rad/s)
alpha23 = [ 0, 0, 6.38024 ] (rad/s^2)
```

B.4 R-RTR-RTR Mechanism

```
% B4
% Velocity and acceleration analysis
% R-RTR-RTR

clear all; clc; close all

AB = 0.15; AC = 0.10; CD = 0.15; %(m)
DF = 0.40; AG = 0.30; % (m)
```

```
phi = pi/6 ;  % (rad)
xA = 0; yA = 0 ; rA = [xA yA 0];
xC = 0; yC = AC; rC = [xC yC 0];
xB = AB*cos(phi); yB = AB*sin(phi); rB = [xB yB 0];
eqnD1='( xDsol - xC )^2 + ( yDsol - yC )^2 = CD^2';
eqnD2='(yB-yC)/(xB-xC)=(yDsol-yC)/(xDsol-xC)';
solD = solve(eqnD1, eqnD2, 'xDsol, yDsol');
xDpositions = eval(solD.xDsol);
yDpositions = eval(solD.yDsol);
xD1=xDpositions(1); xD2=xDpositions(2);
yD1=yDpositions(1); yD2=yDpositions(2);
if(phi>=0&&phi<=pi/2)||(phi>=3*pi/2&&phi<=2*pi)
 if xD1 <= xC xD=xD1; yD=yD1; else xD=xD2; yD=yD2;
 end
   else
 if xD1 >= xC xD=xD1; yD=yD1; else xD=xD2; yD=yD2;
 end
end
rD=[xD yD 0];
phi2=atan((yB-yC)/(xB-xC)); phi3=phi2;
phi4=atan(yD/xD)+pi; phi5=phi4;
xF=xD+DF*cos(phi3); yF=yD+ DF*sin(phi3);
rF=[xF yF 0];
xG=AG*cos(phi5); yG=AG*sin(phi5);
rG = [xG yG 0];

fprintf('Results \n\n')

fprintf('phi = phi1 = %g (degrees) \n',phi*180/pi)
fprintf('rA = [ %g, %g, %g ] (m)\n', rA)
fprintf('rC = [ %g, %g, %g ] (m)\n', rC)
fprintf('rB = [ %g, %g, %g ] (m)\n', rB)
fprintf('rD = [ %g, %g, %g ] (m)\n', rD)
fprintf...
('phi2 = phi3 = %g (degrees) \n',phi2*180/pi)
fprintf...
('phi4 = phi5 = %g (degrees) \n',phi4*180/pi)
fprintf('rF = [ %g, %g, %g] (m)\n', rF)
fprintf('rG = [ %g, %g, %g] (m)\n', rG)

% Graphic of the mechanism
plot([xA,xB],[yA,yB],'k-o',[xD,xF],[yD,yF],'b-o',...
[xA,xG],[yA,yG],'r-o')
xlabel('x (m)'), ylabel('y (m)'),...
title('positions for \phi = 30 (deg)'),...
```

```
text(xA,yA,' A'), text(xB,yB,' B'),...
text(xC,yC,' C'), text(xD,yD,' D'),...
text(xF,yF,' F'), text(xG,yG,' G'),...
axis([-0.3 0.3 -0.1 0.3]), grid

fprintf('\n')
fprintf('Velocity and acceleration analysis \n\n')

n = 50.;
omega1 = [ 0 0 pi*n/30 ]; alpha1 = [0 0 0 ];
fprintf...
('omega1 = [ %g, %g, %g ] (rad/s)\n', omega1)
fprintf...
('alpha1 = [ %g, %g, %g ] (rad/s^2)\n', alpha1)
fprintf('\n')

vA = [0 0 0 ]; aA = [0 0 0 ];

vB1 = vA + cross(omega1,rB); vB2 = vB1;
aB1 = aA + cross(alpha1,rB) - ...
    dot(omega1,omega1)*rB;
vB2 = vB1;
aB2 = aB1;
fprintf...
('vB = vB1 = vB2 = [ %g, %g, %g ] (m/s)\n',vB1)
fprintf...
('aB = aB1 = aB2 = [ %g, %g, %g ] (m/s^2)\n',aB1)
fprintf('\n')

omega3z=sym('omega3z','real');
vB32=sym('vB32','real');
omega3 = [ 0 0 omega3z ];
% omega3z unknown (to be calculated)
% vB32 unknown (to be calculated)
vC = [0 0 0 ]; % C is fixed
% vB3 = vC + omega3 x rCB
% (B3 & C are points on link 3)
vB3 = vC + cross(omega3,rB-rC);
% point B2 is on link 2 and point B3 is on link 3
% vB3 = vB2 + vB3B2
% between the links 2 and 3 there is a
% translational joint B_T
% vB3B2 is the relative velocity of B3 wrt 2
% vB3B2 is parallel to the sliding direcion BC
% vB3B2 is written as a vector
```

```
vB3B2 = vB32*[ cos(phi2) sin(phi2) 0];
% vB3 = vB2 + vB3B2
eqvB = vB3 - vB2 - vB3B2;
% vectorial equation
% the component of the vectorial equation on x-axis
eqvBx = eqvB(1);
% the component of the vectorial equation on y-axis
eqvBy = eqvB(2);
% two equations eqvBx and eqvBy with two unknowns
% solve for omega3z and vB32
solvB = solve(eqvBx,eqvBy);
omega3zs=eval(solvB.omega3z);
vB32s=eval(solvB.vB32);

Omega3 = [0 0 omega3zs]; Omega2 = Omega3;
VB32 = vB32s*[cos(phi2) sin(phi2) 0];

% print the equations for calculating
% omega3 and vB32
fprintf...
('vB3 = vC + omega3 x rCB = vB2 + vB3B2 => \n')
qvBx=vpa(eqvBx,6);
fprintf('x-axis: %s = 0 \n', char(qvBx))
qvBy=vpa(eqvBy,6);
fprintf('y-axis: %s = 0 \n', char(qvBy))
fprintf('=>\n')
fprintf('omega3z =  %g (rad/s)\n', omega3zs)
fprintf('vB32 =  %g (m/s)\n', vB32s)
fprintf('\n')
fprintf...
('omega2=omega3 = [ %g,%g,%g ](rad/s)\n', Omega3)
fprintf('vB3B2 = [ %g, %g, %g ] (m/s)\n\n', VB32)

% Coriolis acceleration
aB3B2cor = 2*cross(Omega3,VB32);

alpha3z=sym('alpha3z','real');
aB32=sym('aB32','real'); % aB32 unknown

alpha3 = [ 0 0 alpha3z ]; % alpha3z unknown
aC = [0 0 0 ]; % C is fixed
% aB3 acceleration of B3
aB3 = aC + cross(alpha3, rB-rC) -...
    dot(Omega3, Omega3)*(rB-rC);
% aB3B2 relative velocity of B3 wrt 2
```

```
% aB3B2 parallel to the sliding direcion BC

aB3B2 = aB32*[ cos(phi2) sin(phi2) 0];
% aB3 = aB2 + aB3B2 + aB3B2cor
eqaB = aB3 - aB2 - aB3B2 - aB3B2cor;
% vectorial equation
eqaBx = eqaB(1); % equation component on x-axis
eqaBy = eqaB(2); % equation component on y-axis
solaB = solve(eqaBx,eqaBy);
alpha3zs=eval(solaB.alpha3z);
aB32s=eval(solaB.aB32);
Alpha3 = [0 0 alpha3zs];
Alpha2 = Alpha3;
AB32 = aB32s*[cos(phi2) sin(phi2) 0];

% print the equations for calculating
% alpha3 and aB32
fprintf...
('aB32cor = [ %g, %g, %d ](m/s^2)\n', aB3B2cor)
fprintf('\n')
fprintf('aB3=aC+alpha3xrCB-(omega3.omega3)rCB\n')
fprintf('aB3=aB2+aB3B2+aB3B2cor =>\n')
qaBx=vpa(eqaBx,6);
fprintf('x-axis:\n')
fprintf('%s = 0\n',char(qaBx))
qaBy=vpa(eqaBy,6);
fprintf('y-axis:\n')
fprintf('%s = 0\n',char(qaBy))
fprintf('=>\n');
fprintf('alpha3z = %g (rad/s)\n', alpha3zs)
fprintf('aB32 = %g (m/s)\n', aB32s)
fprintf('\n')
fprintf...
('alpha2=alpha3 = [ %g,%g,%g ](rad/s^2)\n', Alpha3)
fprintf('aB3B2 = [ %g, %g, %g ] (m/s^2)\n\n', AB32)

% vD3 velocity of D3
% D3 & C points on link 3
vD3 = vC + cross(Omega3,rD-rC);
vD4 = vD3;
fprintf('vD3 = vD4 = [ %g, %g, %g ] (m/s)\n', vD3)
% aD3 acceleration of D3
aD3 = aC + cross(Alpha3, rD-rC)-...
    dot(Omega3, Omega3)*(rD-rC);
aD4 = aD3;
```

```
fprintf('aD3 = aD4 = [ %g, %g, %g ] (m/s^2)\n', aD3)
fprintf('\n')

omega5z = sym('omega5z','real');
% omega5z unknown
vD54 = sym('vD54','real');
% vD54 unknown
omega5 = [ 0 0 omega5z ];
% vD5 velocity of D5
% D5 & A points on link 5
vD5 = vA + cross(omega5,rD);
% vD5D4 relative velocity of D5 wrt 4
% vD5D4 parallel to the sliding direcion DE
vD5D4 = vD54*[ cos(phi5) sin(phi5) 0];
% vD5 = vD4 + vD5D4
eqvD = vD5 - vD4 - vD5D4;
% vectorial equation
eqvDx = eqvD(1); % component on x-axis
eqvDy = eqvD(2); % component on y-axis
solvD = solve(eqvDx,eqvDy);
omega5zs=eval(solvD.omega5z);
vD54s=eval(solvD.vD54);
Omega5 = [0 0 omega5zs];
Omega4 = Omega5;
VD54 = vD54s*[cos(phi5) sin(phi5) 0];

% print the equations for calculating
% omega5 and vD54
fprintf...
('vD5 = vA + omega5 x rD = vD4 + vD5D4 => \n')
qvDx=vpa(eqvDx,6);
fprintf('x-axis: %s = 0 \n', char(qvDx))
qvDy=vpa(eqvDy,6);
fprintf('y-axis: %s = 0 \n', char(qvDy))
fprintf('=>\n')
fprintf('omega5z = %g (rad/s)\n', omega5zs)
fprintf('vD54 = %g (m/s)\n', vD54s)
fprintf('\n')
fprintf...
('omega4=omega5 = [ %g, %g, %g ] (rad/s)\n', Omega5)
fprintf('vD5D4 = [ %g, %g, %g ] (m/s)\n', VD54 )
fprintf('\n')

% Coriolis acceleration
aD5D4cor = 2*cross(Omega5,VD54);
```

```
alpha5z = sym('alpha5z','real');
aD54 = sym('aD54','real');
alpha5 = [ 0 0 alpha5z ]; % alpha5z unknown
aE = [0 0 0 ];

% aD5 acceleration of D5
aD5=aE+cross(alpha5,rD)-dot(Omega5,Omega5)*rD;
% aD5D4 relative velocity of B5 wrt 4
% aD5D4 parallel to the sliding direcion AD
% aD54 unknown
aD5D4 = aD54*[ cos(phi5) sin(phi5) 0];
% aB5 = aB4 + aD5D4+ aD5D4cor
eqaD = aD5 - aD4 - aD5D4 - aD5D4cor;
% vectorial equation
eqaDx = eqaD(1); % component on x-axis
eqaDy = eqaD(2); % component on y-axis
solaD = solve(eqaDx,eqaDy);
alpha5zs = eval(solaD.alpha5z);
aD54s = eval(solaD.aD54);

Alpha5 = [0 0 alpha5zs];
Alpha4 = Alpha5;
AD54 = aD54s*[cos(phi5) sin(phi5) 0];

% print the equations for calculating
% alpha5 and aD54
fprintf...
('aD54cor = [ %g, %g, %g ] (m/s^2)\n', aD5D4cor)
fprintf('\n')
fprintf('aD5=aA+alpha5xrD-(omega5.omega5)rD=\n')
fprintf('aD5=aD4+aD5D4+aD5D4cor =>\n')
qaDx=vpa(eqaDx,6);
fprintf('x-axis: %s = 0 \n', char(qaDx))
qaDy=vpa(eqaDy,6);
fprintf('y-axis: %s = 0 \n', char(qaDy))
fprintf('=>\n')

fprintf('alpha5z = %g (rad/s^2)\n', alpha5zs)
fprintf('aD54 = %g (m/s^2)\n', aD54s)
fprintf('\n')
fprintf...
('alpha4=alpha5 = [ %g,%g,%g ] (rad/s^2)\n', Alpha5)
fprintf('aD54 = [ %g, %g, %g ] (m/s^2)\n', AD54)
% end of program
```

Results:

```
phi = phi1 = 30 (degrees)
rA = [ 0, 0, 0 ] (m)
rC = [ 0, 0.1, 0 ] (m)
rB = [ 0.129904, 0.075, 0 ] (m)
rD = [ -0.147297, 0.128347, 0 ] (m)
phi2 = phi3 = -10.8934 (degrees)
phi4 = phi5 = 138.933 (degrees)
rF = [ 0.245495, 0.0527544, 0] (m)
rG = [ -0.226182, 0.197083, 0] (m)

Velocity and acceleration analysis

omega1 = [ 0, 0, 5.23599 ] (rad/s)
alpha1 = [ 0, 0, 0 ] (rad/s^2)

vB = vB1 = vB2 = [ -0.392699, 0.680175, 0 ] (m/s)
aB = aB1 = aB2 = [ -3.56139, -2.05617, 0 ] (m/s^2)

vB3 = vC + omega3 x rCB = vB2 + vB3B2 =>
x-axis: .250000e-1*omega3z+.392699-.981983*vB32 = 0
y-axis: .129904*omega3z-.680175+.188982*vB32 = 0
=>
omega3z =  4.48799 (rad/s)
vB32 =  0.514164 (m/s)

omega2=omega3 = [ 0,0,4.48799 ](rad/s)
vB3B2 = [ 0.504899, -0.0971678, 0 ] (m/s)

aB32cor = [ 0.872176, 4.53196, 0 ](m/s^2)

aB3=aC+alpha3xrCB-(omega3.omega3)rCB
aB3=aB2+aB3B2+aB3B2cor =>
x-axis:
.250000e-1*alpha3z+.726814e-1-.981983*aB32 = 0
y-axis:
.129904*alpha3z-1.97224+.188982*aB32 = 0
=>
alpha3z = 14.5363 (rad/s)
aB32 = 0.44409 (m/s)

alpha2=alpha3 = [ 0,0,14.5363 ](rad/s^2)
aB3B2 = [ 0.436088, -0.0839252, 0 ] (m/s^2)

vD3 = vD4 = [ -0.127223, -0.661068, 0 ] (m/s)
```

```
aD3 = aD4 = [ 2.5548, -2.71212, 0 ] (m/s^2)

vD5 = vA + omega5 x rD = vD4 + vD5D4 =>
x-axis: -.128347*omega5z+.127223+.753939*vD54 = 0
y-axis: -.147297*omega5z+.661068-.656945*vD54 = 0
=>
omega5z = 2.97887 (rad/s)
vD54 = 0.338367 (m/s)

omega4=omega5 = [ 0, 0, 2.97887 ] (rad/s)
vD5D4 = [ -0.255108, 0.222288, 0 ] (m/s)

aD54cor = [ -1.32434, -1.51987, 0 ] (m/s^2)

aD5=aA+alpha5xrD-(omega5.omega5)rD=
aD5=aD4+aD5D4+aD5D4cor =>
x-axis: -.128347*alpha5z+.766057e-1+.753939*aD54 = 0
y-axis: -.147297*alpha5z+3.09308-.656945*aD54 = 0
=>
alpha5z = 12.1939 (rad/s^2)
aD54 = 1.97423 (m/s^2)

alpha4=alpha5 = [ 0,0,12.1939 ] (rad/s^2)
aD54 = [ -1.48845, 1.29696, 0 ] (m/s^2)
```

B.5 R-RTR-RTR Mechanism: Derivative Method

```
% B5
% Velocity and acceleration analysis
% R-RTR-RTR
% Derivative method

clear all; clc; close all
fprintf('Results \n\n')
fprintf('Velocity and acceleration analysis \n')
fprintf('Derivative method \n\n')
AB = 0.15 ; AC = 0.10 ; CD = 0.15 ; %(m)
DF = 0.40 ; AG = 0.30 ; % (m)
phi = pi/6 ;  % (rad)
xA = 0 ; yA = 0   ; rA = [xA yA 0] ;
xC = 0 ; yC = AC  ; rC = [xC yC 0] ;

n = 50 ; % rpm of the driver link (constant)
```

```
omega = n*pi/30;  % rad/s

% sym  constructs symbolic numbers and variables
t = sym('t','real');
% phi(t) the angle of the diver link
% with the horizontal axis
% phi(t) is a function of time, t
% position of joint B
xB = AB*cos(sym('phi(t)'));
yB = AB*sin(sym('phi(t)'));
% position vector of B in terms of phi(t) - symbolic
rB = [xB yB 0];

% subs(S,OLD,NEW) replaces
% OLD with NEW in the symbolic expression S
xBn = subs(xB,'phi(t)',pi/6); % xB for phi(t)=pi/6
yBn = subs(yB,'phi(t)',pi/6); % yB for phi(t)=pi/6
rBn = subs(rB,'phi(t)',pi/6); % rB for phi(t)=pi/6
fprintf('rB =  [ %g, %g, %g ] (m)\n', rBn)

% velocity of B1=B2
% diff(S,t) differentiates S with respect to t
% vB1=vB2 in terms of phi(t) and diff(phi(t),t)
vB = diff(rB,t);

% calculates numerical value of vB for
% phi(t)=pi/6 and phi'(t)=omega
% creates a list slist for the symbolical variables
% phi''(t), phi'(t), phi(t)
slist = ...
{diff('phi(t)',t,2), diff('phi(t)',t), 'phi(t)'};
% creates a list nlist
% with the numerical values for slist
nlist = {0, omega, pi/6};
% diff('phi(t)',t,2) -> 0
% diff('phi(t)',t) -> omega
% 'phi(t)' -> pi/6

% replaces slist with nlist in vB
vBn = subs(vB,slist,nlist);
% double(S) converts the symbolic object S
%   to a numeric object
% converts the symbolic object vBn
%   to a numeric object
VB  = double(vBn);
```

```
fprintf('vB1=vB2 = [ %g, %g, %g ] (m/s)\n', VB)
% norm(v) is the magnitude of a vector v
VBn = norm(VB);
fprintf('|vB1|=|vB2| = %g (m/s) \n', VBn)

% acceleration of B1=B2
aB  = diff(vB,t);
% numerical value for aB
aBn = double(subs(aB,slist,nlist));
fprintf('aB1=aB2 = [ %g, %g, %g ] (m/s^2)\n', aBn)
ABn = norm(aBn);
fprintf('|aB1|=|aB2| = %g (m/s^2) \n', ABn)
fprintf('\n')

% position of joint D
eqnD1 = '(xDsol - xC)^2 + (yDsol - yC)^2=CD^2 ';
eqnD2 = '(yB-yC)/(xB-xC)=(yDsol-yC)/(xDsol-xC)';
solD = solve(eqnD1, eqnD2, 'xDsol, yDsol');
xDpositions = eval(solD.xDsol);
yDpositions = eval(solD.yDsol);
xD1 = xDpositions(1); xD2 = xDpositions(2);
yD1 = yDpositions(1); yD2 = yDpositions(2);
% select the correct position for D
% for the given input angle
xD1n = subs(xD1,'phi(t)',pi/6);
% xD1 for phi(t)=pi/6
if xD1n < xC
    xD = xD1; yD = yD1;
else
    xD = xD2; yD = yD2;
end
% position vector of D in term of phi(t)-symbolic
rD = [ xD yD 0 ];
xDn = subs(xD,'phi(t)',pi/6);
% xD for phi(t)=pi/6
yDn = subs(yD,'phi(t)',pi/6);
% yD for phi(t)=pi/6
rDn = [ xDn yDn 0 ]; % rD for phi(t)=pi/6
fprintf('rD = [ %g, %g, %g ] (m)\n', rDn)

% velocity of D3=D4
% vD in terms of phi(t) and diff('phi(t)',t)
vD  = diff(rD,t);
% numerical value for vD
vDn = double(subs(vD,slist,nlist));
```

```
fprintf('vD3=vD4 = [ %g, %g, %g ] (m/s)\n', vDn)
fprintf('|vD3|=|vD4| = %g (m/s) \n', norm(vDn))

% acceleration of D3=D4
% aD in terms of phi(t), diff('phi(t)',t),
%    and diff('phi(t)',t,2)
aD  = diff(vD,t);
% numerical value for aD
aDn = double(subs(aD,slist,nlist));
fprintf('aD3=aD4 = [ %g, %g, %g ] (m/s^2)\n', aDn)
fprintf('|aD3|=|aD4| = %g (m/s^2) \n', norm(aDn))
fprintf('\n')

% angular velocities and accelerations

% Link 2
phi2 = atan((yB-yC)/(xB-xC));
% phi2 in terms of phi(t)
phi2n = subs(phi2,'phi(t)',pi/6);
% phi2 for phi(t)=pi/6
fprintf...
('phi2=phi3 = %g (degrees) \n', phi2n*180/pi)

% omega2 in terms of phi(t) and diff('phi(t)',t)
dphi2 = diff(phi2,t);
dphi2nn = subs(dphi2,diff('phi(t)',t),omega);
% numerical value for omega2
dphi2n  = subs(dphi2nn,'phi(t)',pi/6);
fprintf('omega2=omega3 = %g (rad/s) \n', dphi2n)

% alpha2 in terms of phi(t), diff('phi(t)',t),
%        and diff('phi(t)',t,2)
ddphi2  = diff(dphi2,t);
% numerical value for alpha2
ddphi2n = double(subs(ddphi2,slist,nlist));
fprintf('alpha2=alpha3 = %g (rad/s^2) \n', ddphi2n)
fprintf('\n')

% Link 4
phi4  = atan(yD/xD)+pi;
% phi4 in terms of phi(t)
phi4n = subs(phi4,'phi(t)',pi/6);
% numerical value for phi4
fprintf('phi4=phi5 = %g (degrees) \n', phi4n*180/pi)
```

```
dphi4  = diff(phi4,t);
% numerical value for omega4
dphi4n = double(subs(dphi4,slist,nlist));
fprintf('omega4=omega5 = %g (rad/s) \n', dphi4n)

ddphi4  = diff(dphi4,t);
% numerical value for alpha4
ddphi4n = double(subs(ddphi4,slist,nlist));
fprintf('alpha4=alpha5 = %g (rad/s^2)\n', ddphi4n)

plot([xA,xBn],[yA,yBn],'k',...
     [xBn,xDn],[yBn,yDn],'b',...
     [xA,xDn],[yA,yDn],'r')
text(xA,yA,'  A'),...
text(xBn,yBn,'  B'),...
text(xC,yC,'  C'),...
text(xDn,yDn,'  D'),...
grid

% end of program
```

Results :

```
rB =  [ 0.129904, 0.075, 0 ] (m)
vB1=vB2 = [ -0.392699, 0.680175, 0 ] (m/s)
|vB1|=|vB2| = 0.785398 (m/s)
aB1=aB2 = [ -3.56139, -2.05617, 0 ] (m/s^2)
|aB1|=|aB2| = 4.11234 (m/s^2)

rD = [ -0.147297, 0.128347, 0 ] (m)
vD3=vD4 = [ -0.127223, -0.661068, 0 ] (m/s)
|vD3|=|vD4| = 0.673198 (m/s)
aD3=aD4 = [ 2.5548, -2.71212, 0 ] (m/s^2)
|aD3|=|aD4| = 3.72594 (m/s^2)

phi2=phi3 = -10.8934 (degrees)
omega2=omega3 = 4.48799 (rad/s)
alpha2=alpha3 = 14.5363 (rad/s^2)

phi4=phi5 = 138.933 (degrees)
omega4=omega5 = 2.97887 (rad/s)
alpha4=alpha5 = 12.1939 (rad/s^2)
```

B.6 Inverted Slider-Crank Mechanism: Derivative Method

```
% B6
% Velocity and acceleration analysis
% Inverted slider-crank mechanism
% Derivative method

clear all; clc; close all

fprintf('Results \n\n')
fprintf('Velocity and acceleration analysis\n')
fprintf('Derivative method \n\n')

AC = 0.15; BC = 0.20 ; xA = 0; yA = 0;
xC = AC; yC = 0;
n = 30; omega = n*pi/30;

t = sym('t','real');

phi = sym('phi(t)');
xB = sym('xB(t)');
yB = sym('yB(t)');

eqB1 = xB*sin(phi) - yB*cos(phi);
eqB2 = ( xB - xC )^2 + ( yB - yC )^2 - BC^2;

sp = {'phi(t)','xB(t)','yB(t)'};
np = {pi/3,'xBn','yBn'};

eqB1p = subs(eqB1,sp,np);
eqB2p = subs(eqB2,sp,np);

solBp = solve(eqB1p, eqB2p);
xBpositions = eval(solBp.xBn);
yBpositions = eval(solBp.yBn);
xB1 = xBpositions(1); xB2 = xBpositions(2);
yB1 = yBpositions(1); yB2 = yBpositions(2);

if yB1 > 0
    xBp = xB1; yBp = yB1;
else
    xBp = xB2; yBp = yB2;
end
rB = [xBp yBp 0];
fprintf('rB =  [ %g, %g, %g ] (m)\n', rB)
```

```
fp = {pi/3,xBp,yBp};

% velocity of B2=B3
deqB1 = diff(eqB1,t);
deqB2 = diff(eqB2,t);

sv={diff('phi(t)',t),diff('xB(t)',t),...
    diff('yB(t)',t)};

nv = {omega,'vxB','vyB'};

deqB1p=subs(deqB1,sv,nv);
deqB1n=subs(deqB1p,sp,fp);
deqB2p=subs(deqB2,sv,nv);
deqB2n=subs(deqB2p,sp,fp);

solvB = solve(deqB1n, deqB2n);
vBx = eval(solvB.vxB);
vBy = eval(solvB.vyB);
fprintf('vB2 = vB3 = [ %g, %g, %g ] (m/s)\n',...
    [vBx vBy 0])
fprintf('|vB2| = |vB3| = %g (m/s) \n',...
    norm([vBx vBy 0]))

fv = {omega,vBx,vBy};

% acceleration of B2=B3
ddeqB1 = diff(deqB1,t);
ddeqB2 = diff(deqB2,t);

sa = {diff('phi(t)',t,2),diff('xB(t)',t,2),...
      diff('yB(t)',t,2)};
na = {0,'axB','ayB'};

ddeqB1p=subs(ddeqB1,sa,na);
ddeqB1n=subs(ddeqB1p,sv,fv);
ddeqB1f=subs(ddeqB1n,sp,fp) ;

ddeqB2p=subs(ddeqB2,sa,na);
ddeqB2n=subs(ddeqB2p,sv,fv);
ddeqB2f=subs(ddeqB2n,sp,fp);

solaB = solve(ddeqB1f, ddeqB2f);
aBx = eval(solaB.axB);
```

```
aBy = eval(solaB.ayB);
fprintf('aB2 = aB3 = [ %g, %g, %g ] (m/s^2)\n',...
    [aBx aBy 0])
fprintf('|aB2| = |aB3| = %g (m/s^2) \n',...
    norm([aBx aBy 0]))

fa = {0,aBx,aBy};

% angular velocity and acceleration of link 3

phi3 = atan((yB-yC)/(xB-xC));
phi3n = subs(phi3,sp,fp);
fprintf('phi3 = %g (degrees) \n',...
    double(phi3n*180/pi))

dphi3 = diff(phi3,t) ;
dphi3nn = subs(dphi3,sv,fv) ;
dphi3n  = subs(dphi3nn,sp,fp) ;
fprintf('omega3 = %g (rad/s) \n',...
    double(dphi3n))

ddphi3 = diff(dphi3,t);
ddphi3nnn = subs(ddphi3,sa,fa);
ddphi3nn = subs(ddphi3nnn,sv,fv);
ddphi3n  = subs(ddphi3nn,sp,fp);
fprintf('alpha3 = %g (rad/s^2) \n',...
    double(ddphi3n))

plot([xA,xBp],[yA,yBp],'r',...
     [xBp,xC],[yBp,yC],'b'),...
text(xA,yA,' A'),...
text(xBp,yBp,' B'),...
text(xC,yC,' C'), grid
% end of program
```

Results:

```
rB =  [ 0.113535, 0.196648, 0 ] (m)
vB2 = vB3 = [ -0.922477, -0.17106, 0 ] (m/s)
|vB2| = |vB3| = 0.938203 (m/s)
aB2 = aB3 = [ 2.0571, -4.0947, 0 ] (m/s^2)
|aB2| = |aB3| = 4.58238 (m/s^2)
phi3 = -79.4946 (degrees)
omega3 = 4.69102 (rad/s)
alpha3 = -6.38024 (rad/s^2)
```

B.7 R-RTR Mechanism: Derivative Method

```
% B7
% Velocity and acceleration analysis
% R-RTR
% Derivative method

clear all; clc; close all
fprintf('Results \n\n')
fprintf('Velocity and acceleration analysis\n')
fprintf('Derivative method \n\n')

AB = 0.1; AC = 0.1; CD = 0.3;% (m)
phi1 = pi/4; omega = pi;  alpha = 0;

xC = 0; yC = 0;
xA = 0; yA = AC;

t = sym('t','real');
xB1 = xA + AB*cos(sym('phi(t)'));
yB1 = yA + AB*sin(sym('phi(t)'));
% position vector of B function of phi(t) - symbolic
rB = [ xB1 yB1 0 ];

xBn = subs(xB1,'phi(t)',pi/4); % xB for phi(t)=pi/4
yBn = subs(yB1,'phi(t)',pi/4); % yB for phi(t)=pi/4
rBn = subs(rB,'phi(t)',pi/4);  % rB for phi(t)=pi/4
fprintf('rB =  [ %g, %g, %g ] (m)\n', rBn)

% velocity of B1=B2
% differentiate rB with respect to t
vB = diff(rB,t);
% list for the symbolical variables:
% phi''(t), phi'(t), phi(t)
slist = ...
{diff('phi(t)',t,2), diff('phi(t)',t), 'phi(t)'};
% list for the numerical values
% of phi''(t), phi'(t), phi(t)
nlist = {alpha,  omega, phi1};
vBn = double(subs(vB,slist,nlist));
fprintf('vB1 = vB2 = [ %g, %g, %g ] (m/s)\n', vBn)
fprintf('|vB1| = |vB2| = %g (m/s)\n', norm(vBn))
% acceleration of B1=B2
% differentiate vB with respect to t
aB  = diff(vB,t);
```

```
aBn = double(subs(aB,slist,nlist));
fprintf('aB1 = aB2 = [ %g, %g, %g ] (m/s^2)\n', aBn)
fprintf('|aB1| = |aB2| = %g (m/s^2)\n', norm(aBn))
fprintf('\n')

xB = sym('xB(t)'); % xB(t) symbolic
yB = sym('yB(t)'); % yB(t) symbolic
% list for the symbolical variables of B
% xB''(t), yB''(t), xB'(t), yB'(t), xB(t), yB(t)
sB={diff('xB(t)',t,2),diff('yB(t)',t,2),...
    diff('xB(t)',t),diff('yB(t)',t),'xB(t)','yB(t)'};
% three dots (...)
% are used whenever a line break is needed
% list for the numerical values of the sB list
nB={aBn(1),aBn(2),vBn(1),vBn(2),xBn,yBn};

% angular velocity and acceleration of links 2 and 3
phi3 = atan((yB-yC)/(xB-xC));
phi3n = subs(phi3,sB,nB);
fprintf('phi2 = phi3 = %g (degrees)\n',...
    double(phi3n*180/pi))

dphi3 = diff(phi3,t);
dphi3n = subs(dphi3,sB,nB);
fprintf('omega2 = omega3 =  %g (rad/s) \n',...
    double(dphi3n))

ddphi3 = diff(dphi3,t);
ddphi3n = subs(ddphi3,sB,nB) ;
fprintf('alpha2 = alpha3 = %g (rad/s^2) \n',...
    double(ddphi3n))
fprintf('\n')

xD = eval(xC + CD*cos(phi3n));
yD = eval(yC + CD*sin(phi3n));

plot([xA,xBn],[yA,yBn],'r',[xC,xD],[yC,yD],'b'),
text(xA,yA,' A'), text(xBn,yBn,' B'),...
text(xC,yC,' C'), text(xD,yD,' D'), ...
axis([-0.05 0.3 -0.05 0.3])
% end of program
```

Results:

```
rB =  [ 0.0707107, 0.170711, 0 ] (m)
vB1 = vB2 = [ -0.222144, 0.222144, 0 ] (m/s)
```

```
|vB1| = |vB2| = 0.314159 (m/s)
aB1 = aB2 = [ -0.697886, -0.697886, 0 ] (m/s^2)
|aB1| = |aB2| = 0.98696 (m/s^2)

phi2 = phi3 = 67.5 (degrees)
omega2 = omega3 =  1.5708 (rad/s)
alpha2 = alpha3 = 2.40543e-16 (rad/s^2)
```

B.8 R-RRR Mechanism: Derivative Method

```
% B8
% R-RRR
% Velocity and acceleration analysis
% Derivative method
clear all; clc; close all
% Input data
AB=0.15; %(m)
BC=0.35; %(m)
CD=0.30; %(m)
CE=0.15; %(m)
xD=0.30; %(m)
yD=0.30; %(m)
phi = pi/4 ; %(rad)
xA = 0; yA = 0;
rA = [xA yA 0];
rD = [xD yD 0];
xB = AB*cos(phi);
yB = AB*sin(phi);
rB = [xB yB 0];
eqnC1='(xCsol-xB)^2+(yCsol-yB)^2=BC^2';
eqnC2='(xCsol-xD)^2+(yCsol-yD)^2=CD^2';
solC = solve(eqnC1, eqnC2, 'xCsol, yCsol');
xCpositions = eval(solC.xCsol);
yCpositions = eval(solC.yCsol);
xC1 = xCpositions(1); xC2 = xCpositions(2);
yC1 = yCpositions(1); yC2 = yCpositions(2);
if xC1 < xD
     xC = xC1; yC=yC1;
else xC = xC2; yC=yC2; end
rC = [xC yC 0];
eqnE1='(xEsol- xC)^2+(yEsol-yC)^2=CE^2';
eqnE2='(yD-yC)/(xD-xC)=(yEsol-yC)/(xEsol-xC)';
solE = solve(eqnE1, eqnE2, 'xEsol, yEsol');
```

```
xEpositions = eval(solE.xEsol);
yEpositions = eval(solE.yEsol);
xE1 = xEpositions(1); xE2 = xEpositions(2);
yE1 = yEpositions(1); yE2 = yEpositions(2);
if xE1 < xC
     xE = xE1; yE=yE1;
else xE = xE2; yE=yE2; end
rE = [xE yE 0];
phi2 = atan((yB-yC)/(xB-xC));
phi3 = atan((yD-yC)/(xD-xC));
fprintf('Results \n\n')
fprintf('Velocity and acceleration analysis\n')
fprintf('Derivative method \n\n')

t = sym('t','real');
n = 60; % (rpm)  driver link
omega1 = pi*n/30;
alpha1 = eval(diff(omega1, t));
fprintf('omega1 = %3.3f (rad/s)\n', omega1)
fprintf('alpha1 = %3.3f (rad/s^2)\n', alpha1)
fprintf('\n')
phi1 = sym('phi1(t)');
%position vector of B function of phi1(t)-symbolic
rBs = [ AB*cos(phi1) AB*sin(phi1) 0 ];
% velocity of B - symbolic
vB = diff(rBs,t);
slist=...
{diff('phi1(t)',t,2),diff('phi1(t)',t),'phi1(t)'};
%numerical values for slist
nlist={alpha1,  omega1, phi};
vBn = eval(subs(vB,slist,nlist));
fprintf...
('vB1 = vB2 = [ %3.3f, %3.3f, %3.3f ] (m/s)\n',vBn)
fprintf...
('|vB1| = |vB2| = %3.3f (m/s)\n', norm(vBn))
% differentiate vB with respect to t
aB  = diff(vB,t);
aBn = eval(subs(aB,slist,nlist));
fprintf...
('aB1 = aB2 = [ %3.3f, %3.3f, %3.3f ] (m/s^2)\n',aBn)
fprintf('|aB1| = |aB2| = %3.3f (m/s^2)\n', norm(aBn))
fprintf('\n')

xBs = sym('xBs(t)'); % xB  symbolic
yBs = sym('yBs(t)'); % yB  symbolic
```

```
% list for the symbolical variables of B
sB={diff('xBs(t)',t,2),diff('yBs(t)',t,2),...
    diff('xBs(t)',t),diff('yBs(t)',t),...
    'xBs(t)','yBs(t)'};
nB={aBn(1),aBn(2),vBn(1),vBn(2),xB,yB};

xCs = sym('xCs(t)'); % xC  symbolic
yCs = sym('yCs(t)'); % yC  symbolic

pC1=(xCs-xBs)^2+(yCs-yBs)^2-BC^2';
pC2=(xCs-xD )^2+(yCs-yD )^2-CD^2';

dpC1 = diff(pC1,t);
dpC2 = diff(pC2,t);

syms dxC dyC

dpC1s=subs(dpC1,...
{diff('xCs(t)',t),diff('yCs(t)',t)},{dxC, dyC});
dpC2s=subs(dpC2,...
{diff('xCs(t)',t),diff('yCs(t)',t)},{dxC, dyC});

soldC = solve(dpC1s, dpC2s,'dxC, dyC');
vxC = soldC.dxC;
vyC = soldC.dyC;

vC = [ vxC vyC 0 ];
spC = {xCs, yCs};
npC = {xC,  yC };

vCn =eval(subs(subs(vC, sB, nB), spC, npC));
fprintf...
('vC = [ %3.3f, %3.3f, %3.3f ] (m/s)\n', vCn)
fprintf('|vC| = %3.3f (m/s)\n', norm(vCn))

svC={diff('xCs(t)',t),diff('yCs(t)',t),...
    'xCs(t)','yCs(t)'};

nvC={vCn(1), vCn(2), rC(1), rC(2)};

ddpC1 = diff(pC1,t,2);
ddpC2 = diff(pC2,t,2);

syms ddxC ddyC
```

```
ddpC1s = subs(ddpC1,...
 {diff('xCs(t)',t,2),diff('yCs(t)',t,2)},...
  {ddxC, ddyC});
ddpC2s = subs(ddpC2,...
 {diff('xCs(t)',t,2),diff('yCs(t)',t,2)},...
  {ddxC, ddyC});

solddC = solve(ddpC1s, ddpC2s,'ddxC, ddyC');
axC = solddC.ddxC;
ayC = solddC.ddyC;

aC = [ axC, ayC, 0];

aCn =eval(subs(subs(aC, sB, nB), svC, nvC));

fprintf...
('aC = [ %3.3f, %3.3f, %3.3f ] (m/s^2)\n', aCn)
fprintf('|aC| = %3.3f (m/s^2)\n', norm(aCn))

sC={diff('xCs(t)',t,2),diff('yCs(t)',t,2),...
    diff('xCs(t)',t),diff('yCs(t)',t),...
    'xCs(t)','yCs(t)'};
nC={aCn(1),aCn(2),vCn(1),vCn(2),xC,yC};

fprintf('\n')
phi2s=atan((yCs-yBs)/(xCs-xBs));
phi2n = eval(subs(subs(phi2s,sB,nB),sC,nC));
fprintf('phi2 = %3.3f (rad)\n',phi2n)
omega2 = diff(phi2s,t);
omega2n = eval(subs(subs(omega2,sC,nC),sB,nB));
fprintf('omega2 = %3.3f (rad/s)\n', omega2n)
alpha2 = diff(omega2,t);
alpha2n = eval(subs(subs(alpha2,sC,nC),sB,nB));
fprintf('alpha2 = %3.3f (rad/s^2)\n',alpha2n)

fprintf('\n')
phi3s=atan((yCs-yD)/(xCs-xD));
phi3n = double(subs(phi3s,sC,nC));
fprintf('phi3 = %3.3f (rad)\n',phi3n)
omega3 = diff(phi3s,t);
omega3n = double(subs(omega3,sC,nC));
fprintf('omega3 = %3.3f (rad/s)\n',omega3n)
alpha3 = diff(omega3,t);
alpha3n = double(subs(alpha3,sC,nC));
fprintf('alpha3 = %3.3f (rad/s^2)\n',alpha3n)
```

```
fprintf('\n')
xEs = xCs-CE*cos(phi3s);
yEs = yCs-CE*sin(phi3s);

rEs=[xEs yEs 0];
vE = diff(rEs,t);
aE = diff(vE,t);

vEn = double(subs(subs(vE,sC,nC),sB,nB));
fprintf...
('vE = [ %3.3f, %3.3f, %2.3f ] (m/s)\n',vEn)
aEn = double(subs(subs(aE,sC,nC),sB,nB));
fprintf...
('aE = [ %3.3f, %3.3f, %2.3f ] (m/s^2)\n',aEn)

% Graph of the mechanism
plot([xA,xB],[yA,yB],'r-o',...
     [xB,xC],[yB,yC],'b-o',...
     [xD,xE],[yD,yE],'g-o'),...
xlabel('x (m)'), ylabel('y (m)'),...
title('positions for \phi = 45 (deg)'),...
text(xA,yA,'  A'), text(xB,yB,'  B'),...
text(xC,yC,'  C'), text(xD,yD,'  D'),...
text(xE,yE,'  E')
% end of program
```

Results:

```
omega1 = 6.283 (rad/s)
alpha1 = 0.000 (rad/s^2)

vB1 = vB2 = [ -0.666, 0.666, 0.000 ] (m/s)
|vB1| = |vB2| = 0.942 (m/s)
aB1 = aB2 = [ -4.187, -4.187, 0.000 ] (m/s^2)
|aB1| = |aB2| = 5.922 (m/s^2)

vC = [ 0.515, 0.893, 0.000 ] (m/s)
|vC| = 1.031 (m/s)
aC = [ -0.322, -7.654, 0.000 ] (m/s^2)
|aC| = 7.660 (m/s^2)

phi2 = -1.381 (rad)
omega2 = -3.436 (rad/s)
alpha2 = -8.979 (rad/s^2)
```

```
phi3 = -0.523 (rad)
omega3 = -3.436 (rad/s)
alpha3 = 22.640 (rad/s^2)

vE = [ 0.772, 1.340, 0.000 ] (m/s)
aE = [ -0.483, -11.481, 0.000 ] (m/s^2)
```

B.9 R-RTR-RTR Mechanism: Contour Method

```
% B9
% Velocity and acceleration analysis
% R-RTR-RTR
% Contour method

clear all; clc; close all

AB = 0.15 ; AC = 0.10 ; CD = 0.15 ; %(m)
DF = 0.40 ; AG = 0.30 ; % (m)
phi = pi/6 ;  % (rad)
xA = 0 ; yA = 0   ; rA = [xA yA 0] ;
xC = 0 ; yC = AC  ; rC = [xC yC 0] ;
xB = AB*cos(phi); yB = AB*sin(phi); rB = [xB yB 0];
eqnD1=' ( xDsol - xC )^2 + ( yDsol - yC )^2 = CD^2';
eqnD2=' (yB-yC)/(xB-xC)=(yDsol-yC)/(xDsol-xC)';
solD = solve(eqnD1, eqnD2, 'xDsol, yDsol');
xDpositions = eval(solD.xDsol);
yDpositions = eval(solD.yDsol);
xD1=xDpositions(1); xD2=xDpositions(2);
yD1=yDpositions(1); yD2=yDpositions(2);
if(phi>=0&&phi<=pi/2)||(phi>=3*pi/2&&phi<=2*pi)
 if xD1 <= xC xD=xD1; yD=yD1; else xD=xD2; yD=yD2;
 end
   else
 if xD1 >= xC xD=xD1; yD=yD1; else xD=xD2; yD=yD2;
 end
end
rD=[xD yD 0];
phi2=atan((yB-yC)/(xB-xC)); phi3=phi2;
phi4=atan(yD/xD)+pi; phi5=phi4;
xF=xD+DF*cos(phi3); yF=yD+ DF*sin(phi3);
rF=[xF yF 0];
xG=AG*cos(phi5); yG=AG*sin(phi5);
rG = [xG yG 0];
```

```
fprintf('Results \n\n')

fprintf('rA = [ %g, %g, %g ] (m)\n', rA)
fprintf('rC = [ %g, %g, %g ] (m)\n', rC)
fprintf('rB = [ %g, %g, %g ] (m)\n', rB)
fprintf('rD = [ %g, %g, %g ] (m)\n', rD)
fprintf...
('phi2 = phi3 = %g (degrees) \n',phi2*180/pi)
fprintf...
('phi4 = phi5 = %g (degrees) \n',phi4*180/pi)
fprintf('rF = [ %g, %g, %g] (m)\n', rF)
fprintf('rG = [ %g, %g, %g] (m)\n', rG)

% Graphic of the mechanism
plot([xA,xB],[yA,yB],'k-o',[xD,xF],[yD,yF],'b-o',...
[xA,xG],[yA,yG],'r-o')
xlabel('x (m)'), ylabel('y (m)'),...
title('positions for \phi = 30 (deg)'),...
text(xA,yA,' A'), text(xB,yB,' B'),...
text(xC,yC,' C'), text(xD,yD,' D'),...
text(xF,yF,' F'), text(xG,yG,' G'),...
axis([-0.3 0.3 -0.1 0.3]), grid

fprintf('\n')
fprintf('Velocity and acceleration analysis\n\n')
fprintf('Contour method \n')
fprintf('\n')

n = 50.;
omega1 = [ 0 0 pi*n/30 ]; alpha1 = [0 0 0 ];
fprintf...
('omega1 = [ %g, %g, %g ] (rad/s)\n', omega1)
fprintf...
('alpha1 = [ %g, %g, %g ] (rad/s^2)\n', alpha1)
fprintf('\n')

vA = [0 0 0 ]; aA = [0 0 0 ];

vB1 = vA + cross(omega1,rB); vB2 = vB1;
aB1 = aA + cross(alpha1,rB) - ...
    dot(omega1,omega1)*rB;

fprintf...
('vB = vB1 = vB2 = [ %g, %g, %g ] (m/s)\n', vB1)
```

```
aB2 = aB1;
fprintf...
('aB = aB1 = aB2 = [ %g, %g, %g ] (m/s^2)\n', aB1)

fprintf('\n')
fprintf('Contour I  \n\n')
fprintf('Relative velocities \n')
fprintf('\n')

omega10 = omega1;
omega21v = [ 0 0 sym('omega21z','real') ];
omega03v = [ 0 0 sym('omega03z','real') ];
v32v = sym('vB32','real')*[ cos(phi2) sin(phi2) 0];

eqIomega = omega10 + omega21v + omega03v;
eqIvz=eqIomega(3);
eqIv = ...
 cross(rB,omega21v) + cross(rC,omega03v) + v32v;
eqIvx=eqIv(1);
eqIvy=eqIv(2);

Ivz=vpa(eqIvz,6);
fprintf('%s = 0 \n', char(Ivz))
Ivx=vpa(eqIvx,6);
fprintf('%s = 0 \n', char(Ivx))
Ivy=vpa(eqIvy,6);
fprintf('%s = 0 \n', char(Ivy))

solIv=solve(eqIvz,eqIvx,eqIvy);
omega21 = [ 0 0 eval(solIv.omega21z) ];
omega03 = [ 0 0 eval(solIv.omega03z) ];
vB3B2 = eval(solIv.vB32)*[ cos(phi2) sin(phi2) 0];

fprintf...
('omega21 = [ %g, %g, %g ] (rad/s)\n', omega21)
fprintf...
('omega03 = [ %g, %g, %g ] (rad/s)\n', omega03)
fprintf('vB32 = %g (m/s)\n', eval(solIv.vB32))
fprintf('vB3B2 = [ %g, %g, %d ] (m/s)\n', vB3B2)
fprintf('\n')

fprintf('Absolute velocities \n\n')
omega30 = - omega03;
omega20 = omega30;
vC = [0 0 0 ];
```

```
vD3 = vC + cross(omega30,rD-rC);
fprintf...
('omega20=omega30= [%d, %d, %g] (rad/s)\n',omega30)
fprintf('vD3 = vD4 = [ %g, %g, %g ] (m/s)\n', vD3)

fprintf('\n')
fprintf('Relative accelerations \n')
fprintf('\n')

alpha10 = alpha1;
alpha21v = [ 0 0 sym('alpha21z','real') ];
alpha03v = [ 0 0 sym('alpha03z','real') ];
a32v = sym('aB32','real')*[ cos(phi2) sin(phi2) 0];

eqIalpha = alpha10 + alpha21v + alpha03v;
eqIaz=eqIalpha(3);
eqIa=cross(rB,alpha21v)+cross(rC,alpha03v)+...
 a32v+2*cross(omega20,vB3B2)-...
 dot(omega1,omega1)*rB-dot(omega20,omega20)*(rC-rB);
eqIax=eqIa(1);
eqIay=eqIa(2);

Iaz=vpa(eqIaz,6);
fprintf('%s=0 \n',char(Iaz))
Iax=vpa(eqIax,3);
fprintf('%s=0 \n',char(Iax))
Iay=vpa(eqIay,6);
fprintf('%s=0 \n',char(Iay))

solIa=solve(eqIaz,eqIax,eqIay);
alpha21 = [ 0 0 eval(solIa.alpha21z) ];
alpha03 = [ 0 0 eval(solIa.alpha03z) ];
aB3B2 = eval(solIa.aB32)*[ cos(phi2) sin(phi2) 0];

fprintf...
('alpha21 = [ %g, %g, %g ] (rad/s^2)\n', alpha21)
fprintf...
('alpha03 = [ %g, %g, %g ] (rad/s^2)\n', alpha03)
fprintf('aB32 = %g (m/s^2)\n', eval(solIa.aB32))
fprintf('aB3B2 = [ %g, %g, %d ] (m/s^2)\n', aB3B2)
fprintf('\n')

fprintf('Absolute accelerations \n\n')
alpha30 = - alpha03;
alpha20 = alpha30;
```

```
aC = [0 0 0 ];
aD3=aC+cross(alpha30,rD-rC)-...
    dot(omega20,omega20)*(rD-rC);
fprintf...
('alpha20=alpha30=[%d,%d,%g] (rad/s^2)\n',alpha30)
fprintf('aD3 = aD4 = [ %g, %g, %g ] (m/s^2)\n', aD3)

fprintf('\n')
fprintf('Contour II \n\n')
fprintf('Relative velocities \n')
fprintf('\n')

omega43v = [ 0 0 sym('omega43z','real') ];
omega05v = [ 0 0 sym('omega05z','real') ];
v54v = sym('vD54','real')*[ cos(phi4) sin(phi4) 0];

eqIIomega = omega30 + omega43v + omega05v;
eqIIvz=eqIIomega(3);
eqIIv=cross(rC,omega30)+cross(rD,omega43v)+v54v;
eqIIvx=eqIIv(1);
eqIIvy=eqIIv(2);

IIvz=vpa(eqIIvz,6); fprintf('%s = 0 \n', char(IIvz))
IIvx=vpa(eqIIvx,6); fprintf('%s = 0 \n', char(IIvx))
IIvy=vpa(eqIIvy,6); fprintf('%s = 0 \n', char(IIvy))

solIIv=solve(eqIIvz,eqIIvx,eqIIvy);
omega43 = [ 0 0 eval(solIIv.omega43z) ];
omega05 = [ 0 0 eval(solIIv.omega05z) ] ;
vD5D4 = eval(solIIv.vD54)*[ cos(phi4) sin(phi4) 0];

fprintf...
('omega43 = [ %g, %g, %g ] (rad/s)\n', omega43)
fprintf...
('omega05 = [ %g, %g, %g ] (rad/s)\n', omega05)
fprintf('vD54 = %g (m/s)\n', eval(solIIv.vD54))
fprintf('vD5D4 = [ %g, %g, %d ] (m/s)\n', vD5D4)
fprintf('\n')

fprintf('Absolute velocities \n\n')
omega50 = - omega05;
omega40 = omega50;
fprintf...
('omega40=omega50=[%d, %d, %g] (rad/s)\n',omega50)
```

```
fprintf('\n')
fprintf('Relative accelerations \n\n')

alpha43v = [ 0 0 sym('alpha43z','real') ];
alpha05v = [ 0 0 sym('alpha05z','real') ];
a54v = sym('aD54','real')*[ cos(phi4) sin(phi4) 0];

eqIIalpha = alpha30 + alpha43v + alpha05v;
eqIIaz=eqIIalpha(3);
eqIIa=cross(rC,alpha30)+cross(rD,alpha43v)+...
    a54v+2*cross(omega40,vD5D4)-...
    dot(omega30,omega30)*(rD-rC)-...
    dot(omega40,omega40)*(-rD);
eqIIax=eqIIa(1);
eqIIay=eqIIa(2);

IIaz=vpa(eqIIaz,6); fprintf('%s = 0 \n', char(IIaz))
IIax=vpa(eqIIax,6); fprintf('%s = 0 \n', char(IIax))
IIay=vpa(eqIIay,6); fprintf('%s = 0 \n', char(IIay))

solIIa=solve(eqIIaz,eqIIax,eqIIay);
alpha43 = [ 0 0 eval(solIIa.alpha43z) ];
alpha05 = [ 0 0 eval(solIIa.alpha05z) ] ;
aD5D4 = eval(solIIa.aD54)*[ cos(phi4) sin(phi4) 0];

fprintf...
('alpha43 = [ %g, %g, %g ] (rad/s^2)\n', alpha43)
fprintf...
('alpha05 = [ %g, %g, %g ] (rad/s^2)\n', alpha05)
fprintf('aD54 = %g (m/s^2)\n', eval(solIIa.aD54))
fprintf('aD5D4 = [ %g, %g, %d ] (m/s^2)\n', aD5D4)
fprintf('\n')

fprintf('Absolute accelerations \n')
fprintf('\n')
alpha50 = - alpha05;
alpha40 = alpha50;

fprintf...
('alpha40=alpha50=[%d, %d, %g] (rad/s^2)\n',alpha50)

% end of program
```

Results:

```
rA = [ 0, 0, 0 ] (m)
```

```
rC = [ 0, 0.1, 0 ] (m)
rB = [ 0.129904, 0.075, 0 ] (m)
rD = [ -0.147297, 0.128347, 0 ] (m)
phi2 = phi3 = -10.8934 (degrees)
phi4 = phi5 = 138.933 (degrees)
rF = [ 0.245495, 0.0527544, 0] (m)
rG = [ -0.226182, 0.197083, 0] (m)

Velocity and acceleration analysis

Contour method

omega1 = [ 0, 0, 5.23599 ] (rad/s)
alpha1 = [ 0, 0, 0 ] (rad/s^2)

vB = vB1 = vB2 = [ -0.392699, 0.680175, 0 ] (m/s)
aB = aB1 = aB2 = [ -3.56139, -2.05617, 0 ] (m/s^2)


Contour I

Relative velocities

5.23599+omega21z+omega03z = 0
.750000e-1*omega21z+.100000*omega03z+.981983*vB32 = 0
-.129904*omega21z-.188982*vB32 = 0
omega21 = [ 0, 0, -0.747998 ] (rad/s)
omega03 = [ 0, 0, -4.48799 ] (rad/s)
vB32 = 0.514164 (m/s)
vB3B2 = [ 0.504899, -0.0971678, 0 ] (m/s)

Absolute velocities

omega20=omega30= [0, 0, 4.48799] (rad/s)
vD3 = vD4 = [ -0.127223, -0.661068, 0 ] (m/s)

Relative accelerations

alpha21z+alpha03z=0
.750e-1*alpha21z+.100*alpha03z+.980*aB32-.727e-1=0
-.129904*alpha21z-.188982*aB32+1.97224=0
alpha21 = [ 0, 0, 14.5363 ] (rad/s^2)
alpha03 = [ 0, 0, -14.5363 ] (rad/s^2)
aB32 = 0.44409 (m/s^2)
aB3B2 = [ 0.436088, -0.0839252, 0 ] (m/s^2)
```

```
Absolute accelerations

alpha20=alpha30=[0,0,14.5363] (rad/s^2)
aD3 = aD4 = [ 2.5548, -2.71212, 0 ] (m/s^2)

Contour II

Relative velocities

4.48798+omega43z+omega05z = 0
.448798+.128347*omega43z-.753939*vD54 = 0
.147297*omega43z+.656945*vD54 = 0
omega43 = [ 0, 0, -1.50912 ] (rad/s)
omega05 = [ 0, 0, -2.97887 ] (rad/s)
vD54 = 0.338367 (m/s)
vD5D4 = [ -0.255108, 0.222288, 0 ] (m/s)

Absolute velocities

omega40=omega50=[0, 0, 2.97887] (rad/s)

Relative accelerations

14.5363+alpha43z+alpha05z = 0
1.78909+.128347*alpha43z-.753939*aD54 = 0
.147297*alpha43z+.656945*aD54-.951928 = 0
alpha43 = [ 0, 0, -2.3424 ] (rad/s^2)
alpha05 = [ 0, 0, -12.1939 ] (rad/s^2)
aD54 = 1.97423 (m/s^2)
aD5D4 = [ -1.48845, 1.29696, 0 ] (m/s^2)

Absolute accelerations

alpha40=alpha50=[0, 0, 12.1939] (rad/s^2)
```

Appendix C
Programs of Chapter 4: Dynamic Force Analysis

C.1 Slider-Crank (R-RRT) Mechanism: Newton–Euler Method

```
% C1
% Dynamic force analysis
% R-RRT
% Newton-Euler method
clear all; clc; close all
format long
AB = 1; BC = 1; phi = 45*pi/180;
xA = 0; yA = 0; rA = [xA yA 0];
xB = AB*cos(phi); yB = AB*sin(phi); rB = [xB yB 0];
yC = 0; xC = xB+sqrt(BC^2-(yC-yB)^2); rC = [xC yC 0];
phi2 = atan((yB-yC)/(xB-xC));
fprintf('Results \n\n')
fprintf('phi = phi1 = %g (degrees) \n', phi*180/pi)
fprintf('rA = [ %g, %g, %g ] (m)\n', rA)
fprintf('rB = [ %g, %g, %g ] (m)\n', rB)
fprintf('rC = [ %g, %g, %g ] (m)\n', rC)
n = 30/pi; % (rpm)  driver link
omega1 = [ 0 0 pi*n/30 ]; alpha1 = [0 0 0 ];
fprintf...
('alpha1 = [ %g, %g, %g ] (rad/s^2)\n', alpha1)
vA = [0 0 0 ]; aA = [0 0 0 ];
vB1 = vA + cross(omega1,rB); vB2 = vB1;
aB1 = aA + cross(alpha1,rB) - dot(omega1,omega1)*rB;
aB2 = aB1;
omega2z = sym('omega2z','real');
vCx = sym('vCx','real');
omega2 = [ 0 0 omega2z ]; vC = [ vCx 0 0 ];
eqvC = vC - (vB2 + cross(omega2,rC-rB));
```

```
eqvCx = eqvC(1); eqvCy = eqvC(2);
solvC = solve(eqvCx,eqvCy);
omega2zs=eval(solvC.omega2z);
vCxs=eval(solvC.vCx); Omega2 = [0 0 omega2zs];
vCs = [vCxs 0 0];
alpha2z = sym('alpha2z','real');
aCx = sym('aCx','real');
alpha2 = [ 0 0 alpha2z ]; aC = [aCx 0 0 ];
eqaC = aC - (aB1 + cross(alpha2,rC-rB) - ...
    dot(Omega2,Omega2)*(rC-rB));
eqaCx = eqaC(1); eqaCy = eqaC(2);
solaC = solve(eqaCx,eqaCy);
alpha2zs=eval(solaC.alpha2z); aCxs=eval(solaC.aCx);
alpha20 = [0 0 alpha2zs]; aCs = [aCxs 0 0];
fprintf...
('alpha2 = [ %g, %g, %g ] (rad/s^2)\n', alpha20)
alpha30 = [0 0 0];
fprintf...
('alpha3 = [ %g, %g, %g ] (rad/s^2)\n', alpha30)
fprintf('\n')

fprintf...
('Positions and accelerations for mass centers \n')
fprintf('\n')
rC1 = (rA+rB)/2;
fprintf('rC1 = [ %g, %g, %g ] (m)\n', rC1)
rC2 = (rB+rC)/2;
fprintf('rC2 = [ %g, %g, %g ] (m)\n', rC2)
rC3 = rC;
fprintf('rC3 = [ %g, %g, %g ] (m)\n', rC3)

% Graphic of the mechanism
plot([0,xB],[0,yB],'r-o',[xB,xC],[yB,yC],'b-o')
xlabel('x (m)'), ylabel('y (m)'),...
title('positions for \phi = 45 (deg)'),...
text(xA,yA,'  A'),text(xB,yB,'  B'),...
text(xC,yC,'  C=C3'),...
text(rC1(1),rC1(2),' C1'),...
text(rC2(1),rC2(2),' C2'),...
axis([-0.1,1.6,-0.1,1.6])

aC1 = aB1/2;
fprintf('aC1 = [ %g, %g, %g ] (m/s^2)\n', aC1)
aC2 = (aB1+aCs)/2;
fprintf('aC2 = [ %g, %g, %g ] (m/s^2)\n', aC2)
```

```
aC3 = aCs;
fprintf('aC3 = [ %g, %g, %g ] (m/s^2)\n', aC3)

fprintf('\n')
% external force
fe = 100;
Fe = -sign(vCs(1))*[fe 0 0];
fprintf('external force Fe=[ %d, %d, %g ] (N)\n',Fe)

h = 0.01; % height of the bar (m)
d = 0.001; % depth of the bar (m)
hSlider = 0.01; % height of the slider (m)
wSlider = 0.01; % depth of the slider (m)
g = 10.; % gravitational acceleration (m/s^2)

fprintf('\n')
fprintf('Inertia forces and moments \n\n')

fprintf(' link 1 \n')
m1 = 1;
IC1 = m1*(AB^2+h^2)/12;
G1 = [ 0 -m1*g 0 ];
Fin1 = - m1*aC1;
Min1 = - IC1*alpha1;
fprintf('m1 = %g (kg)\n', m1)
fprintf('IC1 = %g (kg m^2)\n', IC1)
fprintf('G1 = [ %d, %g, %d ] (N)\n', G1)
fprintf('Fin1 = [ %g, %g, %d ] (N)\n', Fin1)
fprintf('Min1 = [ %d, %d, %d ] (N m)\n', Min1)

fprintf(' link 2 \n')
m2 = 1;
IC2 = m2*(BC^2+h^2)/12;
G2 = [ 0 -m2*g 0 ];
Fin2 = - m2*aC2;
Min2 = - IC2*alpha20;
fprintf('m2 = %g (kg)\n', m2)
fprintf('IC2 = %g (kg m^2)\n', IC2)
fprintf('G2 = [ %d, %g, %d ] (N)\n', G2)
fprintf('Fin2 = [ %g, %g, %d ] (N)\n', Fin2)
fprintf('Min2 = [ %d, %d, %d ] (N m)\n', Min2)

fprintf(' link 3 \n')
m3 = 1;
IC3 = m3*(hSlider^2+wSlider^2)/12;
```

```
G3 = [ 0  -m3*g  0 ];
Fin3 = - m3*aC3;
Min3 = - IC3*alpha30;
fprintf('m3 = %g (kg)\n', m3)
fprintf('IC3 = %g (kg m^2)\n', IC3)
fprintf('G3 = [ %d, %g, %d ] (N)\n', G3)
fprintf('Fin3 = [ %g, %g, %d ] (N)\n', Fin3)
fprintf('Min3 = [ %d, %d, %d ] (N m)\n', Min3)

fprintf('\n')
fprintf('Dynamic force analysis  \n')
fprintf('Newton-Euler eom  \n\n')

% eom link 3
F03 = [ 0 sym('F03y','real') 0 ];
F23 = [ sym('F23x','real') sym('F23y','real') 0 ];
% sum of the forces for link 3
eqF3 = F03+F23+Fe+G3-m3*aC3;
eqF3x = eqF3(1);
eqF3y = eqF3(2);
fprintf('%s = 0 (1)\n', char(vpa(eqF3x,6)))
fprintf('%s = 0 (2)\n', char(vpa(eqF3y,6)))
% eom link 2
F32 = -F23;
F12 = [ sym('F12x','real') sym('F12y','real') 0 ];
% sum of the forces for link 2
eqF2 = F32+F12+G2-m2*aC2;
eqF2x = eqF2(1);
eqF2y = eqF2(2);
% sum of the moments for link 2 wrt C2
eqM2 = cross(rB-rC2,F12)+cross(rC-rC2,F32)-...
       IC2*alpha20;
eqM2z = eqM2(3);
fprintf('%s = 0 (3)\n', char(vpa(eqF2x,6)))
fprintf('%s = 0 (4)\n', char(vpa(eqF2y,6)))
fprintf('%s = 0 (5)\n', char(vpa(eqM2z,3)))
fprintf('\n')
fprintf...
('Eqs(1)-(5) => F03y, F23x, F23y, F12x, F12y \n')
sol32=solve(eqF3x,eqF3y,eqF2x,eqF2y,eqM2z);
F03ys=eval(sol32.F03y);
F23xs=eval(sol32.F23x);
F23ys=eval(sol32.F23y);
F12xs=eval(sol32.F12x);
F12ys=eval(sol32.F12y);
```

```
F03s = [ 0, F03ys, 0 ];
F23s = [ F23xs, F23ys, 0 ];
F12s = [ F12xs, F12ys, 0 ];

fprintf('F03 = [ %g, %g, %g ] (N)\n', F03s )
fprintf('F23 = [ %g, %g, %g ] (N)\n', F23s )
fprintf('F12 = [ %g, %g, %g ] (N)\n', F12s )
fprintf('\n')

% eom link 1
% sum of the forces for 1
F01=m1*aC1-G1+F12s;
% sum of the moments for 1 wrt A
Mm=IC1*alpha1+cross(rC1,m1*aC1-G1)-cross(rB,-F12s);
fprintf('F01 = [ %g, %g, %g ] (N)\n', F01 )
fprintf('Mm = [ %d, %d, %g ] (N m)\n', Mm )
% end of program
```

Results:

```
phi = phi1 = 45 (degrees)
rA = [ 0, 0, 0 ] (m)
rB = [ 0.707107, 0.707107, 0 ] (m)
rC = [ 1.41421, 0, 0 ] (m)
alpha1 = [ 0, 0, 0 ] (rad/s^2)
alpha2 = [ 0, 0, 0 ] (rad/s^2)
alpha3 = [ 0, 0, 0 ] (rad/s^2)

Positions and accelerations for mass centers

rC1 = [ 0.353553, 0.353553, 0 ] (m)
rC2 = [ 1.06066, 0.353553, 0 ] (m)
rC3 = [ 1.41421, 0, 0 ] (m)
aC1 = [ -0.353553, -0.353553, 0 ] (m/s^2)
aC2 = [ -1.06066, -0.353553, 0 ] (m/s^2)
aC3 = [ -1.41421, 0, 0 ] (m/s^2)

external force Fe=[ 100, 0, 0 ] (N)

Inertia forces and moments

 link 1
m1 = 1 (kg)
IC1 = 0.0833417 (kg m^2)
G1 = [ 0, -10, 0 ] (N)
Fin1 = [ 0.353553, 0.353553, 0 ] (N)
```

```
Min1 = [ 0, 0, 0 ] (N m)
 link 2
m2 = 1 (kg)
IC2 = 0.0833417 (kg m^2)
G2 = [ 0, -10, 0 ] (N)
Fin2 = [ 1.06066, 0.353553, 0 ] (N)
Min2 = [ 0, 0, 0 ] (N m)
 link 3
m3 = 1 (kg)
IC3 = 1.66667e-05 (kg m^2)
G3 = [ 0, -10, 0 ] (N)
Fin3 = [ 1.41421, -0, 0 ] (N)
Min3 = [ 0, 0, 0 ] (N m)

Dynamic force analysis
Newton-Euler eom

F23x+101.414 = 0 (1)
F03y+F23y-10. = 0 (2)
-1.*F23x+F12x+1.06066 = 0 (3)
-1.*F23y+F12y-9.64645 = 0 (4)
-.352*F12y-.352*F12x-.352*F23y-.352*F23x = 0 (5)

Eqs(1)-(5) => F03y, F23x, F23y, F12x, F12y
F03 = [ 0, -87.1213, 0 ] (N)
F23 = [ -101.414, 97.1213, 0 ] (N)
F12 = [ -102.475, 106.768, 0 ] (N)

F01 = [ -102.828, 116.414, 0 ] (N)
Mm = [ 0, 0, 151.492 ] (N m)
```

C.2 Slider-Crank (R-RRT) Mechanism: D'Alembert's Principle

```
% C2
% Dynamic force analysis
% R-RRT
% D'Alembert Principle
clear all; clc; close all
format long
AB = 1; BC = 1;
phi_input = 45; phi = phi_input*(pi/180);
xA = 0; yA = 0; rA = [xA yA 0];
xB = AB*cos(phi); yB = AB*sin(phi);
```

```
rB = [xB yB 0];
yC = 0; xC = xB+sqrt(BC^2-(yC-yB)^2);
rC = [xC yC 0];
phi2 = atan((yB-yC)/(xB-xC));
n = 30/pi; % (rpm)  driver link
omega1 = [ 0 0 pi*n/30 ]; alpha1 = [0 0 0 ];
vA = [0 0 0 ]; aA = [0 0 0 ];
vB1 = vA + cross(omega1,rB); vB2 = vB1;
aB1 = aA + cross(alpha1,rB) - ...
    dot(omega1,omega1)*rB;
aB2 = aB1;
omega2z = sym('omega2z','real');
vCx = sym('vCx','real');
omega2 = [ 0 0 omega2z ]; vC = [ vCx 0 0 ];
eqvC = vC - (vB2 + cross(omega2,rC-rB));
eqvCx = eqvC(1); eqvCy = eqvC(2);
solvC = solve(eqvCx,eqvCy);
omega2zs=eval(solvC.omega2z);
vCxs=eval(solvC.vCx); Omega2 = [0 0 omega2zs];
vCs = [vCxs 0 0];
alpha2z = sym('alpha2z','real');
aCx = sym('aCx','real');
alpha2 = [ 0 0 alpha2z ]; aC = [aCx 0 0 ];
eqaC=aC-(aB1+cross(alpha2,rC-rB)-...
    dot(Omega2,Omega2)*(rC-rB));
eqaCx = eqaC(1); eqaCy = eqaC(2);
solaC = solve(eqaCx,eqaCy);
alpha2zs=eval(solaC.alpha2z);
aCxs=eval(solaC.aCx);
alpha20 = [0 0 alpha2zs]; aCs = [aCxs 0 0];
alpha30 = [0 0 0];

rC1 = (rA+rB)/2;
rC2 = (rB+rC)/2;
rC3 = rC;

% Graphic of the mechanism
plot([0,xB],[0,yB],'r-o',[xB,xC],[yB,yC],'b-o')
xlabel('x (m)'), ylabel('y (m)'),...
title('positions for \phi = 45 (deg)'),...
text(xA,yA,'  A'),text(xB,yB,'  B'),...
text(xC,yC,'  C=C3'),...
text(rC1(1),rC1(2),' C1'),...
text(rC2(1),rC2(2),' C2'),...
axis([-0.1,1.6,-0.1,1.6])
```

```
aC1 = aB1/2;
aC2 = (aB1+aCs)/2;
aC3 = aCs;

h = 0.01; d = 0.01; hSlider = 0.01; wSlider = 0.01;
g = 10.; % gravitational acceleration

m1 = 1;
IC1 = m1*(AB^2+h^2)/12;
G1 = [ 0  -m1*g  0 ];
Fin1 = - m1*aC1;
Min1 = - IC1*alpha1;

m2 = 1;
IC2 = m2*(BC^2+h^2)/12;
G2 = [ 0  -m2*g  0 ];
Fin2 = - m2*aC2;
Min2 = - IC2*alpha20;

m3 = 1 ;
IC3 = m3*(hSlider^2+wSlider^2)/12;
G3 = [ 0  -m3*g  0 ];
Fin3 = - m3*aC3;
Min3 = - IC3*alpha30;

% external force
fe = 100; Fe = -sign(vCs(1))*[fe 0 0];

fprintf('Results \n\n')
fprintf('Dynamic force analysis  \n')
fprintf('D Alembert Principle  \n\n')

% eom link 3
F03 = [ 0 sym('F03y','real') 0 ];
F23 = [ sym('F23x','real') sym('F23y','real') 0 ];
% sum of the forces for link 3
eqF3 = F03+F23+Fe+G3+Fin3 ;
eqF3x = eqF3(1);
eqF3y = eqF3(2);
fprintf('%s = 0 (1)\n', char(vpa(eqF3x,6)))
fprintf('%s = 0 (2)\n', char(vpa(eqF3y,6)))

% eom link 2
F32 = -F23;
```

```
F12 = [ sym('F12x','real') sym('F12y','real') 0 ];
% sum of the forces for link 2
eqF2 = F32+F12+G2+Fin2;
eqF2x = eqF2(1);
eqF2y = eqF2(2);
% sum of the moments for link 2 wrt C2
eqM2 = cross(rB-rC2,F12)+cross(rC-rC2,F32)+Min2;
eqM2z = eqM2(3) ;
fprintf('%s = 0 (3)\n', char(vpa(eqF2x,6)))
fprintf('%s = 0 (4)\n', char(vpa(eqF2y,6)))
fprintf('%s = 0 (5)\n', char(vpa(eqM2z,3)))

% eom link 1
F01 = [ sym('F01x','real') sym('F01y','real') 0 ];
Mm = [ 0 0 sym('Mmz','real') ];
% sum of the forces for 1
eqF1 = F01+Fin1+G1-F12;
eqF1x = eqF1(1);
eqF1y = eqF1(2);
% sum of the moments for 1 wrt C1
eqM1 = cross(rB-rC1,-F12)+cross(rA-rC1,F01)+...
    Min1+Mm;
eqM1z = eqM1(3) ;
fprintf('%s = 0 (6)\n', char(vpa(eqF1x,6)))
fprintf('%s = 0 (7)\n', char(vpa(eqF1y,6)))
fprintf('%s = 0 (8)\n', char(vpa(eqM1z,3)))
fprintf('\n')

fprintf('Eqs(1)-(8) => \n')
fprintf...
('F03y, F23x, F23y, F12x, F12y, F01x, F01y, Mmz \n')


sol321 = solve(eqF3x,eqF3y,eqF2x,eqF2y,eqM2z,...
               eqF1x,eqF1y,eqM1z);

F03ys = eval(sol321.F03y);
F23xs = eval(sol321.F23x);
F23ys = eval(sol321.F23y);
F12xs = eval(sol321.F12x);
F12ys = eval(sol321.F12y);
F01xs = eval(sol321.F01x);
F01ys = eval(sol321.F01y);
Mmzs = eval(sol321.Mmz);
```

```
F03s = [ 0, F03ys, 0 ];
F23s = [ F23xs, F23ys, 0 ];
F12s = [ F12xs, F12ys, 0 ];
F01s = [ F01xs, F01ys, 0 ];
Mms = [ 0, 0, Mmzs ];

fprintf('\n')
fprintf('F03 = [ %g, %g, %g ] (N)\n', F03s)
fprintf('F23 = [ %g, %g, %g ] (N)\n', F23s)
fprintf('F12 = [ %g, %g, %g ] (N)\n', F12s)
fprintf('F01 = [ %g, %g, %g ] (N)\n', F01s)
fprintf('Mm = [ %g, %g, %g ] (N m)\n', Mms)
% end of program
```

Results:

```
F23x+101.414 = 0 (1)
F03y+F23y-10. = 0 (2)
-1.*F23x+F12x+1.06066 = 0 (3)
-1.*F23y+F12y-9.64645 = 0 (4)
-.352*F12y-.352*F12x-.352*F23y-.352*F23x = 0 (5)
F01x+.353552-1.*F12x = 0 (6)
F01y-9.64645-1.*F12y = 0 (7)
-.352*F12y+.352*F12x-.352*F01y+.352*F01x+Mmz = 0 (8)

Eqs(1)-(8) =>
F03y, F23x, F23y, F12x, F12y, F01x, F01y, Mmz

F03 = [ 0, -87.1213, 0 ] (N)
F23 = [ -101.414, 97.1213, 0 ] (N)
F12 = [ -102.475, 106.768, 0 ] (N)
F01 = [ -102.828, 116.414, 0 ] (N)
Mm = [ 0, 0, 151.492 ] (N m)
```

C.3 Slider-Crank (R-RRT) Mechanism: Dyad Method

```
% C3
% Dynamic force analysis
% R-RRT
% Dyad method
clear all; clc; close all
format long
AB = 1; BC = 1; phi = 45*pi/180;
xA = 0; yA = 0; rA = [xA yA 0];
```

```
xB = AB*cos(phi); yB = AB*sin(phi);
rB = [xB yB 0];
yC = 0; xC = xB+sqrt(BC^2-(yC-yB)^2);
rC = [xC yC 0];
phi2 = atan((yB-yC)/(xB-xC));
fprintf('Results \n\n')
fprintf('phi = phi1 = %g (degrees) \n', phi*180/pi)
fprintf('rA =  [ %g, %g, %g ] (m)\n', rA)
fprintf('rB =  [ %g, %g, %g ] (m)\n', rB)
fprintf('rC =  [ %g, %g, %g ] (m)\n', rC)
n = 30/pi; % (rpm)  driver link
omega1 = [ 0 0 pi*n/30 ]; alpha1 = [0 0 0 ];
fprintf...
('alpha1 = [ %g, %g, %g ] (rad/s^2)\n',alpha1)
vA = [0 0 0 ]; aA = [0 0 0 ];
vB1 = vA + cross(omega1,rB); vB2 = vB1;
aB1 = aA + cross(alpha1,rB) - ...
     dot(omega1,omega1)*rB;
aB2 = aB1;
omega2z = sym('omega2z','real');
vCx = sym('vCx','real');
omega2 = [ 0 0 omega2z ]; vC = [ vCx 0 0 ];
eqvC = vC - (vB2 + cross(omega2,rC-rB));
eqvCx = eqvC(1); eqvCy = eqvC(2);
solvC = solve(eqvCx,eqvCy);
omega2zs=eval(solvC.omega2z);
vCxs=eval(solvC.vCx); Omega2 = [0 0 omega2zs];
vCs = [vCxs 0 0];
alpha2z = sym('alpha2z','real');
aCx = sym('aCx','real');
alpha2 = [ 0 0 alpha2z ]; aC = [aCx 0 0 ];
eqaC = aC - (aB1 + cross(alpha2,rC-rB) - ...
     dot(Omega2,Omega2)*(rC-rB));
eqaCx = eqaC(1); eqaCy = eqaC(2);
solaC = solve(eqaCx,eqaCy);
alpha2zs=eval(solaC.alpha2z);
aCxs=eval(solaC.aCx);
alpha20 = [0 0 alpha2zs]; aCs = [aCxs 0 0];
fprintf...
('alpha2 = [ %g, %g, %g ] (rad/s^2)\n',alpha20)
alpha30 = [0 0 0];
fprintf...
('alpha3 = [ %g, %g, %g ] (rad/s^2)\n',alpha30)
fprintf('\n')
```

```
rC1 = (rA+rB)/2;
fprintf('rC1 = [ %g, %g, %g ] (m)\n', rC1)
rC2 = (rB+rC)/2;
fprintf('rC2 = [ %g, %g, %g ] (m)\n', rC2)
rC3 = rC;
fprintf('rC3 = [ %g, %g, %g ] (m)\n', rC3)

% Graphic of the mechanism
plot([0,xB],[0,yB],'r-o',[xB,xC],[yB,yC],'b-o')
xlabel('x (m)'), ylabel('y (m)'),...
title('positions for \phi = 45 (deg)'),...
text(xA,yA,'  A'),text(xB,yB,'  B'),...
text(xC,yC,'  C=C3'),...
text(rC1(1),rC1(2),' C1'),...
text(rC2(1),rC2(2),' C2'),...
axis([-0.1,1.6,-0.1,1.6])

aC1 = aB1/2;
fprintf('aC1 = [ %g, %g, %g ] (m/s^2)\n', aC1)
aC2 = (aB1+aCs)/2;
fprintf('aC2 = [ %g, %g, %g ] (m/s^2)\n', aC2)
aC3 = aCs;
fprintf('aC3 = [ %g, %g, %g ] (m/s^2)\n', aC3)

fprintf('\n')
% external force
fe = 100;
Fe = -sign(vCs(1))*[fe 0 0];
fprintf...
('external force Fe = [ %d, %d, %g ] (N)\n', Fe)

h = 0.01; % height of the bar (m)
d = 0.001; % depth of the bar (m)
hSlider = 0.01; % height of the slider (m)
wSlider = 0.01; % depth of the slider (m)
g = 10.; % gravitational acceleration (m/s^2)

fprintf('\n')
fprintf('Inertia forces and moments \n\n')

fprintf(' link 1 \n');
m1 = 1;
IC1 = m1*(AB^2+h^2)/12;
G1 = [ 0  -m1*g  0 ];
Fin1 = - m1*aC1;
```

```
Min1 = - IC1*alpha1;
fprintf('m1 = %g (kg)\n', m1)
fprintf('IC1 = %g (kg m^2)\n', IC1)
fprintf('G1 = [ %d, %g, %d ] (N)\n', G1)
fprintf('Fin1 = [ %g, %g, %d ] (N)\n', Fin1)
fprintf('Min1 = [ %d, %d, %d ] (N m)\n', Min1)

fprintf(' link 2 \n')
m2 = 1;
IC2 = m2*(BC^2+h^2)/12;
G2 = [ 0  -m2*g  0 ];
Fin2 = - m2*aC2;
Min2 = - IC2*alpha20;
fprintf('m2 = %g (kg)\n', m2)
fprintf('IC2 = %g (kg m^2)\n', IC2)
fprintf('G2 = [ %d, %g, %d ] (N)\n', G2)
fprintf('Fin2 = [ %g, %g, %d ] (N)\n', Fin2)
fprintf('Min2 = [ %d, %d, %d ] (N m)\n', Min2)

fprintf(' link 3 \n')
m3 = 1;
IC3 = m3*(hSlider^2+wSlider^2)/12;
G3 = [ 0  -m3*g  0 ];
Fin3 = - m3*aC3;
Min3 = - IC3*alpha30;
fprintf('m3 = %g (kg)\n', m3)
fprintf('IC3 = %g (kg m^2)\n', IC3)
fprintf('G3 = [ %d, %g, %d ] (N)\n', G3)
fprintf('Fin3 = [ %g, %g, %d ] (N)\n', Fin3)
fprintf('Min3 = [ %d, %d, %d ] (N m)\n', Min3)

fprintf('\n')
fprintf('Dynamic force analysis \n')
fprintf('Dyad method  \n')
fprintf('\n')

% links 2 and 3
F03 = [ 0 sym('F03y','real') 0 ];
F12 = [ sym('F12x','real') sym('F12y','real') 0];
% sum of the forces for links 2 and 3
eqF23 = F03+Fe+G3+F12+G2-m3*aC3-m2*aC2;
eqF23x = eqF23(1);
eqF23y = eqF23(2);
fprintf('%s = 0 (1)\n', char(vpa(eqF23x,6)))
fprintf('%s = 0 (2)\n', char(vpa(eqF23y,6)))
```

```
% sum of the moments for link 2 wrt C
eqM2C = cross(rB-rC,F12)+cross(rC2-rC,G2)-...
    IC2*alpha20-cross(rC2-rC,m2*aC2);
eqM2Cz = eqM2C(3);
fprintf('%s = 0 (3)\n', char(vpa(eqM2Cz,6)))
fprintf('\n');
fprintf('Eqs(1)-(3) => F03y, F12x, F12y \n')
sol32=solve(eqF23x,eqF23y,eqM2Cz);
F03ys=eval(sol32.F03y);
F12xs=eval(sol32.F12x);
F12ys=eval(sol32.F12y);
F03s = [ 0, F03ys, 0 ];
F12s = [ F12xs, F12ys, 0 ];
fprintf('F03 = [ %g, %g, %g ] (N)\n', F03s)
fprintf('F12 = [ %g, %g, %g ] (N)\n', F12s)
fprintf('\n')
F32 = m2*aC2-(F12s+G2);
fprintf('F32 = m2*aC2-(F12+G2) \n')
fprintf('F32 = [ %g, %g, %g ] (N)\n', F32)
fprintf('\n')

% link 1
% sum of the forces for 1
F01=m1*aC1-G1+F12s;
% sum of the moments for 1 wrt A
Mm=IC1*alpha1+cross(rC1,m1*aC1-G1)-cross(rB,-F12s);
fprintf('F01 = [ %g, %g, %g ] (N)\n', F01)
fprintf('Mm = [ %d, %d, %g ] (N m)\n', Mm)
% end of program
```

Results:

```
phi = phi1 = 45 (degrees)
rA =  [ 0, 0, 0 ] (m)
rB =  [ 0.707107, 0.707107, 0 ] (m)
rC =  [ 1.41421, 0, 0 ] (m)
alpha1 = [ 0, 0, 0 ] (rad/s^2)
alpha2 = [ 0, 0, 0 ] (rad/s^2)
alpha3 = [ 0, 0, 0 ] (rad/s^2)

rC1 = [ 0.353553, 0.353553, 0 ] (m)
rC2 = [ 1.06066, 0.353553, 0 ] (m)
rC3 = [ 1.41421, 0, 0 ] (m)
aC1 = [ -0.353553, -0.353553, 0 ] (m/s^2)
aC2 = [ -1.06066, -0.353553, 0 ] (m/s^2)
aC3 = [ -1.41421, 0, 0 ] (m/s^2)
```

```
external force Fe = [ 100, 0, 0 ] (N)

Inertia forces and moments

 link 1
m1 = 1 (kg)
IC1 = 0.0833417 (kg m^2)
G1 = [ 0, -10, 0 ] (N)
Fin1 = [ 0.353553, 0.353553, 0 ] (N)
Min1 = [ 0, 0, 0 ] (N m)
 link 2
m2 = 1 (kg)
IC2 = 0.0833417 (kg m^2)
G2 = [ 0, -10, 0 ] (N)
Fin2 = [ 1.06066, 0.353553, 0 ] (N)
Min2 = [ 0, 0, 0 ] (N m)
 link 3
m3 = 1 (kg)
IC3 = 1.66667e-05 (kg m^2)
G3 = [ 0, -10, 0 ] (N)
Fin3 = [ 1.41421, -0, 0 ] (N)
Min3 = [ 0, 0, 0 ] (N m)

Dynamic force analysis
Dyad method

102.475+F12x = 0 (1)
F03y-19.6464+F12y = 0 (2)
-.707105*F12y-.707105*F12x+3.03552 = 0 (3)

Eqs(1)-(3) => F03y, F12x, F12y
F03 = [ 0, -87.1213, 0 ] (N)
F12 = [ -102.475, 106.768, 0 ] (N)

F32 = m2*aC2-(F12+G2)
F32 = [ 101.414, -97.1213, 0 ] (N)

F01 = [ -102.828, 116.414, 0 ] (N)
Mm = [ 0, 0, 151.492 ] (N m)
```

C.4 Slider-Crank (R-RRT) Mechanism: Contour Method

```
% C4
% Dynamic force analysis
% R-RRT
% Contour method

clear all; clc; close all
format long
AB = 1; BC = 1;
phi_input = 45; phi = phi_input*(pi/180);
xA = 0; yA = 0; rA = [xA yA 0];
xB = AB*cos(phi); yB = AB*sin(phi);
rB = [xB yB 0];
yC = 0; xC = xB+sqrt(BC^2-(yC-yB)^2);
rC = [xC yC 0];
phi2 = atan((yB-yC)/(xB-xC));
n = 30/pi; % (rpm)  driver link
omega1 = [ 0 0 pi*n/30 ]; alpha1 = [0 0 0 ];
vA = [0 0 0 ]; aA = [0 0 0 ];
vB1 = vA + cross(omega1,rB); vB2 = vB1;
aB1 = aA + cross(alpha1,rB) - dot(omega1,omega1)*rB;
aB2 = aB1;
omega2z = sym('omega2z','real');
vCx = sym('vCx','real');
omega2 = [ 0 0 omega2z ]; vC = [ vCx 0 0 ];
eqvC = vC - (vB2 + cross(omega2,rC-rB));
eqvCx = eqvC(1); eqvCy = eqvC(2);
solvC = solve(eqvCx,eqvCy);
omega2zs=eval(solvC.omega2z);
vCxs=eval(solvC.vCx); Omega2 = [0 0 omega2zs];
vCs = [vCxs 0 0];
alpha2z = sym('alpha2z','real');
aCx = sym('aCx','real');
alpha2 = [ 0 0 alpha2z ]; aC = [aCx 0 0 ];
eqaC = aC - (aB1 + cross(alpha2,rC-rB) - ...
    dot(Omega2,Omega2)*(rC-rB));
eqaCx = eqaC(1); eqaCy = eqaC(2);
solaC = solve(eqaCx,eqaCy);
alpha2zs=eval(solaC.alpha2z);
aCxs=eval(solaC.aCx);
alpha20 = [0 0 alpha2zs];
aCs = [aCxs 0 0];
alpha30 = [0 0 0];
```

```
rC1 = (rA+rB)/2;
rC2 = (rB+rC)/2;
rC3 = rC;

% Graphic of the mechanism
plot([0,xB],[0,yB],'r-o',[xB,xC],[yB,yC],'b-o')
xlabel('x (m)'), ylabel('y (m)'),...
title('positions for \phi = 45 (deg)'),...
text(xA,yA,'  A'),text(xB,yB,'  B'),...
text(xC,yC,'  C=C3'),...
text(rC1(1),rC1(2),' C1'),...
text(rC2(1),rC2(2),' C2'),...
axis([-0.1,1.6,-0.1,1.6])

aC1 = aB1/2;
aC2 = (aB1+aCs)/2;
aC3 = aCs;

% Inertia forces and moments

h = 0.01; d = 0.01; hSlider = 0.01; wSlider = 0.01;
g = 10.; % gravitational acceleration

m1 = 1;
IC1 = m1*(AB^2+h^2)/12;
G1 = [ 0  -m1*g  0 ];
Fin1 = - m1*aC1;
Min1 = - IC1*alpha1;

m2 = 1 ;
IC2 = m2*(BC^2+h^2)/12;
G2 = [ 0  -m2*g  0 ];
Fin2 = - m2*aC2;
Min2 = - IC2*alpha20;

m3 = 1;
IC3 = m3*(hSlider^2+wSlider^2)/12;
G3 = [ 0  -m3*g  0 ];
Fin3 = - m3*aC3;
Min3 = - IC3*alpha30;

% external force
fe = 100; Fe = -sign(vCs(1))*[fe 0 0];

fprintf('Results \n\n')
```

```
fprintf('Dynamic force analysis  \n')
fprintf('Contour method \n\n')

% Joint C_T
F03 = [ 0 sym('F03y','real') 0 ];
% sum of the moments for links 3&2 wrt B
eqM32B = cross(rC-rB, F03+G3+Fin3+Fe) +...
    cross(rC2-rB, Fin2+G2) + Min2;
eqM32Bz = eqM32B(3);
fprintf('%s = 0 (1)\n', char(vpa(eqM32Bz,6)))
fprintf('Eq(1) => F03y \n')
solF03=solve(eqM32Bz) ;
F03ys=eval(solF03) ;
F03s=[ 0, F03ys, 0 ];
fprintf('F03 = [ %g, %g, %g ] (N)\n', F03s)
fprintf('\n')

% Joint C_R
F23 = [ sym('F23x','real') sym('F23y','real') 0 ];
% sum of the forces for link 3 projected on x
eqF3 = F23+Fe+G3+Fin3;
eqF3x = eqF3(1);
% sum of the moments for link 2 wrt B
eqM2B = cross(rC-rB,-F23)+cross(rC2-rB,Fin2+G2)+Min2;
eqM2Bz = eqM2B(3);

fprintf('%s = 0 (2)\n', char(vpa(eqF3x,6)))
fprintf('%s = 0 (3)\n', char(vpa(eqM2Bz,6)))
fprintf('Eqs(2)-(3) => F23x, F23y \n')

solF23=solve(eqF3x,eqM2Bz);
F23xs=eval(solF23.F23x);
F23ys=eval(solF23.F23y);
F23s = [ F23xs, F23ys, 0 ];
fprintf('F23 = [ %g, %g, %g ] (N)\n', F23s)
fprintf('\n')

% Joint B_R
F12 = [ sym('F12x','real') sym('F12y','real') 0 ];
% sum of the moments for link 2 wrt C
eqM2C = cross(rB-rC,F12)+cross(rC2-rC,Fin2+G2)+Min2;
eqM2Cz = eqM2C(3);
% sum of the forces for links 2&3 projected on x
eqF23 = (F12+Fin2+G2+G3+Fin3+Fe);
eqF23x = eqF23(1) ;
```

```
fprintf('%s = 0 (4)\n', char(vpa(eqM2Cz,6)))
fprintf('%s = 0 (5)\n', char(vpa(eqF23x,6)))
fprintf('Eqs(4)-(5) => F12x, F12y \n')

solF12=solve(eqM2Cz,eqF23x);
F12xs=eval(solF12.F12x);
F12ys=eval(solF12.F12y);
F12s = [ F12xs, F12ys, 0 ];
fprintf('F12 = [ %g, %g, %g ] (N)\n', F12s)
fprintf('\n')

% Joint A_R
F01 = [ sym('F01x','real') sym('F01y','real') 0 ];
Mm = [ 0 0 sym('Mmz','real') ];
% sum of the moments for link 1 wrt B
eqM1B = cross(-rB,F01)+cross(rC1-rB,Fin1+G1)+Min1+Mm;
eqM1Bz = eqM1B(3);
% sum of the moments for links 1&2 wrt C
eqM12C = cross(-rC, F01)+cross(rC1-rC, Fin1+G1)+...
    Min1+Mm+cross(rC2-rC, Fin2+G2)+Min2;
eqM12Cz = eqM12C(3);
% sum of the forces for links 1&2&3 projected on x
eqF123 = (F01+Fin1+G1+Fin2+G2+Fin3+G3+Fe) ;
eqF123x = eqF123(1) ;

fprintf('%s = 0 (6)\n', char(vpa(eqM1Bz,6)))
fprintf('%s = 0 (7)\n', char(vpa(eqM12Cz,6)))
fprintf('%s = 0 (8)\n', char(vpa(eqF123x,6)))
fprintf('Eqs(6)(7)(8) => F01x, F01y, Mmz \n')

solF01=solve(eqM1Bz,eqM12Cz,eqF123x);
F01xs=eval(solF01.F01x);
F01ys=eval(solF01.F01y);
Mmzs=eval(solF01.Mmz);
F01s = [ F01xs, F01ys, 0 ];
Mms = [ 0, 0, Mmzs ];
fprintf('F01 = [ %g, %g, %g ] (N)\n', F01s)
fprintf('Mm = [ %g, %g, %g ] (N m)\n', Mms)
```

Results:

```
.707105*F03y+61.6039 = 0 (1)
Eq(1) => F03y
F03 = [ 0, -87.1213, 0 ] (N)

F23x+101.414 = 0 (2)
```

```
-.707105*F23y-.707105*F23x-3.03553 = 0 (3)
Eqs(2)-(3) => F23x, F23y
F23 = [ -101.414, 97.1213, 0 ] (N)

-.707105*F12y-.707105*F12x+3.03553 = 0 (4)
F12x+102.475 = 0 (5)
Eqs(4)-(5) => F12x, F12y
F12 = [ -102.475, 106.768, 0 ] (N)

-.707105*F01y+.707105*F01x+3.53552+Mmz = 0 (6)
-1.41421*F01y+13.1421+Mmz = 0 (7)
F01x+102.828 = 0 (8)
Eqs(6)(7)(8) => F01x, F01y, Mmz
F01 = [ -102.828, 116.414, 0 ] (N)
Mm = [ 0, 0, 151.492 ] (N m)
```

C.5 R-RTR-RTR Mechanism: Newton–Euler Method

```
% C5
% Dynamic force analysis
% R-RTR-RTR
% Newton-Euler method
clear all; clc; close all; format long
fprintf('R-RTR-RTR  \n')
fprintf('Dynamic force analysis \n')
fprintf('Newton-Eluer method \n')
fprintf('Results  \n\n')
AB = 0.15 ;
AC = 0.10 ;
CD = 0.15 ;
DF = 0.40 ;
AG = 0.30 ;
phi = pi/6 ;
fprintf('phi = phi1 = %g (degrees) \n', phi*180/pi)
fprintf('\n')
fprintf('Position analysis \n\n')
xA = 0; yA = 0; rA = [xA yA 0];
fprintf('rA = [ %g, %g, %g ] (m)\n', rA)
xC = 0; yC = AC; rC = [xC yC 0];
fprintf('rC = [ %g, %g, %g ] (m)\n', rC)
xB = AB*cos(phi); yB = AB*sin(phi); rB = [xB yB 0];
fprintf('rB = [ %g, %g, %g] (m)\n', rB)
eqnD1 = '( xDsol - xC )^2 + ( yDsol - yC )^2 = CD^2';
```

```
eqnD2 = '(yB-yC)/(xB-xC)=(yDsol-yC)/(xDsol-xC)';
solD = solve(eqnD1, eqnD2, 'xDsol, yDsol');
xDpositions = eval(solD.xDsol);
yDpositions = eval(solD.yDsol);
xD1 = xDpositions(1); xD2 = xDpositions(2);
yD1 = yDpositions(1); yD2 = yDpositions(2);
if(phi>=0&&phi<=pi/2)||(phi >=3*pi/2&&phi<=2*pi)
if ...
xD1 <= xC xD=xD1; yD=yD1; else xD=xD2; yD=yD2;end
   else
if ...
xD1 >= xC xD=xD1; yD=yD1; else xD=xD2; yD=yD2;end
end
rD = [xD yD 0];
fprintf('rD = [ %g, %g, %g ] (m)\n', rD)
phi2 = atan((yB-yC)/(xB-xC)); phi3 = phi2;
phi4 = atan((yD-yA)/(xD-xA))+pi; phi5 = phi4;
xF=xD+DF*cos(phi3); yF=yD+DF*sin(phi3);
rF=[xF yF 0];
fprintf('rF = [ %g, %g, %g ] (m)\n', rF)
xG=xA+AG*cos(phi5); yG=yA+AG*sin(phi5);
rG=[xG yG 0];
fprintf('rG = [ %g, %g, %g ]  (m)\n', rG)
fprintf('phi2 = phi3 = %g (degrees) \n', phi2*180/pi)
fprintf('phi4 = phi5 = %g (degrees) \n', phi4*180/pi)
fprintf('\n')
xC1 = xB/2; yC1 = yB/2; rC1 = [xC1 yC1 0];
fprintf('rC1 = [ %g, %g, %g ] (m)\n', rC1)
rC2 = rB;
fprintf('rC2 = rB = [ %g, %g, %g ] (m)\n', rC2)
xC3 = (xD+xF)/2; yC3 = (yD+yF)/2; rC3 = [xC3 yC3 0];
fprintf('rC3 = [ %g, %g, %g ] (m)\n', rC3)
rC4 = rD;
fprintf('rC4 = rD = [ %g, %g, %g ] (m)\n', rC4)
xC5 = (xA+xG)/2; yC5 = (yA+yG)/2; rC5 = [xC5 yC5 0];
fprintf('rC5 = [ %g, %g, %g ] (m)\n', rC5)

% Graphic of the mechanism
plot([0,xB],[0,yB],'r-o',[xD,xF],[yD,yF],...
     [xA,xG],[yA,yG],'g-o'),...
xlabel('x (m)'), ylabel('y (m)'),...
title('positions for \phi = 30 (deg)'),...
text(xA,yA,'  A'), text(xB,yB,'  B=C2'),...
text(xC,yC,'  C'), text(xD,yD,'  D=C4'),...
text(xF,yF,'  F'),text(xG,yG,'  G'),...
```

```
text(xC1,yC1,'  C1'), text(xC3,yC3,'  C3'),...
text(xC5,yC5,'  C5'), ...
axis([-0.3 0.3 -0.3 0.3]), grid on

fprintf('\n')
fprintf('Velocity and acceleration analysis \n\n')
n = 50.;
omega1 = [ 0 0 pi*n/30 ]; alpha1 = [0 0 0 ];
vA = [0 0 0 ]; aA = [0 0 0 ];
vB1 = vA + cross(omega1,rB); vB2 = vB1;
aB1 = aA + cross(alpha1,rB) -  ...
     dot(omega1,omega1)*rB;
aB2 = aB1;
fprintf ...
('aB1 = aB2 = [ %g, %g, %g ] (m/s^2)\n', aB1)
omega3z=sym('omega3z','real');
alpha3z=sym('alpha3z','real');
vB32=sym('vB32','real');
aB32=sym('aB32','real');
omega3 = [ 0 0 omega3z ];
vC = [0 0 0 ];
vB3 = vC + cross(omega3,rB-rC);
vB3B2 = vB32*[ cos(phi2) sin(phi2) 0];
eqvB = vB3 - vB2 - vB3B2;
eqvBx = eqvB(1); eqvBy = eqvB(2);
solvB = solve(eqvBx,eqvBy);
omega3zs=eval(solvB.omega3z);
vB32s=eval(solvB.vB32);
Omega3 = [0 0 omega3zs]; Omega2 = Omega3;
v32 = vB32s*[cos(phi2) sin(phi2) 0];
vD3 = vC + cross(Omega3,rD-rC); vD4 = vD3;
aB3B2cor = 2*cross(Omega3,v32);
alpha3 = [ 0 0 alpha3z ];
aC = [0 0 0 ];
aB3 = aC + cross(alpha3,rB-rC) - ...
     dot(Omega3,Omega3)*(rB-rC);
aB3B2 = aB32*[ cos(phi2) sin(phi2) 0];
eqaB = aB3 - aB2 - aB3B2 - aB3B2cor;
eqaBx = eqaB(1); eqaBy = eqaB(2);
solaB = solve(eqaBx,eqaBy);
alpha3zs=eval(solaB.alpha3z);
aB32s=eval(solaB.aB32);
Alpha3 = [0 0 alpha3zs]; Alpha2 = Alpha3;
aD3 = aC + cross(Alpha3,rD-rC) - ...
     dot(Omega3,Omega3)*(rD-rC);
```

```
aD4 = aD3;
fprintf('aD3 = aD4 = [ %g, %g, %g ] (m/s^2)\n', aD3)
omega5z=sym('omega5z','real');
alpha5z=sym('alpha5z','real');
vD54=sym('vD54','real');
aD54=sym('aD54','real');
omega5 = [ 0 0 omega5z ];
vD5 = vA + cross(omega5,rD-rA);
vD5D4 = vD54*[ cos(phi5) sin(phi5) 0];
eqvD = vD5 - vD4 - vD5D4;
eqvDx = eqvD(1); eqvDy = eqvD(2);
solvD = solve(eqvDx,eqvDy);
omega5zs = eval(solvD.omega5z);
vD54s = eval(solvD.vD54);
Omega5 = [0 0 omega5zs];
v54 = vD54s*[cos(phi5) sin(phi5) 0];
Omega4 = Omega5;
aD5D4cor = 2*cross(Omega5,v54);
alpha5 = [ 0 0 alpha5z ];
aD5 = aA + cross(alpha5,rD-rA) - ...
    dot(Omega5,Omega5)*(rD-rA);
aD5D4 = aD54*[ cos(phi5) sin(phi5) 0];
eqaD = aD5 - aD4 - aD5D4 - aD5D4cor;
eqaDx = eqaD(1); eqaDy = eqaD(2);
solaD = solve(eqaDx,eqaDy);
alpha5zs = eval(solaD.alpha5z);
aD54s = eval(solaD.aD54);
Alpha5 = [0 0 alpha5zs]; Alpha4 = Alpha5;
aF = aC + cross(Alpha3,rF-rC) - ...
    dot(Omega3,Omega3)*(rF-rC);
aG = aA + cross(Alpha5,rG-rA) - ...
    dot(Omega5,Omega5)*(rG-rA);
fprintf('aF = [ %g, %g, %g ] (m/s^2)\n', aF)
fprintf('aG = [ %g, %g, %g ] (m/s^2)\n', aG)

fprintf ...
('omega4=omega5 = [%g, %g, %g] (rad/s)\n', Omega5)
fprintf('\n')
fprintf ...
('alpha1 = [%g, %g, %g] (rad/s^2)\n', alpha1)
fprintf ...
('alpha2=alpha3 = [%g, %g, %g](rad/s^2)\n', Alpha3)
fprintf ...
('alpha4=alpha5 = [%g, %g, %g](rad/s^2)\n', Alpha5)
fprintf('\n')
```

```
aC1 = aB1/2;
fprintf('aC1 = [ %g, %g, %g ] (m/s^2)\n', aC1)
aC2 = aB2;
fprintf('aC2=aB2 = [ %g, %g, %g ] (m/s^2)\n', aC2)
aC3 = (aD3+aF)/2;
fprintf('aC3 = [ %g, %g, %g ] (m/s^2)\n', aC3)
aC4 = aD3;
fprintf('aC4=aD4 = [ %g, %g, %g ] (m/s^2)\n', aC4)
aC5 = (aA+aG)/2;
fprintf('aC5 = [ %g, %g, %g ] (m/s^2)\n', aC5)

fprintf('\n')
fprintf('Dynamic force analysis  \n')
fprintf('Newton-Euler method  \n\n')

h = 0.01; % height of the bar
d = 0.001; % depth of the bar
hSlider = 0.02; % height of the slider
wSlider = 0.05; % depth of the slider
rho = 8000; % density of the material
g = 9.807; % gravitational acceleration

Me = -sign(Omega5(3))*[0,0,100];
fprintf('Me = [ %d, %d, %g] (N m)\n', Me)
fprintf('\n')
fprintf('Inertia forces and inertia moments\n')
fprintf('\n')

fprintf('Link 1  \n')
m1 = rho*AB*h*d;
Fin1 = -m1*aC1;
G1 = [0,-m1*g,0];
IC1 = m1*(AB^2+h^2)/12;
Min1 = -IC1*alpha1;
fprintf('m1 =  %g (kg)\n', m1)
fprintf('m1 aC1 = [%g, %g, %g] (N)\n', m1*aC1)
fprintf('Fin1 = - m1 aC1 =[%g, %g, %d] (N)\n',Fin1)
fprintf('G1 = - m1 g = [%g, %g, %g] (N)\n', G1)
fprintf('IC1 =  %g (kg m^2)\n', IC1)
fprintf('IC1 alpha1=[%g, %g, %d](N m)\n',IC1*alpha1)
fprintf('Min1=-IC1 alpha1=[%d, %d, %d](N m)\n',Min1)
fprintf('\n')

fprintf('Link 2  \n')
m2 = rho*hSlider*wSlider*d;
```

```
Fin2 = -m2*aC2;
G2 = [0,-m2*g,0];
IC2 = m2*(hSlider^2+wSlider^2)/12;
Min2 = -IC2*Alpha2;
fprintf('m2 =  %g (kg)\n', m2)
fprintf('m2 aC2 = [%g, %g, %g] (N)\n', m2*aC2)
fprintf('Fin2 = - m2 aC2 =[%g, %g, %d] (N)\n',Fin2)
fprintf('G2 = - m2 g = [%g, %g, %g] (N)\n', G2)
fprintf('IC2 =  %g (kg m^2)\n', IC2)
fprintf('IC2 alpha2=[%g, %g, %g](N m)\n',IC2*Alpha2)
fprintf('Min2=-IC2 alpha2=[%d, %d, %g](N m)\n',Min2)
fprintf('\n')

fprintf('Link 3  \n')
m3 = rho*DF*h*d;
Fin3 = -m3*aC3;
G3 = [0,-m3*g,0];
IC3 = m3*(DF^2+h^2)/12;
Min3 = -IC3*Alpha3;
fprintf('m3 =  %g (kg)\n', m3)
fprintf('m3 aC3 = [%g, %g, %g] (N)\n', m3*aC3)
fprintf('Fin3 = - m3 aC3 =[%g, %g, %d] (N)\n',Fin3)
fprintf('G3 = - m3 g = [%g, %g, %g] (N)\n', G3)
fprintf('IC3 =  %g (kg m^2)\n', IC3)
fprintf('IC3 alpha3=[%g, %g, %g](N m)\n',IC3*Alpha3)
fprintf('Min3=-IC3 alpha3=[%d, %d, %g](N m)\n',Min3)
fprintf('\n')

fprintf('Link 4 \n')
m4 = rho*hSlider*wSlider*d;
Fin4 = -m4*aC4;
G4 = [0,-m4*g,0];
IC4 = m4*(hSlider^2+wSlider^2)/12;
Min4 = -IC4*Alpha4;
fprintf('m4 =  %g (kg)\n', m4)
fprintf('m4 aC4 = [%g, %g, %g] (N)\n', m4*aC4)
fprintf('Fin4 = - m4 aC4 =[%g, %g, %d] (N)\n',Fin4)
fprintf('G4 = - m4 g = [%g, %g, %g] (N)\n', G4)
fprintf('IC4 =  %g (kg m^2)\n', IC4)
fprintf('IC4 alpha4=[%g, %g, %g](N m)\n',IC4*Alpha4)
fprintf('Min4=-IC4 alpha4=[%d, %d, %g](N m)\n',Min4)
fprintf('\n')

fprintf('Link 5  \n')
m5 = rho*AG*h*d;
```

```
Fin5 = -m5*aC5;
G5 = [0,-m5*g,0];
IC5 = m5*(AG^2+h^2)/12;
Min5 = -IC5*Alpha5;
fprintf('m5 =  %g (kg)\n', m5)
fprintf('m5 aC5 = [%g, %g, %g] (N)\n', m5*aC5)
fprintf('Fin5 = - m5 aC5 =[%g, %g, %d] (N)\n', Fin5)
fprintf('G5 = - m5 g = [%g, %g, %g] (N)\n', G5)
fprintf('IC5 =  %g (kg m^2)\n', IC5)
fprintf('IC5 alpha5=[%g, %g, %g](N m)\n',IC5*Alpha5)
fprintf('Min5=-IC5 alpha5=[%d, %d, %g](N m)\n',Min5)

fprintf('\n')
fprintf('Joint reactions and equilibrium moment \n')
fprintf('\n')

% link 5
fprintf('eom link 5 \n')
fprintf('\n')
F05x=sym('F05x','real');
F05y=sym('F05y','real');
F45x=sym('F45x','real');
F45y=sym('F45y','real');
xP=sym('xP','real');
yP=sym('yP','real');
F05=[ F05x, F05y, 0]; %unknown joint force of 0 on 5
F45=[ F45x, F45y, 0]; %unknown joint force of 4 on 5
% unknown application point of force F45
rP=[xP, yP, 0];
% point P is on the line ED
eqP=cross(rD-rA,rP-rA);
eqPz=eqP(3); % eq(1)
Pz=vpa(eqPz,6);
fprintf('%s = 0 (1)\n', char(Pz))
% F45 perpendicular to DA
eqF45DA=dot(F45,rD-rA); % eq(2)
F45DA=vpa(eqF45DA,6);
fprintf('%s = 0 (2)\n', char(F45DA))
% Sum of the forces for 5
eqF5=F05+F45+G5-m5*aC5;
eqF5x=eqF5(1); % projection on x-axis  eq(3)
SF5x=vpa(eqF5x,6);
fprintf('%s = 0 (3)\n', char(SF5x))
eqF5y=eqF5(2); % projection on y-axis  eq(4)
SF5y=vpa(eqF5y,6);
```

```
fprintf('%s = 0 (4)\n', char(SF5y))
% Sum of the moments for 5 wrt C5
eqMC5=cross(rA-rC5,F05)+cross(rP-rC5,F45)+Me-...
      IC5*Alpha5;
eqMC5z=eqMC5(3); % projection on z-axis  eq(5)
% print eq(5)
eqMC5I = cross(rA-rC5,F05);
fprintf('%s+\n',char(vpa(eqMC5I(3),6)))
eqMC5II = cross(rP-rC5,F45);
fprintf('%s+\n',char(vpa(eqMC5II(3),6)))
eqMC5III = Me-IC5*Alpha5;
fprintf('%s = 0 (5)\n',char(vpa(eqMC5III(3),6)))

% link 4
fprintf('\n')
fprintf('eom link 4 \n')
fprintf('\n')
F34x=sym('F34x','real');
F34y=sym('F34y','real');
% unknown joint force of 3 on 4
F34=[ F34x, F34y, 0 ] ;
F54=-F45 ; % joint force of 5 on 4
% Sum of the forces for 4
eqF4=F34+F54+G4-m4*aC4;
eqF4x=eqF4(1); % projection on x-axis   eq(6)
SF4x=vpa(eqF4x,6);
fprintf('%s = 0 (6)\n', char(SF4x))
eqF4y=eqF4(2); % projection on y-axis   eq(7)
SF4y=vpa(eqF4y,6);
fprintf('%s = 0 (7)\n', char(SF4y))
% Sum of the moments for 4 wrt C4=D
eqMC4=cross(rP-rC4,F54)-IC4*Alpha4;
eqMC4z=eqMC4(3); % projection on z-axis  eq(8)
SMC4z=vpa(eqMC4z,3);
fprintf('%s = 0 (8)\n',char(SMC4z))

fprintf('\n')
fprintf('Eqs(1)-(8) => \n')
fprintf ...
('F05x, F05y, F45x, F45y, F34x, F34y, xP, yP\n')
sol45=solve(eqF5x,eqF5y,eqMC5z,eqF45DA,eqPz,...
            eqF4x,eqF4y,eqMC4z);
F05xs=eval(sol45.F05x);
F05ys=eval(sol45.F05y);
F05s=[ F05xs, F05ys, 0 ];
```

```
fprintf('F05 = [ %g, %g, %g] (N)\n', F05s)
F45xs=eval(sol45.F45x);
F45ys=eval(sol45.F45y);
F45s=[ F45xs, F45ys, 0 ];
fprintf('F45 = [ %g, %g, %g] (N)\n', F45s)
F34xs=eval(sol45.F34x);
F34ys=eval(sol45.F34y);
F34s=[ F34xs, F34ys, 0 ];
fprintf('F34 = [ %g, %g, %g] (N)\n', F34s)
xPs=eval(sol45.xP);
yPs=eval(sol45.yP);
rPs=[xPs, yPs, 0];
fprintf('rP = [ %g, %g, %g] (m)\n', rPs)

% link 3
fprintf('\n')
fprintf('eom link 3 \n')
fprintf('\n')
F43=-F34s;
F03x=sym('F03x','real');
F03y=sym('F03y','real');
F23x=sym('F23x','real');
F23y=sym('F23y','real');
xQ=sym('xQ','real');
yQ=sym('yQ','real');
%unknown joint force of 0 on 3
F03=[ F03x, F03y, 0];
%unknown joint force of 2 on 3
F23=[ F23x, F23y, 0];
% unknown application point of force F23
rQ=[xQ, yQ, 0];
% point Q is on the line BC
eqQ=cross(rB-rC,rQ-rC);
eqQz=eqQ(3); % eq(9)
Qz=vpa(eqQz,6);
fprintf('%s = 0 (9)\n', char(Qz))
% F23 perpendicular to BC
eqF23BC = dot(F23,rB-rC); % eq(10)
F23BC=vpa(eqF23BC,6);
fprintf('%s = 0 (10)\n', char(F23BC))
% Sum of the forces for 3
eqF3=F43+F03+F23+G3-m3*aC3;
eqF3x=eqF3(1); % projection on x-axis eq(11)
SF3x=vpa(eqF3x,6);
fprintf('%s = 0 (11)\n', char(SF3x))
```

```
eqF3y=eqF3(2); % projection on y-axis eq(12)
SF3y=vpa(eqF3y,6);
fprintf('%s = 0 (12)\n', char(SF3y))
% Sum of the moments for 3 wrt C3
eqMC3=cross(rD-rC3,F43)+cross(rC-rC3,F03)+...
    cross(rQ-rC3,F23)-IC3*Alpha3;
eqMC3z=eqMC3(3); % projection on z-axis  eq(13)
% print eq(13)
eqMC3I = cross(rC-rC3,F03);
fprintf('%s+\n',char(vpa(eqMC3I(3),6)))
eqMC3II = cross(rQ-rC3,F23);
fprintf('%s+\n',char(vpa(eqMC3II(3),6)))
eqMC3III = cross(rD-rC3,F43)-IC3*Alpha3;
fprintf('%s = 0 (13)\n',char(vpa(eqMC3III(3),6)))

% link 2
fprintf('\n')
fprintf('eom link 2 \n')
fprintf('\n')
F12x=sym('F12x','real');
F12y=sym('F12y','real');
% unknown joint force of 1 on 2
F12=[ F12x, F12y, 0];
F32=-F23; % joint force of 3 on 2
% Sum of the forces for 2
eqF2=F32+F12+G2-m2*aC2;
eqF2x=eqF2(1); % projection on x-axis  eq(14)
SF2x=vpa(eqF2x,6);
fprintf('%s = 0 (14)\n', char(SF2x))
eqF2y=eqF2(2); % projection on y-axis  eq(15)
SF2y=vpa(eqF2y,6);
fprintf('%s = 0 (15)\n', char(SF2y))
% Sum of the moments for 4 wrt C2=B
eqMC2=cross(rQ-rC2,F32)-IC2*Alpha2;
eqMC2z=eqMC2(3); % projection on z-axis  eq(16)
SMC2z=vpa(eqMC2z,2);
fprintf('%s = 0 (16)\n', char(SMC2z))

fprintf('\n')
fprintf('Eqs(9)-(16) => \n')
fprintf ...
('F03x, F03y, F23x, F23y, F12x, F12y, xQ, yQ\n')
sol23=solve(eqF3x,eqF3y,eqMC3z,eqF23BC,eqQz,...
            eqF2x,eqF2y,eqMC2z);
F03xs=eval(sol23.F03x);
```

```
F03ys=eval(sol23.F03y);
F03s=[ F03xs, F03ys, 0 ];
fprintf('F03 = [ %g, %g, %g] (N)\n', F03s)
F23xs=eval(sol23.F23x);
F23ys=eval(sol23.F23y);
F23s=[ F23xs, F23ys, 0 ];
fprintf('F23 = [ %g, %g, %g] (N)\n', F23s)
F12xs=eval(sol23.F12x);
F12ys=eval(sol23.F12y);
F12s=[ F12xs, F12ys, 0 ];
fprintf('F12 = [ %g, %g, %g] (N)\n', F12s)
xQs=eval(sol23.xQ);
yQs=eval(sol23.yQ);
rQs=[xQs, yQs, 0];
fprintf('rQ = [ %g, %g, %g] (m)\n', rQs )
fprintf('\n')

% link 1
fprintf('eom link 1 \n')
fprintf('\n')
% Sum of the forces for 1
F01=m1*aC1+F12s-G1;
fprintf('F01 = [ %g, %g, %g] (N)\n', F01)
% Sum of the moments for 1 wrt C1
Mm=IC1*alpha1-cross(rA-rC1,F01)+cross(rB-rC1,F12s);
fprintf('Mm = [ %d, %d, %g] (N m)\n', Mm)

% another way of calculating equilibrium moment Mm
% Sum of the moments for 1 wrt A
Mm1=cross(rB,F12s)+cross(rC1,m1*aC1-G1)+IC1*alpha1;
% fprintf('Mm1 = [ %d, %d, %g] (N m)\n', Mm1);

% end of program
```

Results:

```
phi = phi1 = 30 (degrees)

Position analysis

rA = [ 0, 0, 0 ] (m)
rC = [ 0, 0.1, 0 ] (m)
rB = [ 0.129904, 0.075, 0] (m)
rD = [ -0.147297, 0.128347, 0 ] (m)
rF = [ 0.245495, 0.0527544, 0 ] (m)
rG = [ -0.226182, 0.197083, 0 ]  (m)
```

```
phi2 = phi3 = -10.8934 (degrees)
phi4 = phi5 = 138.933 (degrees)

rC1 = [ 0.0649519, 0.0375, 0 ] (m)
rC2 = rB = [ 0.129904, 0.075, 0 ] (m)
rC3 = [ 0.049099, 0.0905509, 0 ] (m)
rC4 = rD = [ -0.147297, 0.128347, 0 ] (m)
rC5 = [ -0.113091, 0.0985417, 0 ] (m)

Velocity and acceleration analysis

aB1 = aB2 = [ -3.56139, -2.05617, 0 ] (m/s^2)
aD3 = aD4 = [ 2.5548, -2.71212, 0 ] (m/s^2)
aF = [ -4.258, 4.52021, 0 ] (m/s^2)
aG = [ -0.396144, -4.50689, 0 ] (m/s^2)
omega4=omega5 = [0, 0, 2.97887] (rad/s)

alpha1 = [0, 0, 0] (rad/s^2)
alpha2=alpha3 = [0, 0, 14.5363](rad/s^2)
alpha4=alpha5 = [0, 0, 12.1939](rad/s^2)

aC1 = [ -1.78069, -1.02808, 0 ] (m/s^2)
aC2=aB2 = [ -3.56139, -2.05617, 0 ] (m/s^2)
aC3 = [ -0.8516, 0.904041, 0 ] (m/s^2)
aC4=aD4 = [ 2.5548, -2.71212, 0 ] (m/s^2)
aC5 = [ -0.198072, -2.25344, 0 ] (m/s^2)

Dynamic force analysis
Newton-Euler method

Me = [ 0, 0, -100] (N m)

Inertia forces and inertia moments

Link 1
m1 =  0.012 (kg)
m1 aC1 = [-0.0213683, -0.012337, 0] (N)
Fin1 = - m1 aC1 =[0.0213683, 0.012337, 0] (N)
G1 = - m1 g = [0, -0.117684, 0] (N)
IC1 =  2.26e-05 (kg m^2)
IC1 alpha1=[0, 0, 0](N m)
Min1=-IC1 alpha1=[0, 0, 0](N m)

Link 2
m2 =  0.008 (kg)
```

```
m2 aC2 = [-0.0284911, -0.0164493, 0] (N)
Fin2 = - m2 aC2 =[0.0284911, 0.0164493, 0] (N)
G2 = - m2 g = [0, -0.078456, 0] (N)
IC2 =  1.93333e-06 (kg m^2)
IC2 alpha2=[0, 0, 2.81035e-05](N m)
Min2=-IC2 alpha2=[0, 0, -2.81035e-05](N m)

Link 3
m3 =  0.032 (kg)
m3 aC3 = [-0.0272512, 0.0289293, 0] (N)
Fin3 = - m3 aC3 =[0.0272512, -0.0289293, 0] (N)
G3 = - m3 g = [0, -0.313824, 0] (N)
IC3 =  0.000426933 (kg m^2)
IC3 alpha3=[0, 0, 0.00620602](N m)
Min3=-IC3 alpha3=[0, 0, -0.00620602](N m)

Link 4
m4 =  0.008 (kg)
m4 aC4 = [0.0204384, -0.021697, 0] (N)
Fin4 = - m4 aC4 =[-0.0204384, 0.021697, 0] (N)
G4 = - m4 g = [0, -0.078456, 0] (N)
IC4 =  1.93333e-06 (kg m^2)
IC4 alpha4=[0, 0, 2.35748e-05](N m)
Min4=-IC4 alpha4=[0, 0, -2.35748e-05](N m)

Link 5
m5 =  0.024 (kg)
m5 aC5 = [-0.00475373, -0.0540826, 0] (N)
Fin5 = - m5 aC5 =[0.00475373, 0.0540826, 0] (N)
G5 = - m5 g = [0, -0.235368, 0] (N)
IC5 =  0.0001802 (kg m^2)
IC5 alpha5=[0, 0, 0.00219734](N m)
Min5=-IC5 alpha5=[0, 0, -0.00219734](N m)

Joint reactions and equilibrium moment

eom link 5

-.147297*yP-.128347*xP = 0 (1)
-.147297*F45x+.128347*F45y = 0 (2)
F05x+F45x+.475373e-2 = 0 (3)
F05y+F45y-.181285 = 0 (4)
.113091*F05y+.985417e-1*F05x+
(xP+.113091)*F45y-1.*(yP-.985417e-1)*F45x+
-100.002 = 0 (5)
```

```
eom link 4

F34x-1.*F45x-.204384e-1 = 0 (6)
F34y-1.*F45y-.567590e-1 = 0 (7)
-1.*(xP+.147)*F45y+(yP-.128)*F45x-.236e-4 = 0 (8)

Eqs(1)-(8) =>
F05x, F05y, F45x, F45y, F34x, F34y, xP, yP
F05 = [ 336.192, 386.015, 0] (N)
F45 = [ -336.197, -385.834, 0] (N)
F34 = [ -336.176, -385.777, 0] (N)
rP = [ -0.147297, 0.128347, 0] (m)

eom link 3

.129904*yQ-.129904e-1+.250000e-1*xQ = 0 (9)
.129904*F23x-.250000e-1*F23y = 0 (10)
336.203+F03x+F23x = 0 (11)
385.435+F03y+F23y = 0 (12)
-.490990e-1*F03y-.944911e-2*F03x+
(xQ-.490990e-1)*F23y-1.*(yQ-.905509e-1)*F23x+
-88.4776 = 0 (13)

eom link 2

-1.*F23x+F12x+.284911e-1 = 0 (14)
-1.*F23y+F12y-.620067e-1 = 0 (15)
-1.*(xQ-.13)*F23y+(yQ-.75e-1)*F23x-.28e-4 = 0 (16)

Eqs(9)-(16) =>
F03x, F03y, F23x, F23y, F12x, F12y, xQ, yQ
F03 = [ -431.027, -878.152, 0] (N)
F23 = [ 94.8234, 492.717, 0] (N)
F12 = [ 94.7949, 492.779, 0] (N)
rQ = [ 0.129904, 0.075, 0] (m)

eom link 1

F01 = [ 94.7736, 492.884, 0] (N)
Mm = [ 0, 0, 56.9119] (N m)
```

C.6 R-RTR-RTR Mechanism: Dyad Method

```
% C6
% Dynamic force analysis
% R-RTR-RTR
% Dyad method
clear all; clc; close all; format long
fprintf('R-RTR-RTR  \n')
fprintf('Dynamic force analysis \n')
fprintf('Dyad method \n')
fprintf('Results  \n\n')
AB = 0.15 ;
AC = 0.10 ;
CD = 0.15 ;
DF = 0.40 ;
AG = 0.30 ;
phi = pi/6 ;
fprintf('phi = phi1 = %g (degrees)\n', phi*180/pi)
fprintf('\n')
fprintf('Position analysis \n\n')
xA = 0; yA = 0; rA = [xA yA 0];
fprintf('rA = [ %g, %g, %g ] (m)\n', rA)
xC = 0; yC = AC; rC = [xC yC 0];
fprintf('rC = [ %g, %g, %g ] (m)\n', rC)
xB = AB*cos(phi); yB = AB*sin(phi); rB = [xB yB 0];
fprintf('rB = [ %g, %g, %g] (m)\n', rB)
eqnD1 = '( xDsol - xC )^2+( yDsol - yC )^2 = CD^2';
eqnD2 = '(yB-yC)/(xB-xC)=(yDsol-yC)/(xDsol-xC)';
solD = solve(eqnD1, eqnD2, 'xDsol, yDsol');
xDpositions = eval(solD.xDsol);
yDpositions = eval(solD.yDsol);
xD1 = xDpositions(1); xD2 = xDpositions(2);
yD1 = yDpositions(1); yD2 = yDpositions(2);
if ...
(phi>=0 && phi<=pi/2)||(phi >= 3*pi/2 && phi<=2*pi)
   if ...
xD1 <= xC xD=xD1; yD=yD1; else xD=xD2; yD=yD2;end
   else
   if ...
xD1 >= xC xD=xD1; yD=yD1; else xD=xD2; yD=yD2;end
end
rD = [xD yD 0];
fprintf('rD = [ %g, %g, %g ] (m)\n', rD)
phi2 = atan((yB-yC)/(xB-xC)); phi3 = phi2;
phi4 = atan((yD-yA)/(xD-xA))+pi; phi5 = phi4;
```

```
xF=xD+DF*cos(phi3); yF=yD+DF*sin(phi3);
rF=[xF yF 0];
fprintf('rF = [ %g, %g, %g ] (m)\n', rF)
xG=xA+AG*cos(phi5); yG=yA+AG*sin(phi5);
rG=[xG yG 0];
fprintf('rG = [ %g, %g, %g ]  (m)\n', rG)
fprintf('phi2 = phi3 = %g (degrees) \n', phi2*180/pi)
fprintf('phi4 = phi5 = %g (degrees) \n', phi4*180/pi)
fprintf('\n')
xC1 = xB/2; yC1 = yB/2; rC1 = [xC1 yC1 0];
fprintf('rC1 = [ %g, %g, %g ] (m)\n', rC1)
rC2 = rB;
fprintf('rC2 = rB = [ %g, %g, %g ] (m)\n', rC2)
xC3 = (xD+xF)/2; yC3 = (yD+yF)/2; rC3 = [xC3 yC3 0];
fprintf('rC3 = [ %g, %g, %g ] (m)\n', rC3)
rC4 = rD;
fprintf('rC4 = rD = [ %g, %g, %g ] (m)\n', rC4)
xC5 = (xA+xG)/2; yC5 = (yA+yG)/2; rC5 = [xC5 yC5 0];
fprintf('rC5 = [ %g, %g, %g ] (m)\n', rC5)

% Graphic of the mechanism
plot([0,xB],[0,yB],'r-o',[xD,xF],[yD,yF],...
    [xA,xG],[yA,yG],'g-o'),...
xlabel('x (m)'), ylabel('y (m)'), ...
title('positions for \phi = 30 (deg)'),...
text(xA,yA,'  A'),text(xB,yB,'  B=C2'),...
text(xC,yC,'  C'),text(xD,yD,'  D=C4'),...
text(xF,yF,'  F'),text(xG,yG,'  G'),...
text(xC1,yC1,'  C1'),text(xC3,yC3,'  C3'),...
text(xC5,yC5,'  C5'),...
axis([-0.3 0.3 -0.3 0.3]), grid on

fprintf('\n')
fprintf('Velocity and acceleration analysis\n\n')
n = 50.;
omega1 = [ 0 0 pi*n/30 ]; alpha1 = [0 0 0 ];
vA = [0 0 0 ]; aA = [0 0 0 ];
vB1 = vA + cross(omega1,rB); vB2 = vB1;
aB1 = aA+cross(alpha1,rB)-dot(omega1,omega1)*rB;
aB2 = aB1;
fprintf('aB1=aB2 = [ %g, %g, %g ] (m/s^2)\n',aB1)
omega3z=sym('omega3z','real');
alpha3z=sym('alpha3z','real');
vB32=sym('vB32','real');
aB32=sym('aB32','real');
```

```
omega3 = [ 0 0 omega3z ];
vC = [0 0 0 ];
vB3 = vC + cross(omega3,rB-rC);
vB3B2 = vB32*[ cos(phi2) sin(phi2) 0];
eqvB = vB3 - vB2 - vB3B2;
eqvBx = eqvB(1); eqvBy = eqvB(2);
solvB = solve(eqvBx,eqvBy);
omega3zs=eval(solvB.omega3z);
vB32s=eval(solvB.vB32);
Omega3 = [0 0 omega3zs]; Omega2 = Omega3;
v32 = vB32s*[cos(phi2) sin(phi2) 0];
vD3 = vC + cross(Omega3,rD-rC); vD4 = vD3;
aB3B2cor = 2*cross(Omega3,v32);
alpha3 = [ 0 0 alpha3z ];
aC = [0 0 0 ];
aB3 = aC + cross(alpha3,rB-rC) - ...
      dot(Omega3,Omega3)*(rB-rC);
aB3B2 = aB32*[ cos(phi2) sin(phi2) 0];
eqaB = aB3 - aB2 - aB3B2 - aB3B2cor;
eqaBx = eqaB(1); eqaBy = eqaB(2);
solaB = solve(eqaBx,eqaBy);
alpha3zs=eval(solaB.alpha3z);
aB32s=eval(solaB.aB32);
Alpha3 = [0 0 alpha3zs]; Alpha2 = Alpha3;
aD3 = aC + cross(Alpha3,rD-rC) - ...
      dot(Omega3,Omega3)*(rD-rC);
aD4 = aD3;
fprintf('aD3=aD4 = [ %g, %g, %g ] (m/s^2)\n',aD3)
omega5z=sym('omega5z','real');
alpha5z=sym('alpha5z','real');
vD54=sym('vD54','real');
aD54=sym('aD54','real');
omega5 = [ 0 0 omega5z ];
vD5 = vA + cross(omega5,rD-rA);
vD5D4 = vD54*[ cos(phi5) sin(phi5) 0];
eqvD = vD5 - vD4 - vD5D4;
eqvDx = eqvD(1); eqvDy = eqvD(2);
solvD = solve(eqvDx,eqvDy);
omega5zs=eval(solvD.omega5z);
vD54s=eval(solvD.vD54);
Omega5 = [0 0 omega5zs];
v54 = vD54s*[cos(phi5) sin(phi5) 0];
Omega4 = Omega5;
aD5D4cor = 2*cross(Omega5,v54);
alpha5 = [ 0 0 alpha5z ];
```

```
aD5=aA+cross(alpha5,rD-rA)-...
    dot(Omega5,Omega5)*(rD-rA);
aD5D4 = aD54*[ cos(phi5) sin(phi5) 0];
eqaD = aD5 - aD4 - aD5D4 - aD5D4cor;
eqaDx = eqaD(1); eqaDy = eqaD(2);
solaD = solve(eqaDx,eqaDy);
alpha5zs=eval(solaD.alpha5z);
aD54s=eval(solaD.aD54);
Alpha5 = [0 0 alpha5zs]; Alpha4 = Alpha5;
aF=aC+cross(Alpha3,rF-rC)-...
   dot(Omega3,Omega3)*(rF-rC);
aG=aA+cross(Alpha5,rG-rA)-...
   dot(Omega5,Omega5)*(rG-rA);
fprintf('aF = [ %g, %g, %g ] (m/s^2)\n', aF)
fprintf('aG = [ %g, %g, %g ] (m/s^2)\n', aG)

fprintf ...
('omega4=omega5=[ %g, %g, %g ] (rad/s)\n',Omega5)
fprintf('\n')
fprintf ...
('alpha1=[ %g, %g, %g ] (rad/s^2)\n', alpha1)
fprintf ...
('alpha2=alpha3=[ %g, %g, %g ](rad/s^2)\n',Alpha3)
fprintf ...
('alpha4=alpha5=[ %g, %g, %g ](rad/s^2)\n',Alpha5)
fprintf('\n')
aC1 = aB1/2;
fprintf('aC1 = [ %g, %g, %g ] (m/s^2)\n', aC1)
aC2 = aB2;
fprintf('aC2=aB2 = [ %g, %g, %g ] (m/s^2)\n', aC2)
aC3 = (aD3+aF)/2;
fprintf('aC3 = [ %g, %g, %g ] (m/s^2)\n', aC3)
aC4 = aD3;
fprintf('aC4=aD4 = [ %g, %g, %g ] (m/s^2)\n', aC4)
aC5 = (aA+aG)/2;
fprintf('aC5 = [ %g, %g, %g ] (m/s^2)\n', aC5)

fprintf('\n')
fprintf('Dynamic force analysis  \n')
fprintf('Dyad method  \n\n')

h = 0.01; % height of the bar
d = 0.001; % depth of the bar
hSlider = 0.02; % height of the slider
wSlider = 0.05; % depth of the slider
```

```
rho = 8000; % density of the material
g = 9.807; % gravitational acceleration

Me = -sign(Omega5(3))*[0,0,100];
fprintf('Me = [%d, %d, %g] (N m)\n', Me)
fprintf('\n')
fprintf('Inertia forces and inertia moments\n\n')

fprintf('Link 1  \n')
m1 = rho*AB*h*d;
Fin1 = -m1*aC1;
G1 = [0,-m1*g,0];
IC1 = m1*(AB^2+h^2)/12;
Min1 = -IC1*alpha1;
fprintf('m1 = %g (kg)\n', m1)
fprintf('m1 aC1 = [%g, %g, %g] (N)\n', m1*aC1)
fprintf('Fin1= - m1 aC1 =[%g, %g, %d] (N)\n', Fin1)
fprintf('G1 = - m1 g = [%g, %g, %g] (N)\n', G1)
fprintf('IC1 = %g (kg m^2)\n', IC1)
fprintf('IC1 alpha1=[%g, %g, %d](N m)\n',IC1*alpha1)
fprintf('Min1=-IC1 alpha1=[%d, %d, %d](N m)\n',Min1)
fprintf('\n')

fprintf('Link 2 \n')
m2 = rho*hSlider*wSlider*d;
Fin2 = -m2*aC2;
G2 = [0,-m2*g,0];
IC2 = m2*(hSlider^2+wSlider^2)/12;
Min2 = -IC2*Alpha2;
fprintf('m2 = %g (kg)\n', m2)
fprintf('m2 aC2 = [%g, %g, %g] (N)\n', m2*aC2)
fprintf('Fin2= - m2 aC2 =[%g, %g, %d] (N)\n', Fin2)
fprintf('G2 = - m2 g = [%g, %g, %g] (N)\n', G2)
fprintf('IC2 = %g (kg m^2)\n', IC2)
fprintf('IC2 alpha2=[%g, %g, %g](N m)\n',IC2*Alpha2)
fprintf('Min2=-IC2 alpha2=[%d, %d, %g](N m)\n',Min2)
fprintf('\n')

fprintf('Link 3 \n')
m3 = rho*DF*h*d;
Fin3 = -m3*aC3;
G3 = [0,-m3*g,0];
IC3 = m3*(DF^2+h^2)/12;
Min3 = -IC3*Alpha3;
fprintf('m3 = %g (kg)\n', m3)
```

```
fprintf('m3 aC3 = [%g, %g, %g] (N)\n', m3*aC3)
fprintf('Fin3 = - m3 aC3 =[%g, %g, %d] (N)\n',Fin3)
fprintf('G3 = - m3 g = [%g, %g, %g] (N)\n', G3)
fprintf('IC3 = %g (kg m^2)\n', IC3)
fprintf('IC3 alpha3=[%g, %g, %g](N m)\n',IC3*Alpha3)
fprintf('Min3=-IC3 alpha3=[%d, %d, %g](N m)\n',Min3)
fprintf('\n')

fprintf('Link 4 \n')
m4 = rho*hSlider*wSlider*d;
Fin4 = -m4*aC4;
G4 = [0,-m4*g,0];
IC4 = m4*(hSlider^2+wSlider^2)/12;
Min4 = -IC4*Alpha4;
fprintf('m4 = %g (kg)\n', m4)
fprintf('m4 aC4 = [%g, %g, %g] (N)\n', m4*aC4)
fprintf('Fin4 = - m4 aC4 =[%g, %g, %d] (N)\n',Fin4)
fprintf('G4 = - m4 g = [%g, %g, %g] (N)\n', G4)
fprintf('IC4 = %g (kg m^2)\n', IC4)
fprintf('IC4 alpha4=[%g, %g, %g](N m)\n',IC4*Alpha4)
fprintf('Min4=-IC4 alpha4=[%d, %d, %g](N m)\n',Min4)
fprintf('\n')

fprintf('Link 5 \n')
m5 = rho*AG*h*d;
Fin5 = -m5*aC5;
G5 = [0,-m5*g,0];
IC5 = m5*(AG^2+h^2)/12;
Min5 = -IC5*Alpha5;
fprintf('m5 = %g (kg)\n', m5)
fprintf('m5 aC5 = [%g, %g, %g] (N)\n', m5*aC5)
fprintf('Fin5 = - m5 aC5 =[%g, %g, %d] (N)\n', Fin5)
fprintf('G5 = - m5 g = [%g, %g, %g] (N)\n', G5)
fprintf('IC5 = %g (kg m^2)\n', IC5)
fprintf('IC5 alpha5=[%g, %g, %g](N m)\n',IC5*Alpha5)
fprintf('Min5=-IC5 alpha5=[%d, %d, %g](N m)\n',Min5)

fprintf('\n')
fprintf('Joint reactions and equilibrium moment\n\n')

% Links 5 and 4 - dyad 4 & 5 (RTR)
fprintf('Dyad 4 & 5 \n\n')

F05x=sym('F05x','real');
F05y=sym('F05y','real');
```

```
F34x=sym('F34x','real');
F34y=sym('F34y','real');
F05=[F05x, F05y, 0]; % unknown joint force of 0 on 5
F34=[F34x, F34y, 0]; % unknown joint force of 3 on 4

% Sum of the forces for 5 and 4
eqF45=F05+G5+G4+F34-m5*aC5-m4*aC4;
eqF45x=eqF45(1); % projection on x-axis  eq(1)
eqF45y=eqF45(2); % projection on y-axis  eq(2)

SF45x=vpa(eqF45x,3);
fprintf('%s = 0  (1)\n', char(SF45x))
SF45y=vpa(eqF45y,3);
fprintf('%s = 0  (2)\n', char(SF45y))

% Sum of the moments for 5 and 4 wrt D
eqMD45=cross(rA-rD,F05)+cross(rC5-rD,G5-m5*aC5)...
    -IC5*Alpha5-IC4*Alpha4+Me;
eqMD45z=eqMD45(3); % projection on z-axis  eq(3)

SMD45z=vpa(eqMD45z,3);
fprintf('%s = 0  (3)\n', char(SMD45z))

% Sum of the forces for 4 projected on ED
eqF4DA=dot(F34+G4-m4*aC4,rD-rA); % eq(4)
F4DA=vpa(eqF4DA,6);
fprintf('%s = 0  (4)\n', char(F4DA))

% eqs(1)-(4) => F05x, F05y, F34x, F34y
fprintf('Eqs(1)-(4) => F05x, F05y, F34x, F34y\n')
solDI=solve(eqF45x, eqF45y , eqMD45z, eqF4DA);
F05xs=eval(solDI.F05x);
F05ys=eval(solDI.F05y);
F34xs=eval(solDI.F34x);
F34ys=eval(solDI.F34y);
F05s=[ F05xs, F05ys, 0 ];
F34s=[ F34xs, F34ys, 0 ];
fprintf('F05 = [%g, %g, %g] (N)\n', F05s)
fprintf('F34 = [%g, %g, %g] (N)\n', F34s)
fprintf('\n')

% Sum of the forces for 5: F45+F05+G5=m5*aC5 => F45
F45=m5*aC5-G5-F05s;
fprintf('F45 = [ %g, %g, %g] (N)\n', F45)
fprintf('verify: F45 perp to DE \n')
```

```
fprintf('F45.DA = %g \n',dot(F45,rD-rA))
fprintf('\n')

xP=sym('xP','real');
yP=sym('yP','real');
rP=[xP, yP, 0]; % application point of force F45
eqP=cross(rD-rA,rP-rA); % P is on the line ED
eqPz=eqP(3); % eq(5)
Pz=vpa(eqPz,6);
fprintf('%s = 0 (5)\n', char(Pz))
% Sum of the moments for 4 wrt C4=D
eqM4=cross(rP-rC4,-F45)-IC4*Alpha4;
eqM4z=eqM4(3); % eq(6)
M4z=vpa(eqPz,6);
fprintf('%s = 0 (6)\n', char(M4z))
fprintf('Eqs(5)-(6) => xP, yP \n');
% eqs(5)-(6) => xP, yP
solP=solve(eqPz,eqM4z);
xPs=eval(solP.xP);
yPs=eval(solP.yP);
rPs=[xPs, yPs, 0];
fprintf('rP = [%g, %g, %g] (m)\n', rPs)
fprintf('rP - rD = [%g, %g, %g] (m)\n', rPs-rD)
fprintf ...
('because IC4*Alpha4 is a very small number\n')
fprintf('-IC4*Alpha4(3) = %g \n',-IC4*Alpha4(3))
fprintf('\n')

% Links 2 and 3 - dyad 2 & 3 (RTR)
fprintf('Dyad 2 & 3 \n')

F03x=sym('F03x','real');
F03y=sym('F03y','real');
F12x=sym('F12x','real');
F12y=sym('F12y','real');
F03=[F03x, F03y, 0]; % unknown joint force of 0 on 3
F12=[F12x, F12y, 0]; % unknown joint force of 1 on 2
F43=-F34s;
% Sum of the forces for 2 and 3
eqF23=F43+F03+G3-m3*aC3+G2-m2*aC2+F12;
eqF23x=eqF23(1); % projection on x-axis  eq(7)
SF23x=vpa(eqF23x,6);
fprintf('%s = 0 (7)\n', char(SF23x))
eqF23y=eqF23(2); % projection on y-axis  eq(8)
SF23y=vpa(eqF23y,6);
```

```
fprintf('%s = 0 (8)\n', char(SF23y))
% Sum of the moments for 2 and 3 wrt B
eqMB3=cross(rD-rB,F43)+cross(rC-rB,F03)+...
      cross(rC3-rB,G3-m3*aC3);
eqMB2=-IC3*Alpha3-IC2*Alpha2;
eqMB23=eqMB3+eqMB2;
eqMB23z=eqMB23(3); % eq(9)
SMB23z=vpa(eqMB23z,6);
fprintf('%s = 0 (9)\n', char(SMB23z))
% Sum of the forces for 2 projected on BC
eqF2BC=dot(F12+G2-m2*aC2, rC-rB); % eq(10)
F2BC=vpa(eqF2BC,6);
fprintf('%s = 0 (10)\n', char(F2BC))
% eqs(7)-(10) => F03x, F03y, F12x, F12y
fprintf('Eqs(7)-(10)=> F03x, F03y, F12x, F12y\n')
solDII = solve(eqF23x, eqF23y , eqMB23z, eqF2BC);
F03xs=eval(solDII.F03x);
F03ys=eval(solDII.F03y);
F12xs=eval(solDII.F12x);
F12ys=eval(solDII.F12y);
F03s=[ F03xs, F03ys, 0 ];
F12s=[ F12xs, F12ys, 0 ];
fprintf('F03 = [%g, %g, %g] (N)\n', F03s)
fprintf('F12 = [%g, %g, %g] (N)\n', F12s)

fprintf('\n')
F32=m2*aC2-G2-F12s;
fprintf('F32 = [%g, %g, %g] (N)\n', F32)
fprintf('verify: F32 perp to BC \n')
fprintf('F32.BC = %g \n',dot(F32,rC-rB))
fprintf('\n')

xQ=sym('xQ','real');
yQ=sym('yQ','real');
rQ=[xQ, yQ, 0]; % application point of force F32
eqQ=cross(rC-rB,rQ-rC); % Q is on the line BC
eqQz=eqQ(3); % eq(11)
Qz=vpa(eqQz,6);
fprintf('%s = 0 (11)\n', char(Qz))
% Sum of the moments for 2 wrt C2=B
eqM2=cross(rQ-rC2,F32)-IC2*Alpha2;
eqM2z=eqM2(3); % eq(12)
SM2z=vpa(eqM2z,6);
fprintf('%s = 0 (12)\n', char(SM2z))
fprintf('Eqs(11)-(12) => xQ, yQ \n')
```

```
% eqs(11)-(12) => xQ, yQ
solQ=solve(eqQz,eqM2z);
xQs=eval(solQ.xQ);
yQs=eval(solQ.yQ);
rQs=[xQs, yQs, 0];
fprintf('rQ = [%g, %g, %g] (m)\n', rQs)
fprintf('rQ - rB = [%g, %g, %g] (m)\n', rQs-rB)
fprintf ...
('because IC2*Alpha2 is a very small number\n')
fprintf('-IC2*Alpha2(3) = %g \n',-IC2*Alpha2(3))
fprintf('\n');

% Link 1
fprintf('Link 1 \n\n')
% Sum of the forces for 1
F01=m1*aC1-G1+F12s;
fprintf('F01 = [ %g, %g, %g] (N)\n', F01)
% Sum of the moments for 1 wrt A
Mm=-cross(rB,-F12s)-cross(rC1,G1-m1*aC1)-IC1*alpha1;
fprintf('Mm = [ %d, %d, %g] (N m)\n', Mm)
% end of program
```

Results:

```
phi = phi1 = 30 (degrees)

Position analysis

rA = [ 0, 0, 0 ] (m)
rC = [ 0, 0.1, 0 ] (m)
rB = [ 0.129904, 0.075, 0] (m)
rD = [ -0.147297, 0.128347, 0 ] (m)
rF = [ 0.245495, 0.0527544, 0 ] (m)
rG = [ -0.226182, 0.197083, 0 ]  (m)
phi2 = phi3 = -10.8934 (degrees)
phi4 = phi5 = 138.933 (degrees)

rC1 = [ 0.0649519, 0.0375, 0 ] (m)
rC2 = rB = [ 0.129904, 0.075, 0 ] (m)
rC3 = [ 0.049099, 0.0905509, 0 ] (m)
rC4 = rD = [ -0.147297, 0.128347, 0 ] (m)
rC5 = [ -0.113091, 0.0985417, 0 ] (m)

Velocity and acceleration analysis

aB1=aB2 = [ -3.56139, -2.05617, 0 ] (m/s^2)
```

```
aD3=aD4 = [ 2.5548, -2.71212, 0 ] (m/s^2)
aF = [ -4.258, 4.52021, 0 ] (m/s^2)
aG = [ -0.396144, -4.50689, 0 ] (m/s^2)
omega4=omega5=[ 0, 0, 2.97887 ] (rad/s)

alpha1=[ 0, 0, 0 ] (rad/s^2)
alpha2=alpha3=[ 0, 0, 14.5363 ](rad/s^2)
alpha4=alpha5=[ 0, 0, 12.1939 ](rad/s^2)

aC1 = [ -1.78069, -1.02808, 0 ] (m/s^2)
aC2=aB2 = [ -3.56139, -2.05617, 0 ] (m/s^2)
aC3 = [ -0.8516, 0.904041, 0 ] (m/s^2)
aC4=aD4 = [ 2.5548, -2.71212, 0 ] (m/s^2)
aC5 = [ -0.198072, -2.25344, 0 ] (m/s^2)

Dynamic force analysis
Dyad method

Me = [0, 0, -100] (N m)

Inertia forces and inertia moments

Link 1
m1 = 0.012 (kg)
m1 aC1 = [-0.0213683, -0.012337, 0] (N)
Fin1= - m1 aC1 =[0.0213683, 0.012337, 0] (N)
G1 = - m1 g = [0, -0.117684, 0] (N)
IC1 = 2.26e-05 (kg m^2)
IC1 alpha1=[0, 0, 0](N m)
Min1=-IC1 alpha1=[0, 0, 0](N m)

Link 2
m2 = 0.008 (kg)
m2 aC2 = [-0.0284911, -0.0164493, 0] (N)
Fin2= - m2 aC2 =[0.0284911, 0.0164493, 0] (N)
G2 = - m2 g = [0, -0.078456, 0] (N)
IC2 = 1.93333e-06 (kg m^2)
IC2 alpha2=[0, 0, 2.81035e-05](N m)
Min2=-IC2 alpha2=[0, 0, -2.81035e-05](N m)

Link 3
m3 = 0.032 (kg)
m3 aC3 = [-0.0272512, 0.0289293, 0] (N)
Fin3 = - m3 aC3 =[0.0272512, -0.0289293, 0] (N)
G3 = - m3 g = [0, -0.313824, 0] (N)
```

```
IC3 = 0.000426933 (kg m^2)
IC3 alpha3=[0, 0, 0.00620602](N m)
Min3=-IC3 alpha3=[0, 0, -0.00620602](N m)

Link 4
m4 = 0.008 (kg)
m4 aC4 = [0.0204384, -0.021697, 0] (N)
Fin4 = - m4 aC4 =[-0.0204384, 0.021697, 0] (N)
G4 = - m4 g = [0, -0.078456, 0] (N)
IC4 = 1.93333e-06 (kg m^2)
IC4 alpha4=[0, 0, 2.35748e-05](N m)
Min4=-IC4 alpha4=[0, 0, -2.35748e-05](N m)

Link 5
m5 = 0.024 (kg)
m5 aC5 = [-0.00475373, -0.0540826, 0] (N)
Fin5 = - m5 aC5 =[0.00475373, 0.0540826, 0] (N)
G5 = - m5 g = [0, -0.235368, 0] (N)
IC5 = 0.0001802 (kg m^2)
IC5 alpha5=[0, 0, 0.00219734](N m)
Min5=-IC5 alpha5=[0, 0, -0.00219734](N m)

Joint reactions and equilibrium moment

Dyad 4 & 5

F05x+F34x-.157e-1 = 0 (1)
F05y-.238+F34y = 0 (2)
.147*F05y+.128*F05x-100. = 0 (3)
-.147297*F34x-.427435e-2+.128347*F34y = 0 (4)
Eqs(1)-(4) => F05x, F05y, F34x, F34y
F05 = [336.192, 386.015, 0] (N)
F34 = [-336.176, -385.777, 0] (N)

F45 = [ -336.197, -385.834, 0] (N)
verify: F45 perp to DE
F45.DA = 0

-.147297*yP-.128347*xP = 0 (5)
-.147297*yP-.128347*xP = 0 (6)
Eqs(5)-(6) => xP, yP
rP = [-0.147297, 0.128347, 0] (m)
rP - rD = [3.47312e-08, -3.0263e-08, 0] (m)
because IC4*Alpha4 is a very small number
-IC4*Alpha4(3) = -2.35748e-05
```

```
Dyad 2 & 3
336.232+F03x+F12x = 0 (7)
385.373+F03y+F12y = 0 (8)
-124.851-.129904*F03y-.250000e-1*F03x = 0 (9)
-.129904*F12x-.525127e-2+.250000e-1*F12y = 0 (10)
Eqs(7)-(10)=> F03x, F03y, F12x, F12y
F03 = [-431.027, -878.152, 0] (N)
F12 = [94.7949, 492.779, 0] (N)

F32 = [-94.8234, -492.717, 0] (N)
verify: F32 perp to BC
F32.BC = -7.10543e-15

-.129904*yQ+.129904e-1-.250000e-1*xQ = 0 (11)
-492.717*xQ+56.8940+94.8234*yQ = 0 (12)
Eqs(11)-(12) => xQ, yQ
rQ = [0.129904, 0.075, 0] (m)
rQ - rB = [-5.50007e-08, 1.05849e-08, 0] (m)
because IC2*Alpha2 is a very small number
-IC2*Alpha2(3) = -2.81035e-05

Link 1

F01 = [ 94.7736, 492.884, 0] (N)
Mm = [ 0, 0, 56.9119] (N m)
```

C.7 R-RTR-RTR Mechanism: Contour Method

```
% C7
% Dynamic force analysis
% R-RTR-RTR
% Contour method
clear all; clc; close all; format long
AB = 0.15 ;
AC = 0.10 ;
CD = 0.15 ;
DF = 0.40 ;
AG = 0.30 ;
phi = pi/6 ;
fprintf('phi = phi1 = %g (degrees)\n', phi*180/pi)
fprintf('\n')
fprintf('Position analysis \n\n')
```

```
xA = 0; yA = 0; rA = [xA yA 0];
fprintf('rA = [ %g, %g, %g ] (m)\n', rA)
xC = 0; yC = AC; rC = [xC yC 0];
fprintf('rC = [ %g, %g, %g ] (m)\n', rC)
xB = AB*cos(phi); yB = AB*sin(phi); rB = [xB yB 0];
fprintf('rB = [ %g, %g, %g ] (m)\n', rB)
eqnD1 = '(xDsol-xC)^2 + (yDsol-yC)^2 = CD^2';
eqnD2 = '(yB-yC)/(xB-xC)=(yDsol-yC)/(xDsol-xC)';
solD = solve(eqnD1, eqnD2, 'xDsol, yDsol');
xDpositions = eval(solD.xDsol);
yDpositions = eval(solD.yDsol);
xD1 = xDpositions(1); xD2 = xDpositions(2);
yD1 = yDpositions(1); yD2 = yDpositions(2);
if ...
(phi>=0 && phi<=pi/2)||(phi >= 3*pi/2 && phi<=2*pi)
   if ...
xD1 <= xC xD=xD1; yD=yD1; else xD=xD2; yD=yD2;end
   else
   if ...
xD1 >= xC xD=xD1; yD=yD1; else xD=xD2; yD=yD2;end
end
rD = [xD yD 0];
fprintf('rD = [ %g, %g, %g ] (m)\n', rD);
phi2 = atan((yB-yC)/(xB-xC)); phi3 = phi2;
phi4 = atan((yD-yA)/(xD-xA))+pi; phi5 = phi4;
xF=xD+DF*cos(phi3); yF=yD+DF*sin(phi3);
rF=[xF yF 0];
fprintf('rF = [ %g, %g, %g ] (m)\n', rF)
xG=xA+AG*cos(phi5); yG=yA+AG*sin(phi5);
rG=[xG yG 0];
fprintf('rG = [ %g, %g, %g ]  (m)\n', rG)
fprintf('phi2=phi3 = %g (degrees)\n', phi2*180/pi)
fprintf('phi4=phi5 = %g (degrees)\n', phi4*180/pi)
fprintf('\n')
xC1 = xB/2; yC1 = yB/2; rC1 = [xC1 yC1 0];
fprintf('rC1 = [ %g, %g, %g ] (m)\n', rC1)
rC2 = rB;
fprintf('rC2 = rB = [ %g, %g, %g ] (m)\n', rC2)
xC3 = (xD+xF)/2; yC3 = (yD+yF)/2; rC3 = [xC3 yC3 0];
fprintf('rC3 = [ %g, %g, %g ] (m)\n', rC3)
rC4 = rD;
fprintf('rC4 = rD = [ %g, %g, %g ] (m)\n', rC4)
xC5 = (xA+xG)/2; yC5 = (yA+yG)/2; rC5 = [xC5 yC5 0];
fprintf('rC5 = [ %g, %g, %g ] (m)\n', rC5)
```

```
% Graphic of the mechanism
plot([0,xB],[0,yB],'r-o',[xD,xF],[yD,yF],...
    [xA,xG],[yA,yG],'g-o'),...
xlabel('x (m)'), ylabel('y (m)'),...
title('positions for \phi = 30 (deg)'),...
text(xA,yA,'  A'),text(xB,yB,'  B=C2'),...
text(xC,yC,'  C'),text(xD,yD,'  D=C4'),...
text(xF,yF,'  F'),text(xG,yG,'  G'),...
text(xC1,yC1,'  C1'), text(xC3,yC3,'  C3'),...
text(xC5,yC5,'  C5'),...
axis([-0.3 0.3 -0.3 0.3]), grid on

fprintf('\n')
fprintf('Velocity and acceleration analysis\n\n')
n = 50.;
omega1 = [ 0 0 pi*n/30 ]; alpha1 = [0 0 0 ];
vA = [0 0 0 ]; aA = [ 0 0 0 ];
vB1 = vA + cross(omega1,rB); vB2 = vB1;
aB1 = aA+cross(alpha1,rB)-dot(omega1,omega1)*rB;
aB2 = aB1;
fprintf('aB1=aB2 = [ %g, %g, %g ] (m/s^2)\n',aB1)
omega3z=sym('omega3z','real');
alpha3z=sym('alpha3z','real');
vB32=sym('vB32','real');
aB32=sym('aB32','real');
omega3 = [ 0 0 omega3z ];
vC = [0 0 0 ];
vB3 = vC + cross(omega3,rB-rC);
vB3B2 = vB32*[ cos(phi2) sin(phi2) 0];
eqvB = vB3 - vB2 - vB3B2;
eqvBx = eqvB(1); eqvBy = eqvB(2);
solvB = solve(eqvBx,eqvBy);
omega3zs=eval(solvB.omega3z);
vB32s=eval(solvB.vB32);
Omega3 = [0 0 omega3zs]; Omega2 = Omega3;
v32 = vB32s*[cos(phi2) sin(phi2) 0];
vD3 = vC + cross(Omega3,rD-rC); vD4 = vD3;
aB3B2cor = 2*cross(Omega3, v32);
alpha3 = [ 0 0 alpha3z ];
aC = [0 0 0 ];
aB3 = aC + cross(alpha3,rB-rC) - ...
      dot(Omega3,Omega3)*(rB-rC);
aB3B2 = aB32*[ cos(phi2) sin(phi2) 0];
eqaB = aB3 - aB2 - aB3B2 - aB3B2cor;
eqaBx = eqaB(1); eqaBy = eqaB(2);
```

```
solaB = solve(eqaBx,eqaBy);
alpha3zs=eval(solaB.alpha3z);
aB32s=eval(solaB.aB32);
Alpha3 = [0 0 alpha3zs]; Alpha2 = Alpha3;
aD3 = aC + cross(Alpha3,rD-rC) - ...
      dot(Omega3,Omega3)*(rD-rC);
aD4 = aD3;
fprintf('aD3=aD4 = [ %g, %g, %g ] (m/s^2)\n',aD3)
omega5z=sym('omega5z','real');
alpha5z=sym('alpha5z','real');
vD54=sym('vD54','real');
aD54=sym('aD54','real');
omega5 = [ 0 0 omega5z ];
vD5 = vA + cross(omega5,rD-rA);
vD5D4 = vD54*[ cos(phi5) sin(phi5) 0];
eqvD = vD5 - vD4 - vD5D4;
eqvDx = eqvD(1); eqvDy = eqvD(2);
solvD = solve(eqvDx,eqvDy);
omega5zs=eval(solvD.omega5z);
vD54s=eval(solvD.vD54);
Omega5 = [0 0 omega5zs];
v54 = vD54s*[cos(phi5) sin(phi5) 0];
Omega4 = Omega5;
aD5D4cor = 2*cross(Omega5,v54);
alpha5 = [ 0 0 alpha5z ];
aD5 = aA + cross(alpha5,rD-rA) - ...
      dot(Omega5,Omega5)*(rD-rA);
aD5D4 = aD54*[ cos(phi5) sin(phi5) 0];
eqaD = aD5 - aD4 - aD5D4 - aD5D4cor;
eqaDx = eqaD(1); eqaDy = eqaD(2);
solaD = solve(eqaDx,eqaDy);
alpha5zs=eval(solaD.alpha5z);
aD54s=eval(solaD.aD54);
Alpha5 = [0 0 alpha5zs]; Alpha4 = Alpha5;
aF = aC + cross(Alpha3,rF-rC) - ...
     dot(Omega3,Omega3)*(rF-rC);
aG = aA + cross(Alpha5,rG-rA) - ...
     dot(Omega5,Omega5)*(rG-rA);
fprintf('aF = [ %g, %g, %g ] (m/s^2)\n', aF)
fprintf('aG = [ %g, %g, %g ] (m/s^2)\n', aG)

fprintf ...
('omega4=omega5=[ %g, %g, %g ] (rad/s)\n',Omega5)
fprintf('\n')
fprintf ...
```

```
('alpha1= [ %g, %g, %g ] (rad/s^2)\n', alpha1)
fprintf ...
('alpha2=alpha3=[%g, %g, %g] (rad/s^2)\n',Alpha3)
fprintf ...
('alpha4=alpha5=[%g, %g, %g] (rad/s^2)\n',Alpha5)
fprintf('\n')
aC1 = aB1/2;
fprintf('aC1 = [ %g, %g, %g ] (m/s^2)\n', aC1)
aC2 = aB2;
fprintf('aC2=aB2 = [ %g, %g, %g ] (m/s^2)\n',aC2)
aC3 = (aD3+aF)/2;
fprintf('aC3 = [ %g, %g, %g ] (m/s^2)\n', aC3)
aC4 = aD3;
fprintf('aC4=aD4 = [ %g, %g, %g ] (m/s^2)\n',aC4)
aC5 = (aA+aG)/2;
fprintf('aC5 = [ %g, %g, %g ] (m/s^2)\n', aC5)

fprintf('\n')
fprintf('Dynamic force analysis  \n')
fprintf('Newton-Euler method \n\n')

h = 0.01; % height of the bar
d = 0.001; % depth of the bar
hSlider = 0.02; % height of the slider
wSlider = 0.05; % depth of the slider
rho = 8000; % density of the material
g = 9.807; % gravitational acceleration

Me = -sign(Omega5(3))*[0,0,100];
fprintf('Me = [%d, %d, %g] (N m)\n', Me)
fprintf('\n');
fprintf('Inertia forces and inertia moments\n\n')

fprintf('Link 1 \n')
m1 = rho*AB*h*d;
Fin1 = -m1*aC1;
G1 = [0,-m1*g,0];
IC1 = m1*(AB^2+h^2)/12;
Min1 = -IC1*alpha1;
fprintf('m1 = %g (kg)\n', m1)
fprintf('m1 aC1 = [%g, %g, %g] (N)\n', m1*aC1)
fprintf('Fin1 = -m1 aC1 = [%g, %g, %d] (N)\n',Fin1)
fprintf('G1 = - m1 g = [%g, %g, %g] (N)\n', G1)
fprintf('IC1 = %g (kg m^2)\n', IC1)
fprintf('IC1 alpha1=[%g, %g, %d](N m)\n',IC1*alpha1)
```

```
fprintf('Min1=-IC1 alpha1=[%d, %d, %d](N m)\n',Min1)
fprintf('\n')

fprintf('Link 2 \n');
m2 = rho*hSlider*wSlider*d;
Fin2 = -m2*aC2;
G2 = [0,-m2*g,0];
IC2 = m2*(hSlider^2+wSlider^2)/12;
Min2 = -IC2*Alpha2;
fprintf('m2 = %g (kg)\n', m2)
fprintf('m2 aC2 = [%g, %g, %g] (N)\n', m2*aC2)
fprintf('Fin2 = -m2 aC2 = [%g, %g, %d] (N)\n',Fin2)
fprintf('G2 = - m2 g = [%g, %g, %g] (N)\n', G2)
fprintf('IC2 = %g (kg m^2)\n', IC2)
fprintf('IC2 alpha2=[%g, %g, %g](N m)\n',IC2*Alpha2)
fprintf('Min2=-IC2 alpha2=[%d, %d, %g] (N m)\n',Min2)
fprintf('\n')

fprintf('Link 3 \n')
m3 = rho*DF*h*d;
Fin3 = -m3*aC3;
G3 = [0,-m3*g,0];
IC3 = m3*(DF^2+h^2)/12;
Min3 = -IC3*Alpha3;
fprintf('m3 = %g (kg)\n', m3)
fprintf('m3 aC3 = [ %g, %g, %g] (N)\n', m3*aC3)
fprintf('Fin3 = -m3 aC3 = [%g, %g, %d] (N)\n',Fin3)
fprintf('G3 = - m3 g = [%g, %g, %g] (N)\n', G3 )
fprintf('IC3 = %g (kg m^2)\n', IC3);
fprintf('IC3 alpha3=[%g, %g, %g](N m)\n',IC3*Alpha3)
fprintf('Min3=-IC3 alpha3=[%d, %d, %g](N m)\n',Min3)
fprintf('\n')

fprintf('Link 4 \n')
m4 = rho*hSlider*wSlider*d;
Fin4 = -m4*aC4;
G4 = [0,-m4*g,0];
IC4 = m4*(hSlider^2+wSlider^2)/12;
Min4 = -IC4*Alpha4;
fprintf('m4 = %g (kg)\n', m4)
fprintf('m4 aC4 = [ %g, %g, %g] (N)\n', m4*aC4)
fprintf('Fin4 = -m4 aC4 = [%g, %g, %d] (N)\n',Fin4)
fprintf('G4 = - m4 g = [%g, %g, %g] (N)\n', G4)
fprintf('IC4 =  %g (kg m^2)\n', IC4)
fprintf('IC4 alpha4=[%g, %g, %g](N m)\n',IC4*Alpha4)
```

```
fprintf('Min4=-IC4 alpha4=[%d, %d, %g](N m)\n',Min4)
fprintf('\n')

fprintf('Link 5 \n')
m5 = rho*AG*h*d;
Fin5 = -m5*aC5;
G5 = [0,-m5*g,0];
IC5 = m5*(AG^2+h^2)/12;
Min5 = -IC5*Alpha5;
fprintf('m5 = %g (kg)\n', m5)
fprintf('m5 aC5 = [%g, %g, %g] (N)\n', m5*aC5)
fprintf('Fin5 = -m5 aC5 = [%g, %g, %d] (N)\n',Fin5)
fprintf('G5 = - m5 g = [%g, %g, %g] (N)\n', G5)
fprintf('IC5 = %g (kg m^2)\n', IC5)
fprintf('IC5 alpha5=[%g, %g, %g](N m)\n',IC5*Alpha5)
fprintf('Min5=-IC5 alpha5=[%d, %d, %g](N m)\n',Min5)

fprintf('\n')
fprintf('Joint reactions\n')
fprintf('Contour method \n\n')

fprintf('Joint A_rotation (5 and 0)\n\n')
F05=[sym('F05x','real'), sym('F05y','real'), 0];
% Sum of the forces for 5 projected on DA
eqAR1=dot(F05+G5+Fin5,rA-rD);
AR1=vpa(eqAR1,6);
fprintf('%s = 0 \n', char(AR1))
% Sum of the moments for 5 & 4 wrt D
eqAR2=cross(rA-rD,F05)+cross(rC5-rD,G5+Fin5)+...
      Me+Min4+Min5;
eqAR2z=eqAR2(3);
AR2=vpa(eqAR2z,6);
fprintf('%s = 0 \n', char(AR2))
solF05=solve(eqAR1,eqAR2z);
F05s=[eval(solF05.F05x), eval(solF05.F05y), 0];
fprintf('F05 = [ %g, %g, %g] (N)\n', F05s)
fprintf('\n')

fprintf('Joint D_translation \n\n')
F45=[sym('F45x','real'), sym('F45y','real'), 0];
F54=-F45;
rP=[sym('xP','real'), sym('yP','real'), 0];
% Sum of the moments for 5 wrt A
eqDT1=cross(rP-rA,F45)+cross(rC5-rA,G5+Fin5)+...
      Me+Min5;
```

```
eqDT1z=eqDT1(3);
DT1=vpa(eqDT1z,6);
fprintf('%s = 0 \n', char(DT1))
% Sum of the moments for 4 wrt D
eqDT2=cross(rP-rD,F54)+Min4;
eqDT2z=eqDT2(3);
DT2=vpa(eqDT2z,3);
fprintf('%s = 0 \n', char(DT2))
% F45 perpendicular to DA
eqF45DA=dot(F45,rD-rA); %
F45DA=vpa(eqF45DA,6);
fprintf('%s = 0 \n', char(F45DA))
% point P is on the line AD
eqP=cross(rD-rA,rP-rA);
eqPz=eqP(3); % eq(4)
Pz=vpa(eqPz,6);
fprintf('%s = 0 \n', char(Pz))
solF45=solve(eqDT1z,eqDT2z,F45DA,eqPz);
F45s=[eval(solF45.F45x), eval(solF45.F45y), 0];
rPs=[eval(solF45.xP), eval(solF45.yP), 0];
fprintf('F45 = [%g, %g, %g] (N)\n', F45s)
fprintf('rP = [%g, %g, %g] (m)\n', rPs)
fprintf('\n');

fprintf('Joint D_rotation \n\n');
F34=[sym('F34x','real'), sym('F34y','real'), 0];
F43=-F34;
% Sum of the forces for 4 projected on AD
eqDR1=dot(F34+G4+Fin4,rD-rA);
DR1=vpa(eqDR1,6);
fprintf('%s = 0 \n', char(DR1));
% Sum of the moments for 4 & 5 wrt A
eqDR24=cross(rC4-rA,G4+Fin4)+cross(rD-rA,F34)+Min4;
eqDR25=cross(rC5-rA,G5+Fin5)+Me+Min5;
eqDR2=eqDR24+eqDR25;
eqDR2z=eqDR2(3);
DR2=vpa(eqDR2z,6);
fprintf('%s = 0 \n', char(DR2))
solF34=solve(eqDR1,eqDR2z);
F34s=[eval(solF34.F34x), eval(solF34.F34y), 0];
fprintf('F34 = [%g, %g, %g] (N)\n', F34s )
fprintf('\n');

fprintf('Joint C_rotation \n\n')
F03=[sym('F03x','real'), sym('F03y','real'), 0];
```

```
% Sum of the forces for 3 projected on CD
eqCR1=dot(F03-F34s+G3+Fin3,rD-rC);
CR1=vpa(eqCR1,6);
fprintf('%s = 0 \n', char(CR1))
% Sum of the moments for 3 & 2 wrt B
eqCR2=cross(rC3-rB,G3+Fin3)+cross(rC-rB,F03)+...
      cross(rD-rB,-F34s)+Min2+Min3;
eqCR2z=eqCR2(3);
CR2=vpa(eqCR2z,6);
fprintf('%s = 0 \n', char(CR2))
solF03=solve(eqCR1,eqCR2z);
F03s=[eval(solF03.F03x), eval(solF03.F03y), 0];
fprintf('F03 = [%g, %g, %g] (N)\n', F03s)
fprintf('\n')

fprintf('Joint B_translation \n\n');
F23=[sym('F23x','real'), sym('F23y','real'), 0];
F32=-F23;
rQ=[sym('xQ','real'), sym('yQ','real'), 0];
% Sum of the moments for 3 wrt C
eqBT1=cross(rQ-rC,F23)+cross(rC3-rC,G3+Fin3)+...
      cross(rD-rC,-F34s)+Min3;
eqBT1z=eqBT1(3);
BT1=vpa(eqBT1z,6);
fprintf('%s = 0 \n', char(BT1))
% Sum of the moments for 2 wrt B
eqBT2=cross(rQ-rB,F32)+Min2;
eqBT2z=eqBT2(3);
BT2=vpa(eqBT2z,3);
fprintf('%s = 0 \n', char(BT2))
% F23 perpendicular to BC
eqF23BC=dot(F23,rC-rB); %
F23BC=vpa(eqF23BC,6);
fprintf('%s = 0 \n', char(F23BC));
% point Q is on the line BC
eqQ=cross(rB-rC,rQ-rC);
eqQz=eqQ(3); % eq(4)
Qz=vpa(eqQz,6);
fprintf('%s = 0 \n', char(Qz))
solF23=solve(eqBT1z,eqBT2z,F23BC,eqQz);
F23s=[eval(solF23.F23x), eval(solF23.F23y), 0];
rQs=[eval(solF23.xQ), eval(solF23.yQ), 0];
fprintf('F23 = [%g, %g, %g] (N)\n', F23s)
fprintf('rQ = [%g, %g, %g] (N)\n', rQs)
fprintf('\n')
```

```
fprintf('Joint B_rotation \n\n')
F12=[sym('F12x','real'), sym('F12y','real'), 0];
F21=-F12;
% Sum of the forces for 2 projected on BC
eqBR1=dot(F12+G2+Fin2,rC-rB);
BR1=vpa(eqBR1,6);
fprintf('%s = 0 \n', char(BR1))
% Sum of the moments for 2 & 3 wrt C
eqBR2=cross(rB-rC,F12)+cross(rC2-rC,G2+Fin2)+Min2...
+cross(rC3-rC,G3+Fin3)+cross(rD-rC,-F34s)+Min3;
eqBR2z=eqBR2(3);
BR2=vpa(eqBR2z,6);
fprintf('%s = 0 \n', char(BR2))
solF12=solve(eqBR1,eqBR2z);
F12s=[eval(solF12.F12x), eval(solF12.F12y), 0];
fprintf('F12 = [ %g, %g, %g] (N)\n', F12s)
fprintf('\n')

% Sum of the moments for 1 wrt A
M1m=-(cross(rB,-F12s)+cross(rC1,G1+Fin1)+Min1);
fprintf('Mm = [%d, %d, %g] (N m)\n', M1m)
fprintf('\n')

fprintf('Joint A_rotation (1 and 0) \n\n')
F01=[sym('F01x','real'), sym('F01y','real'), 0];
% Sum of the moments for 1 wrt B
eqAAR1=cross(-rB,F01)+cross(rC1-rB,G1+Fin1)+Min1+M1m;
eqAAR1z=eqAAR1(3);
AAR1=vpa(eqAAR1z,6);
fprintf('%s = 0 \n', char(AAR1))
% Sum of the forces for 1 & 2 projected on BC
eqAAR2=dot(F01+G1+Fin1+G2+Fin2,rC-rB);
AAR2=vpa(eqAAR2,6);
fprintf('%s = 0 \n', char(AAR2))
solF01=solve(eqAAR1z,eqAAR2);
F01s=[ eval(solF01.F01x), eval(solF01.F01y), 0 ];
fprintf('F01 = [%g, %g, %g] (N)\n', F01s)
fprintf('\n')

% end of program
```

Results:

```
phi = phi1 = 30 (degrees)
```

```
Position analysis

rA = [ 0, 0, 0 ] (m)
rC = [ 0, 0.1, 0 ] (m)
rB = [ 0.129904, 0.075, 0 ] (m)
rD = [ -0.147297, 0.128347, 0 ] (m)
rF = [ 0.245495, 0.0527544, 0 ] (m)
rG = [ -0.226182, 0.197083, 0 ]  (m)
phi2=phi3 = -10.8934 (degrees)
phi4=phi5 = 138.933 (degrees)

rC1 = [ 0.0649519, 0.0375, 0 ] (m)
rC2 = rB = [ 0.129904, 0.075, 0 ] (m)
rC3 = [ 0.049099, 0.0905509, 0 ] (m)
rC4 = rD = [ -0.147297, 0.128347, 0 ] (m)
rC5 = [ -0.113091, 0.0985417, 0 ] (m)

Velocity and acceleration analysis

aB1=aB2 = [ -3.56139, -2.05617, 0 ] (m/s^2)
aD3=aD4 = [ 2.5548, -2.71212, 0 ] (m/s^2)
aF = [ -4.258, 4.52021, 0 ] (m/s^2)
aG = [ -0.396144, -4.50689, 0 ] (m/s^2)
omega4=omega5=[ 0, 0, 2.97887 ] (rad/s)

alpha1= [ 0, 0, 0 ] (rad/s^2)
alpha2=alpha3=[0, 0, 14.5363] (rad/s^2)
alpha4=alpha5=[0, 0, 12.1939] (rad/s^2)

aC1 = [ -1.78069, -1.02808, 0 ] (m/s^2)
aC2=aB2 = [ -3.56139, -2.05617, 0 ] (m/s^2)
aC3 = [ -0.8516, 0.904041, 0 ] (m/s^2)
aC4=aD4 = [ 2.5548, -2.71212, 0 ] (m/s^2)
aC5 = [ -0.198072, -2.25344, 0 ] (m/s^2)

Dynamic force analysis
Newton-Euler method

Me = [0, 0, -100] (N m)

Inertia forces and inertia moments

Link 1
m1 = 0.012 (kg)
m1 aC1 = [-0.0213683, -0.012337, 0] (N)
```

```
Fin1 = -m1 aC1 = [0.0213683, 0.012337, 0] (N)
G1 = - m1 g = [0, -0.117684, 0] (N)
IC1 = 2.26e-05 (kg m^2)
IC1 alpha1=[0, 0, 0](N m)
Min1=-IC1 alpha1=[0, 0, 0](N m)

Link 2
m2 = 0.008 (kg)
m2 aC2 = [-0.0284911, -0.0164493, 0] (N)
Fin2 = -m2 aC2 = [0.0284911, 0.0164493, 0] (N)
G2 = - m2 g = [0, -0.078456, 0] (N)
IC2 = 1.93333e-06 (kg m^2)
IC2 alpha2=[0, 0, 2.81035e-05](N m)
Min2=-IC2 alpha2=[0, 0, -2.81035e-05] (N m)

Link 3
m3 = 0.032 (kg)
m3 aC3 = [ -0.0272512, 0.0289293, 0] (N)
Fin3 = -m3 aC3 = [0.0272512, -0.0289293, 0] (N)
G3 = - m3 g = [0, -0.313824, 0] (N)
IC3 = 0.000426933 (kg m^2)
IC3 alpha3=[0, 0, 0.00620602](N m)
Min3=-IC3 alpha3=[0, 0, -0.00620602](N m)

Link 4
m4 = 0.008 (kg)
m4 aC4 = [ 0.0204384, -0.021697, 0] (N)
Fin4 = -m4 aC4 = [-0.0204384, 0.021697, 0] (N)
G4 = - m4 g = [0, -0.078456, 0] (N)
IC4 =  1.93333e-06 (kg m^2)
IC4 alpha4=[0, 0, 2.35748e-05](N m)
Min4=-IC4 alpha4=[0, 0, -2.35748e-05](N m)

Link 5
m5 = 0.024 (kg)
m5 aC5 = [-0.00475373, -0.0540826, 0] (N)
Fin5 = -m5 aC5 = [0.00475373, 0.0540826, 0] (N)
G5 = - m5 g = [0, -0.235368, 0] (N)
IC5 = 0.0001802 (kg m^2)
IC5 alpha5=[0, 0, 0.00219734](N m)
Min5=-IC5 alpha5=[0, 0, -0.00219734](N m)

Joint reactions
Contour method
```

```
Joint A_rotation (5 and 0)

.147297*F05x+.239677e-1-.128347*F05y = 0
.147297*F05y+.128347*F05x-100.008 = 0
F05 = [ 336.192, 386.015, 0] (N)

Joint D_translation

xP*F45y-1.*yP*F45x-99.9822 = 0
-1.*(xP+.147)*F45y+(yP-.128)*F45x-.236e-4 = 0
-.147297*F45x+.128347*F45y = 0
-.147297*yP-.128347*xP = 0
F45 = [-336.196, -385.834, 0] (N)
rP = [-0.147297, 0.128347, 0] (m)

Joint D_rotation

-.147297*F34x-.427435e-2+.128347*F34y = 0
-99.9712-.147297*F34y-.128347*F34x = 0
F34 = [-336.176, -385.777, 0] (N)

Joint C_rotation

-.147297*F03x-38.5957+.283473e-1*F03y = 0
-124.851-.129904*F03y-.250000e-1*F03x = 0
F03 = [-431.027, -878.152, 0] (N)

Joint B_translation

xQ*F23y-1.*(yQ-.100000)*F23x-66.3764 = 0
-1.*(xQ-.130)*F23y+(yQ-.750e-1)*F23x-.281e-4 = 0
-.129904*F23x+.250000e-1*F23y = 0
.129904*yQ-.129904e-1+.250000e-1*xQ = 0
F23 = [94.8233, 492.717, 0] (N)
rQ = [0.129904, 0.075, 0] (N)

Joint B_rotation

-.129904*F12x-.525127e-2+.250000e-1*F12y = 0
.129904*F12y+.250000e-1*F12x-66.3837 = 0
F12 = [ 94.7949, 492.779, 0] (N)

Mm = [0, 0, 56.9119] (N m)

Joint A_rotation (1 and 0)
```

```
-.129904*F01y+.750000e-1*F01x+56.9195 = 0
-.129904*F01x-.106608e-1+.250000e-1*F01y = 0
F01 = [94.7736, 492.884, 0] (N)
```

Appendix D
Programs of Chapter 5: Direct Dynamics

D.1 Compound Pendulum

```
% D1
% Direct Dynamics
% Compound pendulum
clear all; clc; close all
syms L m g t
omega = [0 0 diff('theta(t)',t)];
alpha = diff(omega,t);
% diff(X,'t') or diff(X,sym('t'))
% differentiates a symbolic expression
% X with respect to t
% diff(X,'t',n) and diff(X,n,'t')
% differentiates X n times
% n is a positive integer
c = cos(sym('theta(t)'));
s = sin(sym('theta(t)'));
xC = (L/2)*c;
yC = (L/2)*s;
rC = [xC yC 0];
G = [0 -m*g 0];
IC = m*L^2/12;
IO = IC + m*(L/2)^2;
MO = cross(rC,G);
eq = -IO*alpha+MO;
eqz = eq(3);

eqI = subs(eqz,{L,m,g},{3,12/32.2,32.2});
eqI1 = subs(eqI,diff('theta(t)',t,2),'ddtheta');
eqI2 = subs(eqI1,diff('theta(t)',t),sym('x(2)'));
```

```
eqI3 = subs(eqI2,'theta(t)',sym('x(1)'));

% solve a second order ODE using MATLAB
% ODE45 function
% write the second order equation as a
% system of two first order equations,
% by introducing x(2) = x(1)'
%  x(1)' = x(2)  ==>   x' = g(t,x)
%  x(2)' = f
%
% define the vector x = [x(1); x(2)]
% (column-vector).

% first differential equation
dx1 = sym('x(2)');
% second differential equation
dx2 = solve(eqI3,'ddtheta');

% define right-hand side vector
dx1dt = char(dx1);
dx2dt = char(dx2);

g=inline(sprintf('[%s; %s]',dx1dt,dx2dt),'t','x');

t0 = 0;  % define initial time
tf = 10; % define final time
time = [0 tf];

x0 = [pi/4; 0]; % define initial conditions

[t,xs] = ode45(g, time, x0);
% find t, xs, but don't show
% ode45  solves non-stiff differential equations,
% medium order method.
% [ts,ys] = ode45(f,tspan,y0) integrates
% the system of differential equations
% y' = f(t,y) with initial conditions y0.

x1 = xs(:,1); % extract x1 & x2 components from xs
x2 = xs(:,2);

subplot(3,1,1),plot(t,x1,'r'),...
xlabel('t'),ylabel('\theta'),grid,...
subplot(3,1,2),plot(t,x2),...
xlabel('t'),ylabel('\omega'),grid,...
```

```
subplot(3,1,3),plot(x1,x2),...

xlabel('\theta'),ylabel('\omega'),grid

[ts,xs] = ode45(g, 0:0.5:10, x0);
format short
[ts,xs]
% end of program
```

Results:

```
ans =

         0    0.7854         0
    0.5000   -0.8609   -6.8742
    1.0000   -3.7001   -2.3851
    1.5000   -3.3568    3.9823
    2.0000   -0.1916    5.3803
    2.5000    0.7178   -1.2627
    3.0000   -1.6474   -7.4097
    3.5000   -3.8826   -1.0102
    4.0000   -2.8311    5.7023
    4.5000    0.2980    3.6596
    5.0000    0.5131   -2.6538
    5.5000   -2.4083   -6.6686
    6.0000   -3.9189    0.2451
    6.5000   -2.1153    7.0898
    7.0000    0.6019    2.0283
    7.5000    0.0638   -4.5067
    8.0000   -3.1170   -4.8067
    8.5000   -3.7773    1.7866
    9.0000   -1.1661    7.2102
    9.5000    0.7582    0.4365
   10.0000   -0.6490   -6.4667
```

D.2 Compound Pendulum Using the Function R(t,x)

```
% D2
% Direct Dynamics
% Compound pendulum
% the program uses the function: R(t,x)
% the function is defined in the program R.m

clear all; clc; close all
```

```
tfinal=10;
time=[0 tfinal];
% x(1)(0)=pi/4  x(2)(0)=0
x0=[pi/4 0];

% solve a second order ODE
% using MATLAB ODE45 function

[t,x]=ode45(@R,time,x0);

subplot(3,1,1),...
plot(t,x(:,1),'r'),...
xlabel('t'),ylabel('\theta'),grid,...
subplot(3,1,2),...
plot(t,x(:,2)),...
xlabel('t'),ylabel('\omega'),grid,...
subplot(3,1,3),...
plot(x(:,1),x(:,2)),...
xlabel('\theta'), ylabel('\omega'),grid
% end of program
```

D.3 Double Pendulum

```
% D3
% Direct Dynamics
% Double pendulum
clear all; clc; close all

L1 = 1; L2 = 0.5; m1 = 1; m2 = 1; g = 10;

t = sym('t','real');

xB = L1*cos(sym('q1(t)'));
yB = L1*sin(sym('q1(t)'));
rB = [xB yB 0];
rC1 = rB/2;
% differentiates rC1 with respect to t
vC1 = diff(rC1,t);
aC1 = diff(vC1,t);

xD = xB + L2*cos(sym('q2(t)'));
yD = yB + L2*sin(sym('q2(t)'));
```

```
rD = [xD yD 0];
rC2 = (rB + rD)/2;
vC2 = diff(rC2,t);
aC2 = diff(vC2,t);

omega1 = [0 0 diff('q1(t)',t)];
alpha1 = diff(omega1,t);
omega2 = [0 0 diff('q2(t)',t)];
alpha2 = diff(omega2,t);

G1 = [0 -m1*g 0];
G2 = [0 -m2*g 0];

IC1 = m1*L1^2/12;
IA = IC1 + m1*(L1/2)^2;
IC2 = m2*L2^2/12;

% LINK 2
% Sum F for link 2:
% -m2 aC2 + G2 + (-F21) = 0 => F21 = -m2 aC2 + G2
F21 = -m2*aC2 + G2;

% LINK 1
% Sum M for 1 wrt A:"
% -IA alpha1 + AB x F21 + AC1 x G1 = 0
EqA = -IA*alpha1 + cross(rB, F21) + cross(rC1, G1);

% LINK 2
% Sum M for 2 wrt C2:
% -IC2 alpha2 + C2B x (-F21) = 0
Eq2 = -IC2*alpha2 + cross(rB - rC2, -F21);

% list for the symbolical variables
slist={diff('q1(t)',t,2),diff('q2(t)',t,2),...
 diff('q1(t)',t),diff('q2(t)',t),'q1(t)','q2(t)'};
nlist={'ddq1','ddq2','x(2)','x(4)','x(1)','x(3)'};
% diff('q1(t)',t,2) will be replaced by 'ddq1'
% diff('q2(t)',t,2) will be replaced by 'ddq2'
% diff('q1(t)',t) will be replaced by 'x(2)'
% diff('q2(t)',t) will be replaced by 'x(4)'
% 'q1(t)' will be replaced by 'x(1)'
% 'q2(t)' will be replaced by 'x(3)'

eq1 = subs(EqA(3),slist,nlist);
eq2 = subs(Eq2(3),slist,nlist);
```

```
sol = solve(eq1,eq2,'ddq1, ddq2');

dx1 = sym('x(2)');
dx2 = sol.ddq1;
dx3 = sym('x(4)');
dx4 = sol.ddq2;

dx1dt = char(dx1);
dx2dt = char(dx2);
dx3dt = char(dx3);
dx4dt = char(dx4);

g=inline(sprintf('[%s;%s;%s;%s]',...
    dx1dt,dx2dt,dx3dt,dx4dt),'t','x');

t0 = 0;  tf = 5; time = [0 tf];

x0 = [-pi/4; 0; -pi/3; 0]; % initial conditions

[t,xs] = ode45(g, time, x0);

x1 = xs(:,1);
x2 = xs(:,2);
x3 = xs(:,3);
x4 = xs(:,4);

subplot(2,1,1),plot(t,x1*180/pi,'r'),...
xlabel('t (s)'),ylabel('q1 (deg)'),grid,...
subplot(2,1,2),plot(t,x3*180/pi,'b'),...
xlabel('t (s)'),ylabel('q2 (deg)'),grid

%[ts,xs] = ode45(g,0:1:5,x0);
%[ts,xs]
% end of program
```

D.4 Double Pendulum Using the File RR.m

```
% D4
% Direct Dynamics
% Double pendulum
% the program uses the function: RR(t,x)
% the function is defined in the program RR.m
```

```
clear all; clc; close all
L1 = 1; L2 = 0.5; m1 = 1; m2 = 1; g = 10;
t = sym('t','real');
xB = L1*cos(sym('q1(t)'));
yB = L1*sin(sym('q1(t)'));
rB = [xB yB 0];
rC1 = rB/2;
xD = xB+L2*cos(sym('q2(t)'));
yD = yB+L2*sin(sym('q2(t)'));
rD = [xD yD 0];
rC2 = (rB + rD)/2;
vC2 = diff(rC2,t); aC2 = diff(vC2,t);
omega1 = [0 0 diff('q1(t)',t)];
alpha1 = diff(omega1,t);
omega2 = [0 0 diff('q2(t)',t)];
alpha2 = diff(omega2,t);
G1 = [0 -m1*g 0]; G2 = [0 -m2*g 0];
IC1 = m1*L1^2/12; IA = IC1 + m1*(L1/2)^2;
IC2 = m2*L2^2/12;
F21 = -m2*aC2 + G2;
EqA = -IA*alpha1 + cross(rB, F21) + cross(rC1, G1);
Eq2 = -IC2*alpha2 + cross(rB - rC2, -F21);
slist = {diff('q1(t)',t,2),diff('q2(t)',t,2), ...
 diff('q1(t)',t),diff('q2(t)',t),'q1(t)','q2(t)'};
nlist = ...
{'ddq1', 'ddq2', 'x(2)', 'x(4)', 'x(1)','x(3)'};
eq1 = subs(EqA(3),slist,nlist);
eq2 = subs(Eq2(3),slist,nlist);

sol = solve(eq1,eq2,'ddq1, ddq2');
dx2 = sol.ddq1; dx4 = sol.ddq2;

dx2dt = char(dx2);
dx4dt = char(dx4);

% opens the file 'RR.m' in the mode specified by 'w+'
% (create for read and write)
fid = fopen('RR.m','w+');
fprintf(fid,'function dx = RR(t,x)\n');
fprintf(fid,'dx = zeros(4,1);\n');
fprintf(fid,'dx(1) = x(2);\n');
fprintf(fid,'dx(2) = ');
fprintf(fid,dx2dt);
fprintf(fid,';\n');
fprintf(fid,'dx(3) = x(4);\n');
```

```
fprintf(fid,'dx(4) = ');
fprintf(fid,dx4dt);
fprintf(fid,';');
% closes the file associated with file identifier fid
fclose(fid);
cd(pwd);
% cd changes current working directory
% pwd displays the current working directory

% fid = fopen(FIL,PERM) opens the file FILE in the
% mode specified by PERM. PERM can be:
% 'w'     writes (creates if necessary)
% 'w+'    truncates or creates for read and write

t0 = 0;  tf = 5; time = [0 tf];

x0 = [-pi/4 0 -pi/3 0]; % initial conditions

[t,xs] = ode45(@RR, time, x0);

x1 = xs(:,1);
x3 = xs(:,3);

subplot(2,1,1),plot(t,x1*180/pi,'r'),...
xlabel('t (s)'),ylabel('q1 (deg)'),grid,...
subplot(2,1,2),plot(t,x3*180/pi,'b'),...
xlabel('t (s)'),ylabel('q2 (deg)'),grid

% [ts,xs] = ode45(@RR,0:1:5,x0);
% [ts,xs]

% end of program
```

D.5 One-Link Planar Robot Arm

```
% D5
% Direct Dynamics
% R robot arm
clear all; clc; close all

syms t
L = 1; m = 1; g = 9.81;
```

```
c = cos(sym('theta(t)'));
s = sin(sym('theta(t)'));
xC = (L/2)*c; yC = (L/2)*s;
rC = [xC yC 0];

omega = [0 0 diff('theta(t)',t)];
alpha = diff(omega,t);

G = [0 -m*g 0];

IO = m*L^2/3;

beta = 45;
gamma = 30;
qf = pi/3;

T01z = -beta*diff('theta(t)',t)-...
    gamma*(sym('theta(t)')-qf)+0.5*g*L*m*c;
T01 = [0 0 T01z];

eq = -IO*alpha + cross(rC,G) + T01;
eqz = eq(3);

slist={diff('theta(t)',t,2),diff('theta(t)',t),...
       'theta(t)'};
nlist={'ddtheta', 'x(2)' , 'x(1)'};

eqI = subs(eqz,slist,nlist);

dx1 = sym('x(2)');
dx2 = solve(eqI,'ddtheta');

dx1dt = char(dx1);
dx2dt = char(dx2);

g=inline(sprintf('[%s; %s]', dx1dt, dx2dt), 't', 'x');

time = [0 10];
x0 = [pi/18; 0]; % define initial conditions

[ts,xs] = ode45(g, 0:1:10, x0);

plot(ts,xs(:,1)*180/pi,'LineWidth',1.5),...
xlabel('t (s)'), ylabel('\theta (deg)'),...
grid, axis([0, 10, 0, 70])
```

```
fprintf('Results \n\n')
fprintf('    t(s)     theta(rad) omega(rad/s) \n')
[ts,xs]
% end of program
```

Results:

```
    t(s)     theta(rad) omega(rad/s)

ans =

         0    0.1745         0
    1.0000    0.5984    0.3006
    2.0000    0.8175    0.1539
    3.0000    0.9297    0.0788
    4.0000    0.9871    0.0403
    5.0000    1.0164    0.0206
    6.0000    1.0315    0.0105
    7.0000    1.0391    0.0054
    8.0000    1.0431    0.0028
    9.0000    1.0451    0.0014
   10.0000    1.0461    0.0007
```

D.6 One-Link Planar Robot Arm Using the m-File Function Rrobot.m

```
% D6
% Direct Dynamics
% R robot arm
% the program uses the function: Rrobot(t,x)
% the function is defined in the program Rrobot.m

clear; clc; close all

time = [0 10];

x0 = [pi/18 0];

[ts,xs] = ode45(@Rrobot, 0:1:10, x0);

plot(ts,xs(:,1)*180/pi,'LineWidth',1.5),...
xlabel('t (s)'), ylabel('\theta (deg)'),...
grid, axis([0, 10, 0, 70])
```

```
fprintf('Results \n\n')
fprintf('    t(s)    theta(rad) omega(rad/s) \n')
[ts,xs]
% end of program
```

Results:

```
     t(s)    theta(rad) omega(rad/s)

ans =

         0    0.1745         0
    1.0000    0.5984    0.3006
    2.0000    0.8175    0.1539
    3.0000    0.9297    0.0788
    4.0000    0.9871    0.0403
    5.0000    1.0164    0.0206
    6.0000    1.0315    0.0105
    7.0000    1.0391    0.0054
    8.0000    1.0431    0.0028
    9.0000    1.0451    0.0014
   10.0000    1.0461    0.0007
```

D.7 Two-Link Planar Robot Arm Using the m-File Function RRrobot.m

```
% D7
% Direct Dynamics
% Double pendulum
% the program uses the function: RRrobot(t,x)
% the function is defined in the program RRrobot.m

clear all; clc; close all

L1 = 1; L2 = 1; m1 = 1; m2 = 1; g = 9.81;
t = sym('t','real');
xB = L1*cos(sym('q1(t)'));
yB = L1*sin(sym('q1(t)'));
rB = [xB yB 0];
rC1 = rB/2; vC1 = diff(rC1,t); aC1 = diff(vC1,t);
xD = xB + L2*cos(sym('q2(t)'));
yD = yB + L2*sin(sym('q2(t)'));
rD = [xD yD 0];
```

```
rC2 = (rB + rD)/2;
vC2 = diff(rC2,t); aC2 = diff(vC2,t);
omega1 = [0 0 diff('q1(t)',t)];
alpha1 = diff(omega1,t);
omega2 = [0 0 diff('q2(t)',t)];
alpha2 = diff(omega2,t);

G1 = [0 -m1*g 0]; G2 = [0 -m2*g 0];
IC1 = m1*L1^2/12; IA = IC1 + m1*(L1/2)^2;
IC2 = m2*L2^2/12;

F21 = -m2*aC2 + G2;

b01 = 450; g01 = 300;
b12 = 200; g12 = 300;

q1f = pi/6;
q2f = pi/3;

T01z=-b01*diff('q1(t)',t)-g01*(sym('q1(t)')-q1f)...
     +0.5*g*L1*m1*cos(sym('q1(t)'))+...
      g*L1*m2*cos(sym('q1(t)'));
T01 = [0 0 T01z];

T12z = -b12*diff('q2(t)',t)-g12*(sym('q2(t)')-q2f)...
      +0.5*g*L2*m2*cos(sym('q2(t)'));
T12 = [0 0 T12z];

EqA=-IA*alpha1+cross(rB, F21)+cross(rC1, G1)+T01-T12;

Eq2=-IC2*alpha2 + cross(rB - rC2, -F21) + T12;

slist={diff('q1(t)',t,2),diff('q2(t)',t,2),...
 diff('q1(t)',t),diff('q2(t)',t),'q1(t)','q2(t)'};
nlist= ...
 {'ddq1', 'ddq2', 'x(2)', 'x(4)', 'x(1)','x(3)'};

eq1 = subs(EqA(3),slist,nlist);
eq2 = subs(Eq2(3),slist,nlist);

sol = solve(eq1,eq2,'ddq1, ddq2');

dx2 = sol.ddq1;
dx4 = sol.ddq2;
```

```
dx2dt = char(dx2);
dx4dt = char(dx4);

fid = fopen('RRrobot.m','w+');
fprintf(fid,'function dx = RRrobot(t,x)\n');
fprintf(fid,'dx = zeros(4,1);\n');
fprintf(fid,'dx(1) = x(2);\n');
fprintf(fid,'dx(2) = ');
fprintf(fid,dx2dt);
fprintf(fid,';\n');
fprintf(fid,'dx(3) = x(4);\n');
fprintf(fid,'dx(4) = ');
fprintf(fid,dx4dt);
fprintf(fid,';');
fclose(fid);
cd(pwd);

t0 = 0;
tf = 15;
time = [0 tf];

x0 = [-pi/18 0 pi/6 0];

[t,xs] = ode45(@RRrobot, time, x0);

x1 = xs(:,1);
x2 = xs(:,2);
x3 = xs(:,3);
x4 = xs(:,4);

subplot(2,1,1),plot(t,x1*180/pi,'r'),...
xlabel('t (s)'),ylabel('q1 (deg)'),grid,...

subplot(2,1,2),plot(t,x3*180/pi,'b'),...
xlabel('t (s)'),ylabel('q2 (deg)'),grid

[ts,xs] = ode45(@RRrobot,0:1:5,x0);

fprintf('Results \n\n')

fprintf ...
(' t(s) q1(rad) dq1(rad/s) q2(rad) dq2(rad/s) \n')
[ts,xs]

% end of program
```

Results:

```
        t(s)       q1(rad)   dq1(rad/s) q2(rad)   dq2(rad/s)

    ans =

             0   -0.1745         0    0.5236         0
        1.0000    0.1594    0.2373    0.9304    0.1758
        2.0000    0.3327    0.1220    1.0213    0.0391
        3.0000    0.4217    0.0626    1.0415    0.0087
        4.0000    0.4674    0.0321    1.0460    0.0019
        5.0000    0.4908    0.0165    1.0469    0.0004
```

Appendix E
Programs of Chapter 6: Analytical Dynamics

E.1 Lagrange's Equations for Two-Link Robot Arm

```
% E1
% Analytical Dynamics
% RR robot arm
% Lagrange's e.o.m

clear all; clc; close all

syms t L1 L2 m1 m2 g

q1 = sym('q1(t)');
q2 = sym('q2(t)');

c1 = cos(q1); s1 = sin(q1);
c2 = cos(q2); s2 = sin(q2);

xB = L1*c1; yB = L1*s1;
rB = [xB yB 0];
rC1 = rB/2; vC1 = diff(rC1,t);
xD = xB + L2*c2; yD = yB + L2*s2;
rD = [xD yD 0];
rC2 = (rB + rD)/2; vC2 = diff(rC2,t);
omega1 = [0 0 diff(q1,t)];
omega2 = [0 0 diff(q2,t)];

IA = m1*L1^2/3; IC2 = m2*L2^2/12;

% kinetic energy of the link 1
T1 = IA*omega1*omega1.'/2;
```

```
% .' array transpose
% A.' is the array transpose of A

% kinetic energy of the link 2
T2 = m2*vC2*vC2.'/2 + IC2*omega2*omega2.'/2;
T2 = simple(T2);

% total kinetic energy
T = expand(T1 + T2);
fprintf('T = \n'); pretty(T); fprintf('\n')

%deriv(f,g(t)) differentiates f with respect to g(t)

% dT/d(dq)
Tdq1 = deriv(T, diff(q1,t));
Tdq2 = deriv(T, diff(q2,t));
fprintf('dT/d(dq1) = \n'); pretty(simple(Tdq1));
fprintf('\n')
fprintf('dT/d(dq2) = \n'); pretty(simple(Tdq2));
fprintf('\n')

% d(dT/d(dq))/dt
Tt1 = diff(Tdq1, t);
Tt2 = diff(Tdq2, t);
fprintf('d dT/d(dq1)/dt = \n');
pretty(simple(Tt1));
fprintf('\n')
fprintf('d dT/d(dq2)/dt = \n');
pretty(simple(Tt2));
fprintf('\n')

% dT/dq
Tq1 = deriv(T, q1);
Tq2 = deriv(T, q2);
fprintf('dT/dq1 = \n'); pretty(Tq1);
fprintf('\n')
fprintf('dT/dq2 = \n'); pretty(Tq2);
fprintf('\n')

% left-hand side of Lagrange's eom
LHS1 = Tt1 - Tq1;
LHS2 = Tt2 - Tq2;

% generalized active forces
```

```
G1 = [0 -m1*g 0]; G2 = [0 -m2*g 0];

syms T01z T12z

% contact torque of 0 that acts on link 1
T01 = [0 0 T01z];

% contact torque of link 1 that acts on link 2
T12 = [0 0 T12z];

% partial derivatives
rC1_1 = deriv(rC1, q1); rC2_1 = deriv(rC2, q1);
rC1_2 = deriv(rC1, q2); rC2_2 = deriv(rC2, q2);

w1_1 = deriv(omega1, diff(q1,t));
w2_1 = deriv(omega2, diff(q1,t));
w1_2 = deriv(omega1, diff(q2,t));
w2_2 = deriv(omega2, diff(q2,t));

% generalized active force Q1
Q1 = rC1_1*G1.'+w1_1*T01.'+w1_1*(-T12.')+ ...
     rC2_1*G2.'+w2_1*T12.';

% generalized active force Q2
Q2 = rC1_2*G1.'+w1_2*T01.'+w1_2*(-T12.')+ ...
     rC2_2*G2.'+w2_2*T12.';

fprintf('Q1 = \n'); pretty(simple(Q1));
fprintf('\n')
fprintf('Q2 = \n'); pretty(simple(Q2));
fprintf('\n')

% first Lagrange's equation of motion
Lagrange1 = LHS1-Q1;
% second Lagrange's equation of motion
Lagrange2 = LHS2-Q2;

% control torques
b01 = 450; g01 = 300;
b12 = 200; g12 = 300;

q1f = pi/6;
q2f = pi/3;
```

```
T01zc = -b01*diff(q1,t)-g01*(q1-q1f)+ ...
        0.5*g*L1*m1*c1+g*L1*m2*c1;
T12zc = -b12*diff(q2,t)-g12*(q2-q2f)+ ...
        0.5*g*L2*m2*c2;

tor  = {T01z, T12z};
torf = {T01zc,T12zc};

Lagrang1 = subs(Lagrange1, tor, torf);
Lagrang2 = subs(Lagrange2, tor, torf);

data = {L1, L2, m1, m2, g};
datn = {1 , 1 , 1 , 1 , 9.81};
% L1 = 1; L2 = 1; m1 = 1; m2 = 1; g = 9.81;

Lagran1 = subs(Lagrang1, data, datn);
Lagran2 = subs(Lagrang2, data, datn);

ql = {diff(q1,t,2), diff(q2,t,2),...
    diff(q1,t), diff(q2,t), q1, q2};
qf = ...
{'ddq1', 'ddq2', 'x(2)', 'x(4)', 'x(1)', 'x(3)'};

% ql                        qf
%-----------------------------
% diff('q1(t)',t,2)  -> 'ddq1'
% diff('q2(t)',t,2)  -> 'ddq2'
%   diff('q1(t)',t)  -> 'x(2)'
%   diff('q2(t)',t)  -> 'x(4)'
%           'q1(t)'  -> 'x(1)'
%           'q2(t)'  -> 'x(3)'


Lagra1 = subs(Lagran1, ql, qf);
Lagra2 = subs(Lagran2, ql, qf);

% solve e.o.m. for ddq1, ddq2
sol = solve(Lagra1,Lagra2,'ddq1, ddq2');
Lagr1 = sol.ddq1;
Lagr2 = sol.ddq2;

% system of ODE
dx2dt = char(Lagr1);
dx4dt = char(Lagr2);
```

```
fid = fopen('RR_Lagr.m','w+');
fprintf(fid,'function dx = RR_Lagr(t,x)\n');
fprintf(fid,'dx = zeros(4,1);\n');
fprintf(fid,'dx(1) = x(2);\n');
fprintf(fid,'dx(2) = ');
fprintf(fid,dx2dt);
fprintf(fid,';\n');
fprintf(fid,'dx(3) = x(4);\n');
fprintf(fid,'dx(4) = ');
fprintf(fid,dx4dt);
fprintf(fid,';');
fclose(fid); cd(pwd);

t0 = 0;  tf = 15; time = [0 tf];

x0 = [pi/18 0 pi/6 0];

[t,xs] = ode45(@RR_Lagr, time, x0);

x1 = xs(:,1);
x2 = xs(:,2);
x3 = xs(:,3);
x4 = xs(:,4);

subplot(2,1,1),plot(t,x1*180/pi,'r'),...
xlabel('t (s)'),ylabel('q1 (deg)'),grid,...
subplot(2,1,2),plot(t,x3*180/pi,'b'),...
xlabel('t (s)'),ylabel('q2 (deg)'),grid

[ts,xs] = ode45(@RR_Lagr,0:1:5,x0);

fprintf('Results \n\n')
fprintf ...
('t(s) q1(rad) dq1(rad/s) q2(rad) dq2(rad/s) \n')
[ts,xs]
% end of program
```

Numerical results:

```
     t(s)      q1(rad)   dq1(rad/s) q2(rad)   dq2(rad/s)

ans =

          0     0.1745          0     0.5236          0
     1.0000     0.3387     0.1175     0.9305     0.1757
     2.0000     0.4246     0.0605     1.0213     0.0390
```

```
3.0000      0.4688      0.0311      1.0415      0.0087
4.0000      0.4915      0.0160      1.0459      0.0019
5.0000      0.5032      0.0082      1.0469      0.0004
```

E.2 Two-Link Robot Arm: Inverse Dynamics

```
% E2
% Analytical Dynamics
% RR robot arm
% Lagrange's e.o.m
% Inverse dynamics
clear all; clc; close all
syms t L1 L2  m1 m2 m3 g
q1 = sym('q1(t)'); q2 = sym('q2(t)');
c1 = cos(q1); s1 = sin(q1);
c2 = cos(q2); s2 = sin(q2);
xB = L1*c1; yB = L1*s1;
rB = [xB yB 0];
rC1 = rB/2; vC1 = diff(rC1,t);
xD = xB + L2*c2; yD = yB + L2*s2;
rD = [xD yD 0];
rC2 = (rB + rD)/2; vC2 = diff(rC2,t);
omega1 = [0 0 diff(q1,t)];
omega2 = [0 0 diff(q2,t)];
IA = m1*L1^2/3; IC2 = m2*L2^2/12;
T1 = IA*omega1*omega1.'/2;
T2 = m2*vC2*vC2.'/2 + IC2*omega2*omega2.'/2;
T = expand(T1 + T2);
%deriv(f,g(t)) differentiates f with respect to g(t)
Tdq1 = deriv(T, diff(q1,t));
Tdq2 = deriv(T, diff(q2,t));
Tt1 = diff(Tdq1, t); Tt2 = diff(Tdq2, t);
Tq1 = deriv(T, q1); Tq2 = deriv(T, q2);
LHS1 = Tt1 - Tq1; LHS2 = Tt2 - Tq2;

rC1_1 = deriv(rC1, q1); rC2_1 = deriv(rC2, q1);
rC1_2 = deriv(rC1, q2); rC2_2 = deriv(rC2, q2);
w1_1 = deriv(omega1, diff(q1,t));
w2_1 = deriv(omega2, diff(q1,t));
w1_2 = deriv(omega1, diff(q2,t));
w2_2 = deriv(omega2, diff(q2,t));
G1 = [0 -m1*g 0]; G2 = [0 -m2*g 0];
syms T01z T12z
```

```
% contact torque of 0 that acts on link 1
T01 = [0 0 T01z];
% contact torque of link 1 that acts on link 2
T12 = [0 0 T12z];
Q1=rC1_1*G1.'+w1_1*T01.'+w1_1*(-T12.')+ ...
   rC2_1*G2.'+w2_1*T12.';
Q2=rC1_2*G1.'+w1_2*T01.'+w1_2*(-T12.')+ ...
   rC2_2*G2.'+w2_2*T12.';
Lagrange1 = LHS1-Q1; Lagrange2 = LHS2-Q2;
data = {L1, L2, m1, m2, g};
datn = {1 , 1 , 1 , 1 , 9.81};

Lagr1 = subs(Lagrange1, data, datn);
Lagr2 = subs(Lagrange2, data, datn);

% solve for T01z T12z
sol = solve(Lagr1,Lagr2,'T01z, T12z');
T01zc = sol.T01z;
T12zc = sol.T12z;

% INVERSE DYNAMICS

q1f = pi/6 ; q2f = pi/3;
q1s = pi/18; q2s = pi/6;

Tp=15.;

q1n=q1s+(q1f-q1s)/Tp*(t-Tp/(2*pi)*sin(2*pi/Tp*t));
q2n=q2s+(q2f-q2s)/Tp*(t-Tp/(2*pi)*sin(2*pi/Tp*t));

dq1n = diff(q1n,t);
dq2n = diff(q2n,t);

ddq1n = diff(dq1n,t);
ddq2n = diff(dq2n,t);

ql={diff(q1,t,2),diff(q2,t,2), ...
    diff(q1,t),diff(q2,t),q1,q2};
qn={ddq1n, ddq2n, dq1n, dq2n, q1n, q2n};

% ql                      qn
%---------------------------
% diff('q1(t)',t,2) -> ddq1n
% diff('q2(t)',t,2) -> ddq2n
%   diff('q1(t)',t) ->  dq1n
```

```
%   diff('q2(t)',t)  ->  dq2n
%           'q1(t)'  ->   q1n
%           'q2(t)'  ->   q1n

T01zt = subs(T01zc, ql, qn);
T12zt = subs(T12zc, ql, qn);

% ezplot(f,[min,max]) plots f(t)
% over the domain: min < t < max

subplot(2,1,1), ezplot(T01zt,[0,Tp]),...
xlabel('t (s)'), ylabel('T01z (N m)'),...
title(''), grid,...

subplot(2,1,2),ezplot(T12zt,[0,Tp]),...
xlabel('t (s)'), ylabel('T12z (N m)'),...
title(''), grid

% % another way of plotting T01z and T12z
%
% time = 0:1:Tp;
% T01t = subs(T01zt,'t',time);
% T12t = subs(T12zt,'t',time);
% subplot(2,1,1),plot(time,T01t),...
% xlabel('t (s)'),ylabel('T01z (N m)'),grid,...
% subplot(2,1,2),plot(time,T12t),...
% xlabel('t (s)'),ylabel('T12z (N m)'),grid
%
% q_1t = subs(q1n,'t',time);
% q_2t = subs(q2n,'t',time);
%
% subplot(2,1,1),plot(time,q_1t*180/pi),...
% xlabel('t (s)'),ylabel('q1 (deg)'),grid,...
%
% subplot(2,1,2),plot(time,q_2t*180/pi),...
% xlabel('t (s)'),ylabel('q2 (deg)'),grid
% end of program
```

E.3 RRT Robot Arm

```
% E3
% Analytical Dynamics
% RRT robot arm
```

```
% Lagrange's e.o.m
clear all; clc; close all

syms t L1 L2  m1 m2 m3 g real

q1 = sym('q1(t)');
q2 = sym('q2(t)');
q3 = sym('q3(t)');

c1 = cos(q1);
s1 = sin(q1);
c2 = cos(q2);
s2 = sin(q2);

% transformation matrix from RF1 to RF0
R10 = [[1 0 0]; [0 c1 s1]; [0 -s1 c1]];

% transformation matrix from RF2 to RF1
R21 = [[c2 0 -s2]; [0 1 0]; [s2 0 c2]];

% angular velocity of link 1 in RF0 expressed
% in terms of RF1 {i1,j1,k1}
w10 = [diff(q1,t) 0 0 ];

% angular velocity of link 2 in RF0 expressed
% in terms of RF1 {i1,j1,k1}
w201 = [diff(q1,t) diff(q2,t) 0];

% angular velocity of link 2 in RF0 expressed
% in terms of RF2 {i2,j2,k2}
w20 = w201 * transpose(R21);

% angular acceleration of link 1 in RF0 expressed
% in terms of RF1{i1,j1,k1}
alpha10 = diff(w10,t);

% angular acceleration of link 2 in RF0 expressed
% in terms of RF2{i2,j2,k2}
alpha20 = diff(w20,t);

% position vector of mass center C1 of link 1
% in RF0 expressed in terms of RF1 {i1,j1,k1}
rC1 = [0 0 L1];

% linear velocity of mass center C1 of link 1
```

```
% in RF0 expressed in terms of RF1 {i1,j1,k1}
vC1 = diff(rC1,t) + cross(w10, rC1);

% position vector of mass center C2 of link 2
% in RF0 expressed in terms of RF2 {i2,j2,k2}
rC2 = [0 0 2*L1]*transpose(R21) + [0 0 L2];

% linear velocity of mass center C2 of link 2
% in RF0 expressed in terms of RF2 {i2,j2,k2}
vC2 = simple(diff(rC2,t) + cross(w20,rC2));

% position vector of mass center C3 of link 3 in RF0
% expressed in terms of RF2 {i2,j2,k2}
rC3 = rC2 + [0 0 q3];

% linear velocity of mass center C3 of link 3 in RF0
% expressed in terms of RF2 {i2,j2,k2}
vC3 = simple(diff(rC3,t) + cross(w20,rC3));

% linear velocity of  C32 of link 2 in RF0
% expressed in terms of RF2 {i2,j2,k2}
% C32 of link 2 is superposed with C3 of link 3
vC32 = simple(vC2 + cross(w20,[0 0 q3]));

% another way of computing vC3 is:
% vC3p= vC32+diff([0 0 sym('q3(t)')],t);
% vC3-vC3p

% linear accelerations
aC1 = simple(diff(vC1,t)+cross(w10,vC1));
aC2 = simple(diff(vC2,t)+cross(w20,vC2));
aC3 = simple(diff(vC3,t)+cross(w20,vC3));

% gravitational force that acts on link 1 at C1
% RF0 expressed in terms of RF1 {i1,j1,k1}
G1 = [-m1*g 0 0]

% gravitational force that acts on link 2 at C2
% in RF0 expressed in terms of RF2 {i2,j2,k2}
G2 = [-m2*g 0 0]*transpose(R21)

% gravitational force that acts on link 3 at C3
% in RF0 expressed in terms of RF2 {i2,j2,k2}
G3 = [-m3*g 0 0]*transpose(R21)
```

```
syms T01x T01y T01z T12x T12y T12z F23x F23y F23z

% contact torque of 0 that acts on link 1
% in RF0 expressed in terms of RF1 {i1,j1,k1}
T01 = [T01x T01y T01z];

% contact torque of link 1 that acts on link 2
% in RF0 expressed in terms of RF2 {i2,j2,k2}
T12 = [T12x T12y T12z];

% contact force of link 2 that acts on link 3 at C3
% in RF0 expressed in terms of RF2 {i2,j2,k2}
F23 = [F23x F23y F23z];

%deriv(f,g(t)) differentiates f with respect to g(t)

w1_1 = deriv(w10, diff(q1,t));
w1_2 = deriv(w10, diff(q2,t));
w1_3 = deriv(w10, diff(q3,t));

w2_1 = deriv(w20, diff(q1,t));
w2_2 = deriv(w20, diff(q2,t));
w2_3 = deriv(w20, diff(q3,t));

vC1_1 = deriv(vC1, diff(q1,t));
vC1_2 = deriv(vC1, diff(q2,t));
vC1_3 = deriv(vC1, diff(q3,t));

vC2_1 = deriv(vC2, diff(q1,t));
vC2_2 = deriv(vC2, diff(q2,t));
vC2_3 = deriv(vC2, diff(q3,t));

vC32_1 = deriv(vC32, diff(q1,t));
vC32_2 = deriv(vC32, diff(q2,t));
vC32_3 = deriv(vC32, diff(q3,t));

vC3_1 = deriv(vC3, diff(q1,t));
vC3_2 = deriv(vC3, diff(q2,t));
vC3_3 = deriv(vC3, diff(q3,t));

% generalized active forces

% generalized active force Q1
Q1 = w1_1*T01.' + vC1_1*G1.' + ...
     w1_1*transpose(R21)*(-T12.') + ...
```

```
          w2_1*T12.' + vC2_1*G2.' + vC32_1*(-F23.') + ...
          vC3_1*F23.' + vC3_1*G3.'

% generalized active force Q2
Q2 = w1_2*T01.' + vC1_2*G1.' + ...
          w1_2*transpose(R21)*(-T12.') + ...
          w2_2*T12.' + vC2_2*G2.' + vC32_2*(-F23.') + ...
          vC3_2*F23.' + vC3_2*G3.'

% generalized active force Q3
Q3 = w1_3*T01.' + vC1_3*G1.' + ...
          w1_3*transpose(R21)*(-T12.') + ...
          w2_3*T12.' + vC2_3*G2.' + vC32_3*(-F23.') + ...
          vC3_3*F23.' + vC3_3*G3.'

% kinetic energy

% inertia dyadic

% inertia matrix associated with central inertia
% dyadic for link 1 expressed in terms of RF1
I1 = [m1*(2*L1)^2/12 0 0; 0 m1*(2*L1)^2/12 0; 0 0 0];

% inertia matrix associated with central inertia
% dyadic for link 2 expressed in terms of RF2
I2 = [m2*(2*L2)^2/12 0 0; 0 m2*(2*L2)^2/12 0; 0 0 0];

% inertia matrix associated with central inertia
% dyadic for link 3 expressed in terms of RF2
syms I3x I3y I3z real
I3 = [I3x 0 0; 0 I3y 0; 0 0 I3z];

% kinetic energy of the link 1
T1 = (1/2)*m1*vC1*vC1.' + (1/2)*w10*I1*w10.';

% kinetic energy of the link 2
T2 = (1/2)*m2*vC2*vC2.' + (1/2)*w20*I2*w20.';

% kinetic energy of the link 3
T3 = (1/2)*m3*vC3*vC3.' + (1/2)*w20*I3*w20.';

% .' array transpose
% A.' is the array transpose of A

% total kinetic energy
```

```
T = expand(T1 + T2 + T3);

% dT/d(dq)
Tdq1 = deriv(T, diff(q1,t));
Tdq2 = deriv(T, diff(q2,t));
Tdq3 = deriv(T, diff(q3,t));

% d(dT/d(dq))/dt
Tt1 = diff(Tdq1, t);
Tt2 = diff(Tdq2, t);
Tt3 = diff(Tdq3, t);

% dT/dq
Tq1 = deriv(T, q1);
Tq2 = deriv(T, q2);
Tq3 = deriv(T, q3);

% left-hand side of Lagrange's eom
LHS1 = Tt1 - Tq1;
LHS2 = Tt2 - Tq2;
LHS3 = Tt3 - Tq3;

% first Lagrange's equation of motion
Lagrange1 = LHS1-Q1;
% second Lagrange's equation of motion
Lagrange2 = LHS2-Q2;
% third Lagrange's equation of motion
Lagrange3 = LHS3-Q3;

% control torques and control force
q1f=pi/3; q2f=pi/3; q3f=0.3;
b01=450; g01=300;
b12=200; g12=300;
b23=150; g23=50;

T01xc = -b01*diff(q1,t)-g01*(q1-q1f);
T12yc = -b12*diff(q2,t)-g12*(q2-q2f)+ ...
         g*(m2*L2+m3*(L2+q3))*c2;
F23zc = -b23*diff(q3,t)-g23*(q3-q3f)+g*m3*s2;

tor  = {T01x, T12y, F23z};
torf = {T01xc,T12yc,F23zc};

Lagrang1 = subs(Lagrange1, tor, torf);
Lagrang2 = subs(Lagrange2, tor, torf);
```

```
Lagrang3 = subs(Lagrange3, tor, torf);

data = {L1, L2, I3x, I3y, I3z, m1, m2, m3, g};
datn = {0.4, 0.4, 5, 4, 1, 90, 60, 40, 9.81};

Lagran1 = subs(Lagrang1, data, datn);
Lagran2 = subs(Lagrang2, data, datn);
Lagran3 = subs(Lagrang3, data, datn);

ql = {diff(q1,t,2), diff(q2,t,2), diff(q3,t,2), ...
   diff(q1,t), diff(q2,t), diff(q3,t), q1, q2, q3};
qf = {'ddq1', 'ddq2',  'ddq3',...
   'x(2)', 'x(4)', 'x(6)', 'x(1)', 'x(3)', 'x(5)'};

% ql                          qf
%-----------------------------
% diff('q1(t)',t,2)  -> 'ddq1'
% diff('q2(t)',t,2)  -> 'ddq2'
% diff('q3(t)',t,2)  -> 'ddq3'
%   diff('q1(t)',t)  -> 'x(2)'
%   diff('q2(t)',t)  -> 'x(4)'
%   diff('q3(t)',t)  -> 'x(6)'
%           'q1(t)'  -> 'x(1)'
%           'q2(t)'  -> 'x(3)'
%           'q3(t)'  -> 'x(5)'

Lagra1 = subs(Lagran1, ql, qf);
Lagra2 = subs(Lagran2, ql, qf);
Lagra3 = subs(Lagran3, ql, qf);

% solve e.o.m. for ddq1, ddq2, ddq3
sol=solve(Lagra1,Lagra2,Lagra3,'ddq1,ddq2,ddq3');
Lagr1 = sol.ddq1;
Lagr2 = sol.ddq2;
Lagr3 = sol.ddq3;

% system of ODE
dx2dt = char(Lagr1);
dx4dt = char(Lagr2);
dx6dt = char(Lagr3);

fid = fopen('RRT_Lagr.m','w+');
fprintf(fid,'function dx = RRT_Lagr(t,x)\n');
fprintf(fid,'dx = zeros(6,1);\n');
fprintf(fid,'dx(1) = x(2);\n');
```

```
fprintf(fid,'dx(2) = ');
fprintf(fid,dx2dt);
fprintf(fid,';\n');
fprintf(fid,'dx(3) = x(4);\n');
fprintf(fid,'dx(4) = ');
fprintf(fid,dx4dt);
fprintf(fid,';\n');
fprintf(fid,'dx(5) = x(6);\n');
fprintf(fid,'dx(6) = ');
fprintf(fid,dx6dt);
fprintf(fid,';');
fclose(fid); cd(pwd);

t0 = 0;  tf = 15; time = [0 tf];

x0 = [pi/18 0 pi/6 0 0.25 0];

[t,xs] = ode45(@RRT_Lagr, time, x0);

x1 = xs(:,1);
x2 = xs(:,2);
x3 = xs(:,3);
x4 = xs(:,4);
x5 = xs(:,5);
x6 = xs(:,6);

subplot(3,1,1),plot(t,x1*180/pi,'r'),...
xlabel('t (s)'),ylabel('q1 (deg)'),grid,...
subplot(3,1,2),plot(t,x3*180/pi,'b'),...
xlabel('t (s)'),ylabel('q2 (deg)'),grid,...
subplot(3,1,3),plot(t,x5,'g'),...
xlabel('t (s)'),ylabel('q3 (m)'),grid

[ts,xs] = ode45(@RRT_Lagr,0:1:5,x0);
fprintf('Results \n\n')
fprintf...
('    t(s)        q1(rad)    q2(rad)    q3(m)\n')
[ts, xs(:,1), xs(:,3), xs(:,5)]
% end of program
```

Results:

```
G1 =

[ -m1*g,      0,      0]
```

```
G2 =

[ -m2*g*cos(q2(t)), 0, -m2*g*sin(q2(t))]


G3 =

[ -m3*g*cos(q2(t)), 0, -m3*g*sin(q2(t))]


Q1 =

T01x


Q2 =

T12y-L2*m2*g*cos(q2(t))-(L2+q3(t))*m3*g*cos(q2(t))


Q3 =

F23z-m3*g*sin(q2(t))

     t(s)        q1(rad)    q2(rad)    q3(m)

ans =

         0    0.1745    0.5236    0.2500
    1.0000    0.5217    0.9225    0.3060
    2.0000    0.8333    1.0206    0.3344
    3.0000    0.9640    1.0418    0.3296
    4.0000    1.0146    1.0463    0.3213
    5.0000    1.0344    1.0471    0.3149
```

E.4 RRT Robot Arm: Inverse Dynamics

```
% E4
% Analytical Dynamics
% RRT robot arm
% Lagrange's e.o.m
% Inverse Dynamics
clear all; clc; close all
syms t L1 L2  m1 m2 m3 g real
q1=sym('q1(t)'); q2=sym('q2(t)'); q3=sym('q3(t)');
c1=cos(q1); s1=sin(q1); c2=cos(q2); s2=sin(q2);
R10 = [[1 0 0]; [0 c1 s1]; [0 -s1 c1]];
R21 = [[c2 0 -s2]; [0 1 0]; [s2 0 c2]];
w10 = [diff(q1,t) 0 0 ];
w201 = [diff(q1,t) diff(q2,t) 0];
w20 = w201 * transpose(R21);
rC1 = [0 0 L1];
vC1 = diff(rC1,t) + cross(w10, rC1);
rC2 = [0 0 2*L1]*transpose(R21) + [0 0 L2];
vC2 = simple(diff(rC2,t) + cross(w20,rC2));
rC3 = rC2 + [0 0 q3];
vC3 = simple(diff(rC3,t) + cross(w20,rC3));
vC32 = simple(vC2 + cross(w20,[0 0 q3]));
G1 = [-m1*g 0 0];
G2 = [-m2*g 0 0]*transpose(R21);
G3 = [-m3*g 0 0]*transpose(R21);

syms T01x T01y T01z T12x T12y T12z F23x F23y F23z
T01 = [T01x T01y T01z];
T12 = [T12x T12y T12z];
F23 = [F23x F23y F23z];

% deriv(f, g(t)) differentiates f
% with respect to g(t)
w1_1 = deriv(w10, diff(q1,t));
w2_1 = deriv(w20, diff(q1,t));
w1_2 = deriv(w10, diff(q2,t));
w2_2 = deriv(w20, diff(q2,t));
w1_3 = deriv(w10, diff(q3,t));
w2_3 = deriv(w20, diff(q3,t));

vC1_1 = deriv(vC1, diff(q1,t));
vC2_1 = deriv(vC2, diff(q1,t));
vC1_2 = deriv(vC1, diff(q2,t));
vC2_2 = deriv(vC2, diff(q2,t));
```

```
vC1_3 = deriv(vC1, diff(q3,t));
vC2_3 = deriv(vC2, diff(q3,t));

vC32_1 = deriv(vC32, diff(q1,t));
vC3_1  = deriv(vC3, diff(q1,t)) ;
vC32_2 = deriv(vC32, diff(q2,t));
vC3_2  = deriv(vC3, diff(q2,t)) ;
vC32_3 = deriv(vC32, diff(q3,t));
vC3_3  = deriv(vC3, diff(q3,t)) ;

% generalized active forces
Q1 = w1_1*T01.' + vC1_1*G1.' + ...
     w1_1*transpose(R21)*(-T12.') + ...
     w2_1*T12.' + vC2_1*G2.' + vC32_1*(-F23.') + ...
     vC3_1*F23.' + vC3_1*G3.';

Q2 = w1_2*T01.' + vC1_2*G1.' + ...
     w1_2*transpose(R21)*(-T12.') + ...
     w2_2*T12.' + vC2_2*G2.' + vC32_2*(-F23.') + ...
     vC3_2*F23.' + vC3_2*G3.';

Q3 = w1_3*T01.' + vC1_3*G1.' + ...
     w1_3*transpose(R21)*(-T12.') + ...
     w2_3*T12.' + vC2_3*G2.' + vC32_3*(-F23.') + ...
     vC3_3*F23.' + vC3_3*G3.';

I1 = [m1*(2*L1)^2/12 0 0; 0 m1*(2*L1)^2/12 0; 0 0 0];
I2 = [m2*(2*L2)^2/12 0 0; 0 m2*(2*L2)^2/12 0; 0 0 0];
syms I3x I3y I3z
I3 = [I3x 0 0; 0 I3y 0; 0 0 I3z];

T1 = (1/2)*m1*vC1*vC1.' + (1/2)*w10*I1*w10.';
T2 = (1/2)*m2*vC2*vC2.' + (1/2)*w20*I2*w20.';
T3 = (1/2)*m3*vC3*vC3.' + (1/2)*w20*I3*w20.';
T = expand(T1 + T2 + T3);

Tdq1 = deriv(T, diff(q1,t));
Tdq2 = deriv(T, diff(q2,t));
Tdq3 = deriv(T, diff(q3,t));

Tt1=diff(Tdq1,t); Tt2=diff(Tdq2,t); Tt3=diff(Tdq3,t);

Tq1=deriv(T, q1); Tq2=deriv(T, q2); Tq3=deriv(T, q3);

LHS1=Tt1 - Tq1; LHS2=Tt2 - Tq2; LHS3=Tt3 - Tq3;
```

```
Lagrange1 = LHS1 - Q1;
Lagrange2 = LHS2 - Q2;
Lagrange3 = LHS3 - Q3;

data = {L1, L2, I3x, I3y, I3z, m1, m2, m3, g};
datn = {0.4, 0.4, 5, 4, 1, 90, 60, 40, 9.81};

Lagra1 = subs(Lagrange1, data, datn);
Lagra2 = subs(Lagrange2, data, datn);
Lagra3 = subs(Lagrange3, data, datn);

sol = solve(Lagra1,Lagra2,Lagra3,'T01x,T12y,F23z');
T01xc = simple(sol.T01x);
T12yc = simple(sol.T12y);
F23zc = simple(sol.F23z);

q1s = pi/18; q2s = pi/6; q3s = 0.25;
q1f = pi/3 ; q2f = pi/3; q3f = 0.3;

Tp=15.;

q1t=q1s+(q1f-q1s)/Tp*(t-Tp/(2*pi)*sin(2*pi/Tp*t));
q2t=q2s+(q2f-q2s)/Tp*(t-Tp/(2*pi)*sin(2*pi/Tp*t));
q3t=q3s+(q3f-q3s)/Tp*(t-Tp/(2*pi)*sin(2*pi/Tp*t));

dq1t = diff(q1t,t);
dq2t = diff(q2t,t);
dq3t = diff(q3t,t);

ddq1t = diff(dq1t,t);
ddq2t = diff(dq2t,t);
ddq3t = diff(dq3t,t);

ql = {diff(q1,t,2), diff(q2,t,2), diff(q3,t,2), ...
   diff(q1,t), diff(q2,t), diff(q3,t), q1, q2, q3};

qn = ...
  {ddq1t,ddq2t,ddq3t, dq1t,dq2t,dq3t, q1t,q2t,q3t};

T01xt = subs(T01xc, ql, qn);
T12yt = subs(T12yc, ql, qn);
F23zt = subs(F23zc, ql, qn);

time = 0:1:Tp;
```

```
T01t = subs(T01xt,'t',time);
T12t = subs(T12yt,'t',time);
F23t = subs(F23zt,'t',time);

subplot(3,1,1),plot(time,T01t),...
xlabel('t (s)'),ylabel('T01x (N m)'),grid,...

subplot(3,1,2),plot(time,T12t),...
xlabel('t (s)'),ylabel('T12y (N m)'),grid,...

subplot(3,1,3),plot(time,F23t),...
xlabel('t (s)'),ylabel('F23z (N)'),grid

fprintf('Results \n\n')

fprintf ...
('    t(s)     T01x(Nm)  T12y(Nm)  F23z(N)\n')

[time' T01t' T12t' F23t']

% another way of plotting T01x, T12y, and F23z
%
% subplot(3,1,1), ezplot(T01xt,[0,Tp]),...
% title(''), xlabel('t(s)'), ylabel('T01x (N m)'),...
%
% subplot(3,1,2), ezplot(T12yt,[0,Tp]),...
% title(''), xlabel('t(s)'), ylabel('T12y (N m)'),...
%
% subplot(3,1,3), ezplot(F23zt,[0,Tp]),...
% title(''), xlabel('t(s)'), ylabel('F23z (N)')
% end of program
```

Results :

```
     t(s)     T01x(Nm)  T12y(Nm)  F23z(N)

ans =

         0   -0.0000  424.7855  196.2000
    1.0000    1.7703  424.7713  196.5650
    2.0000    3.2150  423.4736  198.8939
    3.0000    4.0454  419.7485  204.7523
    4.0000    4.0801  412.6265  214.9893
    5.0000    3.3036  401.3600  229.5450
    6.0000    1.8890  385.5925  247.4489
```

```
      7.0000    0.1625  365.6353  267.0360
      8.0000   -1.4946  342.6925  286.3519
      9.0000   -2.7690  318.8276  303.6290
     10.0000   -3.4945  296.5656  317.6617
     11.0000   -3.6461  278.2350  327.9492
     12.0000   -3.2804  265.3101  334.6139
     13.0000   -2.4792  258.0103  338.2170
     14.0000   -1.3396  255.2728  339.6061
     15.0000   -0.0000  255.0600  339.8284
```

E.5 RRT Robot Arm: Kane's Dynamical Equations

```
% E5
% Analytical Dynamics
% RRT robot arm
% Kane's dynamical equations
clear all; clc; close all
syms t L1 L2  m1 m2 m3 g I3x I3y I3z ...
 T01x T01y T01z T12x T12y T12z F23x F23y F23z real

% generalized coordinates q1, q2, q3
q1 = sym('q1(t)');
q2 = sym('q2(t)');
q3 = sym('q3(t)');
% generalized speeds u1, u2, u3
u1 = sym('u1(t)');
u2 = sym('u2(t)');
u3 = sym('u3(t)');
% expressing q1', q2', q3' in terms of
% generalized speeds u1, u2, u3
dq1 = u1;
dq2 = u2;
dq3 = u3;

qt = {diff(q1,t), diff(q2,t), diff(q3,t)};
qu = {dq1, dq2, dq3};

c1=cos(q1); s1=sin(q1); c2=cos(q2); s2=sin(q2);

R10 = [[1 0 0]; [0 c1 s1]; [0 -s1 c1]];
R21 = [[c2 0 -s2]; [0 1 0]; [s2 0 c2]];

w10  = [dq1, 0, 0 ];
```

```
w201 = [dq1, dq2, 0];
w20  = w201 * transpose(R21);
alpha10 = diff(w10,t);
alpha20 = subs(diff(w20,t), qt, qu);

rC1 = [0 0 L1];
vC1 = diff(rC1,t) + cross(w10, rC1);
rC2 = [0 0 2*L1]*transpose(R21) + [0 0 L2];
vC2 = subs(diff(rC2, t), qt, qu) + cross(w20,rC2);

rC3 = rC2 + [0 0 q3];
vC3 = subs(diff(rC3, t), qt, qu) + cross(w20,rC3);
vC32 = vC2 + cross(w20,[0 0 q3]);

aC1 = diff(vC1,t) + cross(w10,vC1);
aC2 = diff(vC2,t) + cross(w20,vC2);
aC3 = subs(diff(vC3,t), qt, qu) + cross(w20,vC3);

% the velocities and accelerations are functions of
% q1, q2, q3, u1, u2, u3, and u1', u2', u3'

G1 = [-m1*g 0 0];
G2 = [-m2*g 0 0]*transpose(R21);
G3 = [-m3*g 0 0]*transpose(R21);

T01 = [T01x T01y T01z];
T12 = [T12x T12y T12z];
F23 = [F23x F23y F23z];

%deriv(f,g(t)) differentiates f with respect to g(t)
% partial velocities
w1_1 = deriv(w10, u1); w2_1 = deriv(w20, u1);
w1_2 = deriv(w10, u2); w2_2 = deriv(w20, u2);
w1_3 = deriv(w10, u3); w2_3 = deriv(w20, u3);

vC1_1 = deriv(vC1, u1); vC2_1 = deriv(vC2, u1);
vC1_2 = deriv(vC1, u2); vC2_2 = deriv(vC2, u2);
vC1_3 = deriv(vC1, u3); vC2_3 = deriv(vC2, u3);

vC32_1 = deriv(vC32, u1); vC3_1 = deriv(vC3, u1);
vC32_2 = deriv(vC32, u2); vC3_2 = deriv(vC3, u2);
vC32_3 = deriv(vC32, u3); vC3_3 = deriv(vC3, u3);

% generalized active forces
Q1 = w1_1*T01.' + vC1_1*G1.' + ...
```

```
      w1_1*transpose(R21)*(-T12.') + ...
      w2_1*T12.' + vC2_1*G2.' + vC32_1*(-F23.') + ...
      vC3_1*F23.' + vC3_1*G3.';

Q2 = w1_2*T01.' + vC1_2*G1.' + ...
      w1_2*transpose(R21)*(-T12.') + ...
      w2_2*T12.' + vC2_2*G2.' + vC32_2*(-F23.') + ...
      vC3_2*F23.' + vC3_2*G3.';

Q3 = w1_3*T01.' + vC1_3*G1.' + ...
      w1_3*transpose(R21)*(-T12.') + ...
      w2_3*T12.' + vC2_3*G2.' + vC32_3*(-F23.') + ...
      vC3_3*F23.' + vC3_3*G3.';

I1 = [m1*(2*L1)^2/12 0 0; 0 m1*(2*L1)^2/12 0; 0 0 0];
I2 = [m2*(2*L2)^2/12 0 0; 0 m2*(2*L2)^2/12 0; 0 0 0];
I3 = [I3x 0 0; 0 I3y 0; 0 0 I3z];

% Kane's dynamical equations

% inertia forces

% inertia force for link 1
% expressed in terms of RF1 {i1,j1,k1}
Fin1= -m1*aC1;
% inertia force for link 2
% expressed in terms of RF2 {i2,j2,k2}
Fin2= -m2*aC2;
% inertia force for link 3
% expressed in terms of RF2 {i2,j2,k2}
Fin3= -m3*aC3;

% inertia moments

% inertia moment for link 1
% expressed in terms of RF1 {i1,j1,k1}
Min1 = -alpha10*I1-cross(w10,w10*I1);
% inertia moment for link 2
% expressed in terms of RF2 {i2,j2,k2}
Min2 = -alpha20*I2-cross(w20,w20*I2);
% inertia moment for link 3
% expressed in terms of RF2 {i2,j2,k2}
Min3 = -alpha20*I3-cross(w20,w20*I3);

% generalized inertia forces
```

```
% generalized inertia forces corresponding to q1
Kin1 = w1_1*Min1.' + vC1_1*Fin1.' + ...
       w2_1*Min2.' + vC2_1*Fin2.' + ...
       w2_1*Min3.' + vC3_1*Fin3.';

% generalized inertia forces corresponding to q2
Kin2 = w1_2*Min1.' + vC1_2*Fin1.' + ...
       w2_2*Min2.' + vC2_2*Fin2.' + ...
       w2_2*Min3.' + vC3_2*Fin3.';

% generalized inertia forces corresponding to q3
Kin3 = w1_3*Min1.' + vC1_3*Fin1.' + ...
       w2_3*Min2.' + vC2_3*Fin2.' + ...
       w2_3*Min3.' + vC3_3*Fin3.';

% Kane's dynamical equations
% first Kane's dynamical equation
Kane1 = Kin1 + Q1;
% second Kane's dynamical equation
Kane2 = Kin2 + Q2;
% third Kane's dynamical equation
Kane3 = Kin3 + Q3;

% control torques and control force
q1f=pi/3; q2f=pi/3; q3f=0.3;
b01=450; g01=300;
b12=200; g12=300;
b23=150; g23=50;

T01xc = -b01*dq1-g01*(q1-q1f);
T12yc = -b12*dq2-g12*(q2-q2f)+ ...
         g*(m2*L2+m3*(L2+q3))*c2;
F23zc = -b23*dq3-g23*(q3-q3f)+g*m3*s2;

tor  = {T01x, T12y, F23z};
torf = {T01xc,T12yc,F23zc};

Kan1 = subs(Kane1, tor, torf);
Kan2 = subs(Kane2, tor, torf);
Kan3 = subs(Kane3, tor, torf);

data = {L1, L2, I3x, I3y, I3z, m1, m2, m3, g};
datn = {0.4, 0.4, 5, 4, 1, 90, 60, 40, 9.81};
```

```
Ka1 = subs(Kan1, data, datn);
Ka2 = subs(Kan2, data, datn);
Ka3 = subs(Kan3, data, datn);

ql = {diff(u1,t), diff(u2,t), diff(u3,t) ...
      u1, u2, u3, q1, q2, q3};
qx = {'du1', 'du2',  'du3',...
'x(4)', 'x(5)', 'x(6)', 'x(1)', 'x(2)', 'x(3)'};

% ql                    qx
%----------------------------
% diff('u1(t)',t)  -> 'du1'
% diff('u2(t)',t)  -> 'du2'
% diff('u3(t)',t)  -> 'du3'
%          'u1(t)'  -> 'x(4)'
%          'u2(t)'  -> 'x(5)'
%          'u3(t)'  -> 'x(6)'
%          'q1(t)'  -> 'x(1)'
%          'q2(t)'  -> 'x(2)'
%          'q3(t)'  -> 'x(3)'

Du1 = subs(Ka1, ql, qx);
Du2 = subs(Ka2, ql, qx);
Du3 = subs(Ka3, ql, qx);

% solve for du1, du2, du3
sol = solve(Du1, Du2, Du3,'du1, du2, du3');
sdu1 = sol.du1;
sdu2 = sol.du2;
sdu3 = sol.du3;

% system of ODE
dx1 = char('x(4)');
dx2 = char('x(5)');
dx3 = char('x(6)');
dx4 = char(sdu1);
dx5 = char(sdu2);
dx6 = char(sdu3);

fid = fopen('RRT_Kane.m','w+');
fprintf(fid,'function dx = RRT_Kane(t,x)\n');
fprintf(fid,'dx = zeros(6,1);\n');
fprintf(fid,'dx(1) = '); fprintf(fid,dx1);
fprintf(fid,';\n');
fprintf(fid,'dx(2) = '); fprintf(fid,dx2);
```

```
fprintf(fid,';\n');
fprintf(fid,'dx(3) = '); fprintf(fid,dx3);
fprintf(fid,';\n');
fprintf(fid,'dx(4) = '); fprintf(fid,dx4);
fprintf(fid,';\n');
fprintf(fid,'dx(5) = '); fprintf(fid,dx5);
fprintf(fid,';\n');
fprintf(fid,'dx(6) = '); fprintf(fid,dx6);
fprintf(fid,';  ');
fclose(fid); cd(pwd);

t0 = 0;  tf = 15; time = [0 tf];

x0 = [pi/18 pi/6 0.25 0 0 0];

[t,xs] = ode45(@RRT_Kane, time, x0);

x1 = xs(:,1);
x2 = xs(:,2);
x3 = xs(:,3);
x4 = xs(:,4);
x5 = xs(:,5);
x6 = xs(:,6);

subplot(3,1,1),plot(t,x1*180/pi,'r'),...
xlabel('t (s)'),ylabel('q1 (deg)'),grid,...
subplot(3,1,2),plot(t,x2*180/pi,'b'),...
xlabel('t (s)'),ylabel('q2 (deg)'),grid,...
subplot(3,1,3),plot(t,x3,'g'),...
xlabel('t (s)'),ylabel('q3 (m)'),grid

% end of program
```

E.6 RRTR Robot Arm

```
% E6
% Analytical Dynamics
% RRTR robot arm
% Kane's dynamical equations
clear all; clc; close all

L1=0.1; L2=0.1; L3=0.7;
m1=9.; m2=6.; m3=4.; m4=1.;
```

```
I1x=0.01; I1y=0.02; I1z=0.01;
I2x=0.06; I2y=0.01; I2z=0.05;
I3x=0.4; I3y=0.01; I3z=0.4;
I4x=0.0005; I4y=0.001; I4z=0.001;
k1=3.; k2=5.; k3=1.; k4=3.; k5=0.3;
k6=0.6; k7=30.; k8=41.; g=9.8;
q1f=pi/3; q2f=pi/3; q3f=pi/3; q4f=0.1;

syms t real

q1 = sym('q1(t)');
q2 = sym('q2(t)');
q3 = sym('q3(t)');
q4 = sym('q4(t)');

c2 = cos(q2); s2 = sin(q2);
c3 = cos(q3); s3 = sin(q3);

% transformation matrix from RF2 to RF1
R21 = [[1 0 0]; [0 c2 s2]; [0 -s2 c2]];

% transformation matrix from RF4 to RF2
R42 = [[c3 0 -s3]; [0 1 0]; [s3 0 c3]];

% expressing q1',q2',q3',q4' in terms
% of generalized speeds u1,u2,u3,u4

u1 = sym('u1(t)');
u2 = sym('u2(t)');
u3 = sym('u3(t)');
u4 = sym('u4(t)');

dq1 = (u1*s3-u3*c3)/s2;
dq2 = u1*c3+u3*s3;
dq3 = u2+(u3*c3-u1*s3)*c2/s2;
dq4 = u4;

qt={diff(q1,t),diff(q2,t),diff(q3,t),diff(q4,t),...
    diff(u1,t),diff(u2,t),diff(u3,t),diff(u4,t)};
ut={dq1,dq2,dq3,dq4, 'du1','du2','du3','du4'};

% qt                    ut
%-----------------------
% diff('q1(t)',t)  ->  dq1
% diff('q2(t)',t)  ->  dq2
```

```
% diff('q3(t)',t) ->  dq3
% diff('q4(t)',t) ->  dq4
% diff('u1(t)',t) -> 'du1'
% diff('u2(t)',t) -> 'du2'
% diff('u3(t)',t) -> 'du3'
% diff('u4(t)',t) -> 'du4'

% Angular velocities

% Angular velocities of each link 1, 2, 3, 4,
% in RF0, involving the generalized speeds,
% are expressed using a vector basis
% fixed in the body under consideration

% angular velocity of link 1 with respect to (wrt)
% RF0 expressed in terms of RF1{i1,j1,k1}
w101 = [0, dq1, 0];

% angular velocity of link 1 wrt RF0 expressed in
% terms of RF2{i2,j2,k2}
w102 = w101*transpose(R21);

% angular velocity of link 2 wrt RF1 expressed in
% terms of RF1{i1,j1,k1}
w211 = [dq2, 0, 0];

% angular velocity of link 2 wrt RF1 expressed in
% terms of RF2{i2,j2,k2}
w212 = [dq2, 0, 0];

% angular velocity of link 2 wrt RF0 expressed in
% terms of RF2{i2,j2,k2}
w202 = w102 + w212;

% angular velocity of link 2 wrt RF0 expressed in
% terms of RF4{i4,j4,k4}
w204 = w202*transpose(R42);

% angular velocity of link 3 wrt RF0 expressed in
% terms of RF2{i2,j2,k2}==RF3
w302 = w202;

% angular velocity of link 4 wrt RF2 expressed in
% terms of RF2{i2,j2,k2}
w422 = [0, dq3, 0];
```

```
% angular velocity of link 4 wrt RF2 expressed in
% terms of RF4{i4,j4,k4}
w424 = w422*transpose(R42);

% anglar velocity of link 4 wrt RF0 expressed in
% terms of RF4{i4,j4,k4}
w404 = w424 + w204;

% angular velocity of link 4 wrt RF0 expressed in
% terms of RF2{i2,j2,k2}
w402 = w404*R42;

% Angular accelerations

% angular acceleration of link 1 wrt RF0 expressed
% in terms of RF1 {i1,j1,k1}
alpha101 = subs(diff(w101, t), qt, ut);

% angular acceleration of link 2 wrt RF0 expressed
% in terms of RF2{i2,j2,k2}
alpha202 = subs(diff(w202, t), qt, ut);

% angular acceleration of link 3 wrt RF0 expressed
% in terms of RF2{i2,j2,k2}
alpha302 = alpha202;

% angular acceleration of link 4 wrt RF0 expressed
% in terms of RF4{i4,j4,k4}
alpha404 = subs(diff(w404, t), qt, ut);

% linear velocity of mass center C1 of link 1
% wrt RF0 expressed in terms of RF1{i1,j1,k1} is
% zero since C1 is fixed in RF0
vC101 = [0, 0, 0];

%position vector from C1, mass center of link 1,
% to C2, mass center of link 2, expressed in terms
% of RF1{i1,j1,k1}
rC121 = [L1, L2, 0];

% linear velocity of mass center C2 of link 2 wrt
% RF0 expressed in terms of RF1{i1,j1,k1}
vC201 = diff(rC121, t) + cross(w101, rC121);
```

```
% Remark: velocity of C2, which is "fixed" wrt
% RF1{i1,j1,k1}, can be computed as
% vC201=cross(w101,[L1,0,0]);

% linear velocity of mass center C2 of link 2 wrt
% RF0 expressed in terms of RF2{i2,j2,k2}
vC202 = vC201*transpose(R21);

% position vector from C1, mass center of link 1,
% to C3, mass center of link 3, expressed in terms
% of RF2{i2,j2,k2}
rC132 = [L1, L2, 0]*transpose(R21) + [0, q4, 0];

% linear velocity of mass center C3 of link 3 wrt
% RF0 expressed in terms of RF2{i2,j2,k2}
vC302=subs(diff(rC132,t),qt,ut)+cross(w202,rC132);

% Remark: another way of computing vC302
% vC302=vC202+diff([0,q4,0],t]+cross(w202,[0,q4,0]);

% linear velocity of point C32 of link 2
% expressed in terms of RF2{i2,j2,k2}
% C32, of link 2, is superposed with C3, of link 3
vC3202 = vC202 + cross(w202, [0, q4, 0]);

% linear velocity of mass center C4 of link 4 wrt
% RF0 expressed in terms RF2 {i2,j2,k2}
vC402 = vC302 + cross(w202, [0, L3, 0]);

% Linear accelerations

% linear acceleration of mass center C1 of link 1 wrt
% RF0 expressed in terms of RF1{i1,j1,k1}
aC101 = [0, 0, 0];

% linear acceleration of mass center C2 of link 2 wrt
% RF0 expressed in terms of RF1{i1,j1,k1}
aC201 = subs(diff(vC201,t),qt,ut)+cross(w101,vC201);

% linear acceleration of mass center C3 of link 3 wrt
% RF0 expressed in terms of RF2{i2,j2,k2}
aC302 = subs(diff(vC302,t),qt,ut)+cross(w202,vC302);

% linear acceleration of mass center C4 of link 4 wrt
% RF0 expressed in terms of RF2{i2,j2,k2}
```

```
aC402 = subs(diff(vC402,t),qt,ut)+cross(w402,vC402);

%deriv(f,g(t)) differentiates f with respect to g(t)
% partial velocities

w1_1 = deriv(w101, u1); w2_1 = deriv(w202, u1);
w1_2 = deriv(w101, u2); w2_2 = deriv(w202, u2);
w1_3 = deriv(w101, u3); w2_3 = deriv(w202, u3);
w1_4 = deriv(w101, u4); w2_4 = deriv(w202, u4);

w4_1 = deriv(w404, u1);
w4_2 = deriv(w404, u2);
w4_3 = deriv(w404, u3);
w4_4 = deriv(w404, u4);

vC1_1 = deriv(vC101, u1); vC2_1 = deriv(vC201, u1);
vC1_2 = deriv(vC101, u2); vC2_2 = deriv(vC201, u2);
vC1_3 = deriv(vC101, u3); vC2_3 = deriv(vC201, u3);
vC1_4 = deriv(vC101, u4); vC2_4 = deriv(vC201, u4);

vC32_1 = deriv(vC3202, u1); vC3_1 = deriv(vC302, u1);
vC32_2 = deriv(vC3202, u2); vC3_2 = deriv(vC302, u2);
vC32_3 = deriv(vC3202, u3); vC3_3 = deriv(vC302, u3);
vC32_4 = deriv(vC3202, u4); vC3_4 = deriv(vC302, u4);

vC4_1 = deriv(vC402, u1);
vC4_2 = deriv(vC402, u2);
vC4_3 = deriv(vC402, u3);
vC4_4 = deriv(vC402, u4);

% Generalized active forces

% Contact and gravitational force

% rigid link 1
syms F01x F01y F01z T01x T01y T01z real
syms F21x F21y F21z T21x T21y T21z real

% force applied by base 0 to link 1 at C1
% expressed in terms of RF1{i1,j1,k1}
F01 = [F01x F01y F01z];
% torque applied by base 0 to link 1
% expressed in terms of RF1{i1,j1,k1}
T01 = [T01x T01y T01z];
```

```
% force applied by link 2 to link 1 at C2
% expressed in terms of RF2{i2,j2,k2}
F21 = [F21x F21y F21z];
% torque applied by link 2 to link 1
% expressed in terms of RF2{i2,j2,k2}
T21 = [T21x T21y T21z];

% gravitational force that acts on link 1 at C1
% expressed in terms of RF1{i1,j1,k1}
G1 = [0 -m1*g 0];

% generalized active forces for link 1
Q1a1 = w1_1*(T01.'+R21.'*T21.') + ...
    vC1_1*(G1.'+F01.') + vC2_1*R21.'*F21.';
Q1a2 = w1_2*(T01.'+R21.'*T21.') + ...
    vC1_2*(G1.'+F01.') + vC2_2*R21.'*F21.';
Q1a3 = w1_3*(T01.'+R21.'*T21.') + ...
    vC1_3*(G1.'+F01.') + vC2_3*R21.'*F21.';
Q1a4 = w1_4*(T01.'+R21.'*T21.') + ...
    vC1_4*(G1.'+F01.') + vC2_4*R21.'*F21.';

% rigid link 2
syms F32x F32y F32z T32x T32y T32z real

% force applied by link 1 to link 2 at C2
% expressed in terms of RF2{i2,j2,k2}: -F21

% torque applied by link 1 to link 2
% expressed in terms of RF2{i2,j2,k2}: -T21

% force applied by link 3 to link 2 at C32
% expressed in terms of RF2{i2,j2,k2}
F32 = [F32x F32y F32z];

% torque applied by link 3 to link 2
% expressed in terms of RF2{i2,j2,k2}
T32 = [T32x T32y T32z];

% gravitational force that acts on link 2 at C2
% expressed in terms of RF2{i2,j2,k2}
G2 = [0 -m2*g 0];

% generalized active forces for link 2
```

```
Q2a1 = w2_1*(T32.'-T21.') + ...
    vC2_1*(G2.'-R21.'*F21.') + vC32_1*F32.';
Q2a2 = w2_2*(T32.'-T21.') + ...
    vC2_2*(G2.'-R21.'*F21.') + vC32_2*F32.';
Q2a3 = w2_3*(T32.'-T21.') + ...
    vC2_3*(G2.'-R21.'*F21.') + vC32_3*F32.';
Q2a4 = w2_4*(T32.'-T21.') + ...
    vC2_4*(G2.'-R21.'*F21.') + vC32_2*F32.';

% rigid link 3
syms F43x F43y F43z T43x T43y T43z real

% force applied by link 2 to link 3 at C3
% expressed in terms of RF2{i2,j2,k2}: -F32

% torque applied by link 2 to link 3
% expressed in terms of RF2{i2,j2,k2}: -T32

% force applied by link 4 to link 3 at C4
% expressed in terms of RF2{i2,j2,k2}
F43 = [F43x F43y F43z];

% torque applied by link 4 to link 3
% expressed in terms of RF2 {i2,j2,k2}
T43 = [T43x T43y T43z];

% gravitational force that acts on link 3 at C3
% expressed in terms of RF2{i2,j2,k2}*)
G3 = [0 -m3*g 0]*transpose(R21);

% generalized active forces for link 3

Q3a1 = w2_1*(T43.'-T32.') + ...
    vC3_1*(G3.'-F32.') + vC4_1*F43.';
Q3a2 = w2_2*(T43.'-T32.') + ...
    vC3_2*(G3.'-F32.') + vC4_2*F43.';
Q3a3 = w2_3*(T43.'-T32.') + ...
    vC3_3*(G3.'-F32.') + vC4_3*F43.';
Q3a4 = w2_4*(T43.'-T32.') + ...
    vC3_4*(G3.'-F32.') + vC4_4*F43.';

% rigid link 4
% force applied by link 3 to link 4 at C4
% expressed in terms of RF2{i2,j2,k2}: -F43
```

```
% torque applied by link 3 to link 4
% expressed in terms of RF2 {i2,j2,k2}: -T43

% gravitational force that acts on link 4 at C4
% expressed in terms of RF2{i2,j2,k2}
G4 = [0 -m4*g 0]*transpose(R21);

% generalized active forces for link 4

Q4a1 = w4_1*R42*(-T43).' + vC4_1*(G4.'-F43.');
Q4a2 = w4_2*R42*(-T43).' + vC4_2*(G4.'-F43.');
Q4a3 = w4_3*R42*(-T43).' + vC4_3*(G4.'-F43.');
Q4a4 = w4_4*R42*(-T43).' + vC4_4*(G4.'-F43.');

% generalized active forces
Q1 = simple(Q1a1+Q2a1+Q3a1+Q4a1);
Q2 = simple(Q1a2+Q2a2+Q3a2+Q4a2);
Q3 = simple(Q1a3+Q2a3+Q3a3+Q4a3);
Q4 = simple(Q1a4+Q2a4+Q3a4+Q4a4);

% Dyadics

% inertia matrix associated with central inertia
% dyadic of link 1 in RF1{i1,j1,k1}
I1 = [I1x 0 0; 0 I1y 0; 0 0 I1z];
% inertia matrix associated with central inertia
% dyadic of link 2 in RF2{i2,j2,k2}
I2 = [I2x 0 0; 0 I2y 0; 0 0 I2z];
% inertia matrix associated with central inertia
% dyadic of link 3 in RF2{i2,j2,k2}
I3 = [I3x 0 0; 0 I3y 0; 0 0 I3z];
% inertia matrix associated with central inertia
% dyadic of link 4 in  RF4{i4,j4,k4}
I4 = [I4x 0 0; 0 I4y 0; 0 0 I4z];

% inertia forces

% inertia force Fin1 of link1 in RF0 expressed in
% terms of RF1{i1,j1,k1}
Fin1 = -m1*aC101;

% inertia force Fin2 of link2 in RF0 expressed in
% terms of RF1{i1,j1,k1}
Fin2 = -m2*aC201;
```

```
% inertia force Fin3 of link3 in RF0 expressed in
% terms of RF2{i2,j2,k2}
Fin3 = -m3*aC302;

% inertia force Fin4 of link4 in RF0 expressed in
% terms of RF2{i2,j2,k2}
Fin4 = -m4*aC402;

% inertia moments

% inertia moment Min1 of link 1 in RF0 expressed in
% terms of RF1{i1,j1,k1}
Min1 = -alpha101*I1-cross(w101,w101*I1);

% inertia moment Min2 of link 2 in RF0 expressed in
% terms of RF2{i2,j2,k2}
Min2 = -alpha202*I2-cross(w202,w202*I2);

% inertia moment Min3 of link 3 in RF0 expressed in
% terms of RF2{i2,j2,k2}
Min3 = -alpha302*I3-cross(w202,w202*I3);

% inertia moment Min4 of link 4 in RF0 expressed in
% terms of RF4{i4,j4,k4}
Min4 = -alpha404*I4-cross(w404,w404*I4);

% Generalized inertia forces

Kin1 = w1_1*Min1.' + vC1_1*Fin1.' + ....
       w2_1*Min2.' + vC2_1*Fin2.' + ...
       w2_1*Min3.' + vC3_1*Fin3.' + ...
       w4_1*Min4.' + vC4_1*Fin4.' ;

Kin2 = w1_2*Min1.' + vC1_2*Fin1.' + ....
       w2_2*Min2.' + vC2_2*Fin2.' + ...
       w2_2*Min3.' + vC3_2*Fin3.' + ...
       w4_2*Min4.' + vC4_2*Fin4.' ;

Kin3 = w1_3*Min1.' + vC1_3*Fin1.' + ....
       w2_3*Min2.' + vC2_3*Fin2.' + ...
       w2_3*Min3.' + vC3_3*Fin3.' + ...
       w4_3*Min4.' + vC4_3*Fin4.' ;

Kin4 = w1_4*Min1.' + vC1_4*Fin1.' + ....
       w2_4*Min2.' + vC2_4*Fin2.' + ...
```

```
        w2_4*Min3.' + vC3_4*Fin3.' + ...
        w4_4*Min4.' + vC4_4*Fin4.' ;

% Kane's dynamical equations
Kane1 = Q1 + Kin1;
Kane2 = Q2 + Kin2;
Kane3 = Q3 + Kin3;
Kane4 = Q4 + Kin4;

% control
T01yc = k1*(q1f-q1)-k2*dq1;
T21xc = k3*(q2-q2f)+k4*dq2+g*((m3+m4)*q4+m4*L3)*s2;
T43yc = k5*(q3-q3f)+k6*dq3;
F32yc = k7*(q4-q4f)+k8*dq4-g*(m3+m4)*c2;

tor  = {T01y , T21x , T43y , F32y };
torf = {T01yc, T21xc, T43yc, F32yc };

Kan1 = subs(Kane1, tor, torf);
Kan2 = subs(Kane2, tor, torf);
Kan3 = subs(Kane3, tor, torf);
Kan4 = subs(Kane4, tor, torf);

ql = {q1, q2, q3, q4, u1, u2, u3, u4};
qe = {'x(1)', 'x(2)', 'x(3)', 'x(4)', ...
      'x(5)', 'x(6)', 'x(7)', 'x(8)'};

% ql           qe
%------------------
% 'q1(t)' -> 'x(1)'
% 'q2(t)' -> 'x(2)'
% 'q3(t)' -> 'x(3)'
% 'q4(t)' -> 'x(4)'
% 'u1(t)' -> 'x(5)'
% 'u2(t)' -> 'x(6)'
% 'u3(t)' -> 'x(7)'
% 'u4(t)' -> 'x(8)'

e1 = subs(Kan1, ql, qe);
e2 = subs(Kan2, ql, qe);
e3 = subs(Kan3, ql, qe);
e4 = subs(Kan4, ql, qe);

sol = solve(e1, e2, e3, e4, 'du1, du2, du3, du4');
du1c = sol.du1;
```

```
du2c = sol.du2;
du3c = sol.du3;
du4c = sol.du4;

% system of ODE

dx1 = char(subs(dq1, ql, qe));
dx2 = char(subs(dq2, ql, qe));
dx3 = char(subs(dq3, ql, qe));
dx4 = char(subs(dq4, ql, qe));
dx5 = char(du1c);
dx6 = char(du2c);
dx7 = char(du3c);
dx8 = char(du4c);

fid = fopen('RRTR.m','w+');
fprintf(fid,'function dx = RRTR(t,x)\n');
fprintf(fid,'dx = zeros(8,1);\n');
fprintf(fid,'dx(1) = '); fprintf(fid,dx1);
fprintf(fid,';\n');
fprintf(fid,'dx(2) = '); fprintf(fid,dx2);
fprintf(fid,';\n');
fprintf(fid,'dx(3) = '); fprintf(fid,dx3);
fprintf(fid,';\n');
fprintf(fid,'dx(4) = '); fprintf(fid,dx4);
fprintf(fid,';\n');
fprintf(fid,'dx(5) = '); fprintf(fid,dx5);
fprintf(fid,';\n');
fprintf(fid,'dx(6) = '); fprintf(fid,dx6);
fprintf(fid,';\n');
fprintf(fid,'dx(7) = '); fprintf(fid,dx7);
fprintf(fid,';\n');
fprintf(fid,'dx(8) = '); fprintf(fid,dx8);
fprintf(fid,';');
fclose(fid); cd(pwd);

t0 = 0;  tf = 15; time = [0 tf];

q10=pi/6;
q20=pi/12;
q30=pi/10;
q40=0.01;
u10=0; u20=0; u30=0; u40=0;

x0 = [q10 q20 q30 q40 u10 u20 u30 u40];
```

```
[t,xs] = ode45(@RRTR, time, x0);

x1 = xs(:,1);
x2 = xs(:,2);
x3 = xs(:,3);
x4 = xs(:,4);
x5 = xs(:,5);
x6 = xs(:,6);
x7 = xs(:,7);
x8 = xs(:,8);

subplot(4,1,1),plot(t,x1*180/pi,'r'),...
xlabel('t (s)'),ylabel('q1 (deg)'),grid,...
subplot(4,1,2),plot(t,x2*180/pi,'b'),...
xlabel('t (s)'),ylabel('q2 (deg)'),grid,...
subplot(4,1,3),plot(t,x3*180/pi,'g'),...
xlabel('t (s)'),ylabel('q3 (deg)'),grid,...
subplot(4,1,4),plot(t,x4,'black'),...
xlabel('t (s)'),ylabel('q4 (m)'),grid

[ts,xs] = ode45(@RRTR,0:1:10,x0);

fprintf('Results \n\n')
fprintf ...
('   t(s)     q1(rad)    q2(rad)    q3(rad)    q4(m)\n');
[ts,xs(:,1),xs(:,2),xs(:,3),xs(:,4)]
% end of program
```

Results:

```
      t(s)        q1(rad)    q2(rad)    q3(rad)    q4(m)

   ans =

           0      0.5236     0.2618     0.3142     0.0100
      1.0000      0.7555     0.4377     0.6023     0.0559
      2.0000      0.8858     0.6259     0.7776     0.0808
      3.0000      0.9587     0.7605     0.8838     0.0917
      4.0000      0.9995     0.8527     0.9481     0.0964
      5.0000      1.0218     0.9154     0.9872     0.0984
      6.0000      1.0338     0.9579     1.0108     0.0993
      7.0000      1.0402     0.9867     1.0251     0.0997
      8.0000      1.0435     1.0063     1.0338     0.0998
      9.0000      1.0453     1.0195     1.0391     0.0999
     10.0000      1.0462     1.0285     1.0423     0.1000
```

References

1. I.I. Artobolevski, *Mechanisms in Modern Engineering Design*, MIR, Moscow (1977)
2. M. Atanasiu, *Mechanics* (Mecanica), EDP, Bucharest (1973)
3. H. Baruh, *Analytical Dynamics*, WCB/McGraw-Hill, Boston (1999)
4. G. Baumann, *Mathematica for Theoretical Physics: Classical Mechanics and Nonlinear Dynamics*, Springer-Verlag, (2005)
5. G. Baumann, *Mathematica for Theoretical Physics: Electrodynamics, Quantum Mechanics, General Relativity and Fractals*, Springer-Verlag (2005)
6. A. Bedford and W. Fowler, *Dynamics*, Addison Wesley, Menlo Park, CA (1999)
7. M.I. Buculei, *Mechanisms*, University of Craiova Press, Craiova, Romania (1976)
8. M.I. Buculei, D. Bagnaru, G. Nanu, D.B. Marghitu, *Analysis of Mechanisms with Bars*, Scrisul romanesc, Craiova (1986)
9. A.G. Erdman and G. N. Sandor, *Mechanisms Design*, Prentice-Hall, Upper Saddle River, NJ (1984)
10. D. M. Etter and D.C. Kuncicky, *Introduction to MATLAB for Engineers and Scientists*, Prentice Hall, Upper Saddle River, NJ (1996)
11. F. Freudenstein, "An Application of Boolean Algebra to the Motion of Epicyclic Drives," *Transaction of the ASME, Journal of Engineering for Industry*, pp.176–182 (1971)
12. J.H. Ginsberg, *Advanced Engineering Dynamics*, Cambridge University Press, Cambridge (1995)
13. D.T. Greenwood, *Principles of Dynamics*, Prentice-Hall, Englewood Cliffs, NJ (1998)
14. R.C. Hibbeler, *Engineering Mechanics – Statics and Dynamics*, Prentice-Hall, Upper Saddle River, NJ (1995)
15. T.R. Kane, *Analytical Elements of Mechanics*, Vol. 1, Academic Press, New York (1959)
16. T.R. Kane, *Analytical Elements of Mechanics*, Vol. 2, Academic Press, New York (1961)
17. T.R. Kane and D.A. Levinson, "The Use of Kane's Dynamical Equations in Robotics", *MIT International Journal of Robotics Research*, No. 3, pp. 3–21 (1983)
18. T.R. Kane, P.W. Likins, and D.A. Levinson, *Spacecraft Dynamics*, McGraw-Hill, New York (1983)
19. T.R. Kane and D.A. Levinson, *Dynamics*, McGraw-Hill, New York (1985)
20. R. Maeder, *Programming in Mathematica*, Addison–Wesley Publishing Company, Redwood City, CA (1990)
21. N.H. Madsen, *Statics and Dynamics*, class notes, available at http://www.eng.auburn.edu/users/nmadsen/
22. N.I. Manolescu, F. Kovacs, and A. Oranescu, *The Theory of Mechanisms and Machines*, EDP, Bucharest (1972)
23. D.B. Marghitu, *Mechanical Engineer's Handbook*, Academic Press, San Diego, CA (2001)
24. D.B. Marghitu and M.J. Crocker, *Analytical Elements of Mechanisms*, Cambridge University Press, Cambridge (2001)

25. D.B. Marghitu, *Kinematic Chains and Machine Component Design*, Elsevier, Amsterdam (2005)
26. D.B. Marghitu, *Kinematics and Dynamics of Machines and Machine Design*, class notes, available at http://www.eng.auburn.edu/users/marghitu/
27. J.L. Meriam and L. G. Kraige, *Engineering Mechanics: Dynamics*, John Wiley & Sons, New York (1997)
28. R.L. Mott, *Machine Elements in Mechanical Design*, Prentice Hall, Upper Saddle River, NJ (1999)
29. D.H. Myszka, *Machines and Mechanisms*, Prentice-Hall, Upper Saddle River, NJ (1999)
30. R.L. Norton, *Machine Design*, Prentice-Hall, Upper Saddle River, NJ (1996)
31. R.L. Norton, *Design of Machinery*, McGraw-Hill, New York (2004)
32. L.A. Pars, *A Treatise on Analytical Dynamics*, John Wiley & Sons, New York (1965)
33. I. Popescu, *Mechanisms*, University of Craiova Press, Craiova (1990)
34. F. Reuleaux, *The Kinematics of Machinery*, Dover, New York (1963)
35. J.E. Shigley and J.J. Uicker, *Theory of Machines and Mechanisms*, McGraw-Hill, New York (1995)
36. D. Smith, *Engineering Computation with MATLAB*, Pearson Education, Upper Saddle River, NJ (2008)
37. R.W. Soutas-Little and D.J. Inman, *Engineering Mechanics: Statics and Dynamics*, Prentice-Hall, Upper Saddle River, NJ (1999)
38. J. Sticklen and M.T. Eskil, *An Introduction to Technical Problem Solving with MATLAB*, Great Lakes Press, Wildwood, MO (2006)
39. E.D. Stoenescu, *Dynamics and Synthesis of Kinematic Chains with Impact and Clearance*, Ph.D. Dissertation, Mechanical Engineering, Auburn University (2005)
40. The MathWorks: http://www.mathworks.com/
41. Vassaia Scola, *Theory of Mechanisms and Machines (Teoria mehanizmov i masin)*, Minsk, Russia (1970)
42. R. Voinea, D. Voiculescu, and V. Ceausu, *Mechanics* (Mecanica), EDP, Bucharest (1983)
43. K.J. Waldron and G. L. Kinzel, *Kinematics, Dynamics, and Design of Machinery*, John Wiley & Sons, New York (1999)
44. C.E. Wilson and J.P. Sadler, *Kinematics and Dynamics of Machinery*, Harper Collins College Publishers, New York (1991)
45. H.B. Wilson, L.H. Turcotte, and D. Halpern, *Advanced Mathematics and Mechanics Applications Using MATLAB*, Chapman & Hall/CRC (2003)
46. S. Wolfram, *Mathematica*, Wolfram Media/Cambridge University Press, Cambridge (1999)

Index